MW00843568

Performance by Design

PERFORMANCE BY DESIGN

Hydrodynamics for High-Speed Vessels

by
Donald L Blount

In August 1992, GT/MY Destriero *passed under the Verrazano-Narrows Bridge in New York en route to Ambrose Light for the beginning of the non-stop, Atlantic Ocean record transit of 58 hours, 35 minutes to Bishop Rock in England, averaging 53.09 knots, validating the author's first of several "once-in-a-lifetime" experiences as a design team manager.*

Disclaimer

No responsibility is assumed by the author or publisher for any injury and/or damage to persons or property as a matter of products liability, negligence, or otherwise, or from any use or operation of any methods, products, instructions, or ideas contained in the material herein.

Printed in the United States of America

First Printing, 2014

ISBN 0-978-9890837-1-3

Donald L Blount
PO Box 55171
Virginia Beach, VA 23471
dblount2014@gmail.com

Table of Contents

DEDICATION

I dedicate this book to my grandchildren: Heather, Shannon, Davis, Cassidy, and Thomas.

"Pop"

ACKNOWLEDGMENTS

The catalyst which advances marine design knowledge comes from clients who have unique combinations of needs that have not yet been resolved. To these clients who choose anonymity, I give my thanks for the opportunity of addressing their challenging requirements, which made it possible to move the boundaries of performance craft technology.

On a personal note, I have had the good fortune to work side by side with a number of creative colleagues, many of whom coauthored technical papers with me. I wish to highlight a few. First, Eugene (Gene) Clement introduced me to planing technology and the necessity of being a lifelong student. Nadine Hubble, with her astute understanding of planing knowledge and mastery of computer code, analyzed large experimental databases to develop enhancements for design procedures. And to Dave Fox who helped me establish, early on, the level of writing to be of most use by designers of performance craft.

The staff of Donald L. Blount and Associates has been a joy. We regularly have in-house technical lectures sharing knowledge. They have grown, as most of their designs have been built, and I have now become their student learning from their successes. Many thanks to staff members who provided graphic and analysis support, reviewed parts of text and debated technical points.

I must express my personal gratitude to Office Manager Linda Martes who, with a great sense of humor, tolerated my re-editing of text and remained a cheerleader getting me to the end of the book.

To Editor Barbara Jean Walsh, I also give my thanks as she reworded "engineering speak" to make the book a bit more suitable as a "fireside read." Linda and Barbara Jean connected very well to the benefit of the end product.

This being my first book project, I didn't realize the impact on family—a lot of my hours in isolation. For her tolerance, I wish to thank my wife Shirley for her love and support during this period.

PROLOGUE FROM THE PAST

I must beg, however, to enforce on the naval gentlemen of this Society the necessity, not only of telling us so exactly what it is that they do want, that we may know what to set about providing for them, but also of adjusting their wishes and demands to that which is really possible.

All practical naval constructors will agree with me in saying that it is too common for their masters and mine simply to ask impossibilities. An admiral with authority proportioned to his rank will require you to construct for him a ship which shall be fast.

You prepare a design, and he exclaims, that will never do; you have made her so long that she will never steer; he demands 13 knots, and refuses you 250 feet of length; he requires that she shall stand up like a church, and refuses the tonnage of the large beam necessary to keep her upright; he urges the use of high power for speed, and refuses length of body to carry your boilers; he demands coals for a great many days, and limits you to a draught of water that won't carry it; he asks for a ship that will be as handy as a boat, and as quick as a cutter, and refuses you breadth of rudder and length of tiller, or turns of the wheel, to give you the sufficient purchase; he asks for a steady ship, and gives you such top weight as makes her stagger.

These are some of the causes which lead to bad ships, and to worse understandings between builders and users of them. Difficulties of this kind are only to be got over, in the end, by being conquered in the beginning.

The fighter of the ship and her builder must come to a thorough understanding at the outset, and I trust it may be one of the useful results of such a meeting as this, that the naval commander of a future fleet will let us thoroughly know what he wants, and we will tell him, as thoroughly what it is that it is in our power, and within the limits of our professional skill.

We will do everything for him but impossibilities, if he will be content with asking everything but impossibilities.

J. Scott Russell – 1861
Transactions of the Institution of Naval Architects, Volume II

Performance by Design

PREFACE

I came to the marine field during a century when early performance vessels were demonstrating a need for designers to rethink hydrodynamic principles, and I am in awe of the accomplishments made during that period.

With respect to progress evolving since then, this book is my own contribution to the design and engineering processes, considering maturing technology with methods leading to the expedient delivery of performance marine vessels.

The design process should be an orderly path, leading you to solutions that meet pre-established requirements. Or, if those exact requirements cannot be met, the process should still show you how to approach your objectives with trade-off studies.

My intent is to share the technical information, decision criteria, rules of thumb, and opinionated experiences which have helped me in making design choices to develop marine craft intended to operate beyond displacement speeds.

These pages represent the evolution of positive personal efforts as well as my own analysis of both good and unsuccessful experiences. If I have been successful, this book will facilitate your own efforts and will reduce the time you need to establish the "right size" craft to meet the technical aspects of your client's requirements.

The sciences related to the design of performance marine craft are now mature. Although many designers have addressed technical topics and expressed their personal experiences in technical papers, magazine articles, and lectures, this maturity is colored by a lack of two things: ready access to individual reference sources and an organized approach to their connectivity to modern thought.

Here, from an engineering viewpoint, I will undertake to present the design process and those technologies in a way that should bring some connectivity to these diverse topics.

And what do I mean by "performance?" When we talk about craft's performance, does everyone have the same understanding? Not likely! Since I've titled this book *Performance by Design*, I need to give you my definition.

To me, the term "performance" is multi-faceted. It encompasses an interactive combination of exceptional operational characteristics; speed with range, efficiency, seakeeping, maneuverability, and dynamic stability. Performance defines the measure of the ability of a marine craft to carry out a specified task, and it has the capability of executing required tasks in a defined operational environment.

My first paper, co-authored with E.P. "Gene" Clement in 1963, published a large experimental database for the David Taylor Model Basin Series 62 hard-chine hulls. Since then, I've written or co-authored a variety of technical papers and magazine articles, and I've extracted much of that material for this book.

During my years in the marine industry, a great many influential technical works were produced by respected contemporaries. I've had the good fortune to be able to associate with "the great ones" who devoted their careers to establishing, extending, and documenting the technologies of performance marine craft.

This has allowed me to focus on design and engineering applications of these good works. I have also been fortunate to have had more than one successful "once-in-a-lifetime experience," as well as a few failures, all of which proved to be most instructive.

Therefore, I hope you will regard this book as the "notes of an engineer" who has been, and remains, involved with designing performance craft while being alert to opportunities to apply emerging technologies.

From my initial paper with Gene Clement, and considering the respected views of others, I have both tempered and expanded my own understanding of technologies related to the design of boats and craft.

My experiences have evolved over 50-plus years of working as a naval architect. As a result, some of my earlier views are now at variance with my current thinking.

I have prepared this book for design students and for undergraduate naval architects/marine engineers. Students of powerboat design should find this to be a useful resource for identifying references of contributions from many authors.

The design process outlined here will guide you around the conceptual design spiral. My emphasis, however, focuses squarely on topics that are related to hydrodynamic performance of surface craft.

I am leaving the subject of structures to others, although there are some hydrodynamic loads addressed to provide input for the designer of the structure.

In Appendix 7, starting on p. 311, you will find *Nomenclature, Definitions, and Abbreviations* for dimensionless speeds and hull loading, as well as many other tools to help you in understanding the material that I discuss. I recommend that you take a quick glance now so that you know what is there.

I do intend to acknowledge all the references I have made to original works including their sources of publication. Should there be any omissions or errors, I take full responsibility for flaws and apologize in advance.

The information cited here reflects my personal knowledge of design data for marine craft; what I know and what I dare to believe.

Please be reminded of the date of this publication, though. Future contributions and experiences from all sources may alter or expand the technology base contained here.

My opinions and experience, and the decision criteria originated by others may become an important addition to the knowledge base of this technology—or future research could debunk the materials here as erroneous conjecture or misinterpretation of data.

I truly encourage you to make copious notes throughout this book so you can document the ever growing knowledge, experiences, and technology around you and personalize your copy of this book as a current reference source. You will see that I have included ample spaces for you to do just that.

Donald L Blount
April 2014

CHAPTER ONE

Looking Back: Moving Forward

For many years I wondered about the first time that someone really understood the technology for marine craft exceeding hull speed. Was there an "Ah ha!" moment when that person realized there were other ways to attain higher boat speeds than increasing hull length?

Early theories of naval architecture only focused on the concept that waterline length of a vessel was the dominant factor limiting maximum speed potential. These theories stated that maximum achievable speed would occur at "hull speed" when $V/(L)^{1/2} = 1.34$, which is a length Froude number $F_{nL} = 0.40$.

After 1860, though, a number of visionaries began to build, test, and demonstrate boats which broke the hull-speed barrier. By 1890, boats attained what we now call "semi-planing" speeds. The result? Progress.

We know that powered craft began to achieve speeds greater than hull speed in 1863. Shipbuilding records report sea-trial data for high-speed craft during that threshold year, and speeds since then continued to increase as marine technologies evolved.

Steam power, improved propeller knowledge, and reduced structural weight were all catalysts for the evolution of performance marine vessels. Before the1860s, steam power had already met a variety of purposes as builders began to deliver steam launches, yachts, military ships, and commercial service vessels which ran consistently faster than hull speed—even though that was still considered to be an impossibility by many.

A history of sailing ships and underpowered steam-driven vessels had created a mental block for designers and builders who believed they must increase waterline length to achieve higher speed potential for marine vessels.

Prior to 1863, there is no record of a powered craft exceeding hull speed, the year in which John Thornycroft's 40-foot motor launch *Ariel* attained a speed of 12.2 knots ($F_{nL} = 0.57$).

And yet others in the maritime trades still could not believe a speed of 42 percent greater than hull speed. The concept of hydrodynamic lifting forces beginning to significantly augment—and at sufficiently high speeds overwhelm—hydrostatic/buoyancy forces was unheard of.

In his 1897 book *Yacht Architecture,* Dixon Kemp was the first to write about this phenomenon when he said, "...there is relative to the normal water level a sensible lifting of the boat...," based on his study of photographs of a high-speed launch.

In 1872, Sir Frederick Bramwell was retained as an independent consulting engineer to conduct sea trials on the Thornycroft steam yacht *Miranda*. Bramwell caused some excitement during a lecture in which he described how the boat had achieved twice hull speed. Noted naval architect, J. Scott Russell, was in the audience, and after the presentation he exclaimed that this was "...one of the most delicious papers I have ever heard, for I see here a set of phenomena exhibited in direct contradiction to all the received laws of dynamics and hydrodynamics!"

This was the "Ah ha!" moment I had wondered about!

Early design professionals had no knowledge of what happens above hull speed. Only limited hull-resistance information was available in the mid-1800s. Hard-chine hull data, as shown in Figure 1-1 from Series 62, wasn't published until much later, in 1963. The steep slopes of the R/W curves versus speed in the vicinity of F_{nL} = 0.40 suggested that only small incremental gains in speed could be possible with extremely large propulsion increases in sail area or with steam power. Thus, technology and economics during that era tended to favor longer hulls whenever an increase in speed was desired.

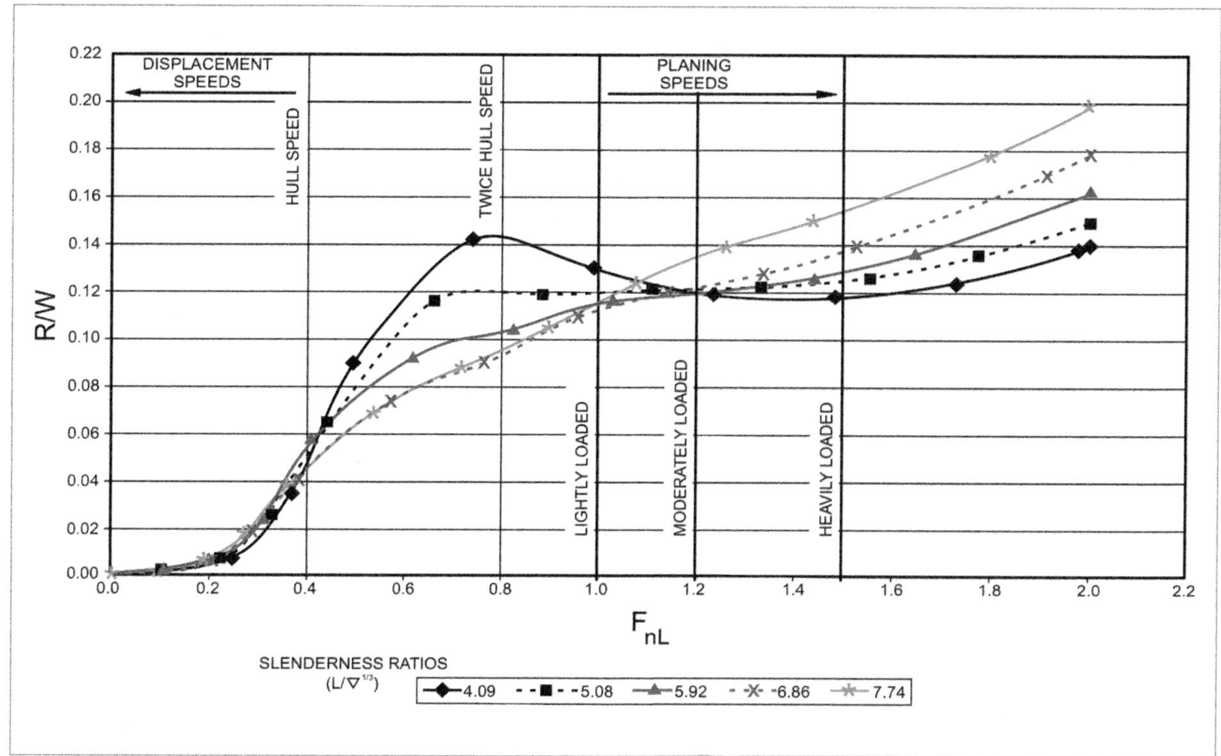

Figure 1-1: Series 62 resistance comparison

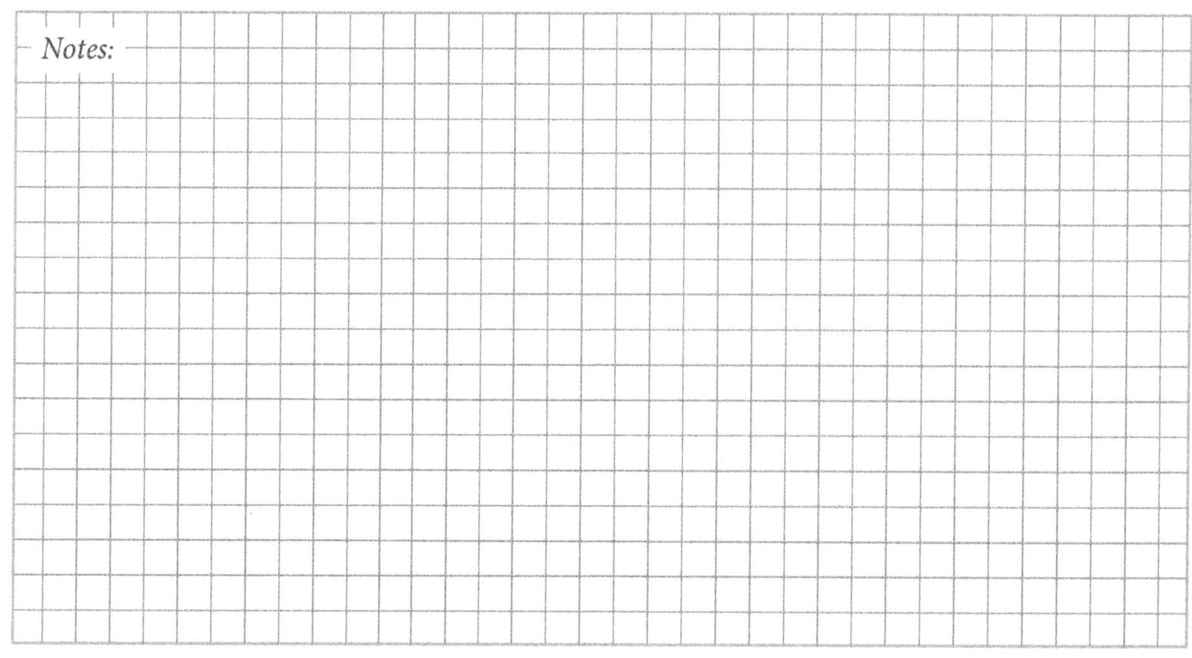

Figure 1-2 is partially a repeat of the data from Figure 1-1 except that the data for three hulls have been removed so as to show R/W trends only with vessels nearly in proportion to the vessels of the late 1800s. Also added is an R/W curve for a modern (circa 1964) round-bilge hull also having proportions of high-speed vessels in the late 1800s. Thus, the round-bilge hull data in this figure depicts the R/W versus speed trends that challenged the will of the early designers/builders to attempt to exceed hull speed.

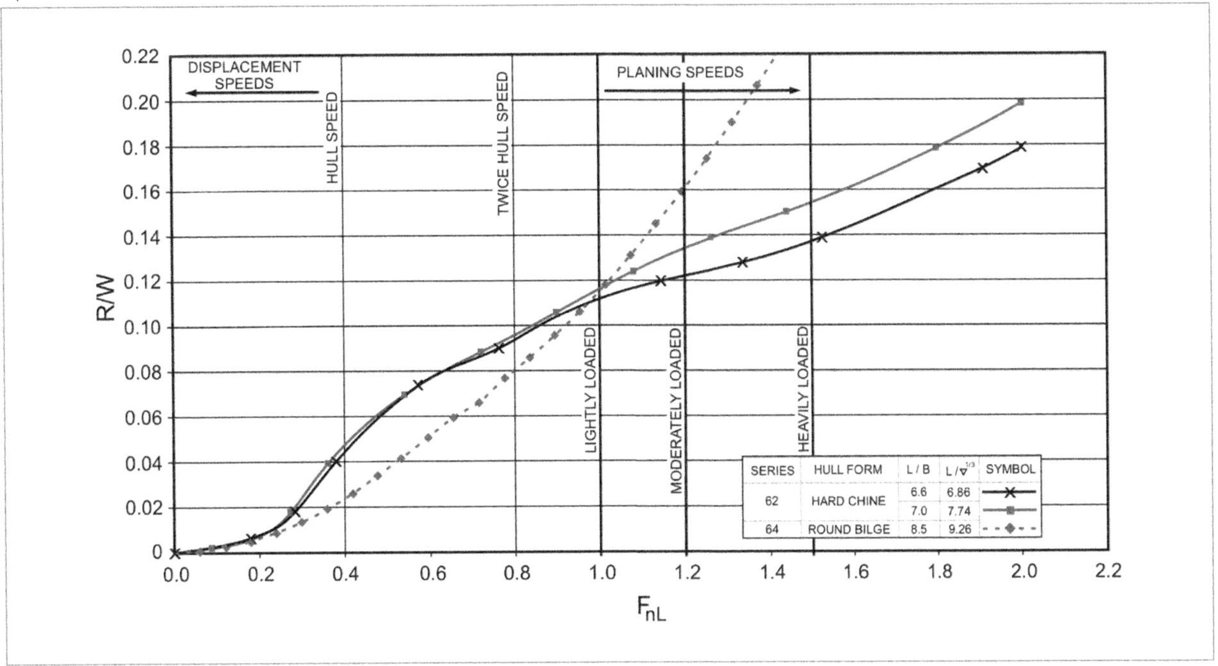

Figure 1-2: Series 62 & Series 64 resistance comparison

We now understand that the support of surface vessel's weight occurs with three distinct and different combinations of buoyancy and dynamic lift whenever there is a broad range of operating speeds.

Below F_{nL} = 0.40 the vessel's weight is supported, for the most part, by buoyancy and with increasing velocity of the hull pushes its shape through the water.

Depending on hull loading for high speeds, F_{nL} > 1.0 to 1.5, a properly shaped hull form is supported largely by dynamic lift with the center of gravity (CG) of the vessel having risen above its static (zero speed) position.

Speeds between F_{nL} = 0.40 and F_{nL} > 1.0 to 1.5 result in a combination of hydrostatic and hydrodynamic lift forces supporting the weight of a vessel. The proportion of weight supported by buoyancy and dynamic lift varies in some relation with speed, hull form, hull loading, and mass distribution.

From the mid-1800s into the early 1900s, relatively small single-cylinder reciprocating steam engines evolved with developments to compound, triple, and quadruple expansion versions.

These steam engine developments, along with a confluence of other technologies, contributed to ever-increasing vessel speeds.

The early 1900s were the waning years of reciprocating steam engines as the propulsion power source of high-speed vessels and the beginning of the transition to internal combustion engines for powering "auto boats," a term related to the source of supply of engines.

The expression "auto boats" was short lived, quickly replaced by the phrase "motor boats."

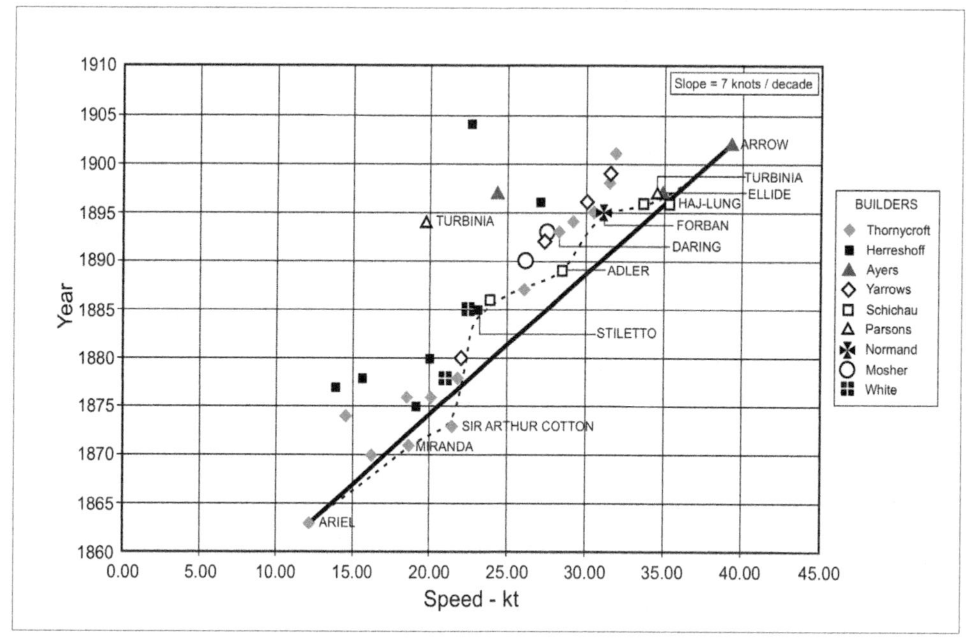

Figure 1-3: Early years of speed gains

The best measurement of performance changes following *Ariel* in 1863 is seen by improvements in speed. Selected high-performance vessels in Figure 1-3 show improving trends during the late 1800s.

On average, the rate of change of these performance data from *Ariel* in 1863 to *Arrow* in 1902 was about seven knots of speed increase for each decade.

Before 1890, demonstrated best dimensionless speed performance remained between 1.5 to 2.0 times greater than hull speed. Designers and builders were mostly European—with the exception of Nathanael Herreshoff.

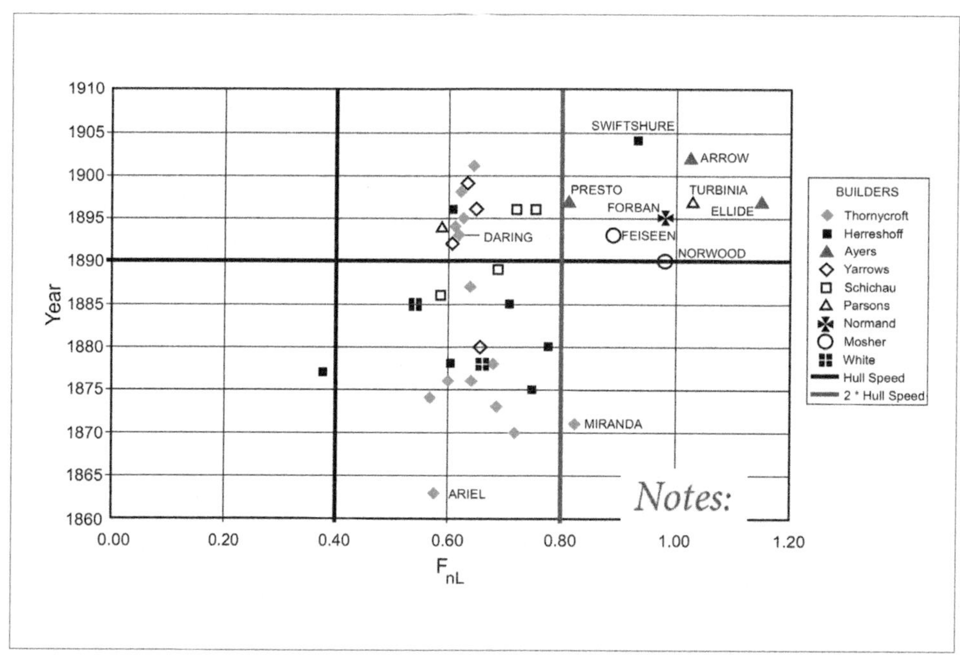

Notes:

Figure 1-4: Year vs. Length Froude number

After 1890, though, the exceptional high-performance vessels exceeded twice the hull speed as shown in Figure 1-4 with the preponderance of designs being developed by American Charles D. Mosher.

Also during this period, Herreshoff produced a design with $F_{nL} = 0.93$. And in England, Charles A. Parsons developed the steam-turbine-powered launch *Turbinia* achieving $F_{nL} = 1.03$.

At the closing of this era of steam-powered craft, the steam launch *Ellide* designed by Mosher was built in 1897 by Ayers in New York.

As shown in Figure 1-5, during sea trials, *Ellide* attained the highest ever dimensionless speed ($F_{nL} = 1.15$) for a steam-powered vessel.

Figure 1-5: S/L Ellide *attained the highest dimensionless speed ($F_{nL} = 1.15$) for a steam-powered vessel (Durand,* Marine Engineering, *December 1898)*

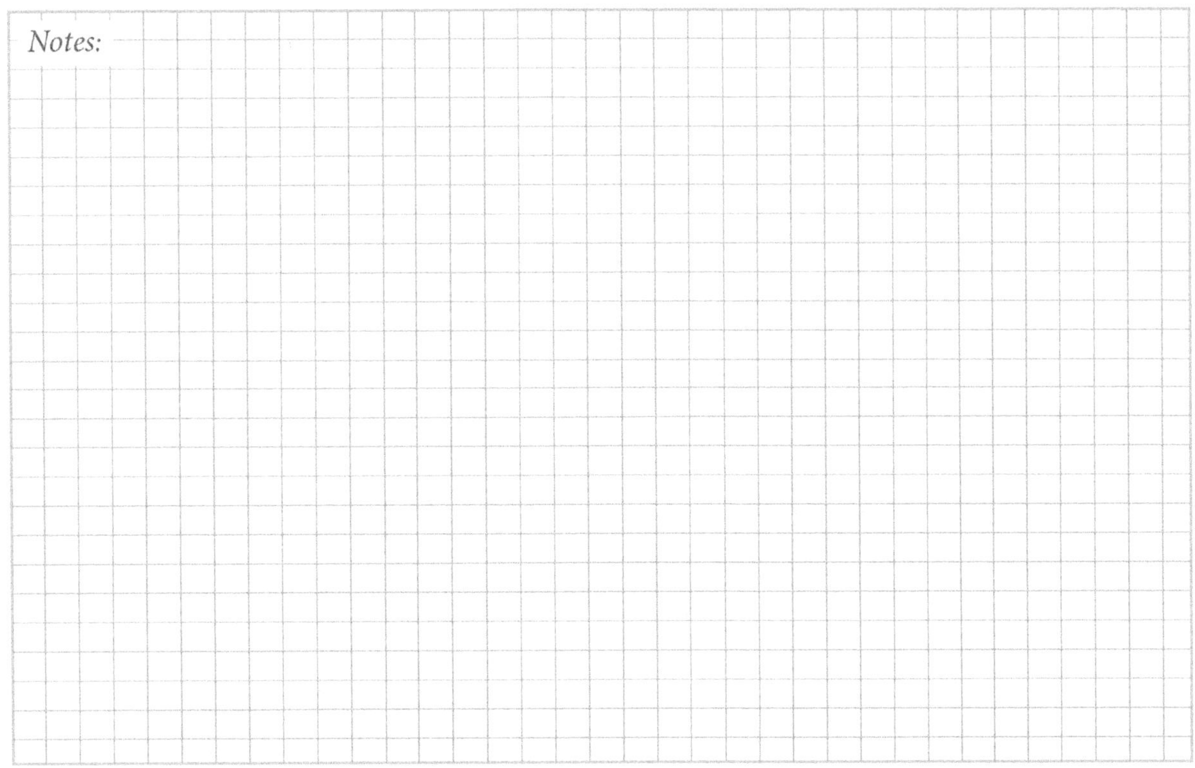

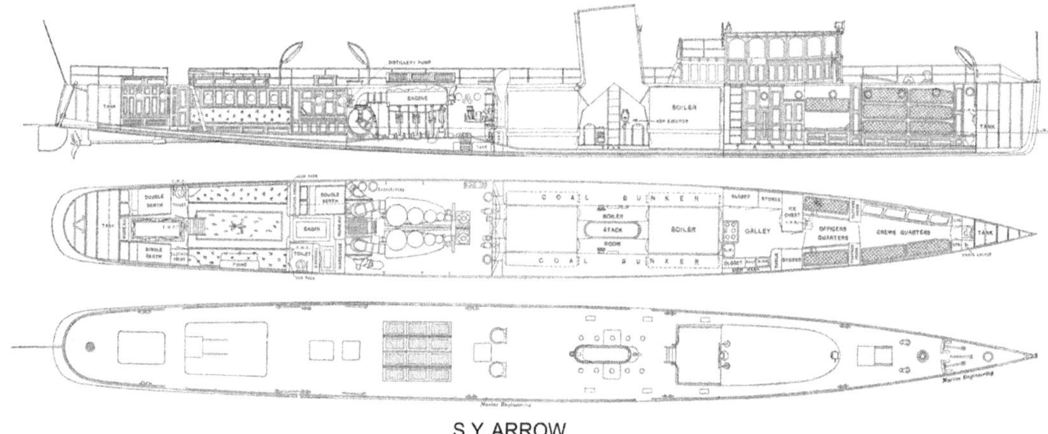

Figure 1-6: S/Y Arrow *attained the highest speed, 39.24 knots, for a steam-powered vessel* (Marine Engineering, *November 1900).*

The steam yacht *Arrow* (Figure 1-6) was also designed by Mosher and built in 1902 by Ayers. On September 7, 1902, *The New York Times* reported that *Arrow* had attained a measured speed of 39.24 knots, just shy of its design goal of 40 knots! *S/Y Arrow* remained in service for 17 years in the New York area under ownership of several different members of the New York Yacht Club.

The concentration of vessels exceeding twice hull speed after 1890 tells us that significant technology advances occurred during that period.

Defining these advances requires some speculation since there are voids in written documents and drawings. We simply don't know much about specific hull geometry of these exceptional boats except that all the hulls had round-bilge transverse sections—no hard chines—LOA/BOA was between 7.5 and 10.4 with an 8.6 average, and slenderness ratios were about 9.3 (varying between 8.3 and 10.3). Few graphics giving transverse sections are available. Some centerline profiles and plan form general arrangements give us a sense of bow and stern waterline endings. All of the vessels were propelled by screw propellers.

If you study these early writings, you'll notice a consistent understanding that ultimate speed is strongly dependent on the ratio of propulsive power to vessel weight. You will also see an interesting relationship between shipbuilders and boiler makers and steam-engine manufacturers. For example, the Simpson, Strickland & Co. Ltd. Catalog first lists steam engines, then boilers, and then lastly marine vessels such as yachts, launches, and tugs. And, from the earliest days, John I. Thorneycroft & Co. Ltd. advertised fast vessels built with engines designed and built by the same company.

A number of noteworthy individuals contributed new pieces to the technology puzzle, and made achievable speed increases possible. Here are a few: In Elbing, Prussia, shipbuilder Ferdinand Schichau was an early manufacturer of triple-expansion reciprocating steam engines in 1884. The Schichau shipyard built high-speed torpedo boats and cruisers for many countries. Designed and built by Schichau for China, the Hai Lung class of torpedo cruisers, were reported by U.S. Naval Intelligence to be the highest speed (35.2 knots) naval vessels in service throughout the world before 1900.

Working for John I. Thornycroft & Co., Ltd., naval architect Sydney Barnaby participated in propeller experiments in the company test facility. In 1885, he published his first edition of the definitive text *Marine Propellers,* and later revised it through six editions, publishing the last version in 1921. In 1894, he published the first critically important cavitation criteria necessary to establish minimum size propellers to absorb engine power.

Charles Parsons' development of the steam turbine was a significant industrial and maritime power source. Parson's knowledge of Barnaby's propeller cavitation criteria was essential for him to size propellers to successfully demonstrate the steam turbine in his launch, *Turbinia*. Parsons then developed the first cavitation tunnel to visually observe and better understand the phenomenon of propeller cavitation. This was highly significant to propeller research.

French designer and ship constructor Augustin Normand developed and published design criteria to minimize and avoid propellers drawing air down from the water surface (which reduces propeller thrust and may cause vibration) on these early high-speed, double-ended hulls.

William F. Durand was an American educator and chronicler of naval architecture and marine engineering who collected and published data on high-speed steam-powered vessels. In 1898, Durand authored and published a very detailed article with sea trial results and technical analysis of *S/L Ellide*, the highest dimensionless speed (F_{nL} = 1.15) yacht powered by reciprocating steam engines.

The enigma of this brief list of contributors was Charles D. Mosher, designer of the exceptional vessels, *S/L Ellide* and *S/Y Arrow*. Little has been written about his life and works.

He mastered the design of quadruple-expansion reciprocating steam engines and boilers, as well as the hydrodynamics of these high-speed, high length-beam ratio round-bilge hulls.

His grasp of high-speed hydrodynamics is best addressed in the following quotation: "…the boat does actually rise out of the water." That's how Mosher described *Ellide*'s sea trials in the 1904 SNAME Transactions. And then he added, "Calculations made from these marks show that the water displaced when underway at high speed was about one-half that displaced when at rest."

Credit for the design of the composite hull structure of *S/Y Arrow* may be attributed to Mosher. The mix of materials was focused on achieving light weight, and the hull materials are most interesting for 1902 as seen in the following list.

Hull Construction List for *S/Y Arrow*

- Steel frames in boiler and engine spaces
- Steel frames below the waterline and aluminum above in other spaces
- Diagonal steel strapping between frames on bottom, sides and deck
- Steel for other major structural elements
- Hull sides double-planked mahogany
- Deck of wood except over boiler space
- Aluminum deck over boiler space
- Aluminum use wherever possible (side & deck stringers, hatch covers & frames, etc.)

In the early 1900s, Clinton Crane was an American naval architect designing both sail and power boats. He is noted for the design of a series of racing boats named *Dixie* winning the International Harmsworth Trophy four times. His significant contributions were demonstrated when marine technology was in rapid transition. Steam power was waning, being replaced in performance vessels by internal combustion engines, and round-bilge hull forms were being overtaken by both hard-chine and stepped hydroplanes in the quest of increased speed.

Crane's round-bilge hull designs had won Harmsworth Trophies in 1907, 1908, and 1910, but the latter race was won only because a much faster hydroplane had an engine failure. Crane's last Harmsworth Trophy was won in 1911 by the hydroplane *Dixie IV*. To Crane's credit, he wanted to better understand the new technology, so he model-tested five stepped-hull designs at the U.S. Navy's Experimental Model Basin managed by Naval Constructor, Admiral David Taylor who showed great interest in these tests. These model resistance tests conducted in a towing tank are likely the first ever for stepped hulls in the U.S.

The speed performance of *Dixie IV* was enhanced by its propeller, which had been designed by Crane in consultation with William Durand, then head of the School of Marine Construction at Cornell University.

Crane shared his accomplishments with dynamically supported boats in the early 1900s in technical papers, discussions presented at SNAME meetings and in articles in *The Rudder* magazine.

The early 1900s was the winding-down period of high length-to-beam ratio, round-bilge hulls powered by steam engines, and the beginning of a new marine performance era. The boating world was in transition to internal combustion engines for propulsive power, and hard-chine, transom stern hulls with length-to-beam ratios trending to lower values, such as those being constructed by William H. Hand, Jr.

Other modes of transportation were evolving at the same time. Automobiles were in their infancy, and the first zeppelin flew in 1900. The Wright Brothers demonstrated controlled flight in 1903, and the Frenchman Henri Fabre designed, constructed, and piloted the first successful seaplane in March 1910 flying a distance of 457 m (1,500 ft) at an altitude of 2 m (6.5 ft). As early as World War I, 55-ft coastal patrol or torpedo boats capable of 40 knots were in military service. Those were heady times!

Through the first half of the century, yacht designers and boatbuilders made incremental improvements in hull form with successive boats coupled with affordable advances in engine power. In Europe and the United States, theoretical and experimental hydrodynamic research was sponsored to increase the success of sea planes taking off and landing on the surface of turbulent waters.

During World War II, countries on both sides of the conflict engaged in battles with large numbers of patrol torpedo boats. As there had been no university programs available, the designers of these boats gained their experience by working with builders of high-performance vessels or as apprentices in design offices.

Proving design guidance when performance requirements included speeds greater than hull speed, individual designers began to publish books about power boats. The following is a list of a few of these books in my library. These, along with many technical papers, articles, and personal experiences have contributed to many of my thoughts.

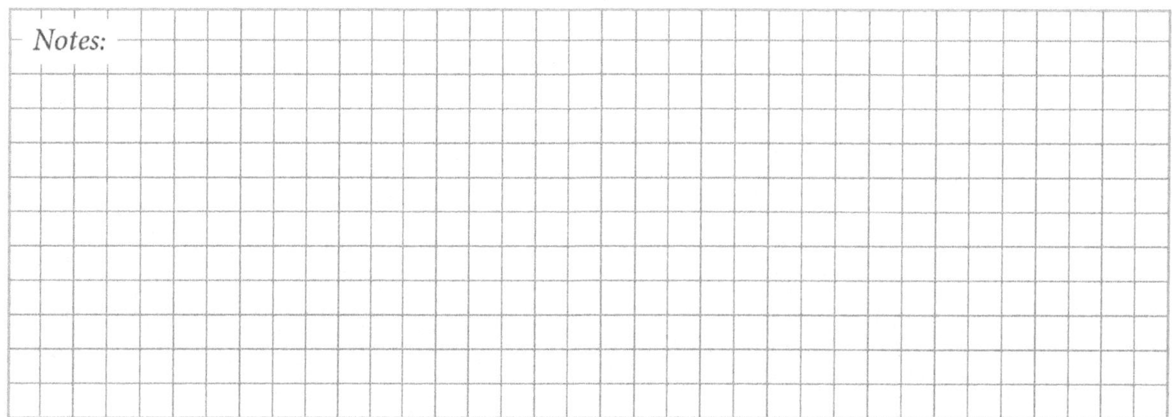

TABLE 1-A PARTIAL LIST OF HISTORICAL BOOKS IN AUTHOR'S LIBRARY
D.W. Taylor, *The Speed and Power of Ships*. (New York: John Wiley & Sons, 1910)
R. Munro Smith, *The Design and Construction of Small Craft*. (New Malden: Vizetelly & Co., Ltd., 1924)
Norman L. Skene, *Elements of Yacht Design*. Third Edition. (New York: Yachting, Inc., 1925)
Henry E. Rossell and Laurence B. Chapman, *Principles of Naval Architecture*, Volumes I and II. Eighth Printing (New York: SNAME, 1949)
Juan Baader, *Cruceros Y Lanchas Veloces*. Spanish language. (Buenos Aires: Nautica Baader, 1951)
Lindsay Lord, *Naval Architecture of Planing Boats*. Third Edition. (Cambridge, MD: Cornell Maritime Press, Inc., 1963)
Jim Stoltz and Joseph G. Koelbel, Jr., *How to Design Planing Hulls*. (New York: Motor Boating, 1963)
Peter DuCane, *High-Speed Small Craft*. First Edition (New York: Cornell Maritime Press, 1950)
Peter DuCane, *High-Speed Small Craft*. Third Edition (New York: Cornell Maritime Press, 1964)
Douglas Phillips-Birt, *Motor Yacht and Boat Design*. Second Edition. (London, in association with Tuckahoe, NY: Adlard Coles Limited, 1966)
Kenneth C. Barnaby, *Basic Naval Architecture*. Third Edition. (London: Hutchinson Scientific and Technical, 1960)
Kenneth C. Barnaby, *Basic Naval Architecture*. Sixth Edition. (London: Hutchinson Scientific and Technical, 1969)
Juan Baader, *Motorkreuzer und Schnelle Sportbotte*. German language. (Bielefeld: Verlag Delius, Klasing & Co., 1970)
Renato (Sonny) Levi, *Dhows to Deltas*. (Lymington: Nautical Publishing Company, 1971)
Andrew G. Hammitt, *Technical Yacht Design*. (New York: Van Nostrand Reinhold Company, 1975)
I.T. Egorov, M.M. Bun'kov and Y.M. Sadovnikov, Propulsive *Performance and Seaworthiness of Planing Vessels*. Russian language. Leningrad, U.S.S.R. Imprint, 1978. English Edition. (Washington, D.C.: NAVSEA Translation No. 1965)
Francis S. Kinney, *Skene's Elements of Yacht Design*. Seventh Edition. (New York: Dodd, Mead & Company, 1962)
Francis S. Kinney, *Skene's Elements of Yacht Design*. Eighth Edition. (New York: Dodd, Mead & Company, 1981)
Peter R. Payne, *Design of High-Speed Boats: Planing*. (Annapolis, MD: Fishergate, Inc., 1988)
Cyrus Hamlin, *Preliminary Design of Boats and Ships*. (Cambridge, MD: Cornell Maritime Press, 1989)
John Teale, *How to Design a Boat*. Third Edition. (Dobbs Ferry, NY: Sheridan House, 2003)

Note: Books are in the English language unless otherwise indicated.

These books can provide you with some "how to do" guidance, but not much design criteria to assist you in making quantified decisions about naval architecture. In order for a design to progress efficiently, you need to make definitive decisions at each step of progress and continue moving forward to a complete and technically correct design.

The value to you in reading these books is in recognizing the technology the authors chose to share, and in seeing the differences in their design philosophies. These books also have merit for documenting the technology available in the public domain at the time they were published.

Let's move on. Further efforts to increase speed of marine craft began to be involved with getting the hull onto the water's surface. Even before 1900, naval architects Dixon Kemp and Charles Mosher observed that when a craft with relatively straight after-body buttocks was underway with increasing speed, it would rise and skim on the surface of water.

Thus, being supported by hydrodynamic or planing lift, at high speeds wetted surface is reduced along with frictional drag. Almost simultaneously, designers were introducing transverse, flow-separating steps in the planing hull's bottom to further reduce wetted surface which mitigates drag at high speeds.

A few years later, in 1906, Italian aircraft designer Enrico Forlanini designed and built a hydrofoil craft which lifted the hull above the water and reached a speed of 37 knots with a 60-horsepower engine driving two air propellers. It wasn't until 1955 that modern development of another dynamically supported craft, air cushion vehicles (ACV) began.

In Figure 1-7, you can see three fundamental means of supporting the weight of marine craft, buoyancy, hydrodynamic lift and mechanically powered lift. This is known as the "Sustention Triangle."

Displacement craft are supported by buoyancy. When underway with increasing speed, planing boats rise and skim on the surface of water supported by dynamic lift. When hulls are fitted with hydrofoil (underwater wings), marine craft can, with sufficient speed, be entirely above the surface of the water surface with dynamic lift. Other classes of high-speed vessels such as ACV and SES are supported above the surface of the water by a low-pressure air cushion generated by mechanically powered equipment.

I am going to concentrate on marine craft with their weight being supported by the combination of buoyancy and hydrodynamic lift of planing, and I will address a design approach, providing supporting technology based on experimental data along with my experiences for developing design criteria.

You will need to differentiate between terminology shared by marine design for both performance vessels and classical naval architecture taught for displacement ships. To be aware of the meaning of terminology used in this book, I encourage you to become familiar with *Appendix 7: Notations, Definitions, and Abbreviations.* It identifies various ways in which dimensionless hull loading and speed might be defined.

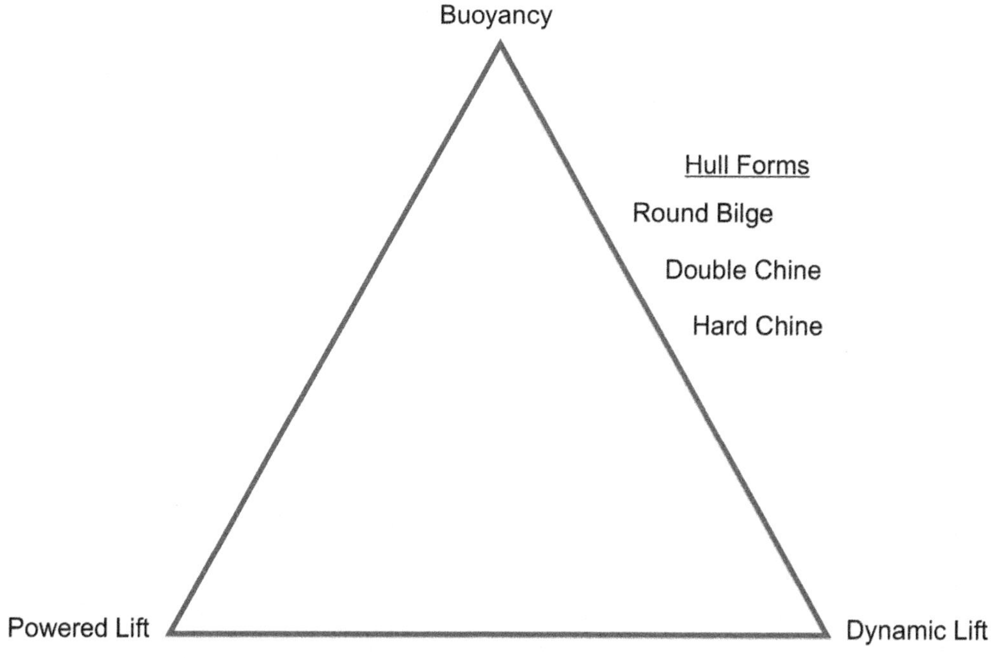

Figure 1-7: Sustention Triangle

CHAPTER TWO

Basics of High-Speed Craft: Hydrodynamics

To ease us through a discussion of the transitional range from speeds of displacement vessels up to dynamically supported planing craft, we need a compilation of information, which I've presented in Figure 2-1. But first, let's consider that the focus on performance vessels is for speeds greater than $F_{nL} = 0.40$; that is, for semi-displacement operation up to high planing speeds.

Personally, for dimensionless speeds, I am of the opinion that length Froude numbers are generally appropriate for comparing hydrodynamic technologies as long as the wetted length of the keel throughout the operational speed range remains not less than 90 percent of its static waterline length. This occurs up to $F_{nL} \approx 0.80$ or $F_{nV} \approx 2.0$. For comparative purposes of hydrodynamic technologies at speeds of $F_{nL} \geq 0.80$ or $F_{nV} \geq 2.0$ should be based on volume Froude numbers since dynamic lift vice buoyancy eventually provides the dominant force supporting a vessel's weight. However, the choice of using F_{nL} vice F_{nV} for specific situations herein depends on technical issues being studied.

The hull of a displacement vessel pushes through the water while a planing vessel at high speed glides along the water's surface. A great many modern vessels have design requirements for operation between $F_{nL} = 0.40$ and $F_{nV} = 3.0$ where the support of the vessel's total weight changes with speed and is apportioned between buoyant and dynamic lift.

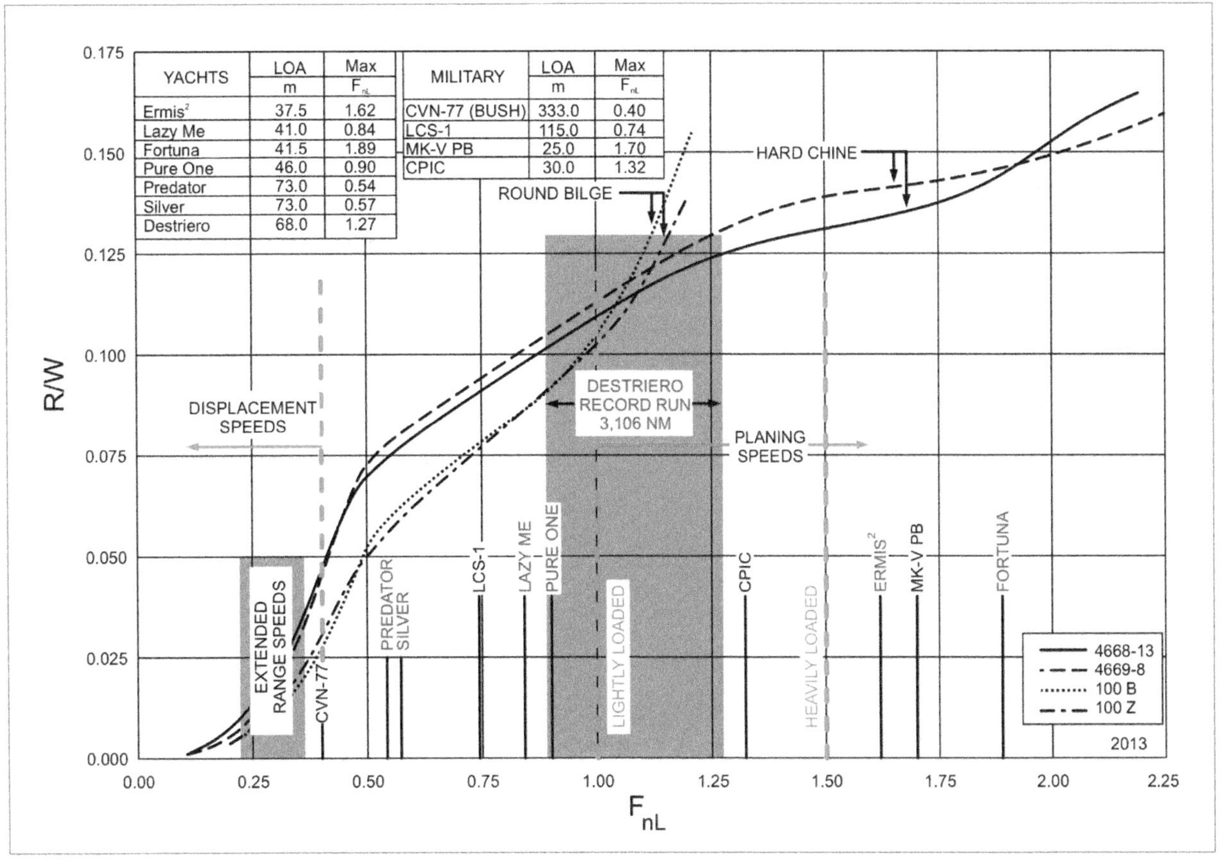

Figure 2-1: Relative calm water performance of significant modern motor yachts and naval vessels

TABLE 2-A: DIMENSIONS OF MODELS IN FIGURE 2-1 SCALED TO A 500 MT VESSEL				
Full Scale	Series 62		NPL	
	4668-13	4669-8	100 B	100 Z
L_{WL}, m	54.0	53.4	51.8	51.8
B_{WL}, m	9.8	7.8	9.6	6.9
L_{WL}/B_{WL}	5.5	7.0	5.4	7.5
$L/\nabla^{1/3}$	6.9	6.8	6.6	6.6
Δ, mt	500	500	500	500
LCG fot, m	19.5	21.8	---	---
LCB fot, m	---	---	22.6	22.6

Locating a Design in the Domain of Dimensionless Speed

In Figure 2-1, look at the R/W curves which are for 500 mt vessels. Notice that the R/W for the round bilge and hard-chine hulls cross near $F_{nL} \approx 1.10$. Hull loading for both hull forms are near, $L/\nabla^{1/3} \approx 6.7$ as shown in Table 2-A. For each hull form, there are two L/Bs of approximately 5.45 and 7.25 for these four tests, indicating that for this loading, L/B has only a secondary influence for speeds below $F_{nL} = 1.10$.

For low speeds, round-bilge hulls have reduced calm-water resistance relative to the hard-chine hulls. At higher speeds, the resistances of the hull forms reverse as the slope of the R/W curves for hard-chine hulls reduce while R/W slopes for the round-bilge hulls increase. Thus, for this simple comparison for $F_{nL} \leq 1.10$, the NPL round-bilge hulls have lower R/W than the Series 62 hard-chine hulls.

Throughout the book, we will discuss preferred hull forms in addition to calm-water resistance and other significant operational factors that affect seakeeping, maneuvering, and dynamic stability for different dimensionless speeds. Let me point out several other noteworthy pieces of information in Figure 2-1. Many large yachts achieve their extended range at speeds of $0.22 \leq F_{nL} \leq 0.35$. You can see displacement speeds highlighted in the figure as extended range speeds.

The maximum speed trends of specific, named large yachts delivered between the years 2000 and 2010 are shown. These vessels have maximum speeds—F_{nL} between 0.54 and 1.89—with the latter currently known to be the highest speed for a recreational vessel longer than 40 m.

For new yachts, trends of increasing overall length and increasing speed continue to command attention. I've also noted in black type a few naval vessels ranging from a nuclear aircraft carrier ($F_{nL} = 0.40$) to a 25m patrol boat ($F_{nL} = 1.70$). It is seen that a privately funded project set the benchmark speed ($F_{nL} = 1.89$) for 40 m-plus operational vessels.

The pink-shaded area in Figure 2-1 bounds the speed range, $0.88 \leq F_{nL} \leq 1.27$, during the Atlantic record made with an average speed of 53.1 knots in Sea State 3+. This also included an unsurpassed 1,403 nm transit during the final 24 hours, averaging 58.4 knots.

A number of other records were established the most important record was establishing the state-of-the-art which defines the relationship between vessel weight, speed, and power discussed in Chapter 8 for transport efficiency in this dimensionless speed range for transport efficiency. Note that this pink-shaded area encompasses the speed ($F_{nL} = 1.10$) where the R/W curves cross for round-bilge and hard-chine hulls. The design decision made for GT/MY *Destriero* was to have a hard-chine hull which favored the highest speeds possible as the displacement reduced as fuel was consumed.

This is an important decision-making principle with regard to performance requirements which include both long range at displacement speeds and a dash capability. The best solution in this situation is achieved with a hull form optimized for the high-speed operation. And last, Figure 2-1 includes blue lines and text to highlight $F_{nL} = 0.40$, the hull speed that I personally considered to be the upper speed for unadulterated displacement technologies.

For higher speed, a vessel's weight shows signs of being supported by a combination of hydrostatic and hydrodynamic lift. The lower speed boundaries for planing in the speed format of F_{nL} are defined by two blue lines, one for a lightly loaded hull $L/\nabla^{1/3} = 9.0$ and another for a heavily loaded hull $L/\nabla^{1/3} = 4.0$ of which both lines equate to $F_{nV} = 3.0$.

Performance vessel design requires naval architects to extend their range of knowledge well beyond fundamentals of displacement-ship technology. Displacement technology provides the necessary foundation only up to hull speed $F_{nL} = 0.40$.

In this book, the measure of heavily or lightly loaded hull forms (dimensionless or relative weight of vessels) is quantified by values of slenderness ration ($L/\nabla^{1/3}$) or area coefficient ($A_P/\nabla^{2/3}$) slenderness ratio is mostly significant for semi-displacement and -planing speeds while area coefficient is generally used for semi-planing and planing speeds. There are a total of five definitions for hull loading found under various naval architectural references included in definitions for "Relative Hull Loading" in Appendix 7. The approximate relationship between $A_P/\nabla^{2/3}$ and $L/\nabla^{2/3}$ is given in Equation 2-8 and in Appendix 7 noting that $L \approx L_p$.

Fortunately, there is a large knowledge base for designing performance vessels for speeds well beyond $F_{nL} = 0.40$. Unfortunately, sources of many of these references may be difficult to locate. Actually finding quantified criteria necessary to make sound design decisions when applying the existing knowledge base is difficult. Considering their accumulated experiences to be proprietary, design professionals often are reluctant to share criteria honed from comparisons of their design process and assumptions with sea trial results.

In August 1992, the 67.7 m, 1,000 mt vessel GT/MY *Destriero* established the record of minimum time (58 hr–34 min–30 sec) for a non-refueled transit of the Atlantic Ocean from New York to England. In less than a day after arrival in Plymouth, I was onboard the vessel to congratulate Captain Odoardo Mancini for commanding *Destriero* on a record crossing for this race horse of the sea. You can imagine my joy at the success of this once-in-a-lifetime experience.

My intent is to bring together reliable experimental resources that designers may use along with the same decision-making criteria that has provided me with many successes.

Donald Blount and Capt. Odoardo Mancini exchange mutual congratulations for GT/MY Destriero*'s record*

Planing Defined

In an academic setting, naval architects study displacement technology representative of low-speed vessels. Performance craft designers, small craft operators, and waterfront wags speak of performance craft and planing speeds from a different mindset, talking about a hull skimming across the water's surface—rather than pushing its way through the water.

So let's pause here for my opinion on conditions for planing, along with the minimum dimensionless speeds for which hydrodynamic lift is likely to provide the major part of the support of the weight of a speeding craft.

Let me quickly debunk the myth that "any hull will plane if enough propulsive power is installed." Ready? A double-ended canoe will not plane. As a canoe's speed increases, it sinks deeper into the water due to negative or suction pressures that developed by the velocity of water flowing along its convex curved buttocks.

In order to have planing occur, a hull bottom must have a three-dimensional shape, which, in reaction to trim attitude and the velocity of water over the wetted shape, develops net positive pressure of a magnitude that results in the center of gravity rising to an elevation that the remaining below-water hull volume is less than 50 percent of its static volume.

Due to all the possible shapes hulls may have, there is no single dimensional or dimensionless speed that can be designated as the threshold for a boat to plane.

In order to have planing occur, a hull bottom must have a three-dimensional shape, which, in reaction to trim attitude and the velocity of water over the wetted shape, develops net positive pressure of a magnitude that results in the center of gravity rising to an elevation that the remaining below-water volume is less than 50 percent of its static volume.

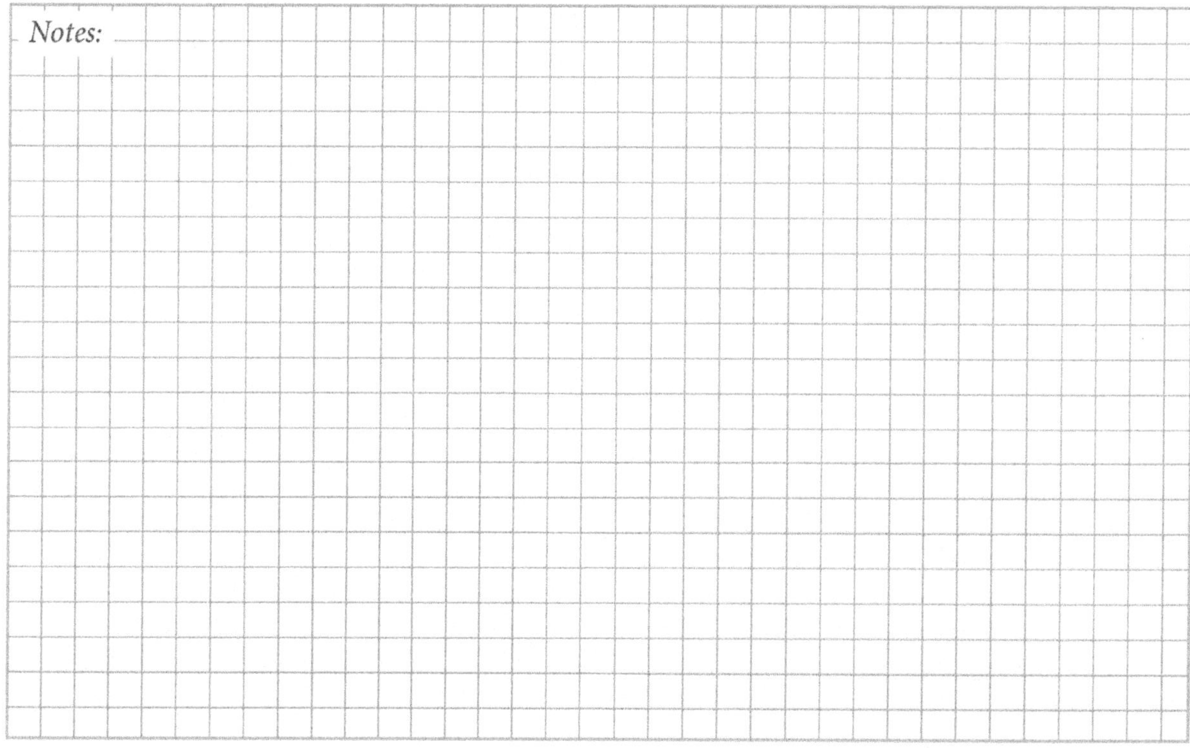

As there cannot be an exact definition of planing based solely on speed, there is no reason to debate the merits of various definitions in print, or even with friends over a drink in a bar.

I will, however, list a few examples of the technical, mythical, and empirical definitions of the threshold planing speed:

1. Speed at which water separates from the hull bottom at the transom $C_V/(L_M/B_T)^{1/2} \approx 1$ (Savitsky, Neidinger, 1954)
2. Speed (without the use of trim control devices) at which the maximum trim angle of the hull begins to reduce
3. Speed-length ratio of 3.5 (based on static waterline length)
4. Length Froude number $F_{nL} = 1.0$ (based on static waterline length)
5. Volume Froude numbers $F_{nV} = 2.8$ or $F_{nV} = 3.0$
6. Speed at which the VCG of the boat begins to rise above its static level
7. Buoyant component of planing load is negligible, that is, for constant trim the wetted length-to-beam ratio is not governed by C_{Lb}.[1]

NOTE:
[1]*Based on formulations for planing lift replicating experimental data which does not have more than a 20 percent contribution of buoyant lift. (Kapryan, Weinstein – 1952) (Locke – 1948) (Weinstein, Kapryan – 1953).*

For a boat with chine length of 69 ft (21 m) and a displacement of 100,000 lb (45.4 mt) based on model tests of the parent hull of Series 62, this table provides a comparison of the definitions listed above:

DEFINITION	APPROXIMATE SPEEDS FOR DEFINITIONS	F_{NV}
1	24.3 kt	2.1
2	31.5 kt	2.8
3	27.9 kt	2.4
4	26.7 kt	2.3
5a (F_{nV} = 2.8) 5b (F_{nV} = 3.0)	32.0 kt 34.3 kt	2.8 3.0
6	20.0 kt	1.7
7	46.8 kt	4.1

You can visualize the differences in these seven definitions in Figure 2-2 depicting resistance-to-weight ratio and trim versus F_{nV}. Each planing speed definition is identified by the number from this table. There is enough scatter in the diagram to indicate that some of the definitions have little hydrodynamic significance.

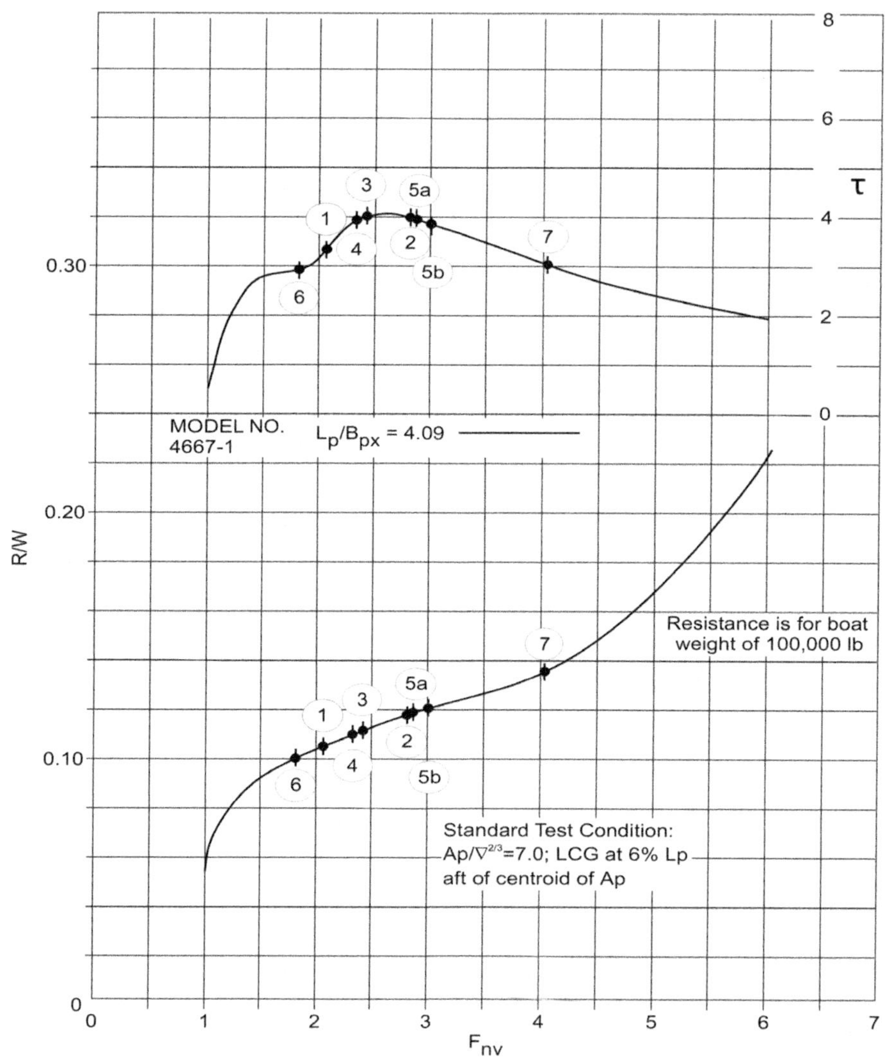

Figure 2-2: R/W and angle of attack versus F_{nV} indicating various definitions of the threshold of planing

UNDERWAY VCG

Both 12.5° deadrise data for Series 62 hull designs and 25° from Delft provide sufficient information to quantify the approximate speeds and hull loading at which the VCG of hard-chine planing boats begin to rise above its static level due to dynamic lift. The round-bilge hulls of the NPL series also exhibit VCG rise above their static level at approximately the same F_{nL} as hard-chine hulls. Following initial squat with low forward speed, with further increase in speed, the hull's VCG rises through its static elevation due to hydrodynamic lift supplementing buoyancy, as shown in Figure 2-3.

The speed at which the VCG returns to the zero speed elevation is approximated by equations 2.1 and 2.3. Experimental data from both Series 62 and NPL data indicate the bandwidth of variation for the VCG to pass through the zero speed elevation is about $0.6 \leq F_{nL} \leq 0.8$.

$$F_{nV} = 0.37 + 0.2167(L/\nabla^{1/3}) \quad \text{Equation 2-1}$$

And by definition

$$F_{nL} = F_{nV}/(L/\nabla^{1/3})^{1/2} \quad \text{Equation 2-2}$$

Therefore Equation 2-1 becomes

$$F_{nL} = 0.37(L/\nabla^{1/3})^{-1/2} + 0.2167(L/\nabla^{1/3})^{1/2} \quad \text{Equation 2-3}$$

The speed at which the VCG passes through its static elevation is sensitive to hull loading for both hard-chine and round-bilge craft. Hull loading defined by slenderness ratio is germane for round-bilge hulls while for planing hulls area coefficient, $A_P/\nabla^{2/3}$ is relevant.

Round-bilge model tests indicate that VCG generally passes through its static elevation for $F_{nL} \leq 0.80$ when $L/\nabla^{1/3} \geq 6.0$. For $L/\nabla^{1/3} \geq 6.0$ (CG Rise)/B tends to reach a maximum of 0.02 near $F_{nL} \approx 1.20$ and if round-bilge hulls are overpowered, VCG descends for further speed increase, possibly leading to dynamic instability. VCG may not return to its static elevation at any speed when $L/\nabla^{1/3} \leq 4.5$.

Figure 2-3 provides experimental boundaries for Series 62 and Delft-type planing hulls for deadrise between 12.5° and 25°. The upper curve (very heavy) is representative of a planing area coefficient of 4.0. The middle curve relates to an area coefficient of 5.5. A typical full-load design condition while the lower curve (very light) would be relevant for a very high-speed monohull.

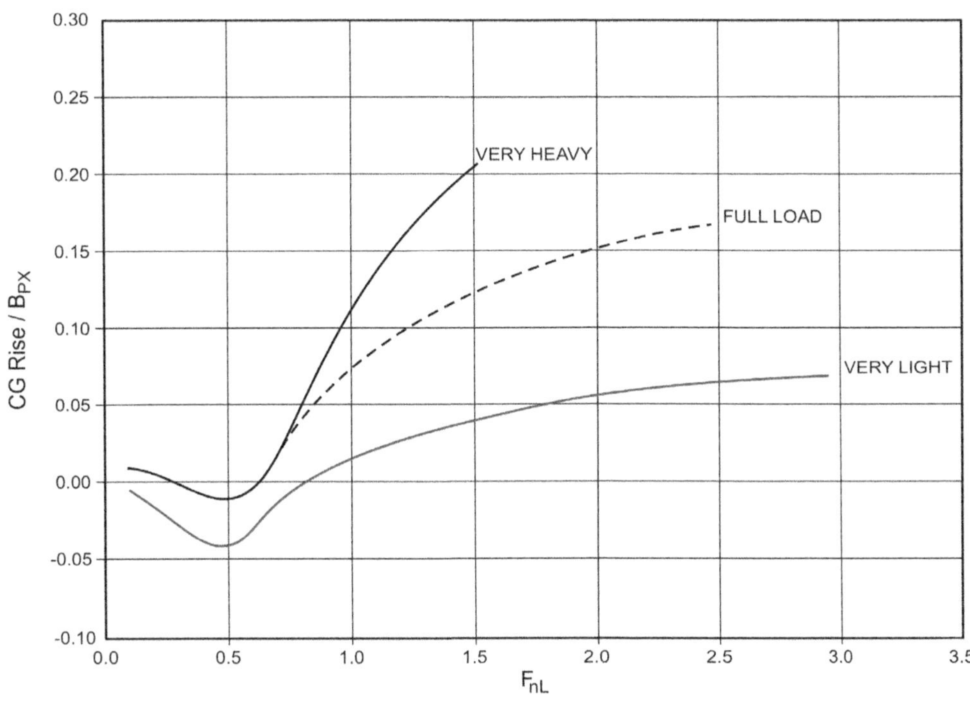

Figure 2-3: Experimental boundaries of CG rise/B_{PX} for Series 62 models

Speeds Defined by Technology

At increasing speed, whereby the CG rises to the same elevation of zero speed, the static weight of the vessel is supported by the combination of buoyancy along with the sum of both positive and negative hydrodynamic forces. However, the LCB of the submerged part of the hull is likely to be shifted aft from its zero speed longitudinal location as the underway trim will have changed. The proportion of static vessel weight supported by buoyancy will be substantial, but less than 100%, when the underway VCG returns to its static elevation.

The range from zero speed up to $F_{nL} = 0.40$ is referred to as displacement speeds, as hydrostatic principles tend to dominate and the hull's draft has increased due to sinkage from speed. At speeds above $F_{nL} \approx 0.40$, vessels essentially become supported by the sum of hydrostatic buoyancy, and both positive and negative hydrodynamic forces. As the VCG of the hull returns to its static elevation near $F_{nL} \approx 0.65$ it is essentially at the upper boundary for semi-displacement speeds as hydrostatic and hydrodynamic forces have equivalent significance. When there is an appropriate hull shape, further increases in speed develop additional hydrodynamic lift, which becomes greater than hydrostatic buoyancy. This results in continued changing of proportions of these lifting components supporting the total weight of the vessel. Then semi-planing speeds will occur from $F_{nL} \approx 0.65$ up to $F_{nV} \approx 3.0$. Note that the lower speed is defined by F_{nL} while the upper speed is defined by F_{nV}.

As I mentioned previously, there cannot be an exact definition for the threshold of planing based *solely* on speed since hull loading constitutes displaced volume as well as dynamic lift factors. For our purposes, however, we will accept $F_{nV} = 3.0$ as the lowest planing speed.

In Figure 2-4, I show general graphic guidelines of my definition of different speed regimes versus volume of displacement for appropriate hydrodynamic technologies for both FPS and SI dimension systems. Varied effects of specific hull geometry, however, cannot be included in this format. The approximate lengths provided at the top of the graph are for $L/\nabla^{1/3} = 6.0$, a fairly typical hull loading for performance vessels.

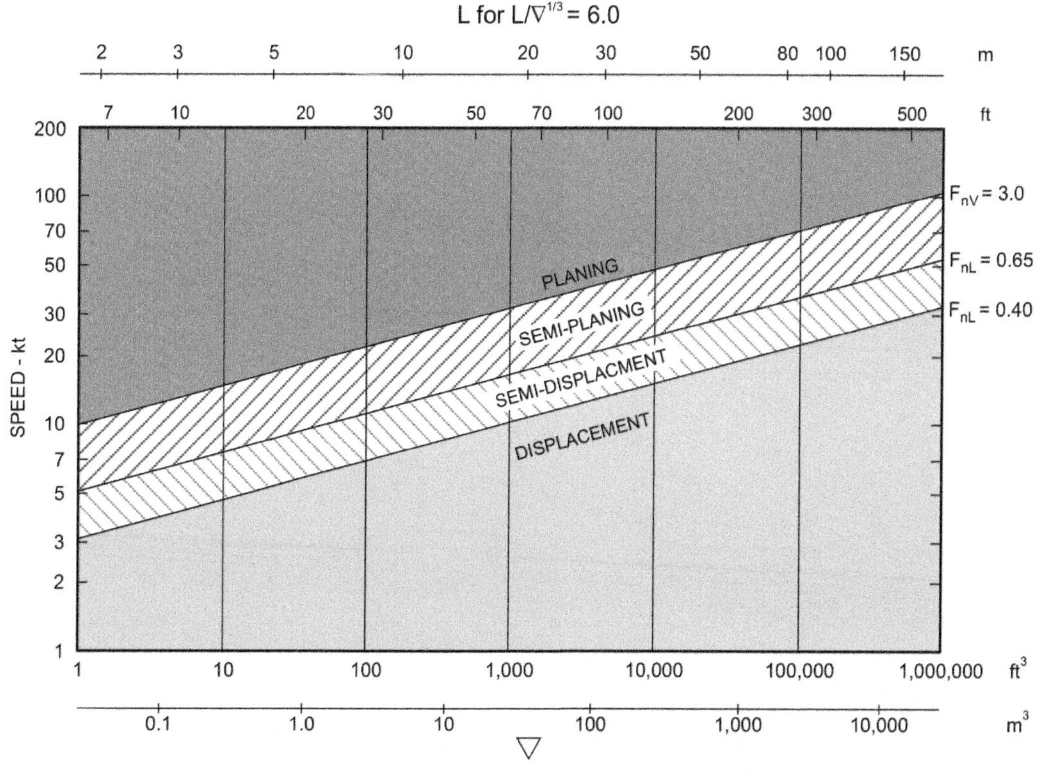

Figure 2-4: General guidelines for different speed regimes

Dry Transom Speed and Wake Hollow

Before leaving low vessel speeds, I want to mention the condition at which flow separates from transom sterns. This condition can be of interest to designers and operators when requirements call for stern launching and retrieving small craft, unmanned surface vessels, towed sensors, and so on, from a vessel when underway.

The water surface behind the hull should be as smooth as it ever will be when the water flow cleanly separates from the hull bottom at the transom. When this condition initially occurs, it is called "dry-transom speed."

It is also referred to as the transom being "fully ventilated" in some studies. Noteworthy references on this topic are Savitsky and Neidinger (1954), Oving (1985), Maki (2005), and Hadler (2007).

In Figure 2-5a, you will find guidance for predicting the minimum deep-water speed for dry transom to be achieved for round-bilge and hard-chine hulls.

The right side of the figure represents higher speeds when the transom is dry. The left side, identified as "transom unwetting," begins at zero speed with the transom immersed to the static waterline.

As the vessel accelerates, the level of turbulent water aft of the transom reduces until flow separates from the hull bottom resulting in a "dry transom".

At speeds above $F_{nL} \approx 0.40$, significant parameters are based on both static transom draft (T_T) and waterline beam (B_T) along with underway transom draft Froude number (F_T).

Although there are some differences in the guidance offered by these references, these variances are not significant from an engineering perspective.

To make an independent assessment, I had a model tested to accurately measure the minimum speed for clean transom flow separation for three displacements with three LCGs at each displacement.

These data points from my tests are plotted in Figure 2-5a and are very close to Oving's results.

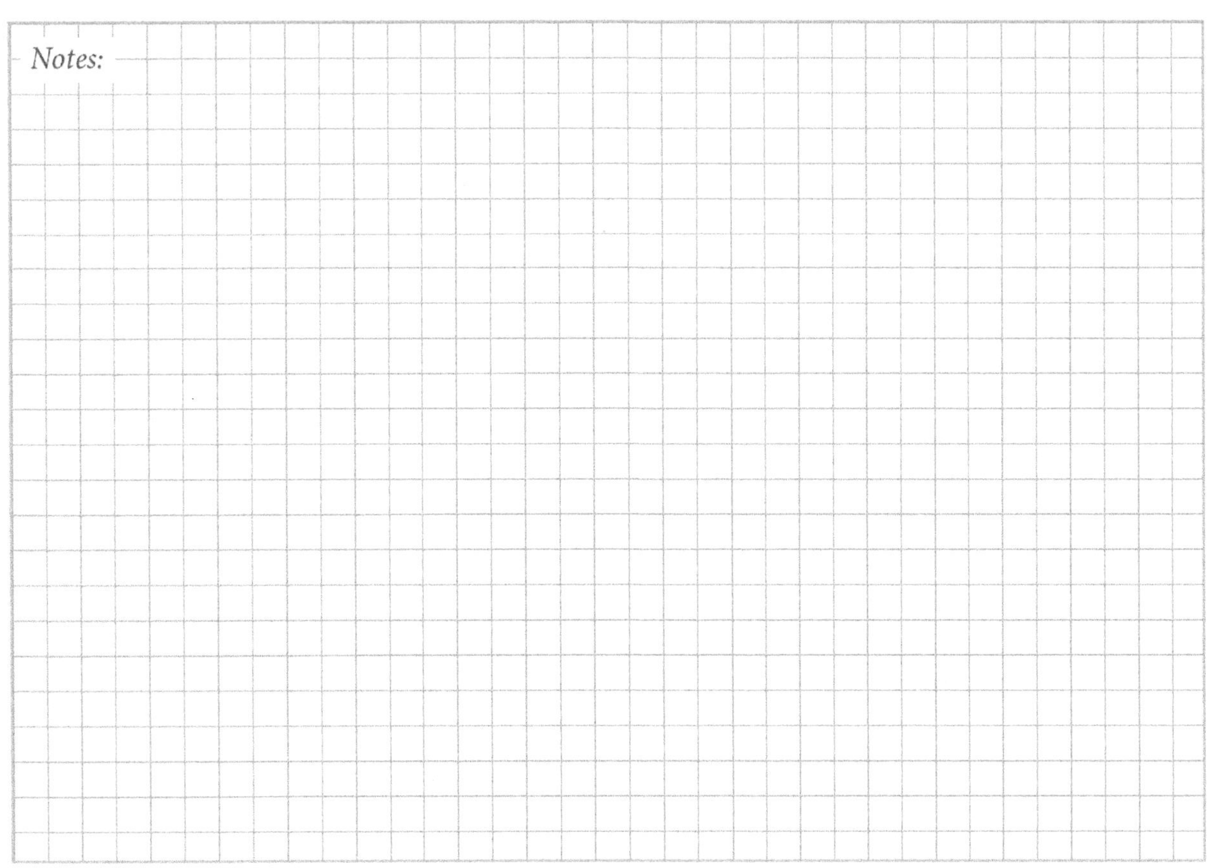

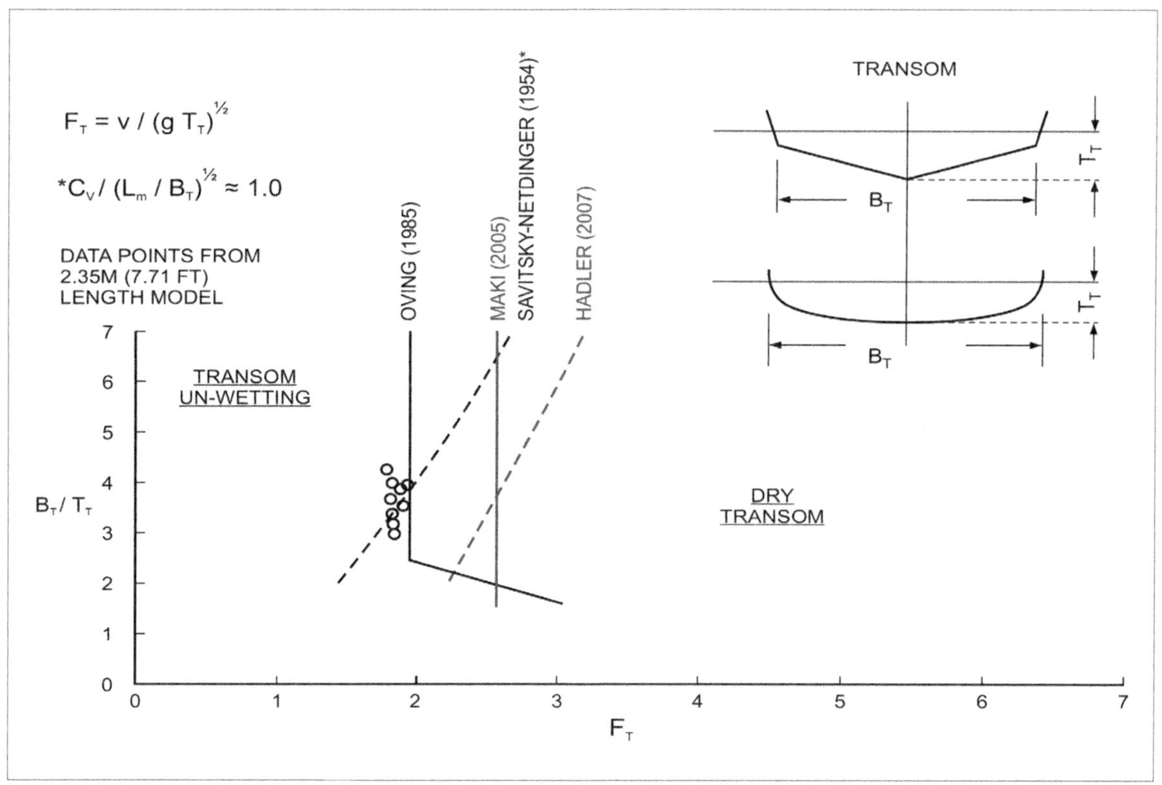

Figure 2-5a: Guidance for predicting the minimum speed for dry transom in deep water

A second confirming set of model tests of a flat-bottomed planing hull reported by Morabito (2013) show that deep-water transom ventilation occurs at an average $F_T = 1.80$ for a wide range of B_T/T_T. I use the Oving criteria of $F_T = 1.95$ for $B_T/T_T \geq 2.50$. More conservative decision makers for $B_T/T_T \geq 2.0$ might use the Maki 2005 Equation 2 (the two coefficient regression curve fit) which suggests $F_T = 2.55$.

In 2012, Taravella et al. published a very simple model (VSM) approximating with about 90 percent accuracy the length and wake hollow depth behind a fully ventilated—or dry transom—of high-speed vessels. When you are locating height and length of transom-mount structures such as dive or swim platform, this VSM can be useful for concept design studies. It can also provide input when you are making an assessment of wave resistance.

The origin of the Cartesian coordinate system for the equations approximating the length and depth of the wake hollow is at the centerline of the still waterline of the hydrodynamic transom with positive longitudinal distances (X) measured aft of the hull. Equation 2-4 estimates the wake length, and Equation 2-5 provides the longitudinal distribution of the depth of the wake hollow aft of a dry transom.

The wake hollow geometry is depicted in Figure 2-5b.

$$L_W/T_T = F_T\,(B_T/T_T)\alpha^{1/2} \qquad \text{Equation 2-4}$$

$$Z(x)/T_T = (1/\alpha)\,(1/F_T^{2})\,(T_T/B_T)^{2}\,X^{2} + (2/\alpha^{1/2})\,(1/F_T)\,(T_T/B_T)\,X - 1$$

$$\text{for } 0 \leq X \leq L_W/T_T \qquad \text{Equation 2-5}$$

The geometry of the transom is represented by the numerical value of α and affects this approximation. For transom shapes which are triangular, α = 1.0; for elliptical shapes, α = π/2; and α = 2.0 for rectangular transoms. This VSM is parabolic downwards (negative Z is measured down from the still water surface), and it shows the flow becoming asymptotic to the water surface at L_W.

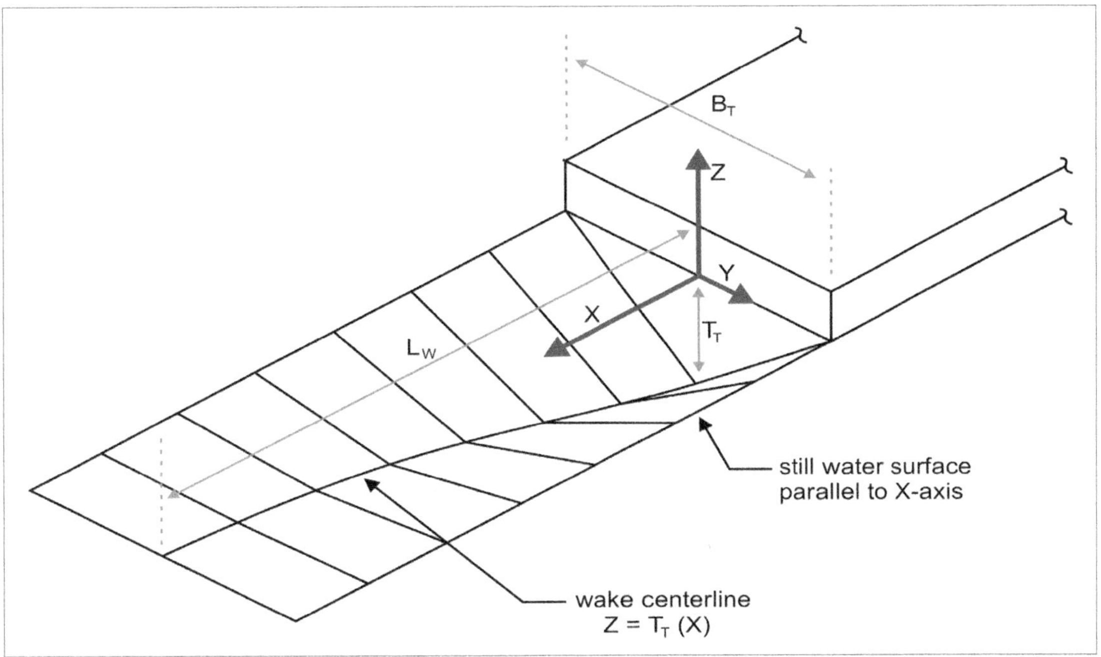

Figure 2-5b: Wake hollow geometry

Semi-Displacement Hulls Powered for Speed Increase

The combination of cruise and maximum speeds, along with increasing hull lengths of motoryachts, commercial and military vessels, has resulted in an increase in the number of vessels having operational Froude numbers (F_{nL}) exceeding 1.2.

In the concentration of yacht projects in 2005, vessels typically had F_{nL} up to 0.6 with a few approaching operational speeds of F_{nL} = 1.2. Hull forms with different transverse sections show variations in resistance characteristics for similar slenderness ratios in this range of Froude numbers.

On occasion, to achieve minimum resistance for traditional round-bilge displacement hull forms in this speed range, designers will employ techniques such as forward LCG and stern wedges with hook in aft buttocks.

These techniques need to be approached with caution. While others have reported instabilities as low as 16 knots, my experience has been that dynamic instabilities generally result from induced bow-down trim moments at speeds greater than 22 to 25 knots.

For example: For round-bilge hulls, flow-separating strakes crossing the bilge radius and extending forward to become spray rails above the water surface may be essential when design maximum speed requirements begin to exceed 22 to 25 knots.

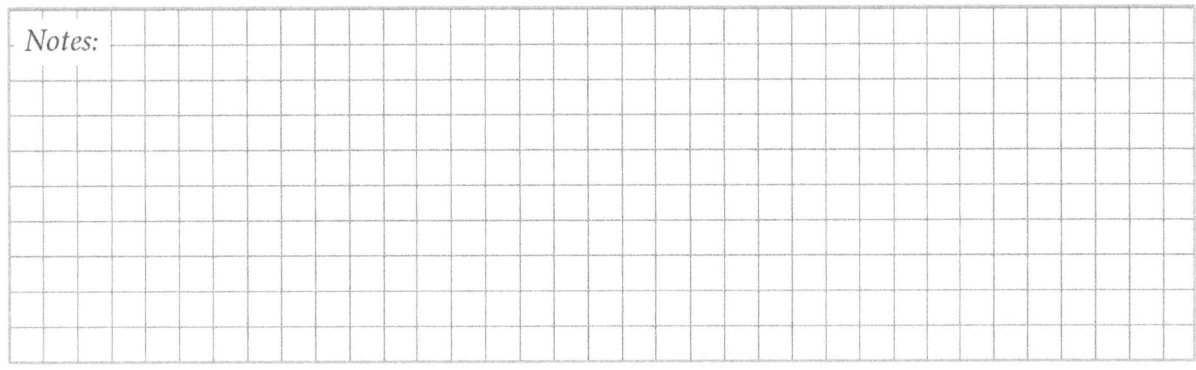

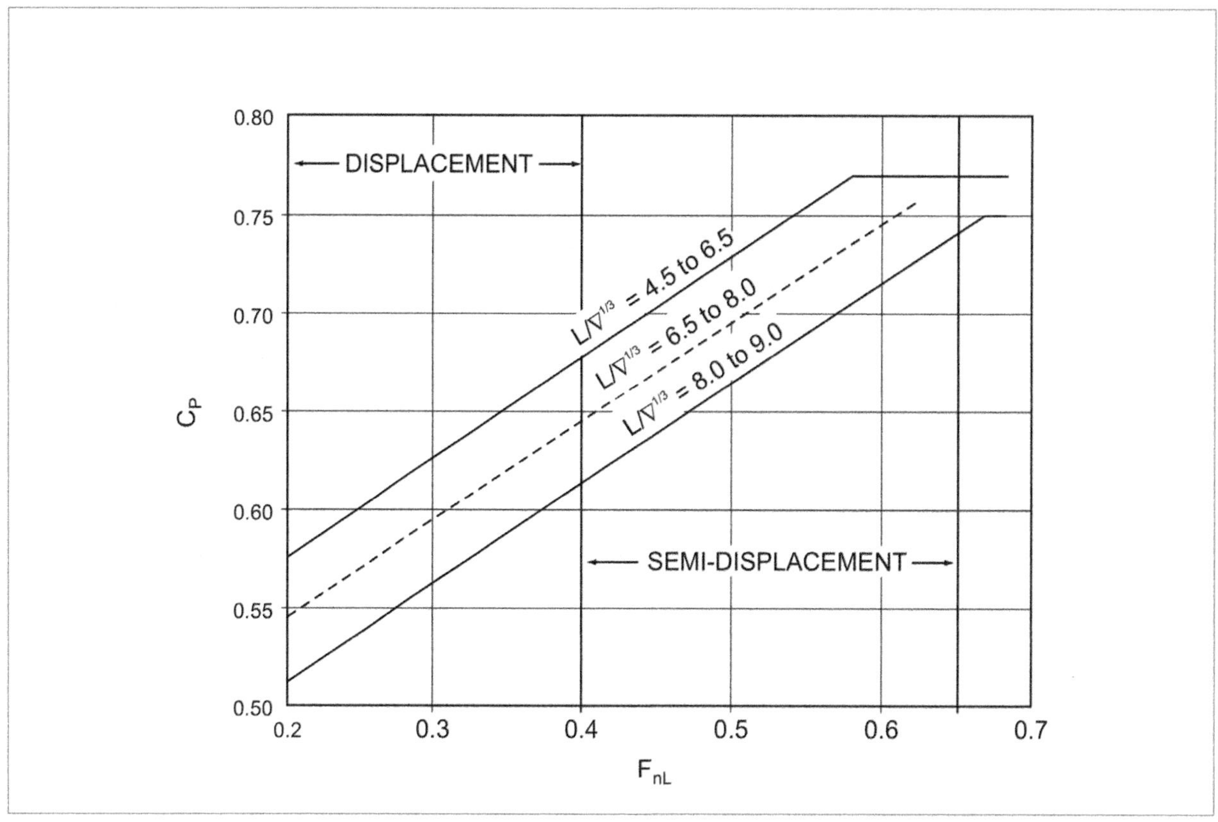

Figure 2-6: C_P versus F_{nL} for minimum R/W

The Waning Influence of Some Displacement Hull Form Coefficients

In the 1860s, the thought-to-be speed barrier for marine vessels—(F_{nL} = 0.40)—was exceeded. As we transition from that period to one where craft consistently attain higher Froude numbers, we should discuss the waning influence of some displacement hull form coefficients.

The Taylor Standard Series (1907-1914) showed that residuary resistance is mainly dependent on prismatic coefficient (C_P) at low Froude numbers while displacement length ratio, a dimensional number, $\Delta/(0.01L)^3$ had dominant influence at high Froude numbers.

Note: *In this book, the dimensionless displacement-hull loading factor, slenderness ratio, is used in preference to displacement-length ratio. The relationship between these two ratios for salt water is:*

$$L/\nabla^{1/3} = 30.57/[\Delta/(0.01L)^3]^{1/3} \qquad \text{Equation 2-6}$$

With additional experimental data supplementing the C_P guidance originally derived from Taylor's Standard Series, optimum C_P for minimum R/W (vice minimum R_r/Δ) has been found to differ somewhat.

This is provided in Figure 2-6 for $0.2 \leq F_{nL} \leq 0.65$. As noted in this figure, $L/\nabla^{1/3}$ also has some influence on optimum C_P. For $F_{nL} > 0.65$, C_P does not seem to have further influence for hull-form optimization with regard to resistance.

Block coefficient (C_B) also has little significant effect for optimizing high-speed hull forms for minimum R/W for $0.40 \leq F_{nL} \leq 0.65$. This finding is supported by round bilge model series data for $0.37 \leq C_B \leq 0.46$. These models had an average C_B of 0.43.

In the speed range of $0.40 \leq F_{nL} \leq 0.65$, the half angle of entrance at the waterline $i_e \leq 11$ degrees has little effect on R/W except for a narrow speed range between 0.50 and 0.60 whereby R/W is increased only by an increment of 0.01 when $i_e > 8.0$ degrees.

The comments about C_P, C_B, and i_e are for calm-water conditions. A Navy study of round-bilge vessels in head seas indicated that pitch is reduced and ride quality could be improved by having LCF as much as 0.11L aft of LCB for $0.38 \leq F_{nL} \leq 0.54$.

One additional point to consider is a bulbous bow when the design includes a long range requirement for $F_{nL} < 0.40$ and maximum speeds up to $F_{nL} \leq 0.65$.

In this speed range bulbous bows can have a significant influence on lowering calm water R/W. However, due consideration for the details of bulb design must be taken into account as both positive and negative effects on pitch in waves are possible. You will find guidance for selecting the size and shape of bulbous bows in Chapter 4.

As we go forward, our discussions of hydrodynamic characteristics will continue to progress from low vessel speeds on to planing conditions. And our study of model test data for fast round-bilge hull forms will lead up to the high speeds of planing craft. This effort will continue to be based on analysis and characteristics of model experiments with some findings related to documented sea trial results of as-built boats and craft.

When it is important to clarify a technical point for contrasting hydrodynamic phenomenon from semi-displacement up to planing operation, I may occasionally deviate from this pattern with arguments supported by analyzed data.

Design decisions cannot be expected to have the same effect on performance at all speeds. You must have a complete understanding of hydrodynamic aspects of hulls, operating from zero to very high planing speeds.

Wetted areas and displaced hull volumes change materially with speed, so take care that no performance-enhancing detail for hull or appendages, focused on intermediate speeds, negatively impacts performance at other speeds.

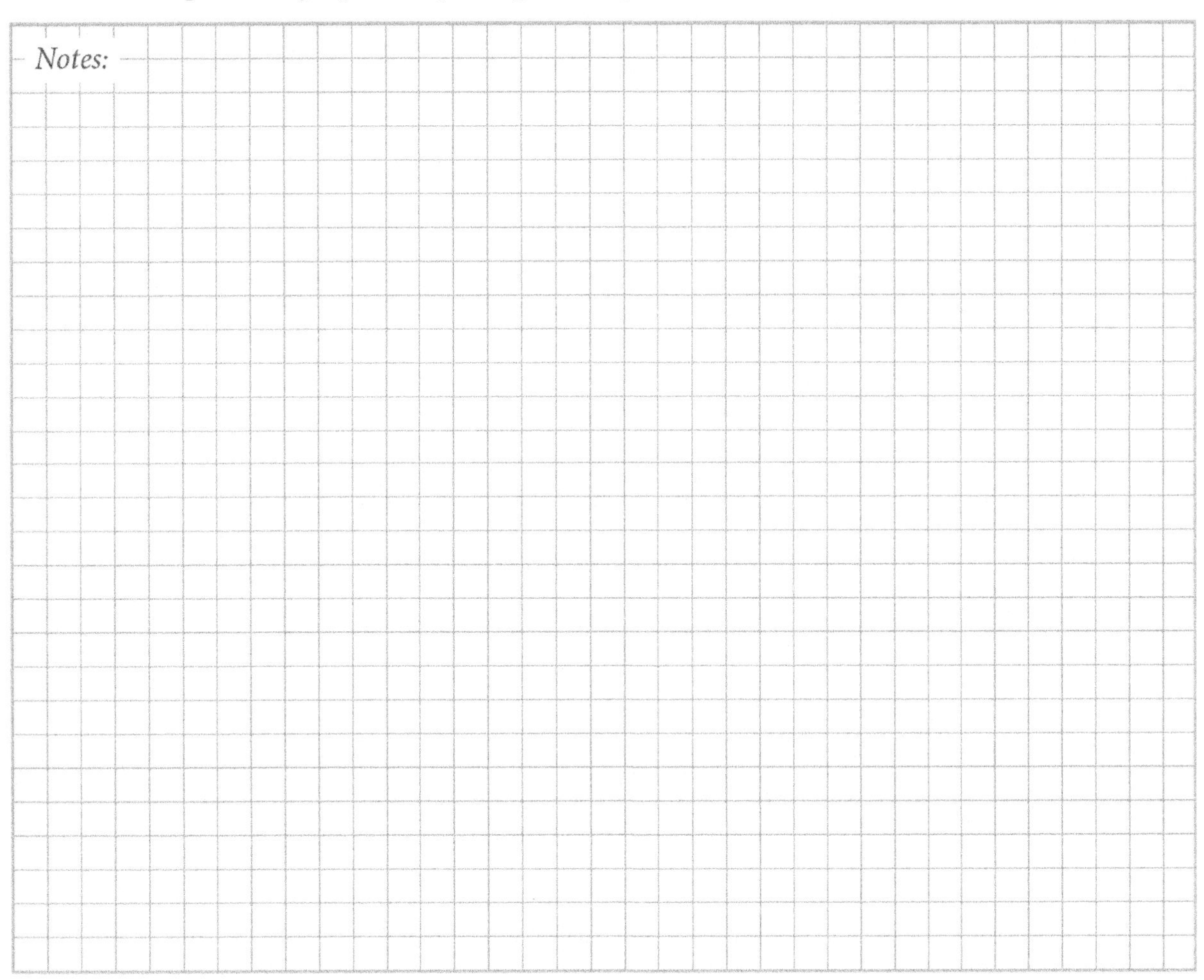

Speed-Related Changes to Flow Patterns

A sequence of underwater photographs of a hard-chine hull best depict the speed-related change of flow patterns. Figure 2-7 provides definition of calm water flow on the hull bottom for a planing speed. L_K and L_C, respectively, are the wetted lengths at the keel and chine.

The black line indicates where the still water surface would intersect the hull in its elevated and trimmed position. The red stagnation line notes the location of solid water contact with the hull's surface: peak hydrodynamic pressure. The region of spray is between the stagnation line and the spray boundary line.

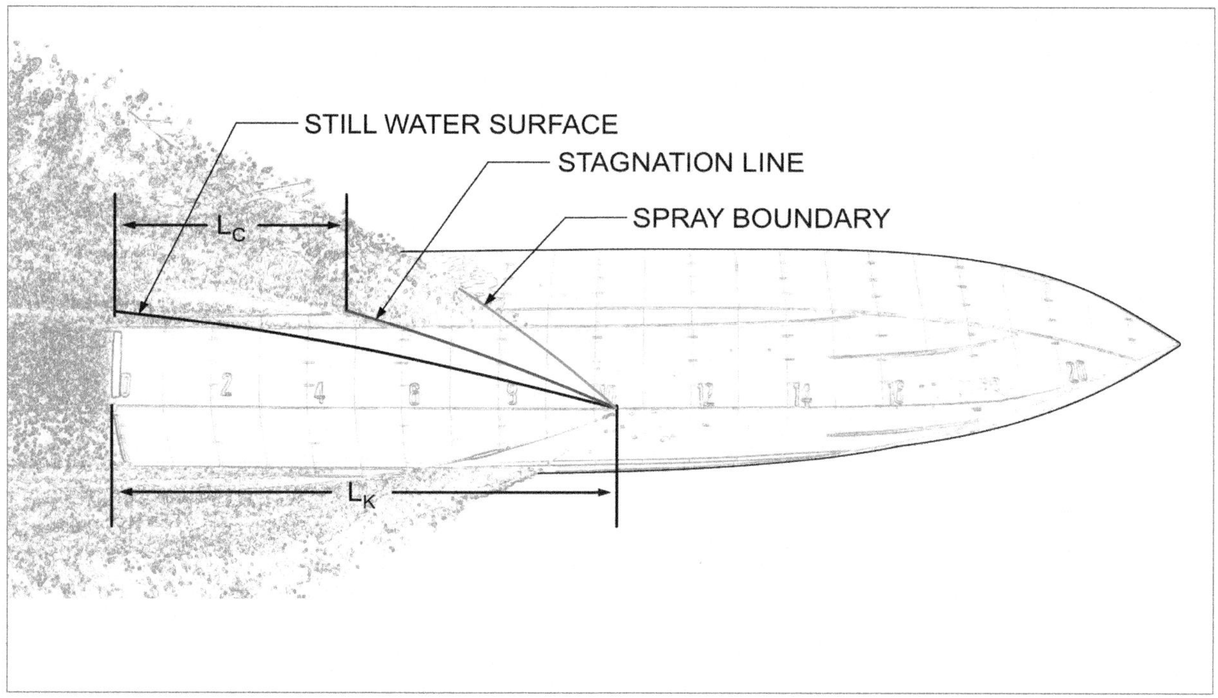

Figure 2-7: Definition of water flow boundaries at a planing speed

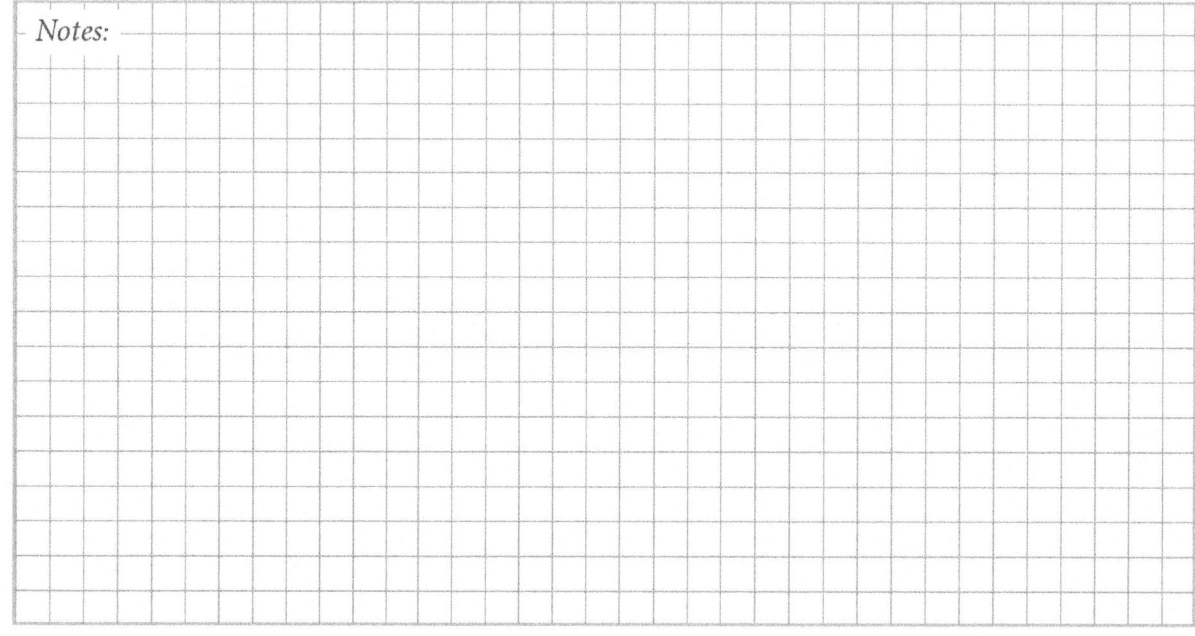

Figure 2-8 provides a sequence of underwater photographs of a hard-chine hull designed for F_{nV} = +5.0. Geometric characteristics for the 100,000 lb (45.4 mt) example are provided in Table 2-B. These photographs show how the area of wetted bottom changes significantly as speed increases from displacement to high planing speeds. In this figure, dimensionless speeds of F_{nL} are based on static waterline length. You should take note of changes of both the location and shape of the stagnation line as well as the direction of spray with increased speed.

Also seen are both a reduction of wetted lifting surface and the centroid of wetted area moving aft. Thus, care must be taken when developing the geometry of the entire hull surface below the chine so that the weight of the craft will be supported in a stable and efficient manner throughout its operational speed range.

And especially be aware of the changing wetted areas so that you can properly locate sea-water intakes and transducers/sensors which must always be submerged in non-aerated water.

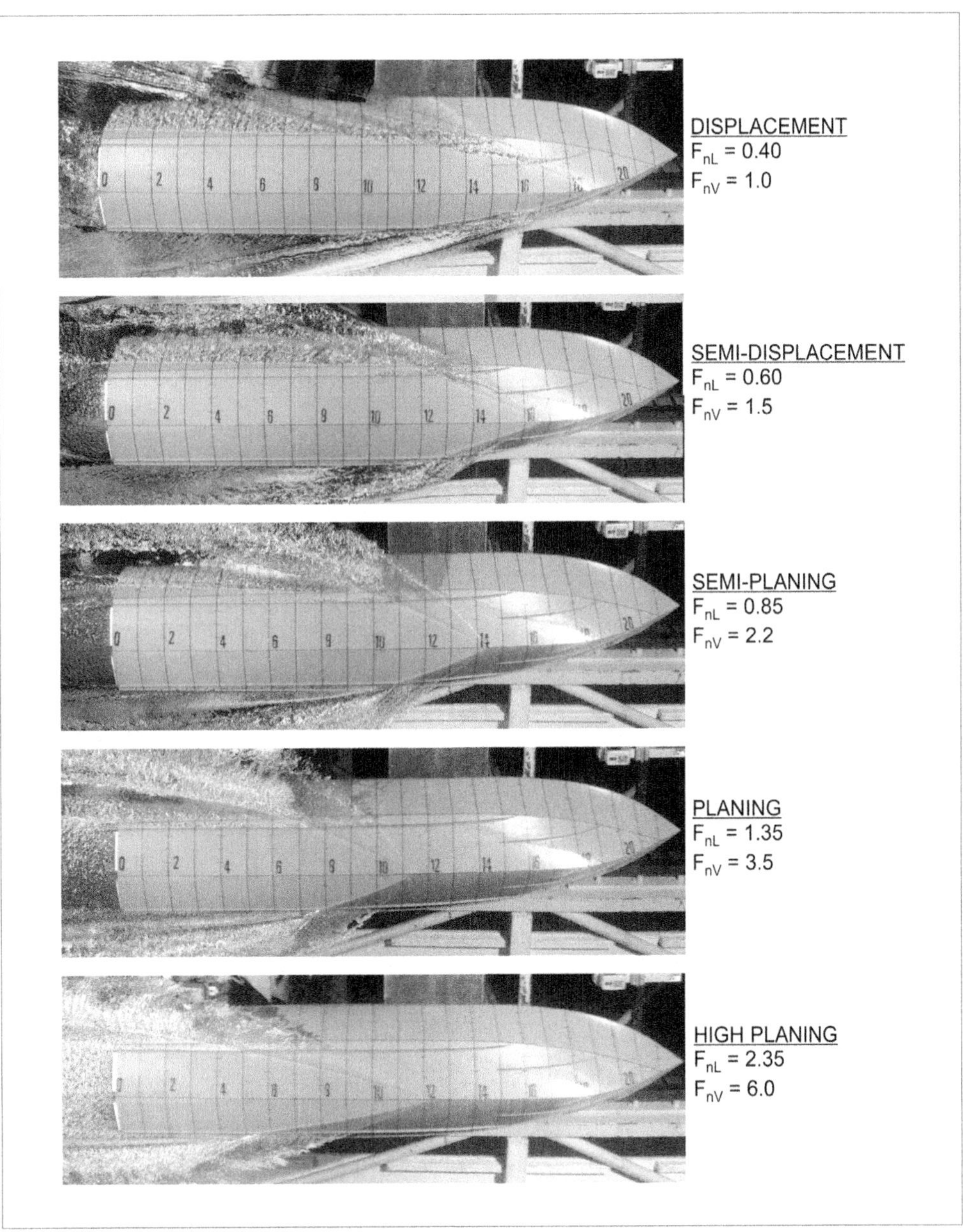

Figure 2-8: Sequence of underwater photographs from displacement to high planing speeds of a hard chine craft

TABLE 2-B PRINCIPAL CHARACTERISTICS OF HARD-CHINE CRAFT		
Measurements	FPS	SI
Displacement	100,000 lb	45.4 mt
L_P	75.3 ft	22.9 m
B_{PX}	13.4 ft	4.1 m
β_{mid}	21.7 deg	21.7 deg
β_T	20 deg	20 deg
Afterbody Warp	0.60 deg/Beam	
L_P/B_{PX}	5.6	
B_{PT}/B_{PX}	0.937	
$A_P/\nabla^{2/3}$	6.3	
$L_P/\nabla^{1/3}$	6.5	
$(CA_P - LCG)\ L_P$	14.3%	

Having examined the variation of wetted bottom area throughout a range of speeds, let's look at what happens to boats in a calm-water environment.

Figure 2-9 provides experimental model data, towed in the free-to-trim-and-heave conditions. (This is the model seen in the underwater photographs in Figure 2-8.) To show the approximate proportion of weight supported by buoyancy as the craft increases speed (F_{nV}), a still-water plane was projected on the profile of the hull so that the submerged volume (∇_V) could be calculated.

The ratio ∇_V/∇ is shown as a function of speed, which indicates that at $F_{nV} = 1.0$ the hull squats (due to negative dynamic pressure) resulting in a potential virtual load on the water of 1.1 times the displacement of the craft.

For speeds greater than $F_{nV} \approx 1.5$, net hydrodynamic lift begins to offset the buoyant support of craft weight. At the nominal threshold of planing, $F_{nV} = 3.0$ as defined in this book, about 50 percent of the craft's volume is below the still water plane and at very high planing speeds, this volume is approximately 15 percent of the static hull volume of displacement.

The volume ratio (∇_v/∇) however, represents only an approximation of the proportion of buoyant lift relative to total craft weight. This can only be an approximation as the transom of this underwater shape is dry, and is devoid of hydrostatic pressure.

Using a combination of experimental data for a rectangular flat plate and analytical pure planing lift equations, Shuford (1958) assumed that the difference between these two values represented the lift due to buoyancy. The range of Shuford's model trim data was from 4° to 20°, and he reported that the calculated volume below the still water surface multiplied by density overestimated the buoyant lift force by a factor of two.

When looking only at these data for trim angles of 4° and 8°—which is in keeping with the trim range of planing craft free-to-trim and heave–the buoyant lift force would be about 2/3 of the calculated volume below the surface multiplied by the density of water.

For $\tau = 4°$ buoyant lift was approximately 90% of the product of water density multiplied by the submerged volume, while for large trim angles ($\tau > 8°$) buoyant lift was only about 40% of the product of density times submerged volume.

Brown (1971) found a factor of 0.62 for calculating static lift from the underwater volume of dry-transom planing vessels. Thus, the static lift force would be found by multiplying underwater hull volume by density and 0.62.

Referring to the graph of ∇_v/∇ in Figure 2-9, I calculated static lift force for dry transom conditions up to the very high speed ($F_{nV} = 6.7$). I found—for the hull design reported in Figure 2-9—that an average multiplying factor of 0.70 times the underwater volume and density makes a reasonable estimate of static lift force.

With this multiplying factor of 0.70, the centroid of underwater volume and therefore the assumed hydrostatic lift has an essentially constant value of 0.28 L_K forward of the transom.

For estimating the static lifting force for a dry-transom condition at recommended planing speeds, I recommend that you use an average multiplying factor of 0.66. Thus, for each speed, static lift would be underwater volume multiplied by density and 0.66 with its effective center of lift located at $0.28L_K$ forward of the transom.

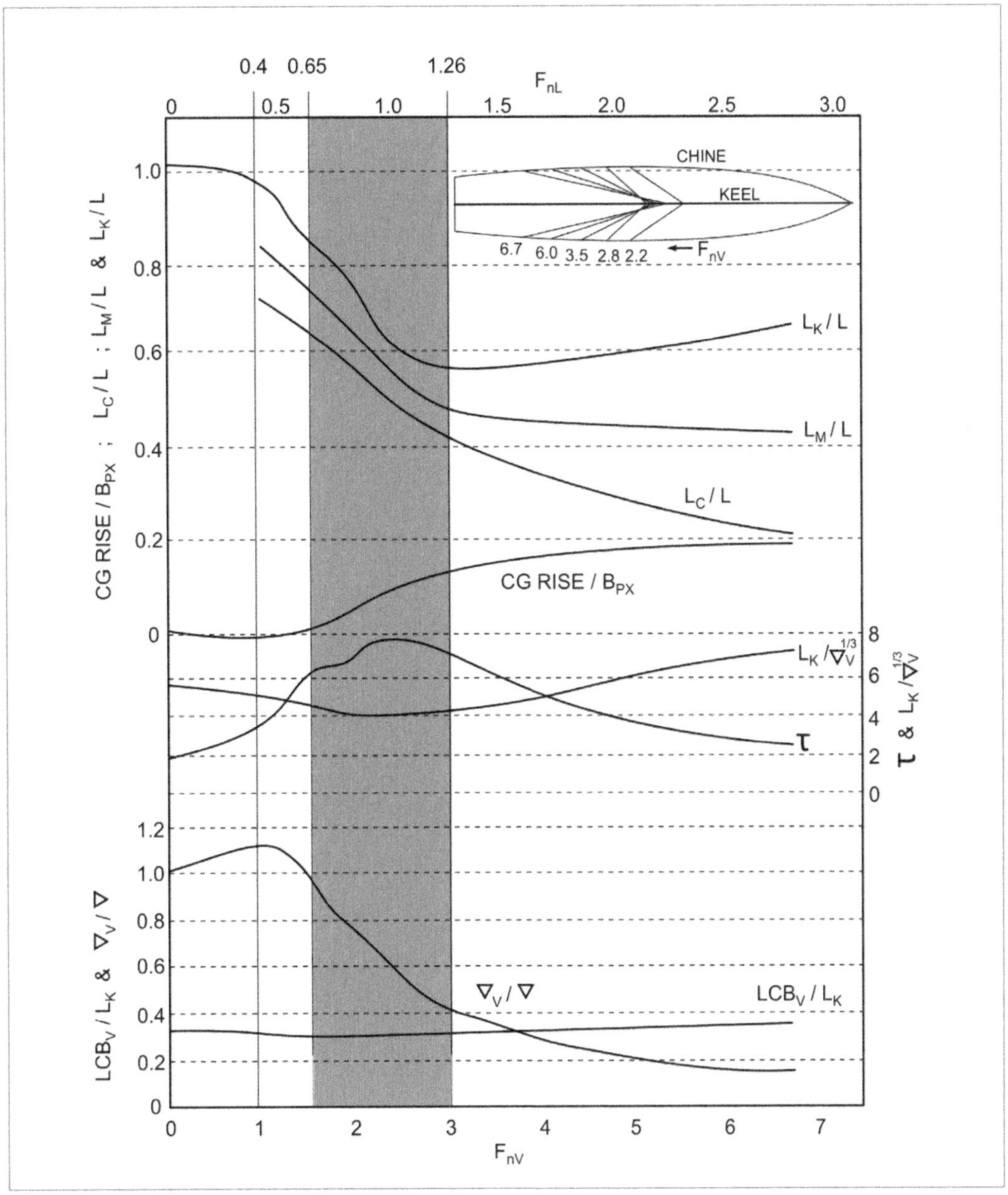

Figure 2-9: Speed-related changes of a hard-chine hull relative to the water surface

Notes:

Figure 2-9 also shows the rise of the hull, presented as a ratio CG/B as well as trim versus speed. Worthy of note is wetted lengths relative to static waterline length for the keel (L_K/L), for the chine (L_C/L) and mean of keel and chine (L_M/L). For this example, the chine wetted length continues to shorten with increasing speed while the wetted length of the keel shortens until F_{nV} = 3.0 and then begins to increase with speed.

The mean wetted length, however, tends to become constant at very high speeds. You can see a graphic of the planing bottom inserted in Figure 2-9 with the stagnation lines given for a range of speeds. At semi- and low-planing speeds, these lines make a constant angle with the centerline. At very high speeds, this angle reduces significantly as L_K begins to increase in length while L_C continues to shorten. During model tests or sea trials, observable measurement of increasing L_K when planing provides an indication of the potential onset of porpoising at higher speeds.

For high planing speeds, the lengthening of L_K while L_C shortens results in the stagnation line making a smaller angle with the centerline. Further increases in speed for this hull (F_{nV} > 6.0) will eventually cause the L_C to move aft past the end of the chine and actually separate at the transom (defined as a chines dry condition) potentially leading to chine walking.

In order to describe hydrodynamics for a full range of vessel speeds from zero to F_{nV} > 6 it was necessary to use model test results of a hard-chine hull. Designers need to have a clear understanding of the dimensionless speed specified by their clients' requirements. This piece of information is an initial indicator of the appropriate hull form.

It also predicts the difficulty of the design task. For requirements of FnV ≤ 1.0, the hull and propulsion hydrodynamics for vessels having normal loads are the least difficult. For dimensionless speeds of 3.0 ≤ FnV ≤ 5.0, hull and propulsion hydrodynamics are manageable design tasks for calm water, but seakeeping and ride quality in seas may dominate the design process as the hull rises relative to the water surface and the bottom responds to the irregular sea surface. The in-between speeds (1.0 ≤ FnV ≤ 3.0) give rise to additional technical complications as a vessel's weight transitions from buoyant to hydrodynamic support as speed increases.

Revisit Figure 2-9 to see the CG changing from a sinkage position to a substantial elevation, the mean wetted length reduces from 85 to 48 percent of the static waterline length, the trim angle almost doubles and buoyant support is reduced to about 50 percent of total weight.

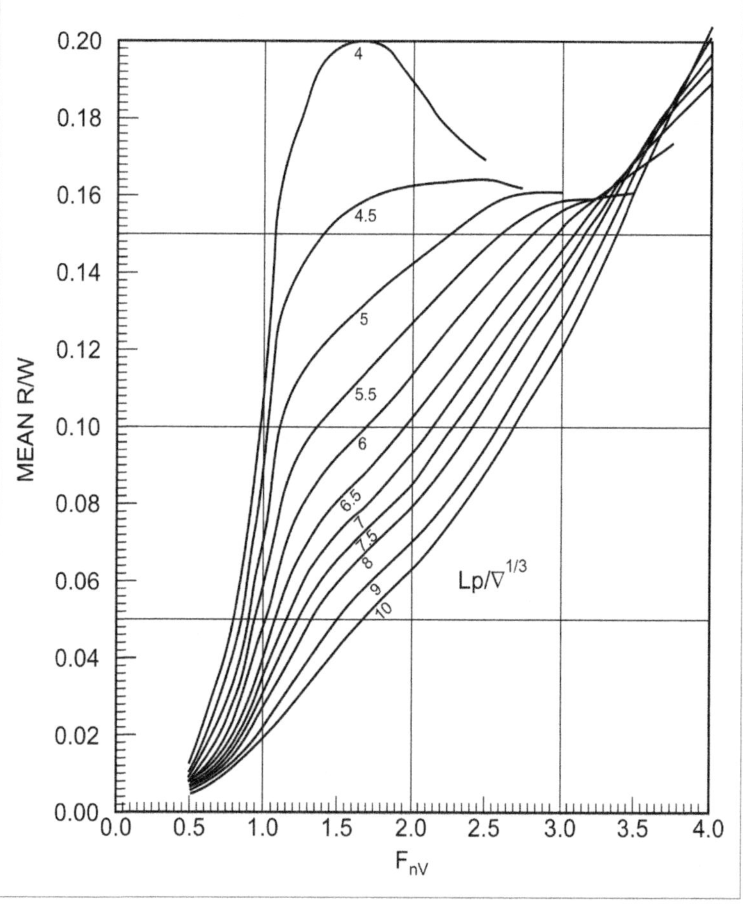

Figure 2-10: Mean R/W in salt water for the Series 62 & 65 regression for displacement = 100,000 lb (45.4 mt) with C_A = 0

In Figure 2-10, you can appreciate how the extreme sensitivity of R/W is to $L/\nabla^{1/3}$ in this speed range which adds another complication to the risk of making accurate speed predictions as well as responsible design decisions for propulsor and power sizing. This figure should grab the attention of designers of vessels having requirements for maximum or cruise speeds between volume Froude numbers of 1.0 and 3.0 as well as planing vessels which must transit through this range. An additional complication relates to increases to resistance when operating in shallow water which we will cover in Chapter 3.

In summary, the successful design of performance craft begins with requirements in which the operational range of dimensionless speeds is defined, a target full-load displacement can be estimated, and the most likely operational sea environment is known. With this information, initial hull form and overall dimensions may be bounded by hull loading. Slenderness ratio ($L/\nabla^{1/3}$) will dominate design decisions for F_{nV} between 1.0 and 3.0. For high-speed vessels/planing craft operating at dimensionless speeds $F_{nV} \geq 3.0$ loading based on hull bottom area ($A_P/\nabla^{2/3}$) tends to control design decisions.

This area coefficient ($A_P/\nabla^{2/3}$) is the "footprint on the water" and has a practical limiting value to allow sustained hydrodynamic lift of a hull at or near the water's surface. Later in this chapter, the planing area coefficient will be discussed further.

Hull Resistance for $0.4 \leq F_{nL} \leq 1.0$

With the technical challenges related to minimizing hull resistance for speeds from F_{nL} = 0.40 to 1.00, a study of public domain data was undertaken by Blount and McGrath (2009) to explore the subtleties necessary for defining criteria for low-drag, dynamically stable hull forms. It is also important, however, to remember that while a goal is achieving low-hull resistance in order to have a fast vessel with good fuel economy, speed in a seaway without adequate dynamic stability and good ride quality can result in an ill-served client. My hope here is to dissect the technology for semi-displacement speeds, and to provide informed guidance for the design process with initial focus on bare hull resistance in calm water.

In studying references for $F_{nL} > 0.30$, I found reports of significant experimental data for models tested free to trim and heave, with systematic variations both in hull geometry and loading. Body plans for parent hulls of the systematic series, along with a brief summary of the variables studies in each series, are provided in Appendix 3. Recent model test data for individual hull designs have also been used to independently validate trends and magnitude of these resistance data. However, the latter body plans are proprietary and are not included in Appendix 3.

For comparative analyses, we chose to use total bare-hull resistance-to-weight ratio for a single displacement which takes into account both frictional and residual resistance. We took this approach in preference to the format of the Taylor Standard Series, which focused only on presenting the hull-form effect of changes to residual resistance which necessitated additional calculations of frictional resistance followed by total resistance before sufficient knowledge is available to make a design decision.

Figure 2-11 brings some reality to the speed range of this section. R/W versus speed is provided from several model tests scaled to 500 mt vessels for calm, deep salt water using ATTC 1947 friction line with zero correlation allowance. This friction line had been used in the majority of public domain information regarding resistance data from available model series tests listed in Appendix 3. All model data were scaled to displacements of 500 mt with full-scale waterline lengths allowed to vary with a resulting mean of 178 ft (54.2 m). Also, full-scale Reynolds number (R_n) values are greater than $1x10^8$ which is in a range, whereby C_F is equivalent for both ATTC 1947 and ITTC 1957 formulations.

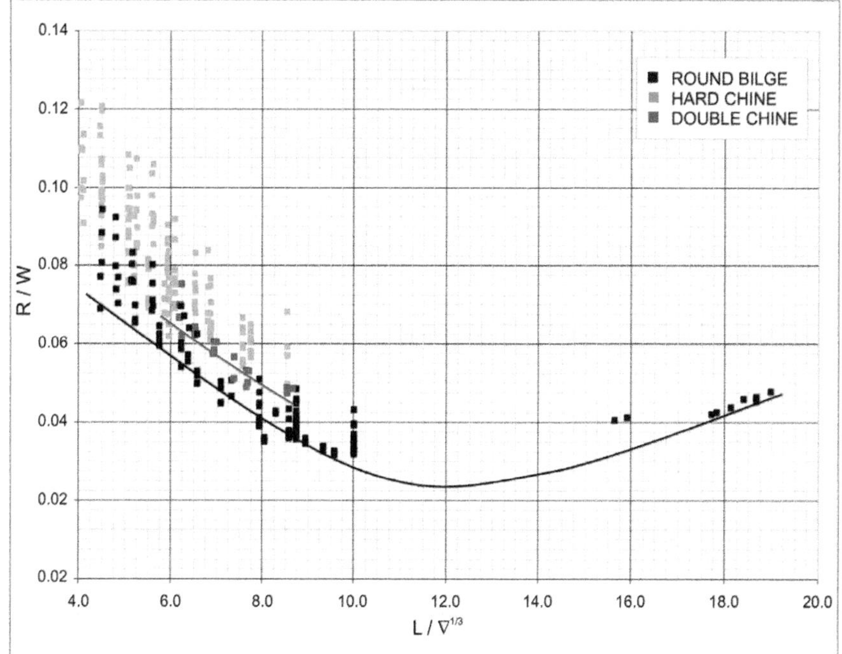

Figure 2-11a: F_{nL} = 0.5

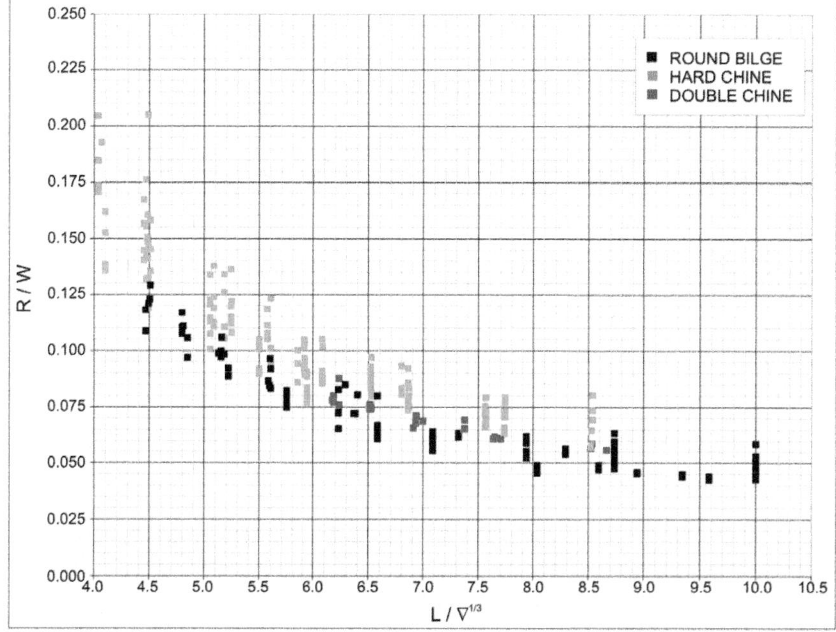

Figure 2-11b: F_{nL} = 0.6

Figure 2-11: Collection of R/W data for 500 mt vessels having round-bilge, double-chine, and hard hulls

Design documentation indicates that a relationship between waterline length and the displacement-slenderness ratio, for example, has a significant influence on hull resistance to weight ratio (R/W) at semi-displacement speeds. There are other factors, such as prismatic coefficient (C_P) which influence variations in R/W in part of this speed range.

Longitudinal center of gravity (LCG) and hull form are also influencing factors on R/W, along with other secondary factors (Blount and McGrath, 2009.)

Systematic experimental study of hull geometry, loading, and distribution of weight exists in the public domain for round-bilge, double-chine, and hard-chine vessels in the range of length Froude numbers from 0.3 to 1.0 and higher. Round-bilge databases typically represent low-speed motoryachts and military applications, double-chine data reflect high-speed monohull ferries and some motoryachts. Hard-chine data have been related to planing boats, military patrol craft, and very high-speed motoryachts.

Beyond displacement speeds, the influence of dynamic hull bottom pressure begins to dominate the effect of buoyancy. From semidisplacement up to semiplaning speed range between $F_{nL} = 0.4$ and 1.0, factors in addition to traditional displacement hull form coefficients have influence on total bare hull resistance. Slenderness ratio is clearly the dominate factor with LCB/L and L/B following in significance, and I will discuss these three factors in more detail below.

Some of the information that I have included here may also have application to the design of demi-hulls of catamarans and amas of multi-hulls excluding the interactions of the transverse and longitudinal proximity (staggering) of hulls.

Hull Loading: Slenderness Ratio

At $F_{nL} = 0.3$, the predominately lowest resistance characteristics are attained with round-bilge, displacement hull forms. There are, however, widely dispersed groups of data at this relatively low dimensionless speed and again at the high speed of $F_{nL} = 1.0$.

These clusters of data in some regards are related to diverse hulls operating outside of their design speed range. Between these low and high speeds, hull-form variants displayed rather orderly trends with changes in speed. With increase of speed some clear demarcation between round-bilge, double-chine and hard-chine hulls begins to be established.

In Figure 2-11, bare hull R/W of all nine model geometric series are plotted with separate symbols for round-bilge, double-chine and hard-chine hull forms. In Figure 2-11b, the trends for each hull form are clearly discernible.

In addition, for speed of $F_{nL} = 0.5$ as in Figure 2-11a, some additional very slender round-bilge hull data were available which indicates that minimum R/W occurs for a slenderness ratio of about 12. In Figure 2-11a, for values of $L/\nabla^{1/3}$ greater than 12, R/W increases for very high values of slenderness ratio, up to about 19, as frictional resistance increases slowly at a greater rate than residual resistance reduces.

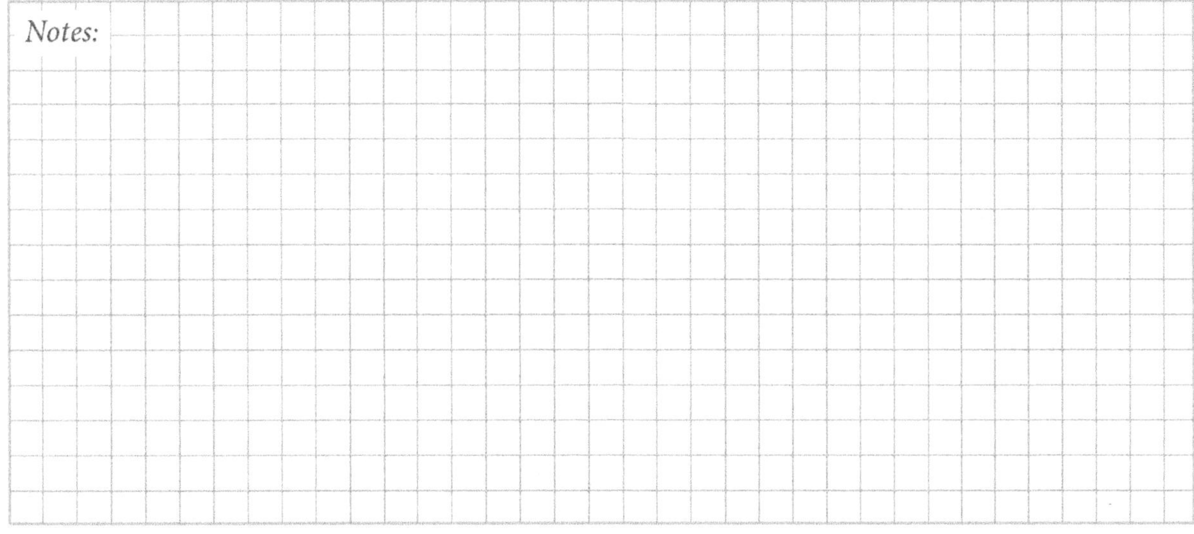

NPL (round bilge) data in Figure 2-12 show the same general trends seen in other series. It is noted that R/W for F_{nL} = 0.3 and 0.4 are relatively constant for wide variation in $L/\nabla^{1/3}$. R/W then begins to increase significantly with reduction in $L/\nabla^{1/3}$ below 7.0 for speeds of F_{nL} = 0.5 and greater; for F_{nL} increases between 0.4 and 0.5 there are especially significant increases of R/W with reduction in $L/\nabla^{1/3}$.

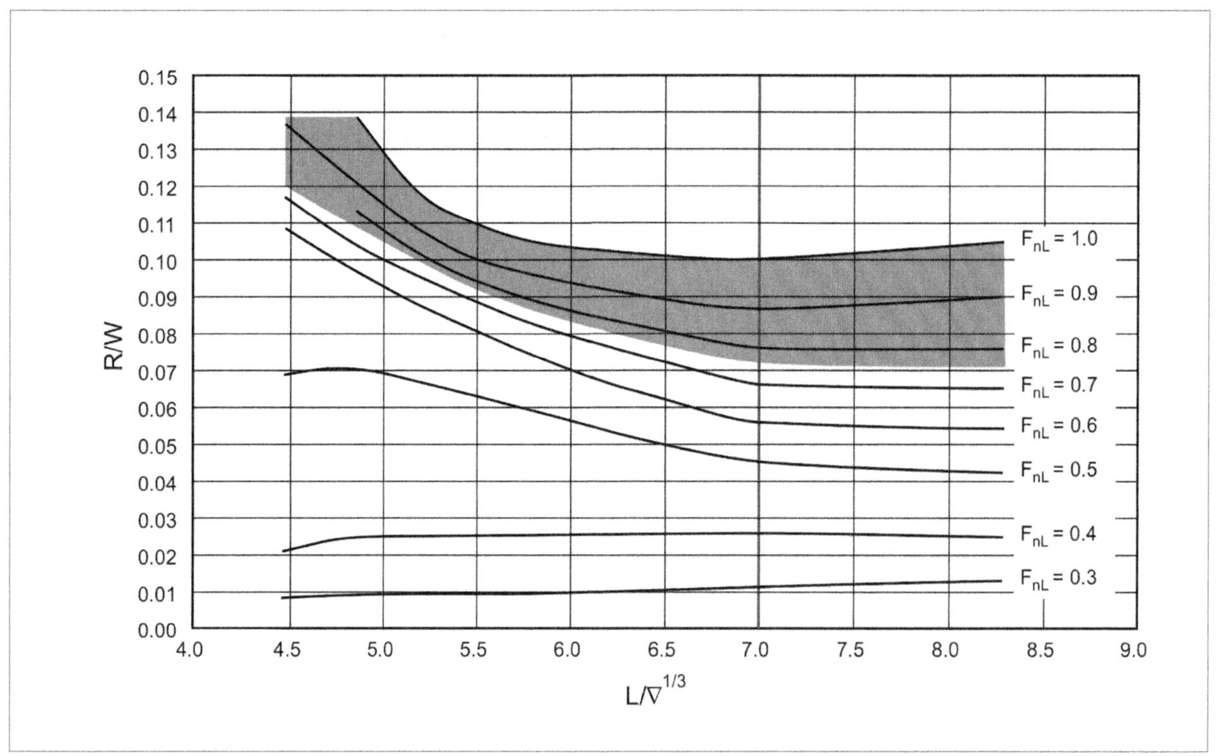

Figure 2-12: R/W data for 500 mt vessels calculated using the NPL round-bilge series.

Note: Some design conditions in the pink-shaded area of Figures 2-12 and 2-13 have exhibited dynamic instabilities.

LCB/L

Not all of the series varied the longitudinal position of center of gravity or buoyancy—most notably Taylor's series—but within three hull forms there were sufficient data to establish trends and/or optimum positions for LCB/L.

Figure 2-13 defines the approximate locus of minimum R/W with a red line. The R/W penalty at high speeds is readily quantified for a forward or aft mis-location of LCB or LCG. At low displacement speeds, the minimum R/W is achieved with the center of weight near midship.

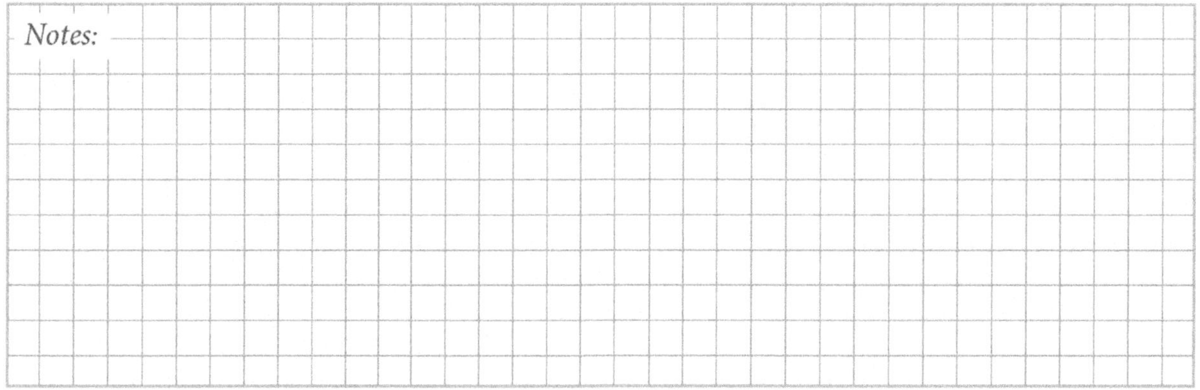

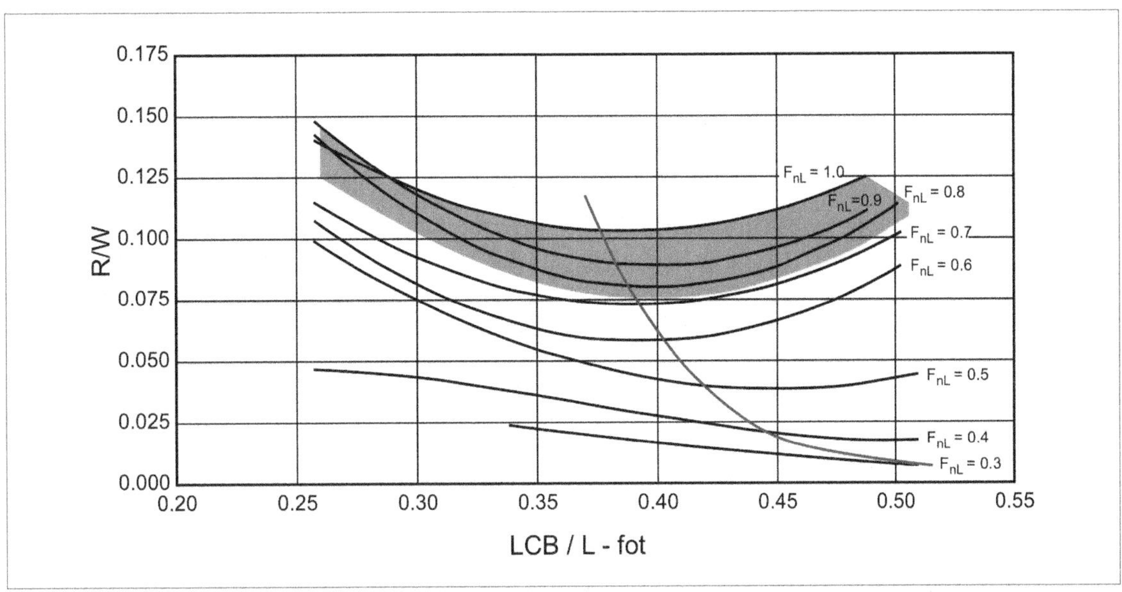

Figure 2-13: Location of LCB/L for minimum R/W for 500 mt vessels having round-bilge, double-chine and hard-chine hull forms

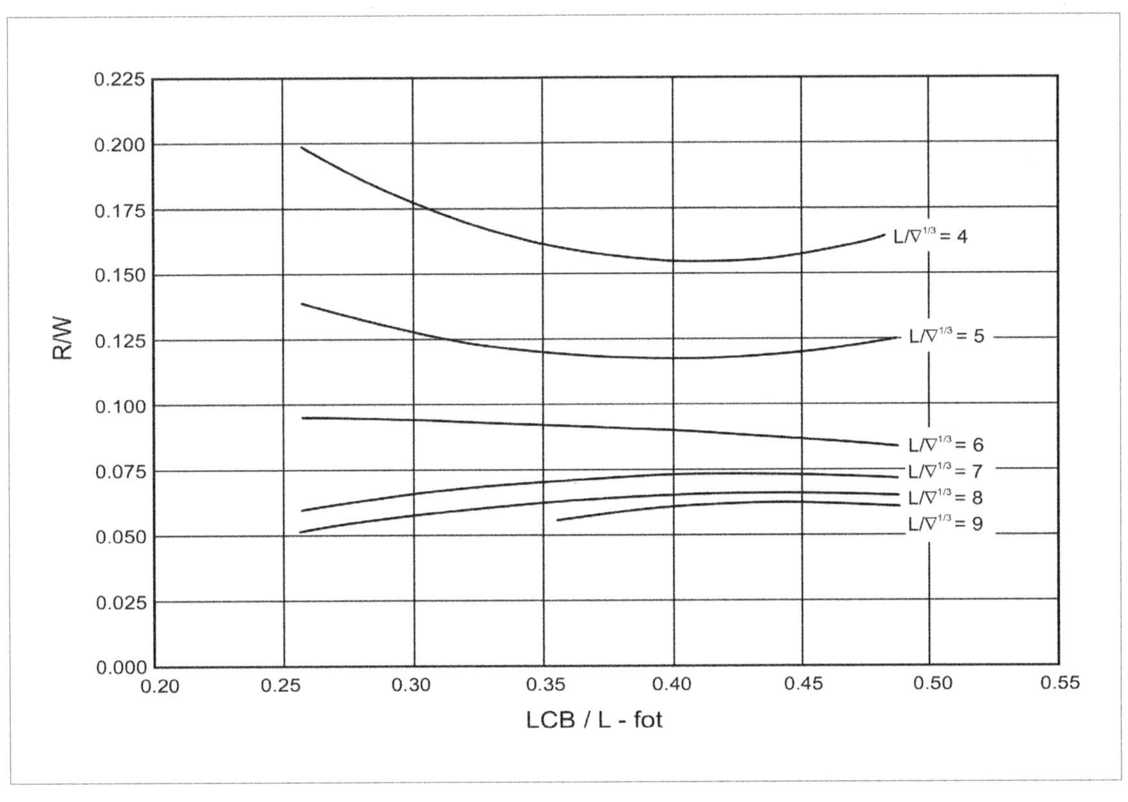

Figure 2-14: Relative insensitivity of R/W to LCB/L at F_{nL} = 0.6

Figure 2-14 shows that R/W at $F_{nL} = 0.6$ is relatively insensitive to LCB/L variation for constant $L/\nabla^{1/3}$ especially for values greater than 6. At this speed R/W is primarily influenced by the magnitude of $L/\nabla^{1/3}$.

By comparing data between Figures 2-13 and 2-14, you can see that the minimum value of R/W at $F_{nL} = 0.6$ corresponds approximately to $L/\nabla^{1/3} = 9.0$ at LCB/L of 40% measured from the transom.

L/B

Length-to-beam ratio is not totally independent of slenderness ratio. Vessels with the same waterline length, but having different waterline beams, most likely will not have the same slenderness ratios due to different displacements. The lesser beam hull does not have the capacity/volume to be densely equipped and/or filled with consumables. Thus, at increasing semi-displacement speeds, L/B becomes a secondary parameter relative to resistance in calm, deep water as seen in Figure 2-15.

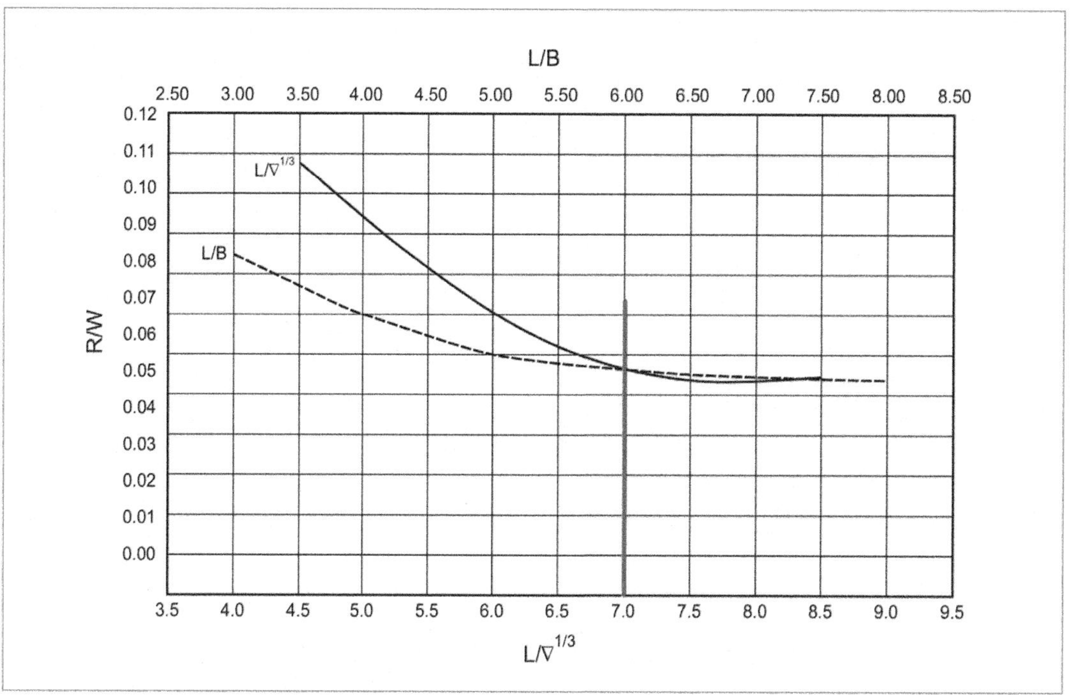

Figure 2-15: Comparison of R/W with L/B and $L/\nabla^{1/3}$ at $F_{nL} = 0.6$ for NPL round bilge series

Minimum Bare-Hull Resistance for $0.40 \leq F_{nL} \leq 1.0$

Appendix 1, Table A1-G provides a table of minimum bare-hull R/W derived for notional 500 mt displacement vessels without bulbous bows for round-bilge, double-chine, and hard-chine hull forms for F_{nL} from 0.4 to 1.0 as a function of slenderness ratio. This table provides a benchmark by which naval architects may evaluate the relative resistance of new hull designs. These tabular results indicate attainable bare-hull resistance for these three hull forms without yet defining other geometric and weight coefficients/characteristics (Blount and McGrath, 2009).

Graphically, these data are shown in Figure 2-16 and depict R/W versus $L/\nabla^{1/3}$ for contours of speed (F_{nL}). These contours represent the lower boundaries of hundreds of available data points from collections of model hull forms: round bilge, double chine, and hard chine. These data show clearly how R/W is affected by $L/\nabla^{1/3}$ and F_{nL} over the range at which these variables had been explored experimentally. Both increasing hull loading—as indicated by reducing $L/\nabla^{1/3}$—and speed results in higher R/W for round-bilge and double-chine hulls for $0.40 \leq F_{nL} \leq 0.80$.

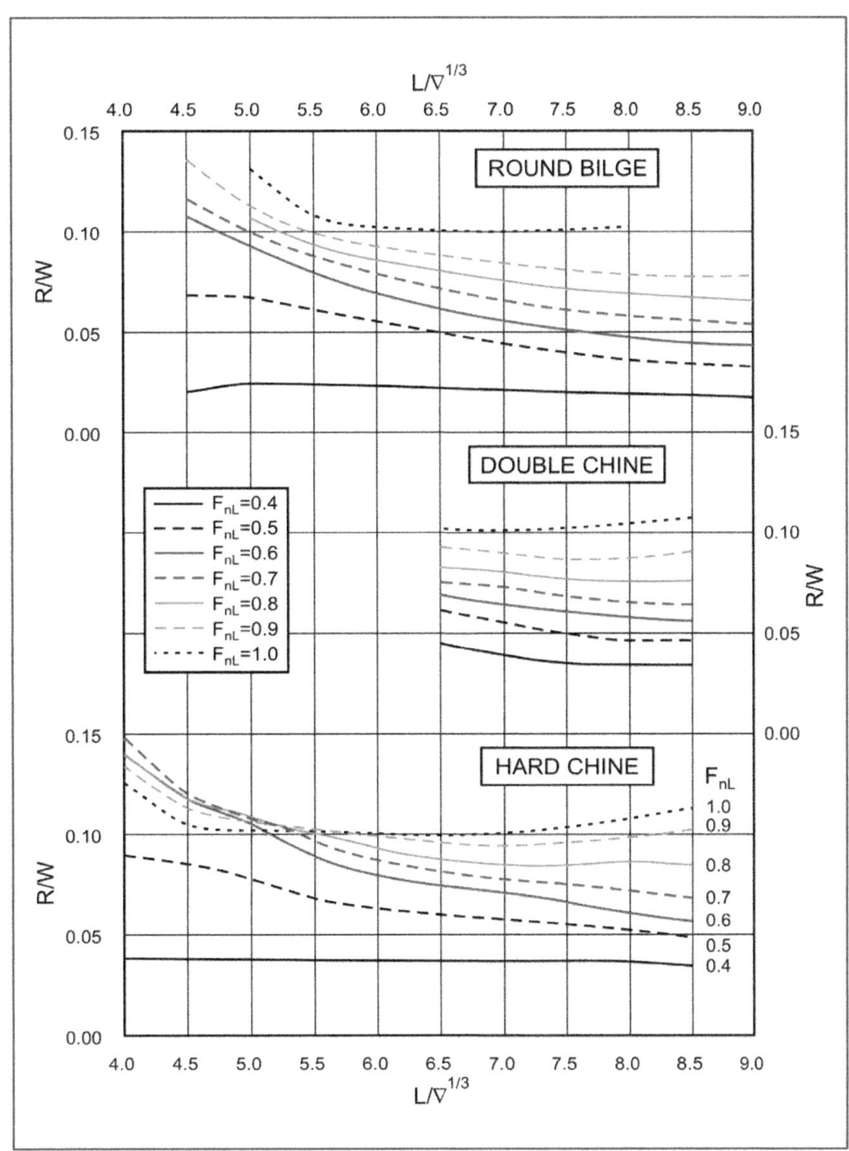

Figure 2-16: Lowest values of R/W found for 500 mt vessels during the 2009 Blount & McGrath study for round-bilge double-chine and hard-chine hull forms

Hard-chine hulls exhibit similar trends at semi-displacement and semi-planing speeds, but as dynamic lift begins to supplement or replace buoyant lift, R/W is affected differently for $L/\nabla^{1/3}$ and F_{nL}.

As speed becomes greater than $F_{nL} > 0.70$, speed contours begin to cross for $L/\nabla^{1/3} < 6.0$. For example, for $F_{nL} = 1.0$, R/W becomes lower for most speeds as $L/\nabla^{1/3}$ becomes smaller.

This overlapping of speed contours is an indication the planing lifting forces have become dominant, resulting in the slope of R/W reducing with speed increase for hard-chine hulls as compared to the resistance characteristics of round-bilge hulls.

You can now begin to envision the difference of resistance implications between displacement and planing vessels. Look at the high-speed contours of the round-bilge and hard-chine hulls with those heavily loaded shown on the left side of Figure 2-16.

This difference is emphasized by plotting R/W data from Figure 2-10 in a different format shown in Figure 2-17. Mean R/W data are plotted versus F_{nL} for $L/\nabla^{1/3}$ from 4.0 to 10.0 for two constant speeds in the format of volume Froude numbers, $F_{nV} = 1.25$ and $F_{nV} = 3.25$, semi-displacement and planing speeds, respectively.

Curve A for a semi-displacement speed shows that R/W is almost vertical for a wide range of hull loading, $L/\nabla^{1/3}$; that is, R/W is strongly linked to $L/\nabla^{1/3}$.

For a planing speed, R/W versus speed (F_{nL}) in Curve B is almost constant for this same wide variation in hull loading ($L/\nabla^{1/3}$); that is, R/W is weakly linked to $L/\nabla^{1/3}$ for this hydrodynamically supported condition.

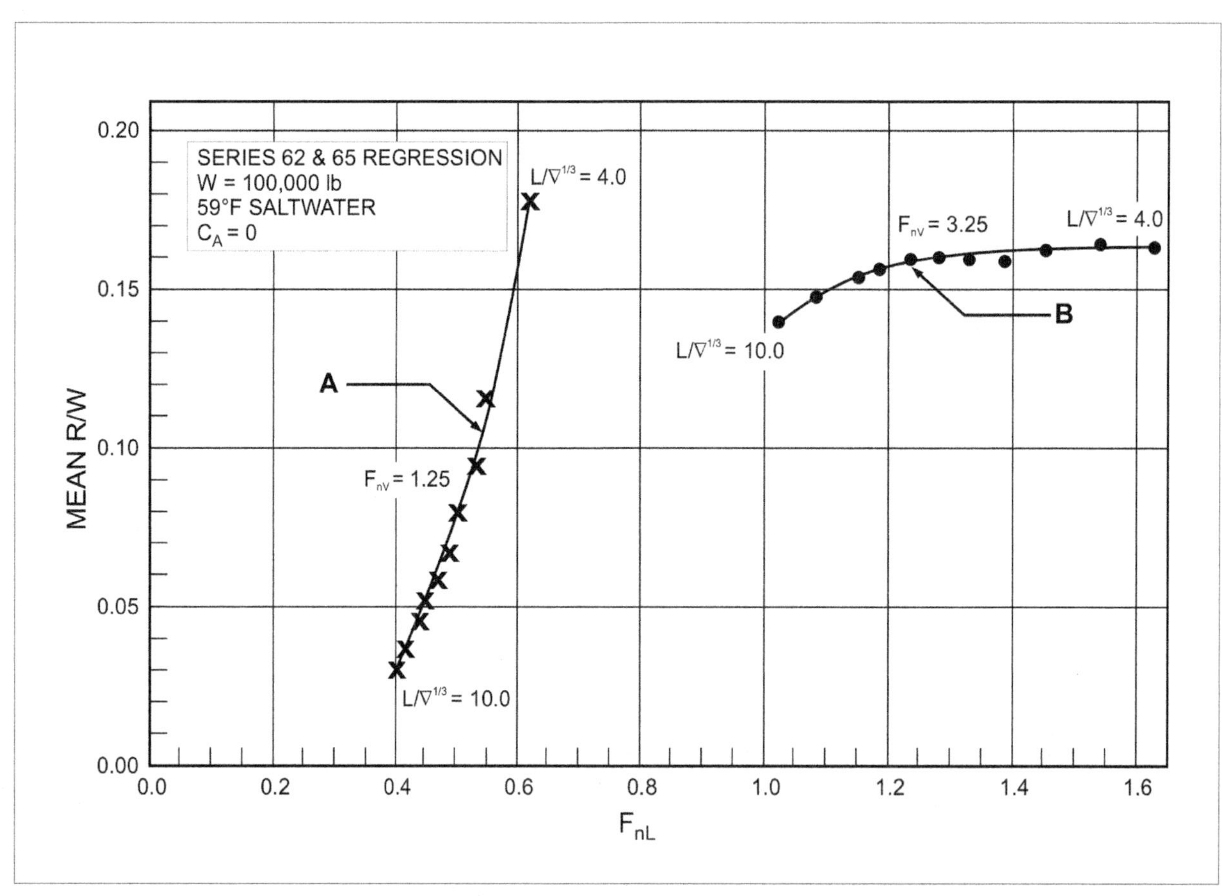

Figure 2-17: Variation of R/W with F_{nL} and $L/\nabla^{1/3}$ for constant $F_{nV} = 1.25$ (semi-displacement speed) and $F_{nV} = 3.25$ (planing speed)

Hydrodynamics at Planing — $F_{nV} \geq 3.0$

Think about this earlier statement: "In order to have planing occur, a hull bottom must have a three-dimensional shape, which, in reaction to trim attitude and the velocity of water over the wetted shape, develops *net positive pressure* of a magnitude that results in the center of gravity rising to an elevation that the remaining below water hull volume is less than 50 percent of its static volume." *Net positive pressure* is an essential condition for planing as hydrodynamic pressures generated by water flow over a surface can locally develop distributed areas of both positive and negative forces, respectively lifting up or sucking down the hull's surface. Thus, the net lifting force of a surface is a function of the projected area A_P and the distributed hydrodynamic pressure.

A significant aspect of planing craft design is controlling where flow separates from the hull when operating in both calm and rough water. This aspect of the design process is represented by the projected longitudinal distribution of the chine beam and transom for the appropriate A_P. Figure 2-18 provides a comparison of plan forms of hard-chine model series parent hulls: one for Series 62 and another for the USCG Series. Notice that in Figure 2-18, the USCG parent has essentially equal transom and maximum chine beams with a relative fine bow. This distribution results in an aft centroid CA_P of the projected area, A_P. This longitudinal chine beam distribution is typical of craft designed for semi-planing and low planing speeds.

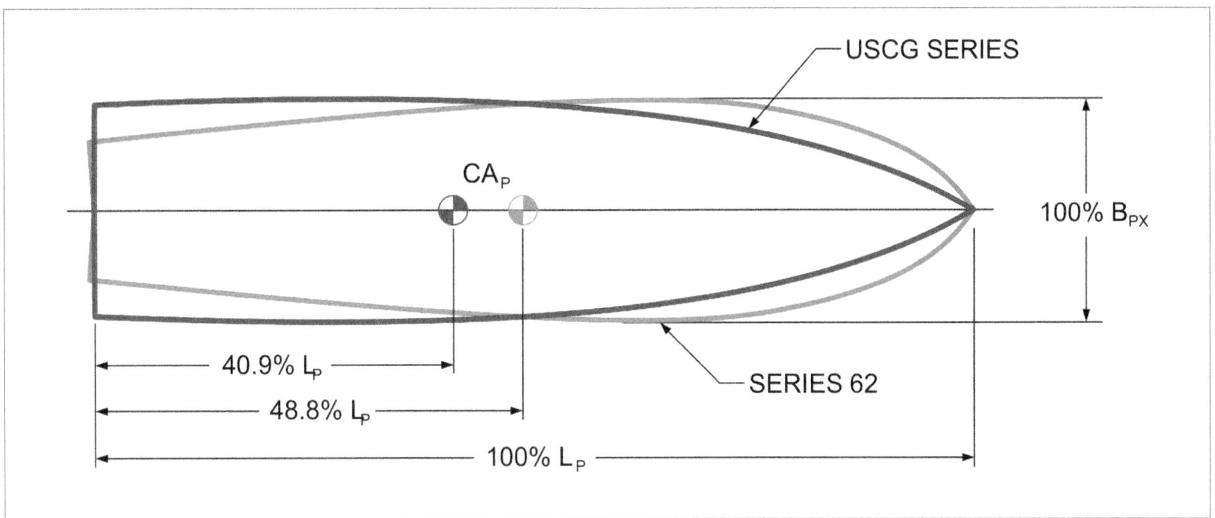

Figure 2-18: Comparison of longitudinal chine beam distribution and centroids of projected area

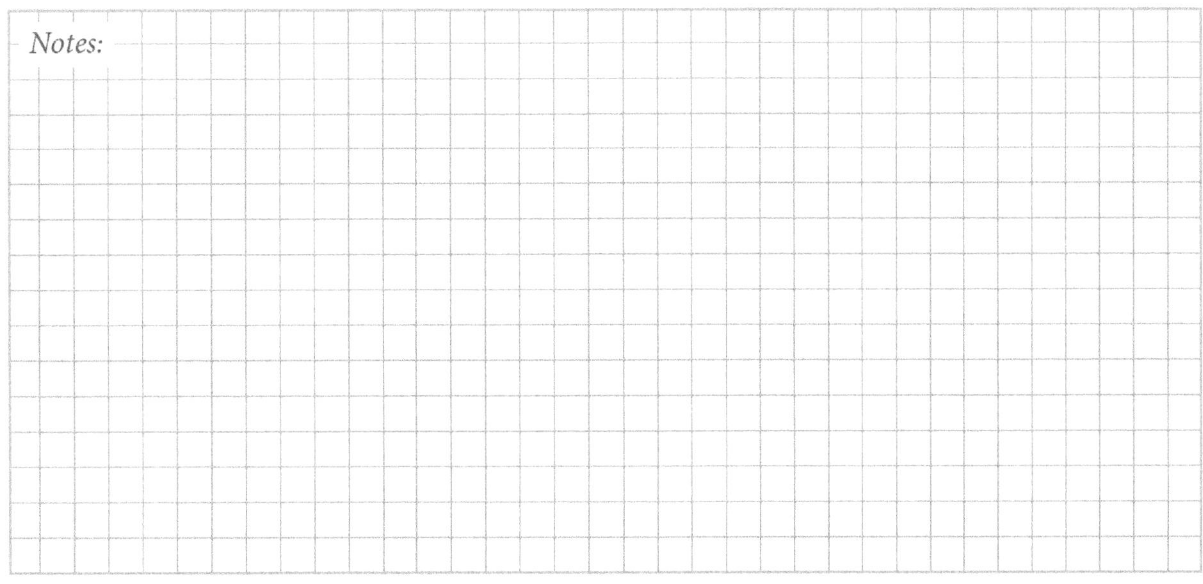

The Series 62 parent developed from PT boat hull lines has a narrow chine beam at the transom with maximum beam well forward, with a relatively full bow resulting in CA_P being near $L_P/2$.

You can see that after establishing design hull loading area coefficient, $A_P/\nabla^{2/3}$ there remains flexibility for a designer to vary longitudinal chine beam distribution to establish the preferred relationship of the LCG relative to the CA_p for best performance in the required operational environment.

This longitudinal chine-beam distribution adopted for Series 62 evolved from pre-WWII designs for hard-chine military craft operating at planing speeds. Over time, improved chine beam distributions with variation in transverse section shape have improved ride quality for rough water operation.

For planing craft, the most important factors affecting design and performance are not the waterline at rest and the shape of the underwater hull as for displacement vessels. Rather, the important factors are those influencing the planing bottom in providing effective dynamic lift (Clement and Kimon, 1957). Thus, there is a similarity of design factors between planing craft and airplanes as the weight of each is supported primarily by dynamic lift.

For an airplane, the projected wing area is of fundamental importance for lift, as the projected bottom area, A_P is the principle source of lift for planing craft. Some designers, however, have argued that the part of A_P forward of each speed's calm-water stagnation line should not be considered as part of design hull loading for calculating the planing area coefficient, $A_P/\nabla^{2/3}$.

This contention assumes that the dry forward area does not always contribute to hydrodynamic lift. This argument, however, ignores the fact that in random waves there are moments of impact when the entire area A_P bounded by the chine and transom is in contact with the sea.

Thus, A_P defines the total lifting surface available to provide recovery from wave entry, as well as just supporting craft weight when operating in calm water at high speed. The chine and transom boundaries of A_P define the edges of desired flow separation when craft are operating in either calm or rough water.

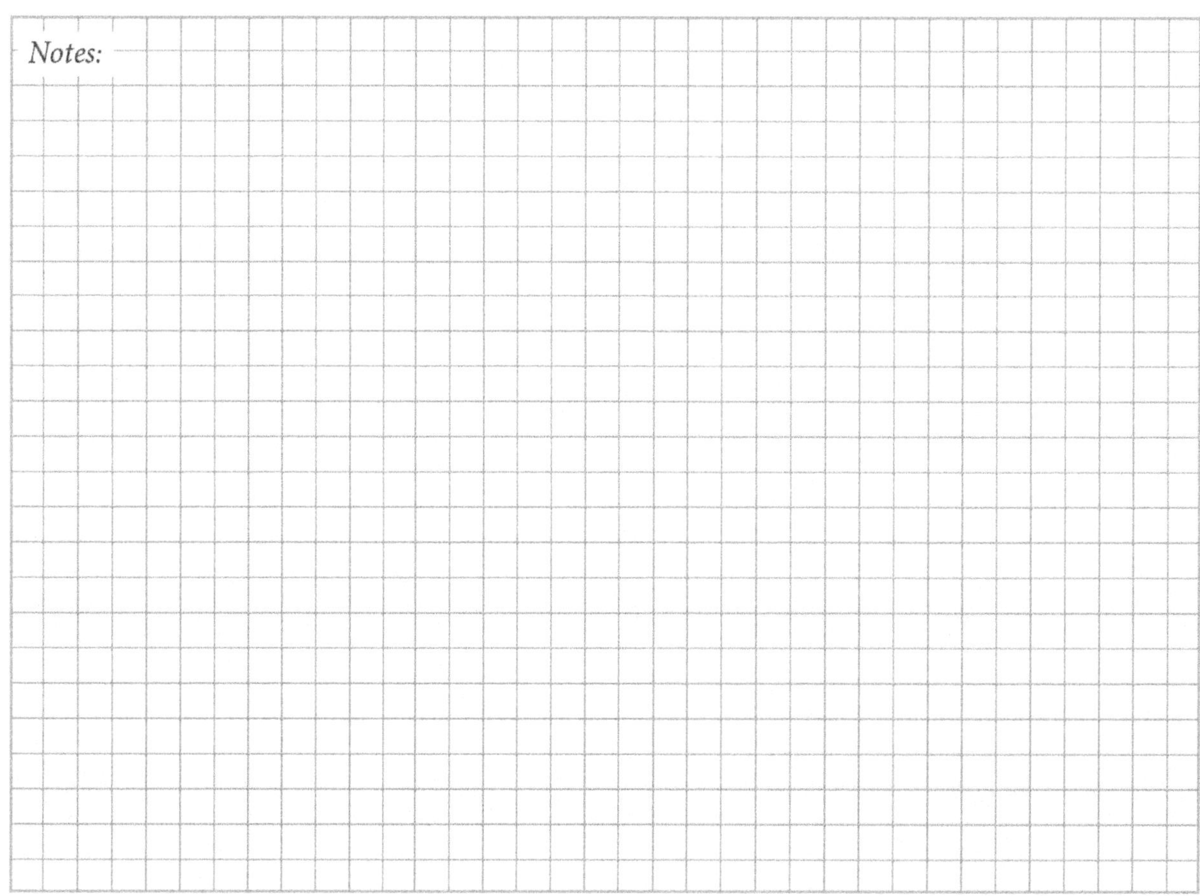

The area coefficient, $A_P/\nabla^{2/3}$ is a dimensionless factor defining the hull loading of a planing hull which is seen to be analogous to that of projected wing area supporting the weight of an aircraft.

The "Principle of Similitude" provides the means of defining comparative relationships which exist between geometrically similar hull forms of different sizes. Thus, the ratio of A_P to $\nabla^{2/3}$ calculated with a consistent dimensional system will be invariant for all sizes of geometrically similar shapes.

You can then streamline the design process by working with dimensionless coefficients during concept and using feasibility studies which define the domain of possible solutions, as well as optimum solutions which meet client's requirements.

An aircraft analogy is also useful in defining a method of LCG location. In figure 2-18 the two plan forms are differentiated by the longitudinal positions of their respective centroids of area (CA_P).

As the LCG of an aircraft is referenced relative to the effective centroid of the wing's lift, you can reasonably expect that the LCG of the craft's weight should be located relative to the centroid of the planing surface.

It does not seem logical that LCGs of these two boat designs in Figure 2-18 would be located at the same percent of L_P forward of the transom, but that should be referenced relative to their respective centroids of area.

This is somewhat the same process as aircraft designers follow who treat the LCG location in terms of the centroid of lifting area relative to the mean aerodynamic chord of the entire wing in order to take into account of such geometric features as sweepback of a wing.

You can achieve a similar effect for various hull shapes for planing boats by referencing the LCG in terms of the distance from the CA_P.

With CA_P established as the reference for measurement of LCG for planing craft, then a dimensionless definition follows for LCG as a percentage of L_P.

$$\% LCG = (CA_P - LCG)100/L_P \qquad \text{Equation 2-7}$$

where CA_P and LCG are distances forward of the transom (fot).

Series 62

Working from these dimensionless definitions for planing craft, Gene Clement and I developed and executed an extensive scientific model test program to establish calm-water hydrodynamic characteristics for a range of small craft geometries, area coefficients, LCG variations and speeds for 12.5 degree deadrise hulls. In 1963, SNAME published the data from this program, known as Series 62.

Series 62 brought a broad spectrum of hydrodynamic science to the forefront. This particular series of model tests, involving variations in length-to-beam ratio, displacements and longitudinal centers of gravity, amounts to such a grand collection of test conditions that designers can look at quantitative data and see their effects in different dimensionless speed ranges.

In contrast with naval architecture of displacement vessels studied at universities, the Series 62 experimental program covered the spectrum of calm water hydrodynamics. Data start at zero speed; go up through and past displacement, semi-displacement, semi-planing and on to high planing speeds.

As test speeds increase within the data of Series 62, there are changing proportions of hydrostatic and hydrodynamic lift. By mining this experimental database of hard-chine models described in Figure 2-19 and Table 2-C, you can discover an abundance of hydrodynamic knowledge.

In contrast with naval architecture of displacement vessels studied at universities, the Series 62 experimental program covered the spectrum of calm-water hydrodynamics.

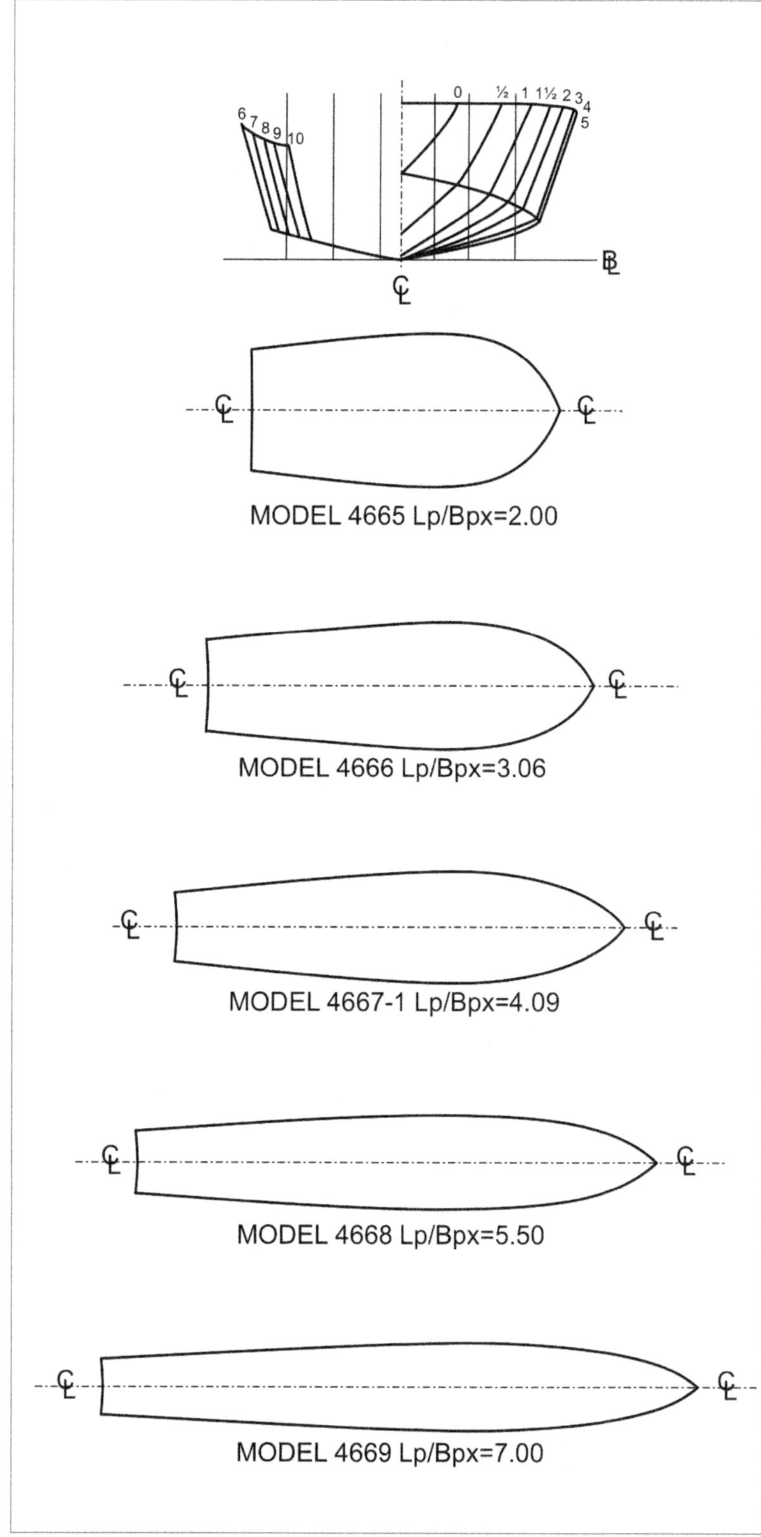

Figure 2-19: Parent body plan and hull plan forms of Series 62 models

It is interesting to follow the calm-water trends of R/W for these hulls with afterbody $\beta = 12.5$ degrees in Figures 2-20a to 2-20c relative to area coefficient $A_P/\nabla^{2/3}$, L_P/B_{PX} and speed F_{nV} for a fairly common LCG four percent L_P aft of CA_P.

These graphics show how R/W varies with L_P/B_{PX} as speed increases from low F_{nV} to high planing speeds.

For $1.0 \le F_{nV} \le 2.0$ low values of L_P/B_{PX} result in R/W being significantly higher than $L_P/B_{PX} = 7.0$.

Then L_P/B_{PX} has little influence on R/W for $2.5 \le F_{nV} \le 5.0$ when area coefficients are in a fairly normal range of $5.5 \le A_P/\nabla^{2/3} \le 7.0$.

At high planing speeds, $F_{nV} = 5.5$ and 6.0 there are hints that minimum R/W may be achieved when $3.5 \le L_P/B_{PX} \le 6.0$.

The values of area coefficients presented in Figures 2-20a to 2-20c cover an extremely wide range of dimensionless hull loading. Table 2-D provides relative examples.

A large population of boats and craft in military, commercial, and recreational service, operating at semi-planing and planing speeds has area coefficients between 4.5 and 7.5, representing the range of full load to light displacement.

It takes a talented naval architect to design an extremely heavy planing craft having, at full load, $A_P/\nabla^{2/3} \le 4.5$, which offers dependable operational characteristics in calm or rough, deep water.

Such a heavily loaded craft may not even attain planing speeds in shallow water.

TABLE 2-C: PARTICULARS OF SERIES 62 MODELS

Particulars	4665	4666	4667-1	4668	4669
A_P ft^2	6.469	9.715	12.800	9.518	7.479
L_P ft	3.912	5.987	8.000	8.000	8.000
B_{PA} ft	1.654	1.623	1.600	1.190	0.935
B_{PX} ft	1.956	1.956	1.956	1.455	1.143
B_{PT} ft	1.565	1.386	1.262	0.934	0.734
L_P/B_{PX}	2.365	3.690	5.000	6.720	8.560
L_P/B_{PX}	2.000	3.060	4.090	5.500	7.000
B_{PX}/B_{PA}	1.180	1.210	1.220	1.220	1.220
B_{PT}/B_{PX}	0.800	0.710	0.640	0.640	0.640
Centroid A_P % L_P forward of transom	47.5	48.2	48.8	48.8	48.8
Angle of afterbody chine in plan view	5.0°	5.0°	5.0°	3.7°	2.9°

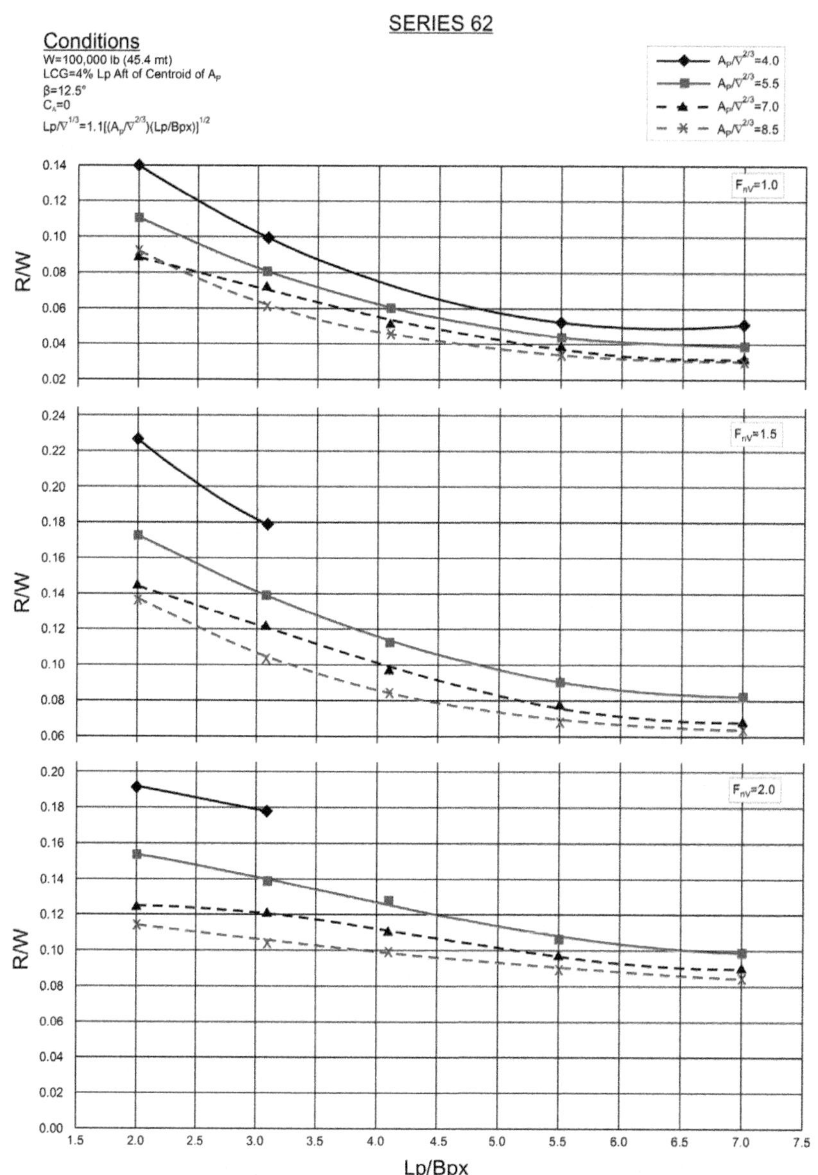

Figure 2-20a: R/W for 100,000 lb displacement in salt water, $C_A = 0$ versus L_P/B_{PX} for speeds F_{nV} and hull loading $A_P/\nabla^{2/3}$

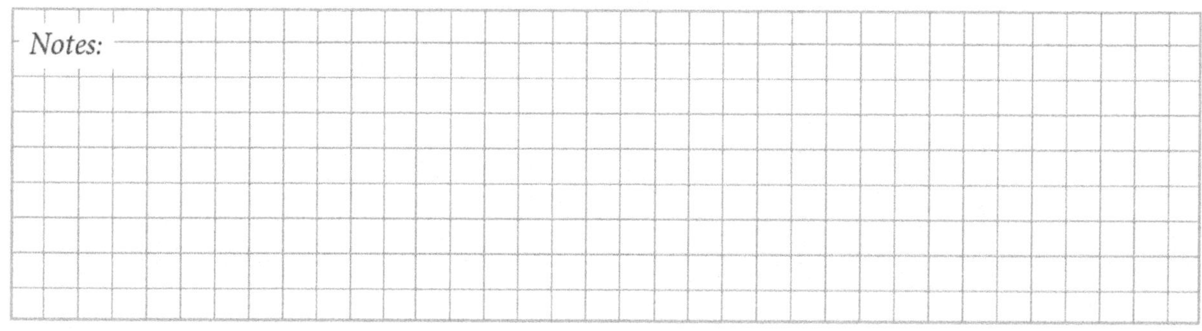

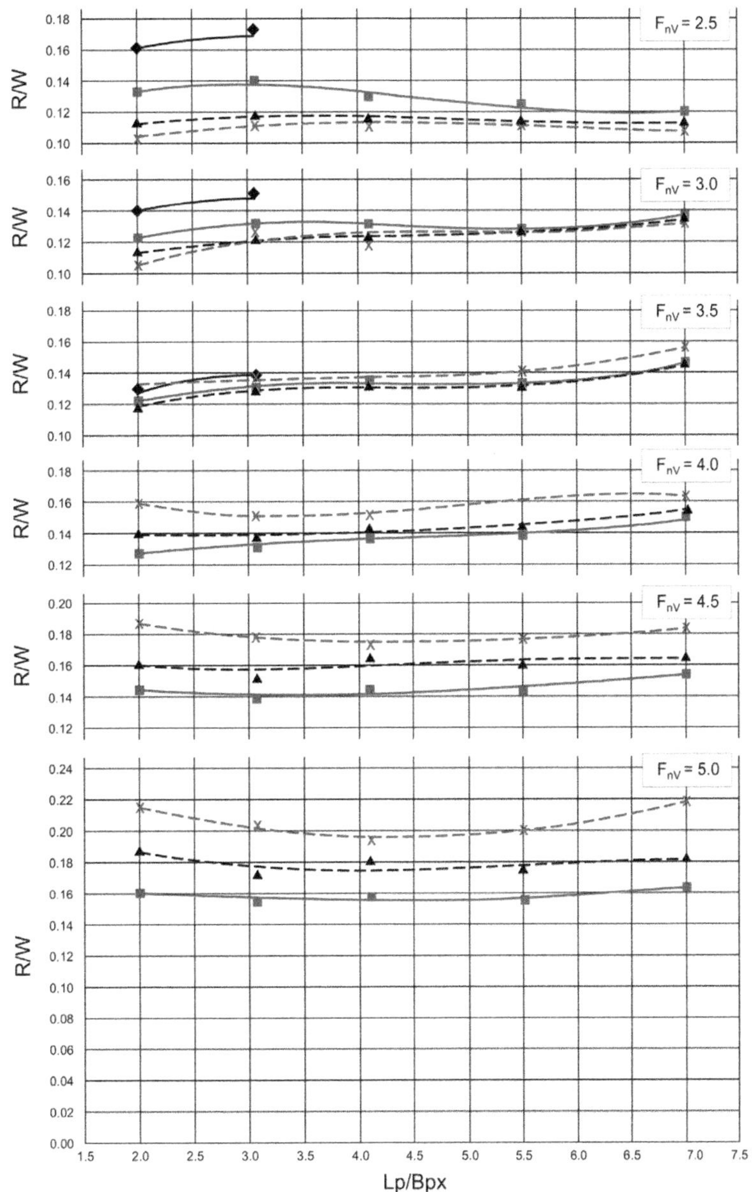

Figure 2-20b(Continued)

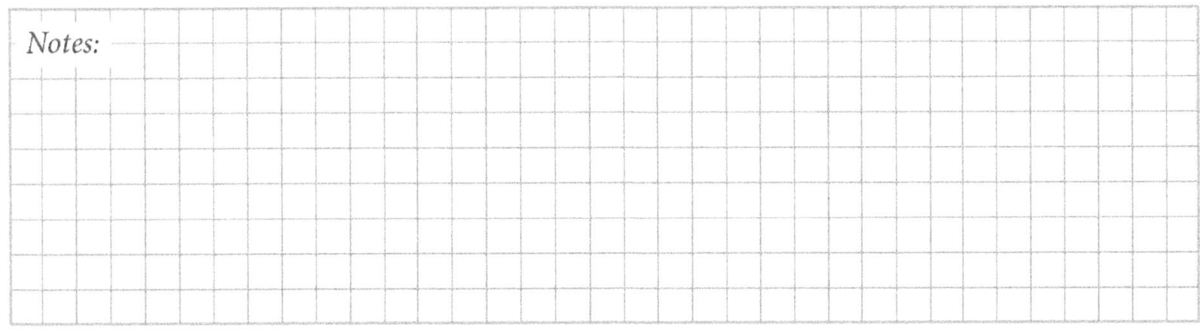

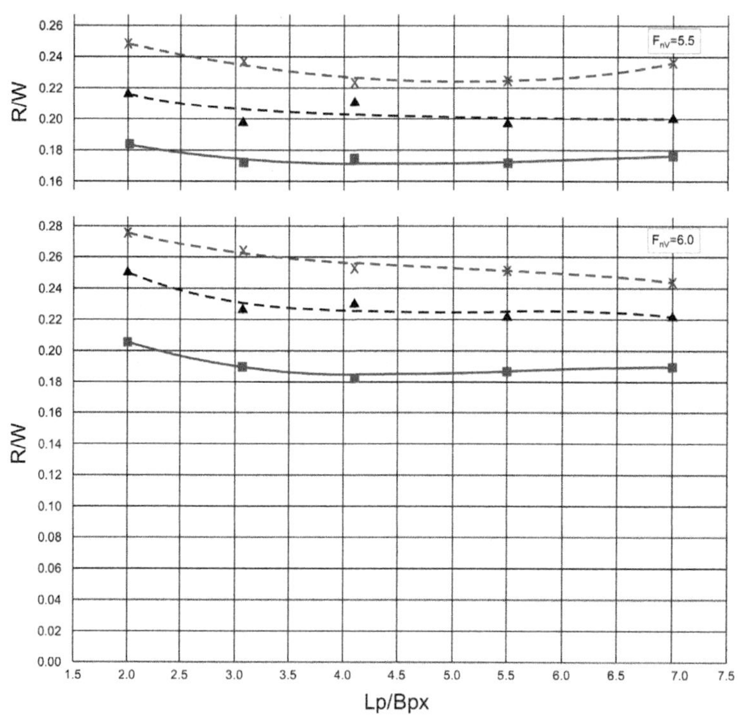

Figure 2-20c (continued)

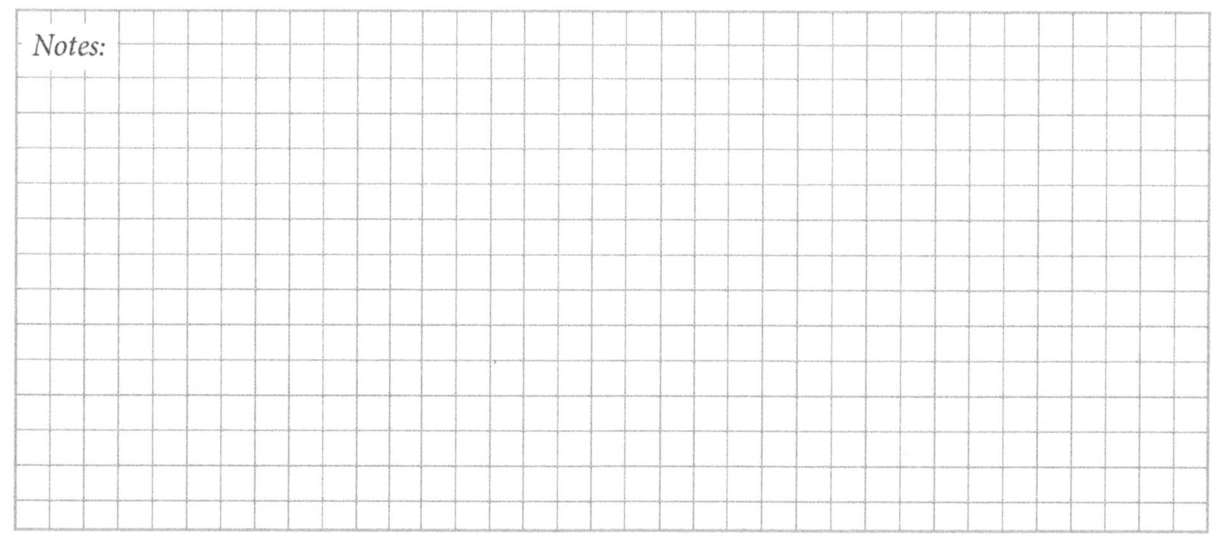

TABLE 2-D: RELATIVE HULL LOADING

$A_P/\nabla^{2/3}$	Description	Examples
4.0	Extremely heavy	Military landing craft loaded with a tank
5.5	Heavy	Fast motoryacht, military patrol boat, offshore sportfishing boat
7.0	Light	Outboard powered recreational craft
8.5	Extremely light	Offshore racing monohull

Take note that in Figures 20a to 20c, R/W data for $A/\nabla^{2/3} = 4.0$ is only provided for the full range of L_P/B_{PX} for $F_{nV} = 1.0$ and for L_P/B_{PX} of 2.0 to 3.06 for speeds up to $F_{nV} = 3.5$. Model-test resistance data for speeds $F_{nV} < 1.0$ are given in Clement and Blount (1963) for $A_P/\nabla^{2/3} = 4.0$.

With regard to the influence of $A_P/\nabla^{2/3}$ on R/W, you will observe an important phenomenon as speed increases. This is shown in Figures 2-20a, 2-20b, and 2-20c. Note that at $F_{nV} = 1.0$, R/W is highest for all values of L_P/B_{PX} when the hull loading is the heaviest, $A_P/\nabla^{2/3} = 4.0$. As speed increases, the R/W curves for each $A_P/\nabla^{2/3}$ become closer together until they essentially coincide somewhere between $F_{nV} = 3.0$ and 3.5.

Clearly, somewhere between these two speeds the trend reverses. That is, at $F_{nV} \geq 3.5$ the lightest loaded hull, $A_P/\nabla^{2/3} = 8.5$ has the highest R/W. At speeds above $F_{nV} = 3.5$, the increments of separation between each R/W curve increases.

This reversal of the relationship of R/W with $A_P/\nabla^{2/3}$ between $F_{nV} = 3.0$ and 3.5 for Series 62 occurs because the running trim angle of lightly loaded planing hulls reduces below the trim angle for minimum resistance at lower speeds than for headily loaded hulls.

Hydrodynamic lift is the primary force supporting the constant weight of a planing boat, and the lift coefficient is related to the reciprocal of speed squared. Thus, a reduction in lift coefficient as planing speed increases is naturally achieved through lowering the equilibrium trim angle.

Whenever the trim angle reduces below that of the minimum resistance, frictional drag (R_F) begins to increase at a higher rate than residual drag (R_R) reduces—thus causing total resistance to increase.

The manner in which R/W varies with L_P/B_{PX} for speeds (F_{nV}) and loading ($A_P/\nabla^{2/3}$) is shown in Figure 2-20. These data are also provided in Appendix 6 in a different format which allows the study of R/W trends due to hull loading ($A_P/\nabla^{2/3}$) over a wide range of speeds for each of the length-to-beam ratios of the Series 62 models. Independent of L_P/B_{PX}, the trends of the slopes of the R/W versus $A_P/\nabla^{2/3}$ curves are consistent with speeds. Below planing, R/W increases as hulls become heavier, and are sensitive to LCG variations at hump speed.

At low planing speeds, R/W is essentially independent of hull loading over a wide range of $A_P/\nabla^{2/3}$. For high planing speeds, R/W reduces with increasing hull loading and becomes sensitive to LCG variations for craft operated in a free-to-trim and heave mode with no operator or automatic trim control.

Considering the formats of Figure 2-20 and Appendix 6 you can readily deduce that for speeds $F_{nV} \geq 2.5$ that hull loading, $A_P/\nabla^{2/3}$ and %LCG have the major influence on R/W throughout the speed range of hard-chine planing craft operating in a free-to-trim and heave mode.

Since CG rise (heave) hardly exceeds 0.15 B_{PX}, this has only minor influence on resistance. Thus, running trim angle has the greatest influence on wetted area and thus on resistance. Operating at the trim angle for minimum resistance at each speed results in the most efficient calm water performance. You will certainly want some means of trim control.

This extensive study was limited, however, to one deadrise angle (β = 12.5°) for the afterbody planing surface. Deadrise angle does have an effect on resistance and trim in calm water as well as ride quality, and added drag due to waves. The original series has since been extended for speeds up to F_{nV} = 3.0 to include test data for Delft Series hulls with β = 25° and β = 30° by Keuning and Gerritsma (1982) and Keuning, Gerritsma, and Van Terwisga (1993), respectively.

The 1993 Delft reference also developed a regression model for estimating calm water resistance, trim, and CG rise for series hulls having 12.5° ≤ β ≤ 30.0° for speeds up to F_{nV} = 3.0. Calm water model resistance tests towed in a free-to-trim and heave manner have not been conducted for Series 62 style hulls having β > 12.5° for 3.0 ≤ F_{nV} ≤ 6.0.

In addition to the Keuning et al. (1993) regression model, I have found the calm water resistance prediction method developed by Hubble (1974 and 1982) to be useful for concept and feasibility studies. Based on the test data of two hard-chine model series programs, Series 62 and Series 65, the Hubble method is described in Appendix 4 for a range of parameters bounded by 22 models.

The geometric parameters of the Series 62 and Series 65 regression by Hubble are bounded by 13° ≤ β ≤ 37° at $L_P/2$, afterbody warp (twist) up to 7.6° per length of B_{PX} and transom beam ratios 0.35 ≤ B_{PT}/B_{PX} ≤ 0.99. This prediction model for speeds 0.5 ≤ F_{nV} ≤ 4.0 assumes that the primary factor affecting resistance is the slenderness ratio ($L/\nabla^{1/3}$) for the pre-planing range seen in Figures 2-10 and A4-4.

Thus, only craft displacement and projected chine length, L_P are necessary to make resistance predictions for mean R/W, minimum R/W or a designer-selected interim value if correlation data are available. This method, however, does not allow for the study of specific craft features such as β, L_P/B_{PX} or LCG, nor is running trim angle predicted. Only resistance is predicted for a range of speeds.

With reference to Figures 2-20a and 2-20b, data indicate that L_P/B_{PX} has little influence on calm-water R/W for 2.5 ≤ F_{nV} ≤ 5.0 and not much of an effect for 1.0 ≤ F_{nV} ≤ 2.0 for L_P/B_{PX} ≥ 5.0. Note that Figure 2-20 uses hull loading defined by $A_P/\nabla^{2/3}$ and the Hubble method is based on $L_P/\nabla^{1/3}$.

$$L_P/\nabla^{1/3} = 1.1[(A_P/\nabla^{2/3})\,(L_P/B_{PX})]^{1/2} \qquad \text{Equation 2-8}$$

This equation provides the approximate relation between $A_P/\nabla^{2/3}$ and $L_P/\nabla^{1/3}$. (It is repeated in *Definitions and General Notes* in Appendix 7 as Equation A7-22.)

If designers accept the notion that high-speed vessel's performance requirements will primarily be at F_{nV} ≥ 3.0, then L_P/B_{PX} is a relatively unimportant factor with regard to resistance for calm-water operation so long as stability requirements are satisfied. Since vessels must reliably accelerate from zero speed up to F_{nV} ≥ 3.0, the complex relationship at low speeds between hull resistance in deep and shallow-water propulsor

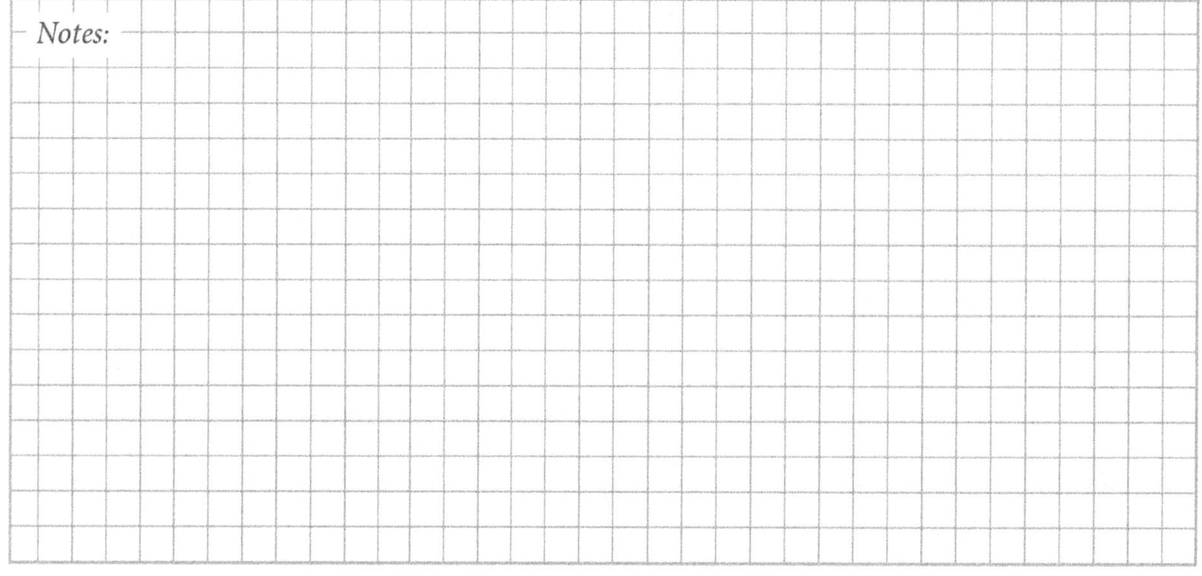

thrust characteristics and low-rpm engine torque characteristics must be addressed, and I will discuss this complex relationship in some detail in Chapters 3 and 6.

I do, however, want to mention here that L_P/B_{PX} can be part of the solution to the conflict which exists between the design of internal general arrangements, hull resistance, and agility to accelerate from low speeds considering both engine and propulsor characteristics, gear ratio, stability, and ride quality in rough seas.

Let's return to the effect of deadrise angle on hull resistance for speeds, $F_{nV} > 3.0$. Other than Series 62, little data for geometrically related model tests of hard chine free-to-trim-and-heave hulls are available in the public domain for studying the effect of variation in deadrise angle at these high planing speeds.

Referring to Clement and Blount's work from 1963, Kimon stated, "The effect of deadrise angle on resistance at high speeds is at least as important as the length-to-beam ratio." Kimon may have missed the mark slightly as far as calm water is concerned since deadrise angle has more significance than L_P/B_{PX}.

The visualization of a prismatic planing surface, meaning constant beam and deadrise, similar to that seen in Figures 2-21 and 2-29, is frequently offered as an illustration of the source of the hydrodynamic lifting force which elevates a high-speed hull toward the water's surface and identifies the resistance component resulting from pressure. In the simplest sense, this illustration ignores the buoyant component of lift and its trimming moment.

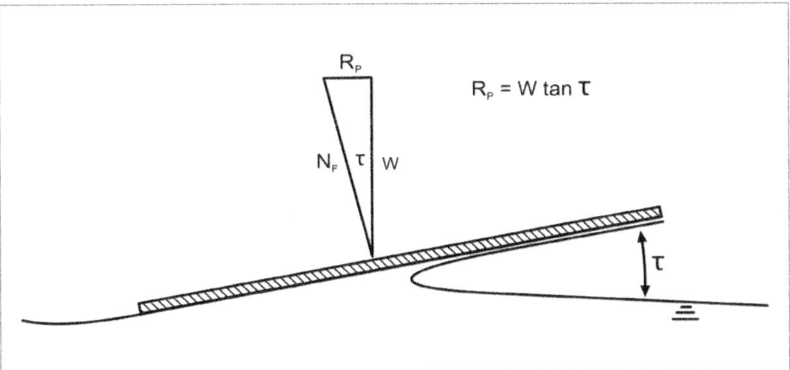

Figure 2-21: Flat plate when planing

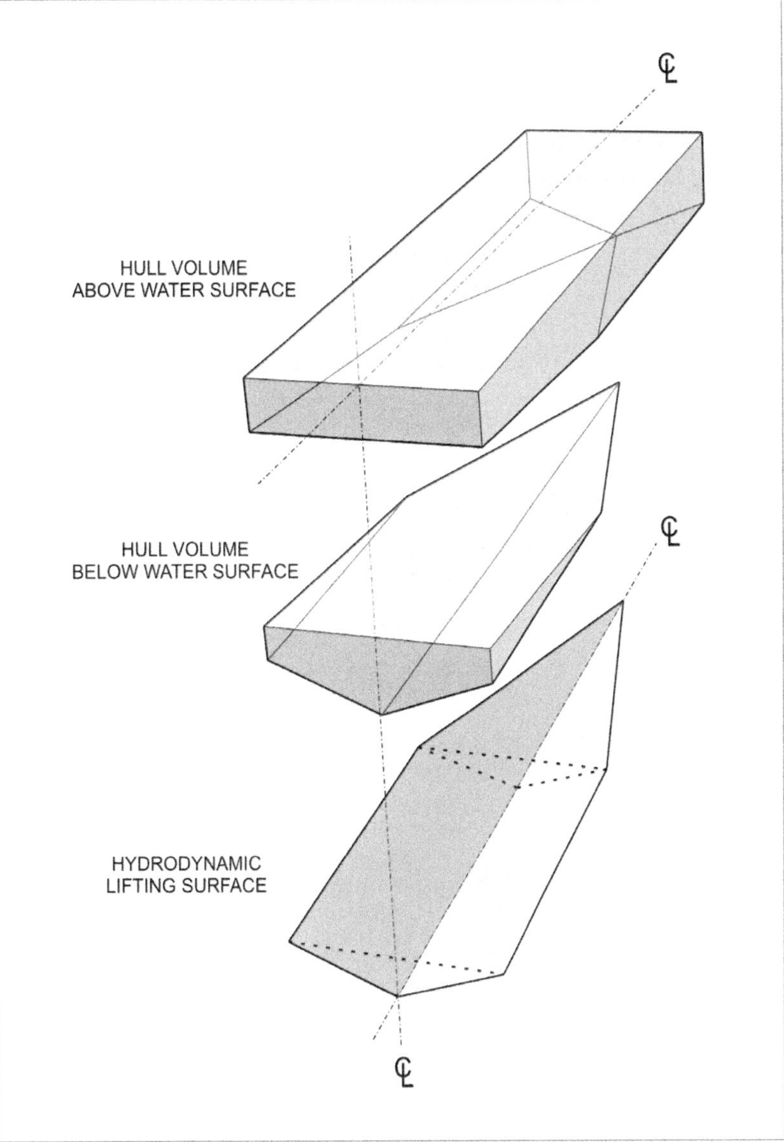

Figure-2-22: Representation of lifting forces of plaing craft—bouyancy and hydrodynamic lift

With regard to the design of planing hull surfaces, this is not representative of reality for speeds when a buoyant component of lift exists or whenever curved buttocks, either forward or aft, are wetted. When convex curved buttocks on the forward hull bottom are wetted, peak dynamic pressure is not the same as that obtained from experiments with prismatic surfaces.

Convex hook or concave rocker in buttocks on the aft hull bottom just forward of the transom, respectively, generate hydrodynamic lifting or depressing pressures affecting bow-down or bow-up trimming moments. Thus, you should consider the depiction in Figure 2-21 only as a special case and as an idealized representation of planing.

Figure 2-22 offers a better representation of the interaction of lifting forces as a combination of hydrostatics of submerged hull volume and hydrodynamics. Planing lift is provided by water flow over the shape of the hull's surface bounded by the transom aft, chines and stagnation line forward, taking note that the stagnation line is elevated above the plane of the water surface due to wave rise.

Thus, hydrodynamic lift is dependent on the aspect ratio of the wetted area of the hull, the longitudinal shape of the buttocks of the hull; that is, the longitudinal straightness or curvature of the buttocks, and the trim angle of the buttocks mostly in the vicinity of the stagnation line, relative to the plane of the water surface.

If the submerged volume of the hull at low speeds has the transom and hull sides wetted, then you must multiply the volume below the plane of the water surface by the density of the water to compute the buoyant lift. In this case, the center of gravity of the buoyant force is located at the centroid of the volume.

However, when water flow has separated from the chines and the intersection of the transom with the hull's bottom, the pressure at the separation boundaries becomes atmospheric, 14.7 psi (101.3 kPa). [*I commented on this earlier, on page 46 of this chapter.*] Then the buoyant lifting force is less than that computed by the hull's volume below the plane of the water surface multiplied by density of the water.

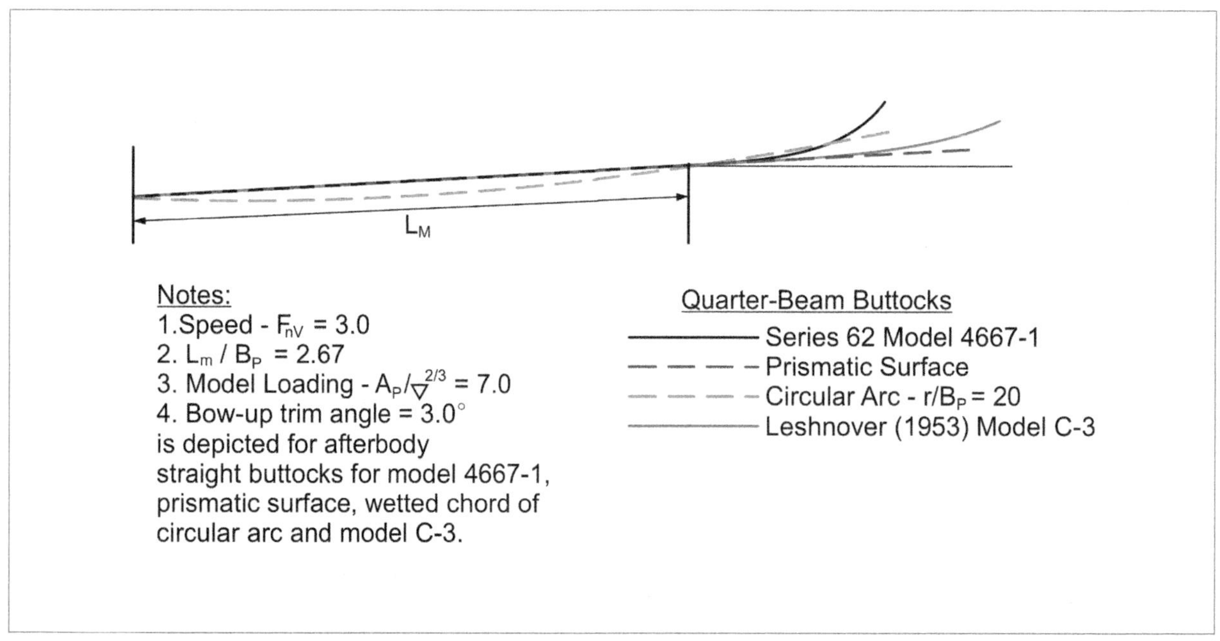

Figure 2-23: Comparison of longitudinal buttock shapes

Convex Buttock Curvature at the Bow

Why should we be concerned with the shape of the bow of planing craft? There's a good reason: It comes into play in both calm and rough water. As a vessel pitches and heaves in waves, the buttock lines, convex forward fairing into a straight line aft, can be fully wetted to possibly momentarily being completely dry.

With operator trim control, the curved portion of bow of a hull can become wetted or with sufficient speed running trim naturally reduces until the forward curved hull bottom is wetted. It is even not so easy to answer the question about what is the influence of bow curvature on a planing boat's longitudinal acceleration from a displacement condition to planing speed. Over time, how does the magnitude of the resultant hydrodynamic lift and drag force vector—and its line of action—vary when both curved and straight buttocks are wetted simultaneously?

That's a tough question.

My current thinking has been influenced by various references for existing experimental planing data for shapes which I could reduce to common format for comparison. These sources excluded buoyant forces and reported lift, drag, and pitching moment for the shapes tested. The geometries found for β = 20° included three boat-like forms (Leshnover, 1953), prismatic surfaces with straight buttocks (Chambliss et al. [1953], Kapryan et al. [1952]), a prismatic surface having constant radius, longitudinal curved buttocks (Mottard 1960), and several unpublished data points with Series 65 model 5239 (average afterbody β = 19°).

There is little experimental data that represents the entire longitudinal shape of the buttocks of planing boats. Typically, a buttock begins at the transom, extends forward as a straight line to about midship where it curves upward until it intersects with the chine. Highlighting these differences, Figure 2-23 shows a buttock comparison for straight lines of a prismatic shape, a longitudinal convex shape having a constant radius, a quarter beam buttock of the parent hull of Series 62, and one from a series of flying-boat forebodies studied for diving tendencies.

These lines are shown at an effective trim angle of 3° between the straight after-bottom buttocks and the calm-water surface except for the circular arc section. Following the method reported by Mottard (1959 and 1960), the chord line of the wetted length of the circular arc has been taken as the effective body axis reference for trim (τ_{ch}) with respect to the water surface.

Mottard (1959 and 1960) conducted calm water experiments with two longitudinal convex surfaces having a radius-to-beam ratio, r/B_P = 20; one had β = 0° and the other with β = 20°. In Figures 2-24 and 2-25, I have compared these data with those for prismatic surfaces.

Hydrodynamic lift coefficients for a range of trim angles are based on projected wetted area versus mean wetted length-to-beam ratios. For both straight and longitudinal curved buttock planing surfaces, L_m/B and L_{mc}/B respectively, are plotted on the same horizontal scale. Look at the pink area between the adjacent pairs of curves to see the reduction in lift coefficients due to convex curvature for trim angles typical of planing craft.

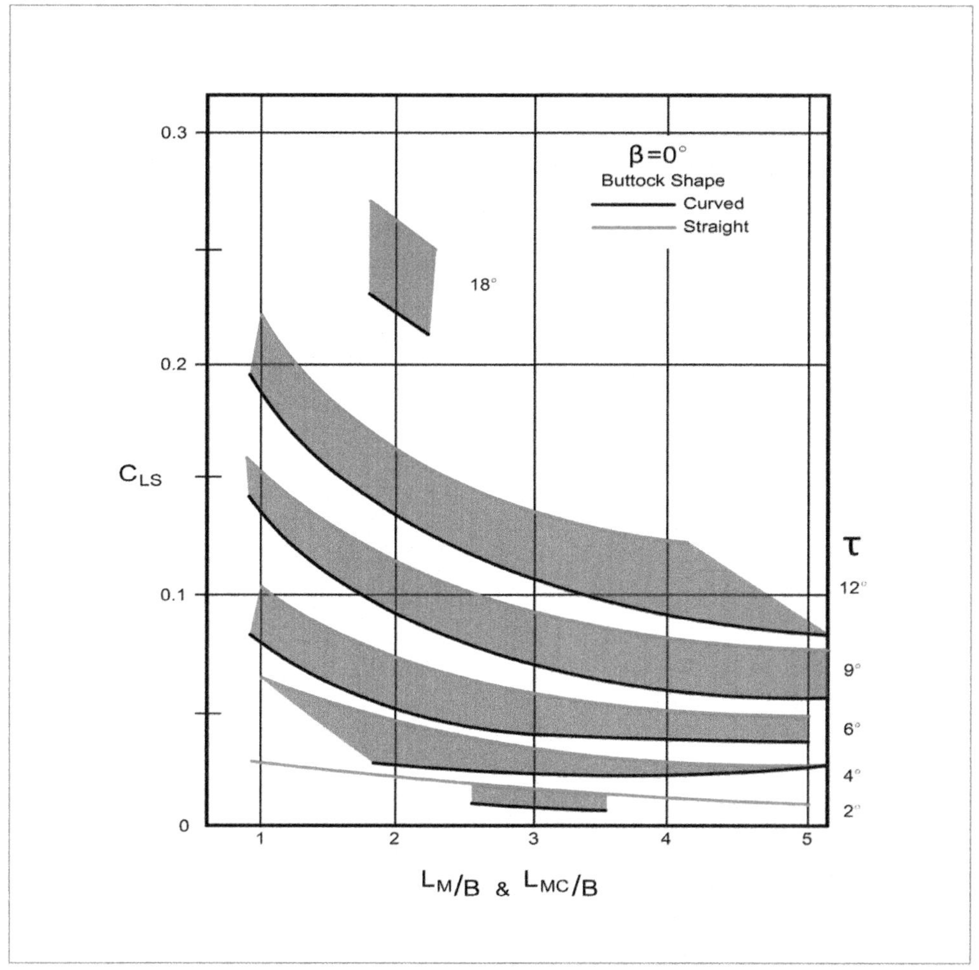

Figure 2-24: Comparison of loss of C_{LS} relative to a prismatic surface due to longitudinal curvature for $\beta = 0°$ planing surfaces

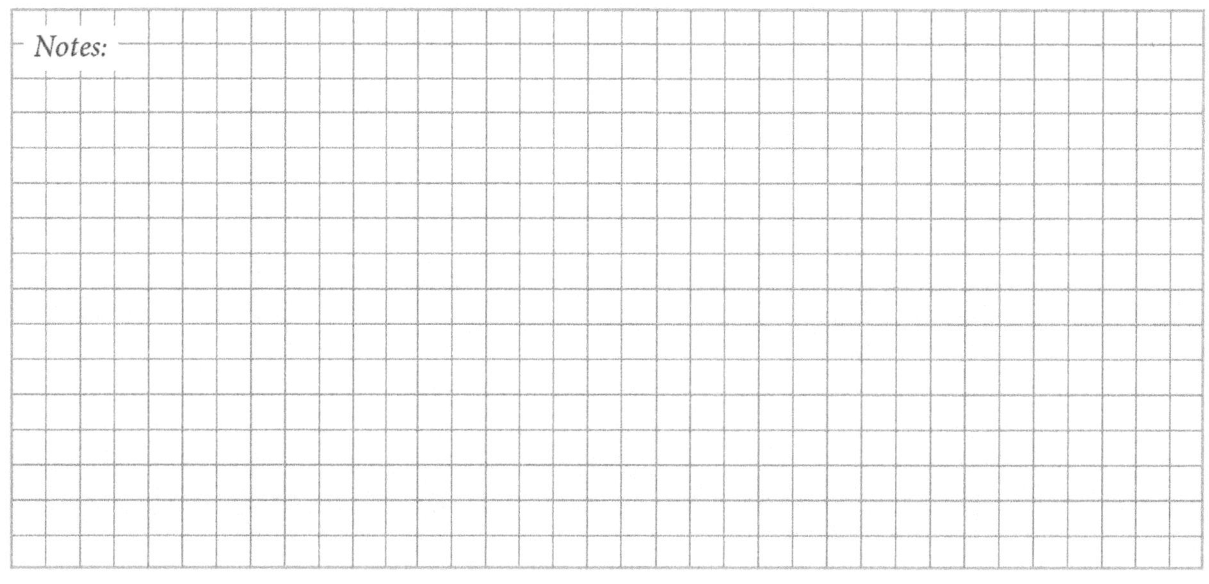

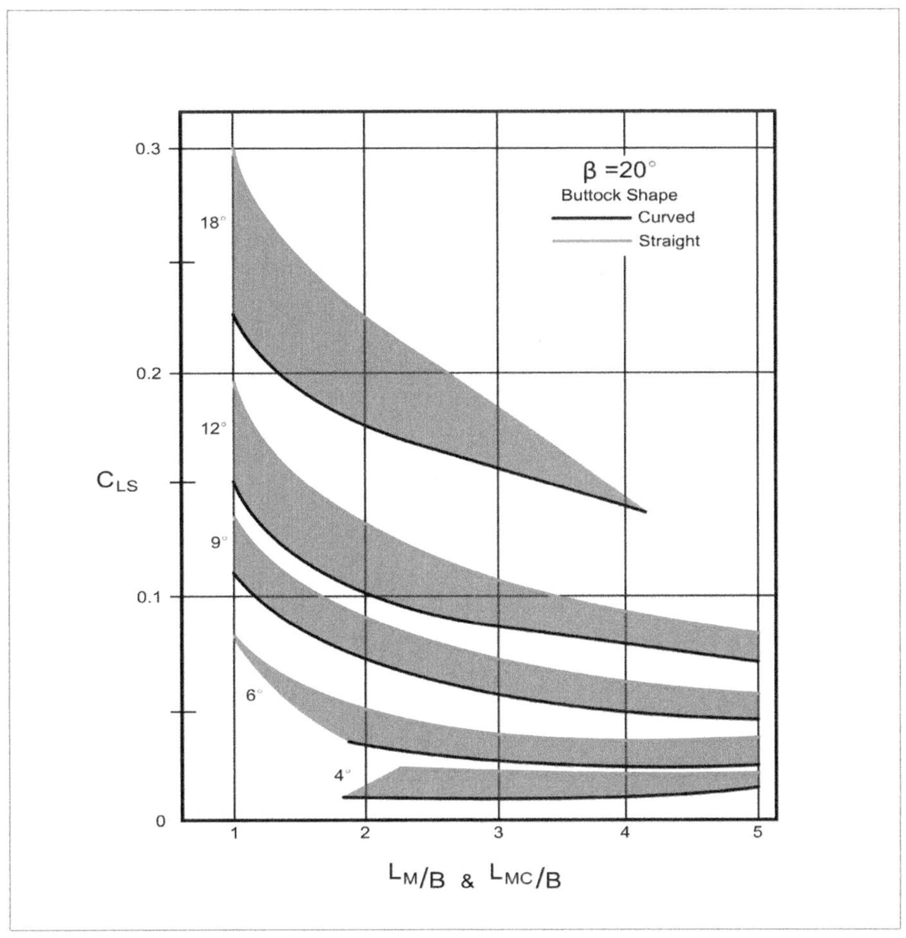

Figure 2-25: Comparison of loss of C_{LS} relative to a prismatic surface due to longitudinal curvature for $\beta = 20°$ planing surfaces

Furthermore, for β = 0° in Figure 2-26, notice how the interactions of both lift and center of pressure are affected by convex curvature. For low-running trim angles, when the convex shape of the forward area of the hull bottom can become wetted both the magnitude of dynamic lift (positive or negative) and its center of pressure change relative to that of an equivalent prismatic surface.

Certainly when trim angles of the planing surface become less than 2° or possibly even 3°, hydrodynamic characteristics based on prismatic data become increasingly unreliable.

For deadrises between β = -10° and β = 70°, a substantial amount of experimental data for prismatic shapes exists in the public domain. The majority of these tests are for trim angles between 4° and 34° with some limited data for τ = 2°. Obtaining accurate data with prismatic shapes at very low trim angles is difficult since the proximity of the model distorts the local water surface. Thus, experimental hydrodynamic lift data is generally extrapolated from the minimum angle tested, often 2° or 4° to be zero lift at zero trim angle.

Data reported by Mottard are available for planing surfaces with convex longitudinal curvature for β = 0° and 20°. With convex curved planing surfaces at the bow, it was possible to conduct experiments at negative trim angles relative to the straight aft buttocks at the transom.

Leshnover (1953) published data for forebodies of flying boats to study diving tendencies during landing also called "stuffing" in planing boat terms. Three of these models with β = 20° were boat-like from a longitudinal perspective.

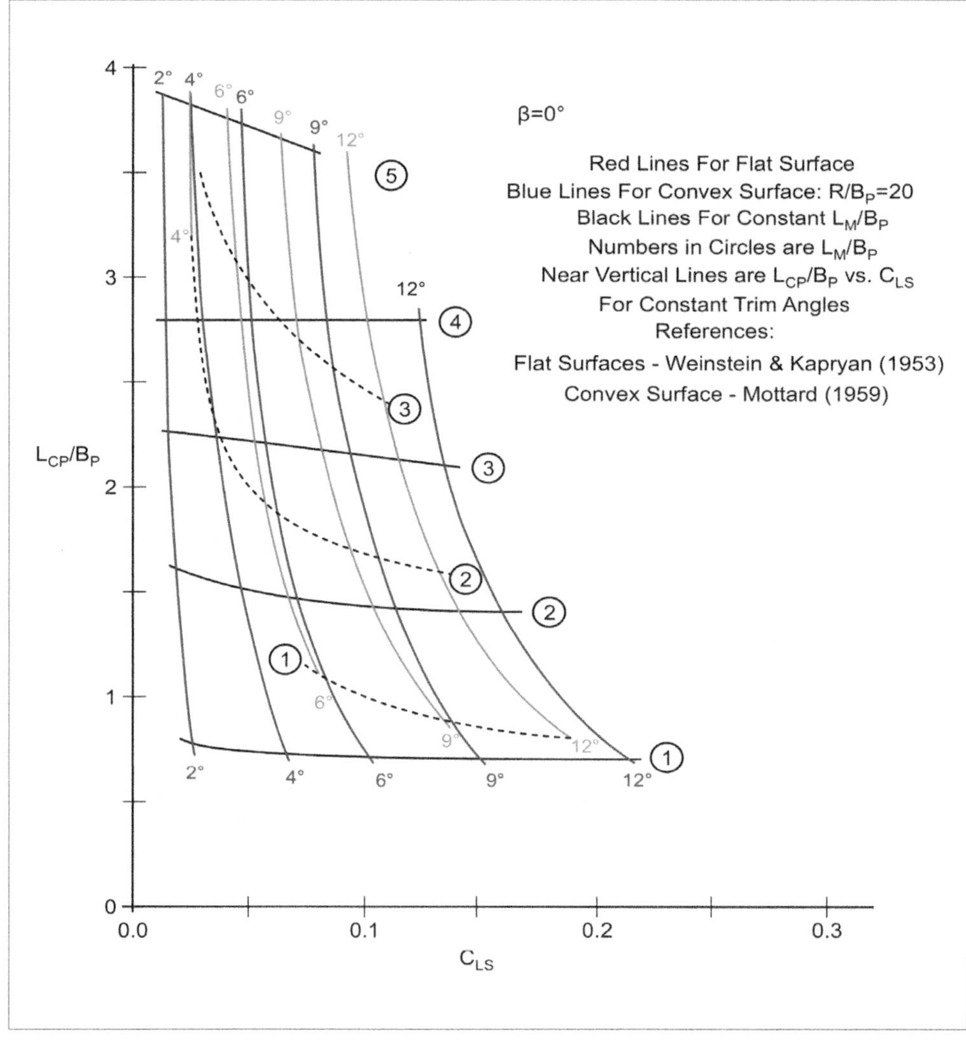

Figure 2-26: The influence for trim angles of longitudinal curvature relative to a prismatic source for $\beta = 0°$ planing surfaces on center of lift versus C_{LS}

The aft buttocks were prismatic with each having different longitudinal buttock curvature over the forward portion.

These models were fixed in heave and tested at constant draft, Leshnover measured force and moment data for three speeds at fixed trim angle between +3° to -5° referenced to the straight buttocks of the afterbody.

I developed two figures from these resources. Figure 2-27 provides, with Shuford's 1958 representation for prismatic surfaces, a comparison of lift coefficients based on projected wetted area (C_{LS}) versus trim angle for the three boat-like models C-1, C-2, and C-3.

Shuford's theory for prismatic surfaces is shown for $L_M/B_P = 2.0$ and 5.0. At 3° trim angle, these Leshnover models had $L_M/B_P = 2.0$, 2.2, and 2.4, respectively. Also shown in Figure 2-27 is the $B_{PX}/4$ buttocks of the three Leshnover models with the height offsets exaggerated to better depict the differences of bow shapes.

The data clearly show that C_{LS} is not zero at $\tau = 0°$ when there is wetted upward buttock curvature at the bow. Actually, the stagnation line is on the curved shape of the bow at low and negative trim for angles referenced to the aft straight buttocks.

The angle of the hull's contact with the water (τ_{ch}) is approximated by extending to the water's surface the tangent to the point on the hull of the intersection of the buttock $B_{PX}/4$ and the stagnation line.

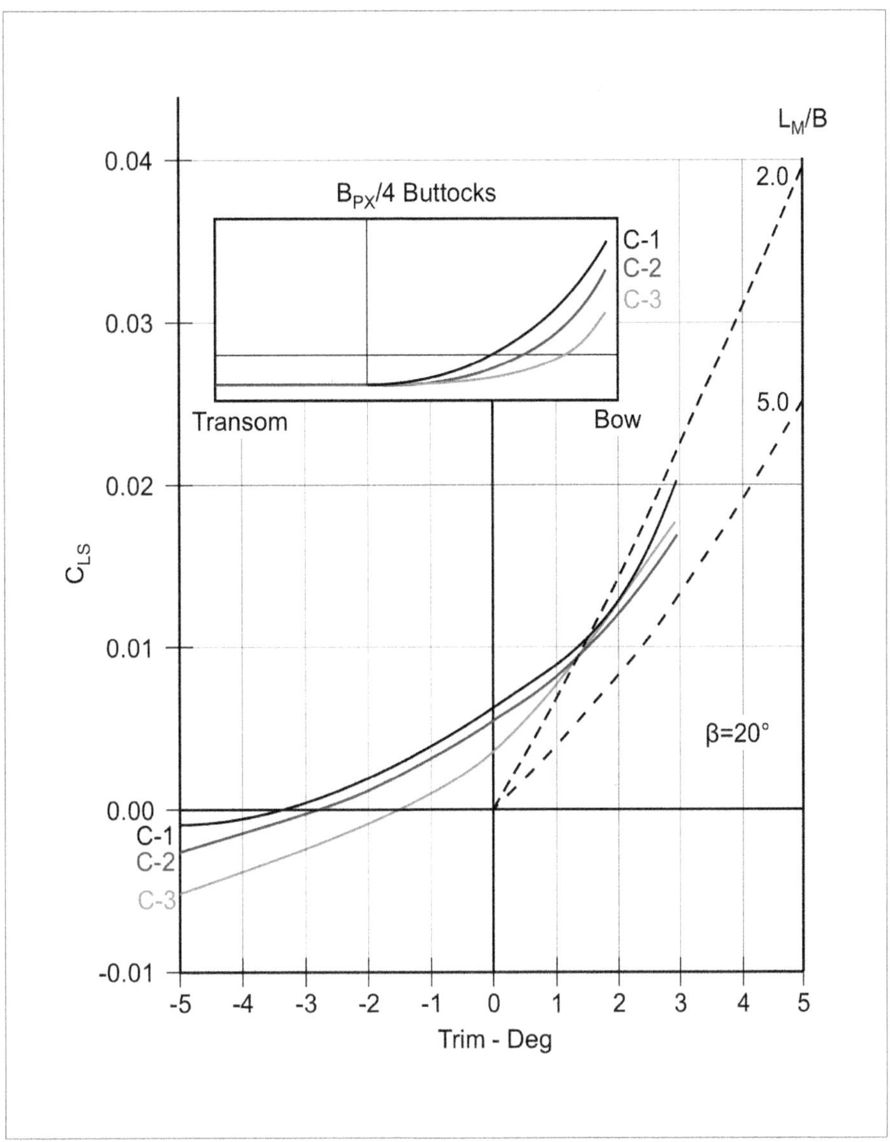

Figure 2-27: Characteristics of C_{LS} for $\beta = 20°$ for both prismatic and boat-like bows with curved buttocks at low and negative trim angles, measured relative to the aft straight buttocks

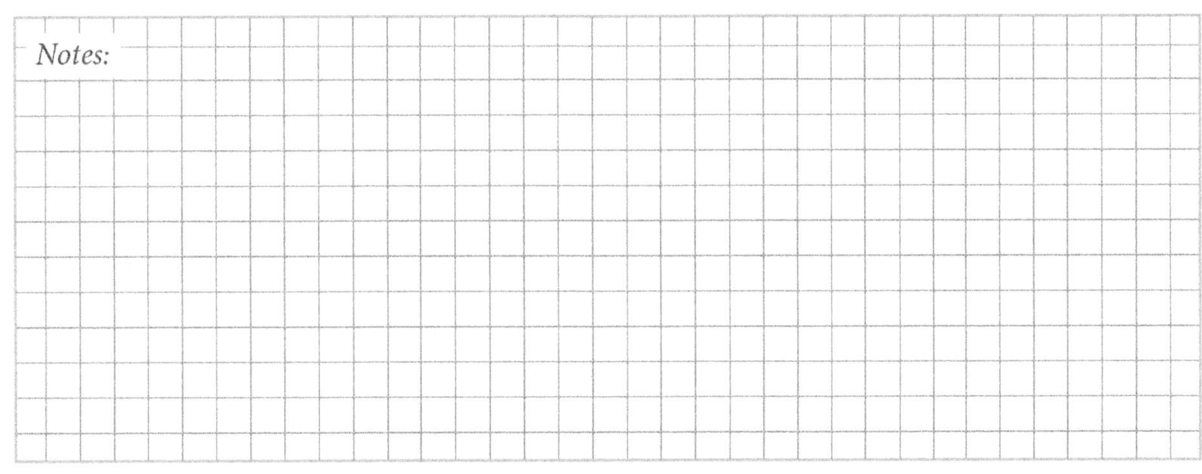

For convex buttocks in the forebody this tangent angle of hull-to-water contact (τ_{ch}) is generally greater than trim measured relative to straight afterbody buttocks, and it is dependent on the local radius of buttock curvature. τ_{ch} may be approximated by Equation 2-9 (Mottard, 1960).

$$\tau_{ch} = \tau + 90/(\pi r)(L_C + L_K) \qquad \text{Equation 2-9}$$

The trim angle from free-to-trim-and-heave model tests of planing hulls results in an effective hydrodynamic trim angle at each speed which is between τ and τ_{ch}.

Keep in mind that randomly designed planing hull surfaces are likely to have different longitudinal positions of tangency where straight buttocks aft connect with convex, curved buttocks forward.

For various speeds, effective hydrodynamic trim angles, measured relative to straight aft buttocks, will be inconsistent from boat to boat, whenever any part of the curved buttocks forward are wetted. It is unlikely that a reliable regression analysis of trim angle data referenced to straight aft buttocks can result from test results of a collection of randomly designed hulls.

It may be better to take a trim reference to be zero when the hull is stationary and measure change of trim with speed. This may provide compliant data which are a candidate for regression analysis to develop an approach for a trim prediction.

In rough seas, the initial impact of the stagnation line occurs on the curved hull buttocks at an angle τ_{ch} which can be very much greater than τ measured relative to the quarter-beam buttock of the afterbody.

As indicated by Equation 2-9, τ_{ch} can easily be 10° to 15° greater than τ when the forward buttock radius (r) in a vertical plane is small. Thus, a designer might hypothesize that buttocks with small r would have higher vertical accelerations in waves than when r is large.

This relationship of acceleration with buttock radius (r) seems to be consistent with results from models used in rough-water tests (Fridsma, 1971) which had rather

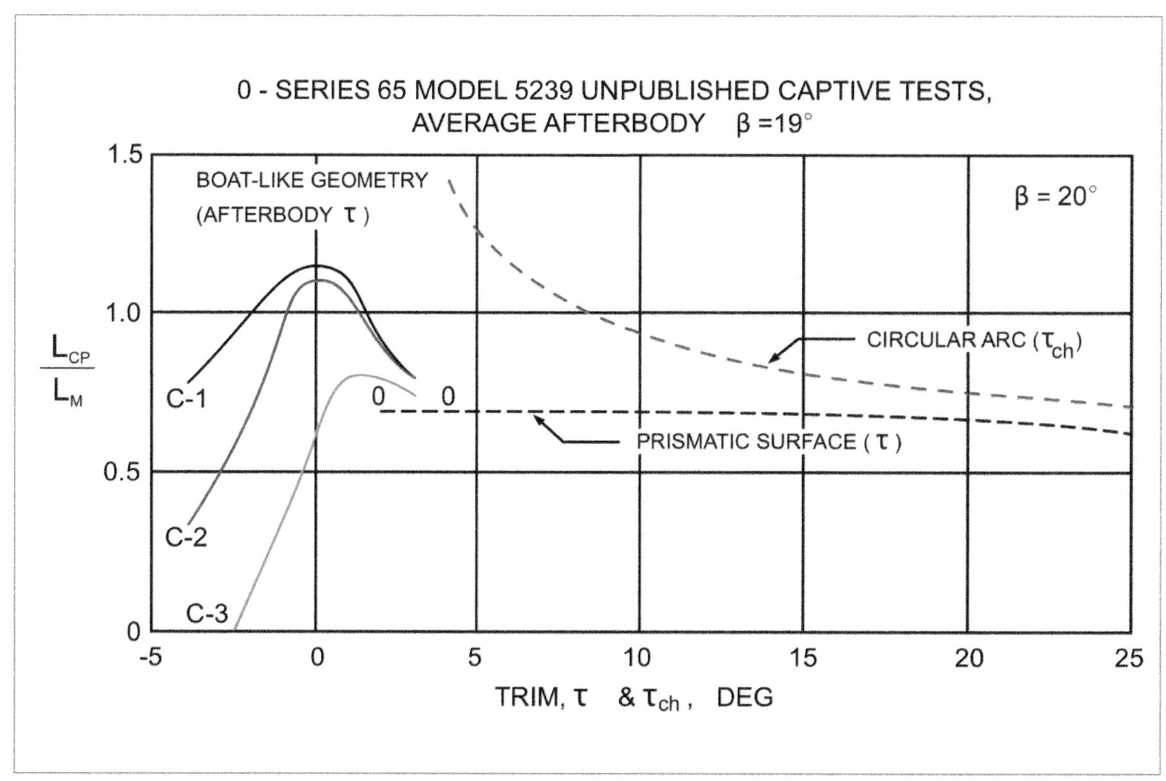

Figure 2-28: Collection of experimental data of L_{CP}/L_M versus trim for β = 20° planing surfaces with constant beam

bluff bows (small buttock radii). These 1971 test results tend to predict higher vertical accelerations than does data from hard chine planing craft having conventional bow geometry.

Once I add Figure 2-28 to this collection of experimental data for $\beta = 20°$ shapes, you can see some effects of buttock curvature on the line of action of the resultant hydrodynamic force vector (L_{CP}/L_M) referenced to the hydrodynamic transom. Some of these effects may be difficult to accept, such as the $L_{CP}/L_M > 1.0$ for low trim angles, measured relative to the tangent of the buttocks at the transom.

This might be a result of the definition of the trim-angle reference or that hydrodynamically L_{CP} could be primarily controlled by stagnation between L_C and L_K. I am showing these data to encourage you to ponder the reality or absurdity of trends.

In 1987, Codega and Lewis reported sea-trial results of a 30-ft surf rescue boat which exhibited dynamic instability at speeds approaching 30 knots. Instrumented tests included a number of flush-mounted pressure transducers to measure bottom pressure distribution and magnitude during these unstable events.

Time varying dynamic pressures on the order of -2.0 to -2.5 psi (suction) were measured mostly on the hull bottom between 20 and 50 percent of the chine length, aft of the bow. In 1992, Codega provided an understanding of the influence of the significance of forward planing surface buttock curvature with regard to both positive and negative dynamic pressures:

"If we look at pressure data from a family of standard airfoils that vary only in thickness/chord (t/c) ratio (Abbott and von Doenhoff, 1959), we find that higher t/c sections develop lower local pressures at low angles of attack than do lower t/c sections.

The extension of the analogy suggests that highly curved underwater buttocks are more prone to developing local low pressure areas with the accompanying destabilizing moments than are less curved buttocks. ...[thus,] a boat with a highly curved buttocks, the equivalent of an airfoil with a high thickness-to-chord ratio, is more prone to develop the local low pressure areas that lead to instabilities."

Leshnover's study concluded the Model C-3 was more likely than C-1 to bow-dive at low and negative trim angles. C-3 had little buttock curvature until well forward transitioning into a relative small radius near the bow. This resulted in negative lift far from the transom which exacerbates a bow-down pitching moment. At a lift coefficient ($C_{LS} = 0$) there could be negative lift at the bow and positive at the stern with a net hydrodynamic lift of zero and the resultant L_{CP} shifting aft.

In summary, I believe that the condition of forward bottom buttock curvature contributes to planing craft becoming dynamically unstable at low trim angles; to suddenly pitch-down, lose directional stability (bow steering), and/or heeling to one side. I also believe that running trim angle and the curvature of the forebody buttock shape are essential considerations for the development of successful high-speed planing boats which will be dynamically stable in rough seas.

Longitudinal Variation of Deadrise: Warp

There are many planing craft which do not have constant deadrise hull bottoms. When a longitudinal variation of deadrise increases from the transom to about mid-length, the hull is referred to as having a warped or twisted planing bottom. In 1975, Peter Ward-Brown conducted tests at the Davidson Laboratory to obtain hydrodynamic characteristics of two constant beam, warped planing surfaces similar to the form seen in Figure 2-29.

The linear measure of warp of the planing surface is the change in deadrise angle over a longitudinal length equal to the maximum projected chine beam. Some of these data were presented by Savitsky and Brown the next year and further analysis of the 3° per beam data was published by Savitsky in 2012. This reference also included a tabulation of remaining unpublished Ward-Brown data for a higher warp of 9° per beam.

The aim of Savitsky in 2012 was to confirm that his widely used planing equations for prismatic hulls from 1964 are also suitable for predicting resistance and trim for warped hulls with certain stipulations. The necessary condition is that the effective deadrise is taken for each speed at the mean wetted length. However, the predicted trim angle relative to the water surface must then be referenced to the quarter beam buttock, $B_{PX}/4$ of the afterbody hull lines.

With only modest changes to computer code following the 1964 procedures of Savitsky resistance and trim predictions for planing craft, having warp may be approximated using his 2012 approach.

Relative to Savitsky's 1964 resistance and trim prediction process, the variation of functions when all forces pass through the CG for warped surfaces are noted in Table 2-E. Other than changes noted in Table 2-E, the remainder of his prediction procedure are to be followed.

Note: The definition of $\lambda = L_M/B_{PX}$ is used only for the Savitsky prediction methods. For the remainder of this book, λ is defined as the linear ratio between two different sized hulls.

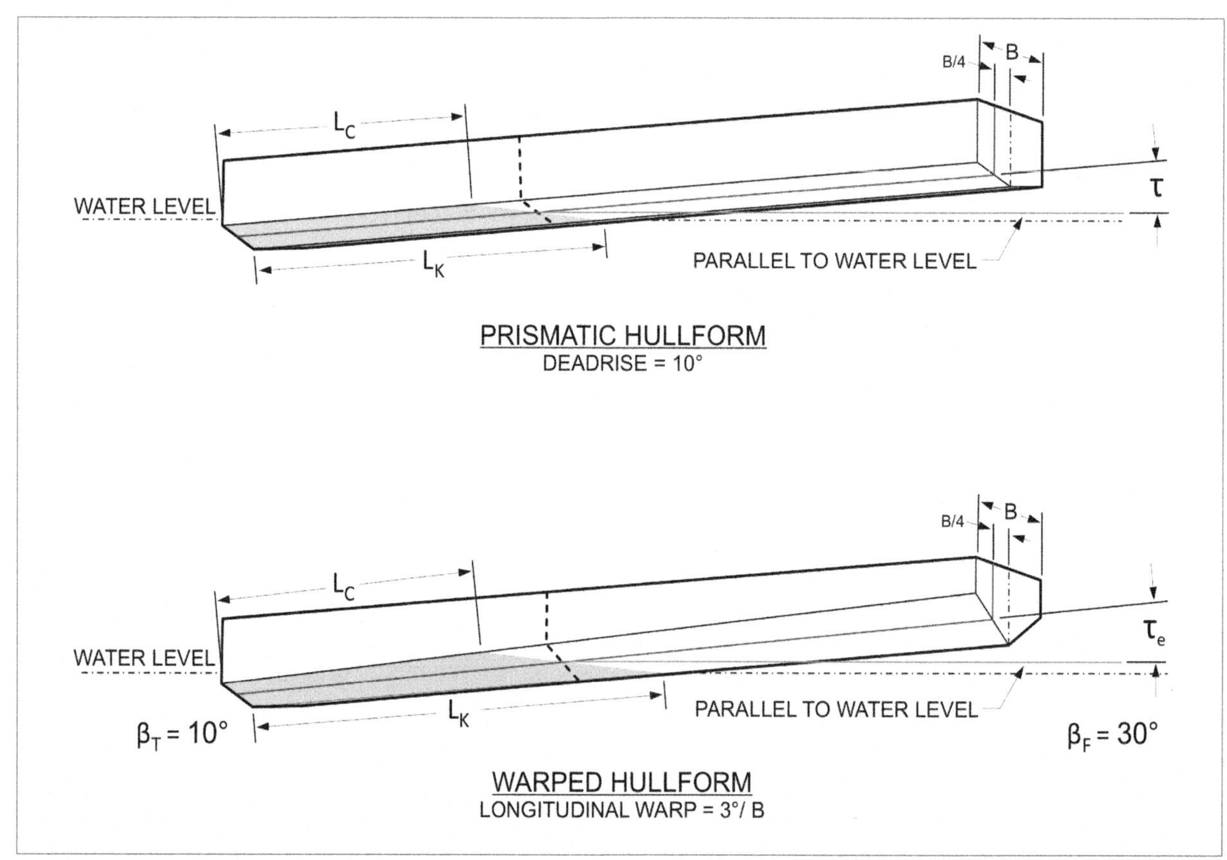

Figure 2-29: Depiction of prismatic and warped hull forms

TABLE 2-E: VARIATION OF FUNCTIONS BETWEEN PRISMATIC AND WARPED SURFACES		
	Prismatic Savitsky (1964)	Warped Savitsky (2012)
Deadrise for each speed (deg)	β	$\beta_e = \beta_\tau + \lambda$ (Warp)
Predicted hydrodynamic trim angle for each speed (deg)	τ referenced to keel	$\tau_e = \tau + \theta_{B/4}$ referenced to $B_{PX}/4$
Center of pressure of total lift forward of the transom. Solve for $\lambda = L_M/B_{PX}$	$CP = L_{cp}/(\lambda B_{PX}) = 0.75 - 1/[(5.21C_V^2/\lambda^2) + 2.39] \approx LCG/L_M$	
Explicit numerical approximation: $\lambda = L_M/B_{PX}$ defined by the equation above for CP	$\lambda = 1.33\,(CP/B_{PX}) + 0.7\,(CP/B_{PX})/[0.43 + \{0.23 + (CP/B_{PX})/C_V\}^{-4}]$	
Where $\theta_{B/4} = (1/4)\,\text{atan}\,[(\tan\beta_{mid} - \tan\beta_T)/(L_P/B_{PX})]$		

Non-constant afterbody deadrise (warp or twist) has been considered by some to result in a higher resistance when compared with a constant-deadrise hull. This opinion is not adequately supported by experimental data when comparisons with constant and warped afterbody hulls are made for the case where both have equal deadrise angles at the center of pressure. For this case there is little difference in relative hull resistance, but there is some difference in dynamic trim. Thus, warping is actually a designer's tool to control dynamic trim in the same fashion that wedges, buttock hook, or when bottom-plate extensions are built into craft. Bottom-plate extensions may conventionally be bent up or down to change dynamic trim as desired after builder's trials.

Some Collected Thoughts

Hydrodynamic pressures can be either positive or negative, with resulting forces directed in three-dimensional space. On this note, Figure 2-30, revised from Müller-Graf (1997), depicts several curved buttock shapes with resulting longitudinal hydrodynamic pressure distributions. Positive pressures are dark blue, negative pressures are light blue, and the red arrow length indicates approximate resultant force and direction as well as longitudinal point of application, assuming force magnitudes are proportional to blue-shaded areas.

Both the magnitude and direction—positive or negative—of hydrodynamic pressure can vary radically with buttock shape and trim, resulting in variation from net positive lift to net negative suction, and the effective center of dynamic pressure moves longitudinally about the boat with the potential for dynamic trimming moments changing from bow up to bow down.

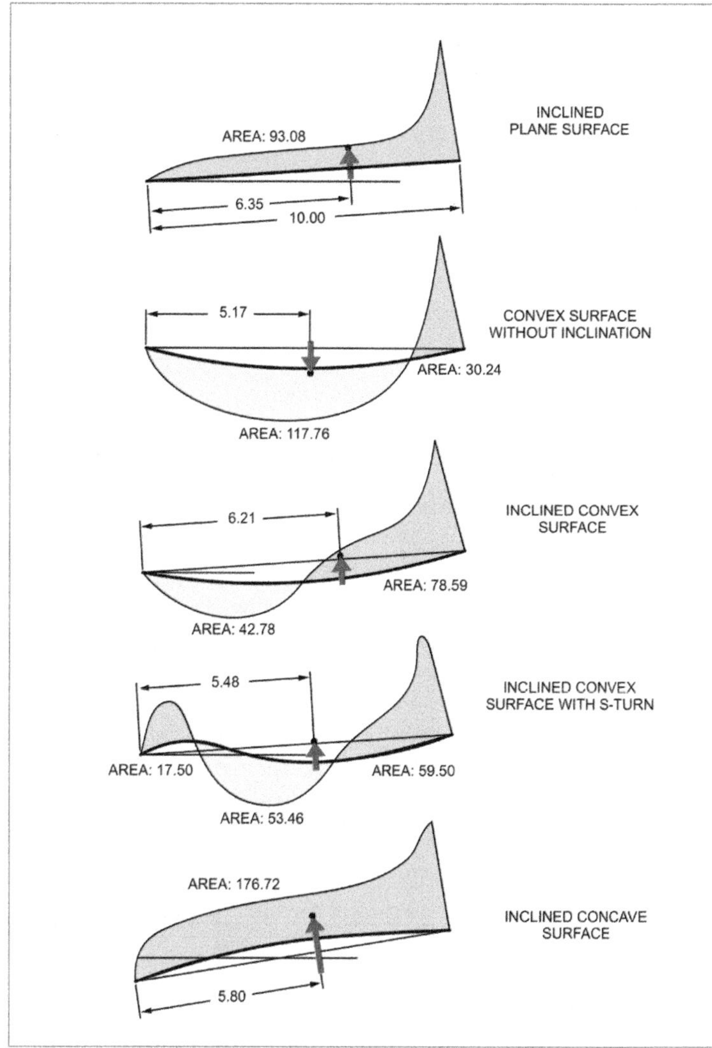

Figure 2-30: Approximate 2-D pressure distribution for various convex and concave buttock shapes (revised from Muller-Graf 1997)

As shown in Figure 2-30, the naval architect's definition of longitudinal distribution of buttock shape can have dramatic influence on hydrodynamic bottom pressure as well as the resultant contribution of buoyant support. And it follows then that the building shipyard has a responsibility to faithfully and accurately reproduce the design shape according to the specified hull geometry.

The figure from Müller-Graf (1997) depicts a variety of hydrodynamic pressure distributions possible by varying the longitudinal buttock shape of the wetted portion of the hull bottom of a planing boat. If you assume that a planing boat's dynamic lift may be represented by an approximate prismatic surface, you may—without careful attention to detail—misrepresent both the magnitude of lift and the effective center of lift.

Yes, these curves may be different, but it is highly likely that a realistic planing hull surface will develop some negative pressure at low trim angles. When curved hull buttocks at the bow are wetted, this will be confirmed by low-speed observations of the bow trimming down. The same thing can occur again at high speeds when, with low trim angles, the curved bow buttocks again become wetted and forward lift can be reduced.

It's important to visualize a planing hull bottom area relative to straight afterbody buttocks and forward curved buttocks. Figure 2-31 depicts the hull bottom from this aspect showing four plan and profile views. Figure 2-31a presents designed bottom geometry, shaded for the area of straight buttocks.

This enhanced area is shown from the transom forward to the location of the tangent at the beginning of forward, upward convex curvature. The curved buttock surface area at bow is white. For the design operating condition (displacement, speed, LCG and trim) model test data provided the location of the stagnation line given in Figure 2-31b. The overlay of the stagnation and tangent lines is quite good; thus, achieving maximum beneficial lift for the wetted area.

When the boat is trimmed down 1.5 degrees with trim tabs, interceptors or a shift forward in load the stagnation line moves as you can see in Figure 2-31c. In this case the stagnation line is in the area of convex buttock curvature and the peak pressure forward is changed due to the flow over curved buttocks.

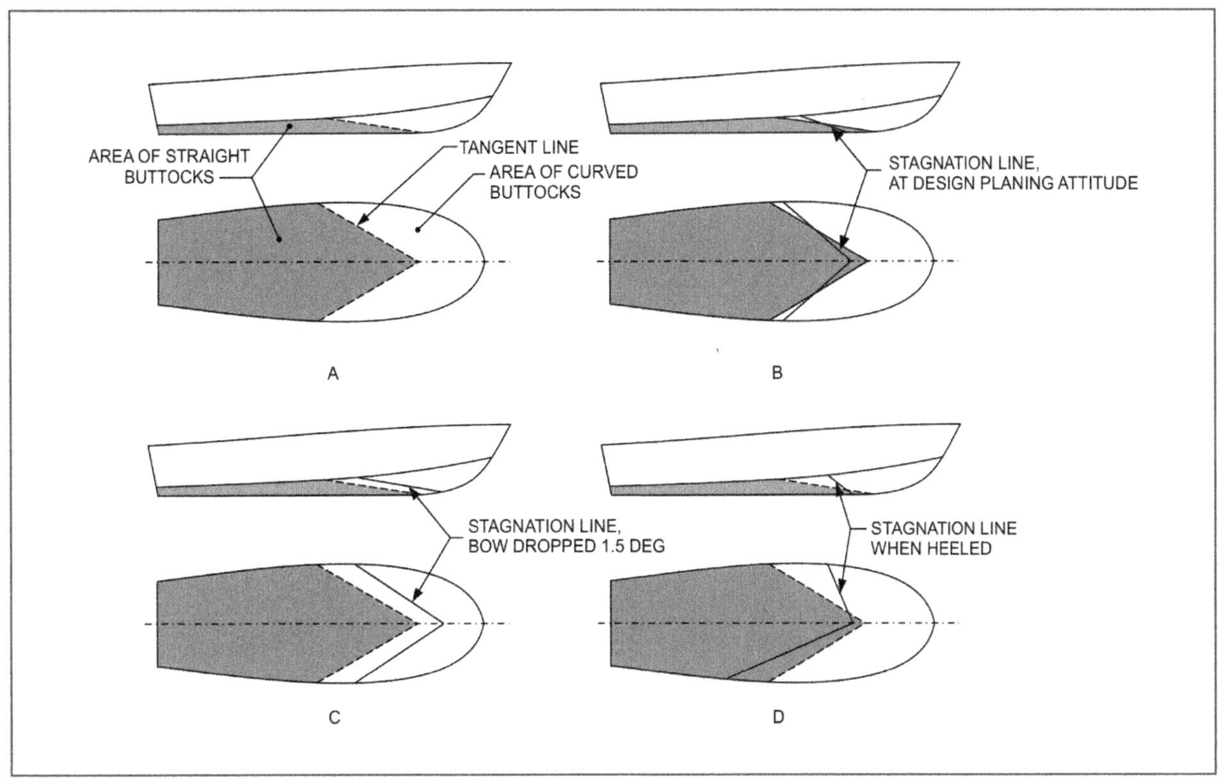

Figure 2-31: Profile and plan view of a Series 62 hull showing straight buttocks (lines) and forward convex curved buttocks (white area)

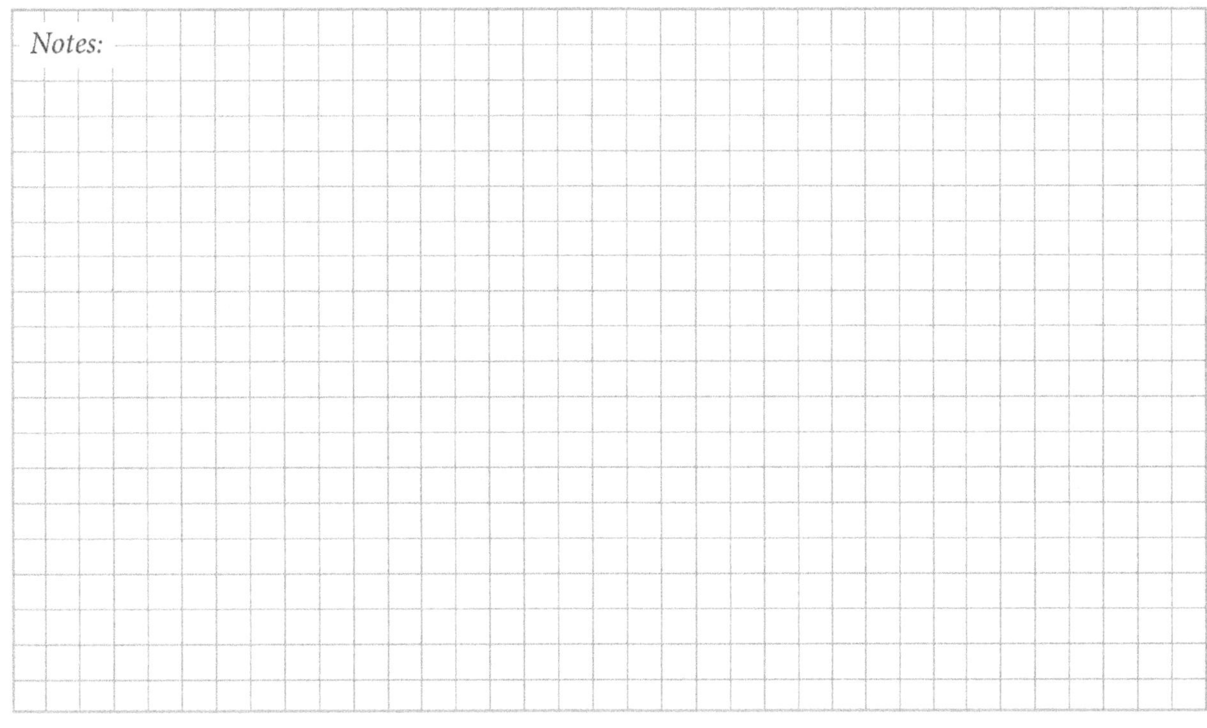

Figure 2-31d depicts a fourth condition: With off-centered weight or in a turn, the boat will operate with a heel angle. For this example, note that the stagnation line is clearly in the region of straight buttocks one side and in the region of curved buttocks on the other side. This asymmetric pressure distribution tends to result in a roll moment producing heel angle down to the side of wetted curved buttocks.

To sum up: Analytical resistance prediction methods using experimental prismatic planing surface data, plus representative displaced volumes for supporting the total weight of a planing boat, may need to be refined to correctly model three-dimensional hull forms. It is likely that you will need to adjust experimental prismatic lift versus trim angle relationship due to bow curvature as it will change the slope of prismatic lift curves. At very low trim angles, it may result in a dynamic bow-down moment.

Or better yet, hydrodynamicists should study velocities and pressures of planing hull shapes using longitudinals streamlines realistically replicating flow along buttock lines.

TABLE 2-F: REFERENCES FOR TESTS OF PRISMATIC FORMS HAVING DEADRISE ANGLES FROM -10° TO 70°

Deadrise Angle β	Reference
-10°	Kimon (1957)
0°	Weinstein et al. (1953)
20° & 40°	Chambliss et al. (1953)
50°	Springston et al. (1955)
70°	Pope (1958)
Analytical representation for 0° ≤ β ≤ 50°	Shuford (1958)

I have also included references in Table 2-F for convenience in documenting experimental data for resistance, lift, center of lift and wetted length, only at positive trim angles, from tests of prismatic forms for deadrise angles from -10° to +70°. These are included in the Reference section of this book.

These reported data represent only dynamic (planing) lift for these experiments when the buoyant component near the transom is only a minimal amount of the measured vertical force. The various analytical planing lift equations developed from these or similar experimental sources are incorporated in many resistance prediction approaches and are appropriate only for very high planing speeds. However, to represent total support of a boat, hydrodynamic contributions from all sources must be coupled with the buoyant component of support and its effective center of lift (LCB).

Over a range of operational speeds, inclusion of all the vertical forces and their effective centers of application are necessary to achieve a significant representation when predicting both total hull resistance and especially running trim angle.

For equilibrium, the weight of a planing boat must be supported by the total of hydrostatic buoyancy and the net sum of both positive and negative hydrodynamic forces as well as influences of appendages and propulsors. Also, for stable performance both the centroid of buoyant forces, as well as all hydrodynamic forces, must combine such that their resultant total is in vertical alignment with the static center of gravity of the boat's weight. Designers must not isolate themselves from understanding the realities of the entire operational speed range of a high-performance craft. Even at zero speed, all surface vessels must float upright and be stable, based on hydrostatic principles.

As speed increases negative dynamic pressures (suction) will increase the draft resulting in the boat virtually getting heavier. Further increase in speed of a craft with appropriate hull lines causes the bow to trim up. After passing peak trim, a hard-chine hull will begin to reduce running trim with further increase in speed.

For a boat designed for high speeds, buoyancy will have provided support for the total weight of the boat at zero speed and total weight of a vessel plus both hydrodynamic positive and negative forces for semi-displacement speeds. For semi-planing speeds, partial support of total weight coupled with trimming moment is attributed to buoyancy. Hydrodynamic pressures begin to lift the CG above its static elevation. With increasing speeds buoyant forces will reduce to a small percentage (but rarely less than ten percent) of total weight as planing lift will dominate total support.

This description of speed-related events is clearly shown by studying the model test results in Figure 2-9. For example, Table 2-G (taken from Figure 2-9) relates how the volume of the hull, below the water surface, relative to static volume changes from zero to high planing speeds. And it is a reminder that at speeds when the transom is dry, the volume of the hull below the still water surface when multiplied by water density alone overestimates buoyant lift of planing craft.

In this chapter I have attempted to point out (for calm, deep-water operation) important aspects of dimensionless speed.

Appropriate hull form and loading are both affected by client requirements and dependent on dimensionless speeds which influence designers when making early decisions in programs developing new craft or modifications to an existing vessel.

TABLE 2-G: APPROXIMATE RATIO OF VOLUME BELOW THE WATER SURFACE RELATIVE TO STATIC HULL VOLUME VERSUS SPEED

Speed		
F_{nL}	F_{nV}	$\nabla v/\nabla$
0.00	0.00	1.00
0.40	1.00	1.10
0.65	1.65	1.00
1.20	3.00	0.45
1.60	4.00	0.30
2.00	5.00	0.20
2.35	6.00	0.15

My goal is to provide technology trends to assist naval architects in making early hydrodynamic choices to quickly go around the design spiral to confirm that client requirements can be achieved.

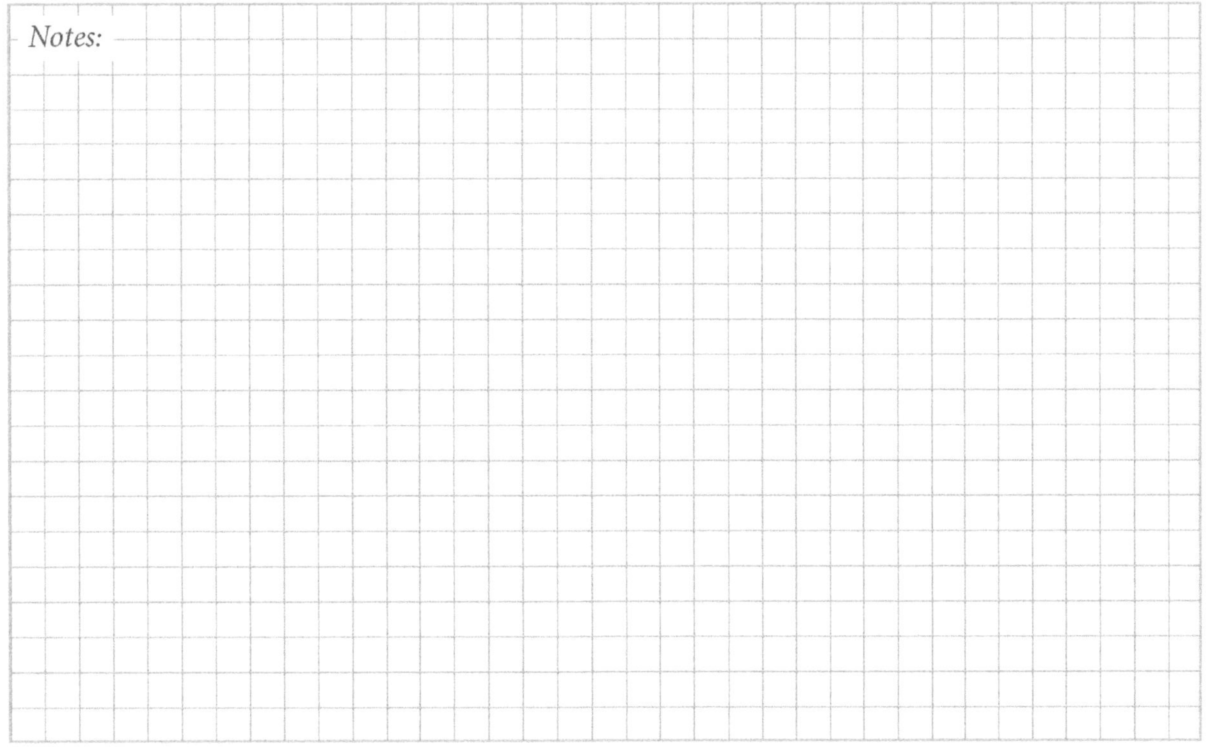

CHAPTER THREE

Calm-Water Hull Resistance Deep and Shallow Water

To be successful, the design of a performance vessel must be based on scientific principles and processes—in addition to experiment and experience. Analytical approaches require experimental validation. Throughout the design process, experience with performance vessels must meet a gut feeling of reality, and designers need to have a sense of the magnitude of expected results from numerical computations. When making design decisions, naval architects must clearly understand the assumptions and boundaries of any computer software they are using before relying on the results.

Of all the diversity of possible client's specifications—whether for military, commercial, or recreational uses—there is one common requirement: attainment of a specified speed. And now we'll proceed to discuss the fundamentals of prediction of hull resistance versus speed.

The hydrodynamics of craft operating at the water surface is more complex than when a body is completely immersed deeply in a fluid. This complexity occurs at the boundary of two fluids of different densities, air and water, as gravity waves are created by craft moving both at the surface and slightly above, whether supported by hydrostatic, hydrodynamic, or aerodynamic forces and also when operating at shallow depths of submergence.

In the 1870s, William Froude undertook the complex task of investigating a method of analyzing the total resistance of vessels operating at water's surface. His work defined the approach we now use for scaling resistance of models to differently-sized vessels having similar dimensionless geometry. That is, the method of expanding model resistance data versus speed to that for full-size ship resistance versus its corresponding speed.

Froude studied wave profiles of different-sized models having the same geometry, and that led to his finding that wave profiles compared well at speeds proportional to the square root of their respective lengths.

$$V \propto L^{1/2} \qquad \text{Equation 3-1}$$

When the "Principle of Similitude" (discussed shortly in this chapter) was accounted for, however, with regard to total resistance versus speed for models of different lengths, Froude found that total resistance coefficients at corresponding dimensionless speeds for different length models did not overlay. The curves had similar shapes, but were displaced. He reasoned that frictional resistance and wave-making resistance versus speed must follow different scaling laws.

Then, to essentially estimate wave-making resistance, Froude conducted a separate study by towing thin planks of various proportions and roughness through water to measure skin-friction resistance. With these experimental data, he developed a representation for skin-frictional resistance based on wetted length and surface area.

He used this formulation to calculate frictional resistance, which he subtracted from model total resistance to yield the remaining residual (mostly wave-making) resistance (R_R). He then reduced the residual resistance to a dimensionless coefficient (C_R), which was found to compare well at the same speed-length ratios, $V/(L)^{1/2}$, of different size and geometrically similar models of a ship form. [*Note: The dimensionless form of $V/(L)^{1/2}$ which is proportional to length Froude number (F_{nL}) is used here, in preference to dimensional notation speed-length ratio.*]

$$R_R = R_T - R_F \qquad \text{Equation 3-2A}$$

$$C_R = R_R/[(1/2\rho S v^2] \qquad \text{Equation 3-2B}$$

Since Froude's work in the 1870s, it has been accepted that:

$R_T = R_F + R_R$ (Dimensional) Equation 3-3A

$C_T = C_F + C_R$ (Dimensionless coefficients) Equation 3-3B

R_F and R_R follow different dimensionless speed laws because R_F is a function of Reynolds number (R_n) taking into account viscosity of water and R_R is a function of Froude number F_{nL} which includes gravity (g) governing wave making. Since R_F can be calculated with reasonable accuracy, one can clearly understand that the primary reason for conducting calm-water resistance tests with models is to experimentally obtain the residual resistance coefficient (C_R).

With R_R and speed extended to dimensionless format of C_R versus F_{nL}, you, as a designer, essentially have acquired invariant characteristics for the hull lines, loading, and LCG tested. C_R versus F_{nL} is, in fact, a "hydrodynamic fingerprint" for the model test conditions. With some forethought in planning an expanded test program to include a matrix of loadings and LCGs, you now have the seeds for building a design database over time, and this will enhance optimization opportunities for your future projects. We'll discuss this more in *Chapter 10 – Model Testing.*

Since the late 1940s, much of the high-performance craft resistance data published in the public domain, when scaled from model to full-scale vessels, has used either the ATTC 1947 or ITTC 1957 formulations to calculate the friction-resistance coefficient.

ATTC 1947 Schoenherr Friction Resistance Coefficient

$0.242/(C_F)1/2 = Log_{10} (R_n \times C_F)$ Equation 3-4

ITTC 1957 Model-Ship Correlation Line

$C_F = 0.075/[(Log_{10} R_n) - 2]^2$ Equation 3-5

In Figure 3-1, you can see curves for both ATTC 1947 and ITTC 1957. The difference between using these two C_F formulations for predicting full-scale hull resistance can result in one percent variation when small-scale models Reynolds numbers are on the order of $1 X 10^7$ and about three percent when model scale $R_n \approx 1 X 10^6$.

In general, I am using the ITTC 1957 formulation for examples in this book, and I recommended it for high-performance craft.

You must, however, remember, that C_R is directly linked to the friction formulation used to calculate C_F.

To put it another way: Do not use C_R derived from model data using C_F from the ATTC 1947 formulation to calculate full-scale C_T by adding C_R to C_F computed using ITTC 1957 formulation or vice versa, unless the slowest model $R_n > 2 X 10^7$.

Take a look at, Appendix 1, Table A1-F. That does provide a method for adjusting C_R obtained with ATTC 1947 to C_R computed with ITTC 1957 and vice versa.

Figure 3-1 shows you how Froude's finding may be visualized graphically.

Friction resistance coefficient formulations for turbulent flow vary with Reynolds number.

Model data are used to calculate total resistance coefficient ($C_{T,m}$) versus $R_{n,m}$ and plotted on the graph. For example, at $R_{n,m} = 3 X 10^6$ vertical lines are indicated for $C_{F,m}$ (ITTC 1957) and C_R which added together represent $C_{T,m}$.

In keeping with Froude's work for model data scaled up to a 500 mt vessel at $F_{nL} = 0.404$, C_{Ff} is computed for the corresponding full-scale $R_{nf} = 3.96 \times 10^8$. Then the full-scale total resistance coefficient is $C_{Tf} - C_{Ff} + C_R$ as seen in Figure 3-1.

Figure 3-1 depicts the following calculation procedure when model resistance data are available for a specific F_{nL}.

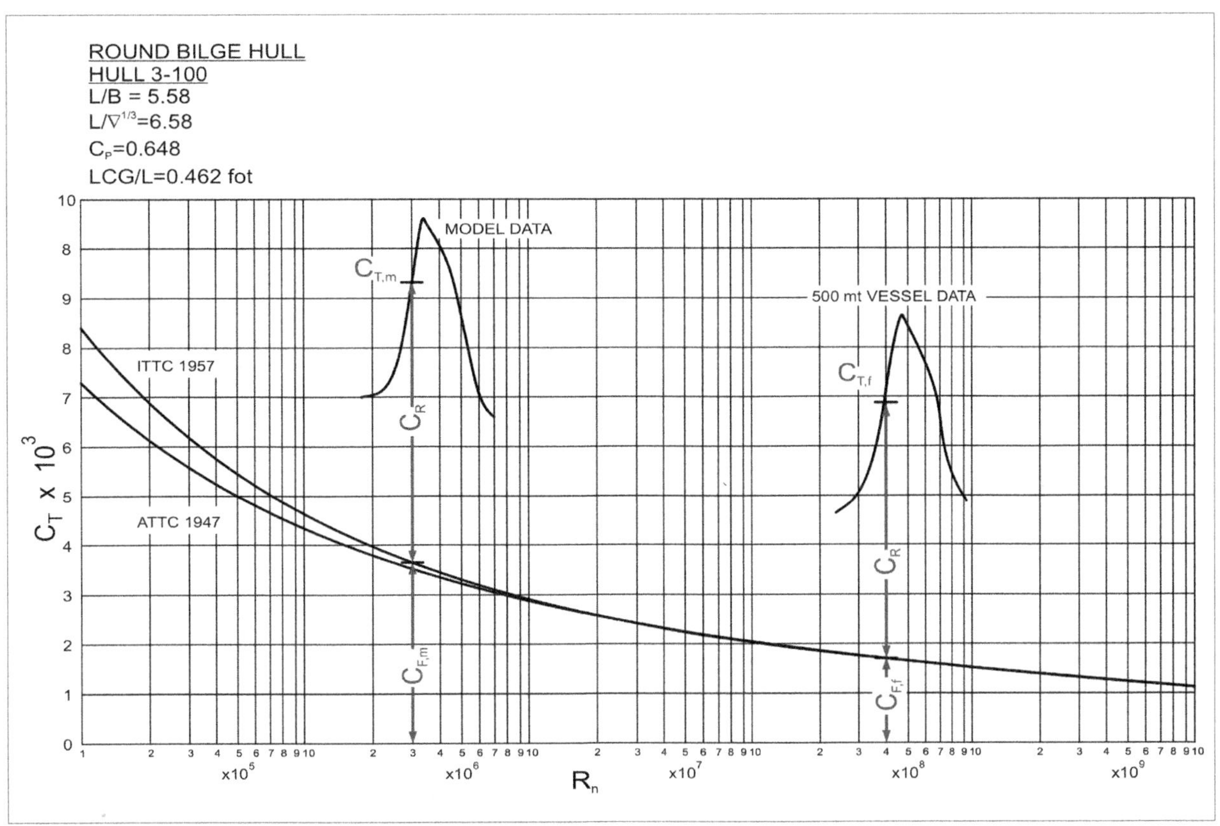

Figure 3-1: ATTC 1947 and ITTC 1957 friction formulations with graphic example of Froude's scaling method

$$C_R = C_{T,m} - C_{F,m}$$ Equation 3-6

For a full-scale vessel having the same F_{nL}

$$C_{T,f} = C_R + C_{F,f}$$ Equation 3-7

The bare-hull resistance of the full-scale vessel at the same F_{nL} is

$$R = C_{T,f}(1/2)\rho_f S_f v_f^2$$ Equation 3-8

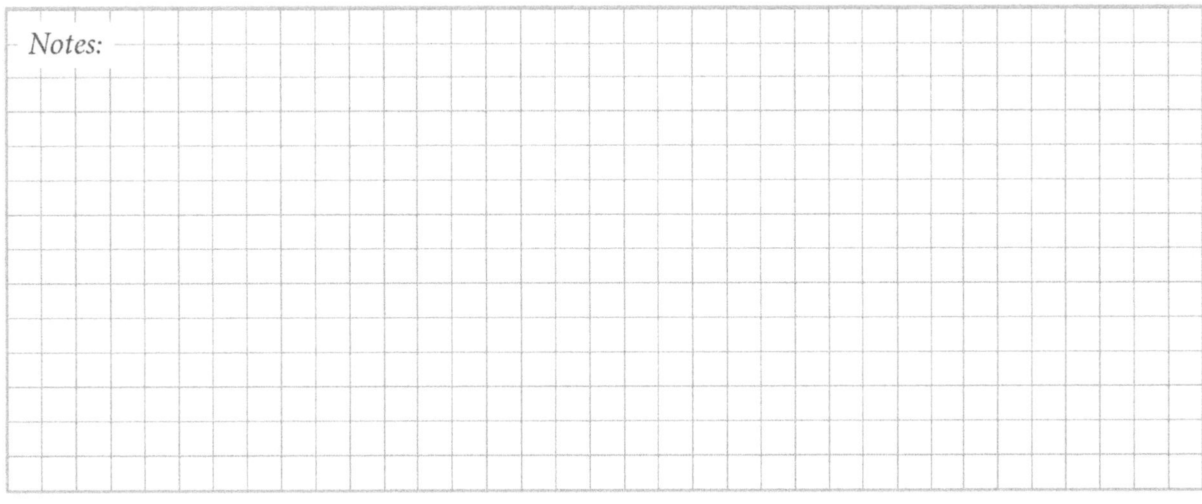

Principle of Similitude

The Principle of Similitude provides means of defining comparative relationships which exist between vessels of various sizes having geometrically similar hull forms, such as a model and a ship, or two geometrically similar ships of different lengths, but having the same dimensionless hull loading.

For your convenience, I've provided Table 3-A to show the case when mass and speed are proportional to volume and the square root of length, respectively. The relationships in the table are functions of the linear scale ratio (λ) of the two length vessels defined as:

$$\lambda = L_f/L_m \qquad \text{Equation 3-9}$$

TABLE 3-A: RELATIONSHIPS BETWEEN TWO SIZE VESSELS

Quantity	From small scale	Multiply by	For full scale
Linear dimensions	L_m	λ	L_f
Area	S_m	λ^2	S_f
Volume	∇_m	λ^3	∇_f
Linear velocity	V_m	$\lambda^{1/2}$	V_f
Reynolds number	$R_{n,m}$	$\lambda^{3/2}(\nu_m/\nu_f)$	$R_{m,f}$
Resistance	R_m*	$\lambda^3(\rho_f/\rho_m)$	R_f
Weight	W_m & Δ_m	$\lambda^3(\rho_f/\rho_m)$	W_f & Δ_f
Thrust	T_m	$\lambda^3(\rho_f/\rho_m)$	T_f
Power	$P_{D,m}$	$\lambda^{3.5}(\rho_f/\rho_m)$	$P_{D,f}$
Trim angle	τ_m	I	τ_f
Subscript m refers to model and f refers to full scale *After $C_{T,m}$ is corrected to $C_{T,f}$.			

Predicting Hull Resistance Using Calm, Deep-Water Model Test Data

At this point, we need a simple definition of deep water to differentiate from shallow water where there is significant influence on hull resistance over a range of speeds. When water depth is greater than 80 percent of length overall (LOA), there is essentially no bottom effect on resistance. (Blount and Hankley, 1976.) We'll address the effect of shallow water on resistance later in this chapter.

The principle for scaling deep-water model test resistance data to full-size vessels applies to displacement and semi-displacement vessels as well as semi-planing and planing craft. There are, however, differences in detailed measurements to be recorded during testing. The weight of displacement and semi-displacement vessels is primarily supported by buoyancy (hydrostatics) and, as such, the wetted length and surface area of the hull has little or no change throughout the operating speed range: L and S are considered to be constant. The Model 3-100 profile photographs of a round-bilge hull, seen in Figure 3-2, show little change in wetted length and surface area for the high-speed range of F_{nL} from 0.55 to 0.91.

For semi-planing and planing craft, weight is supported mostly by dynamic lift that, with increasing speed, tends to elevate the hull relative to the water's surface. Thus, both the wetted length and wetted surface change with speed. To accurately reflect these changes affecting the calculation of the frictional resistance component of total resistance, wetted lengths of keel and chine, CG rise, and running trim-angle measurements must also be made during resistance tests.

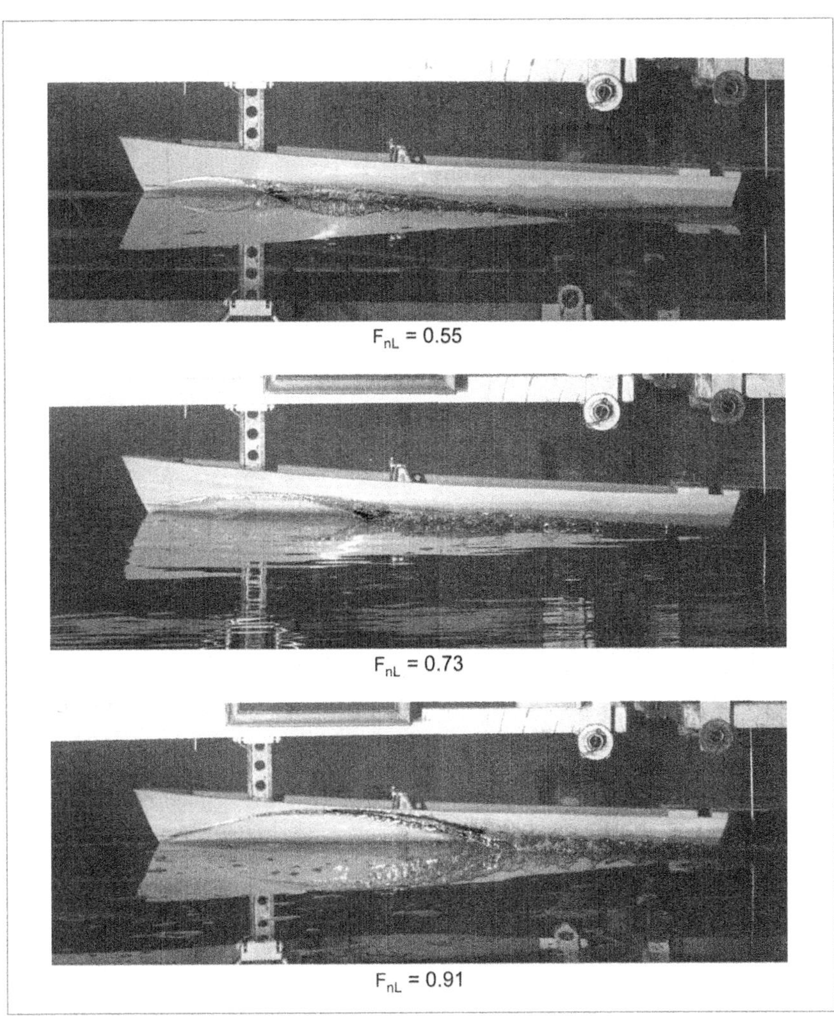

Figure 3-2: Wave profile photographs of high-speed round-bilge hull

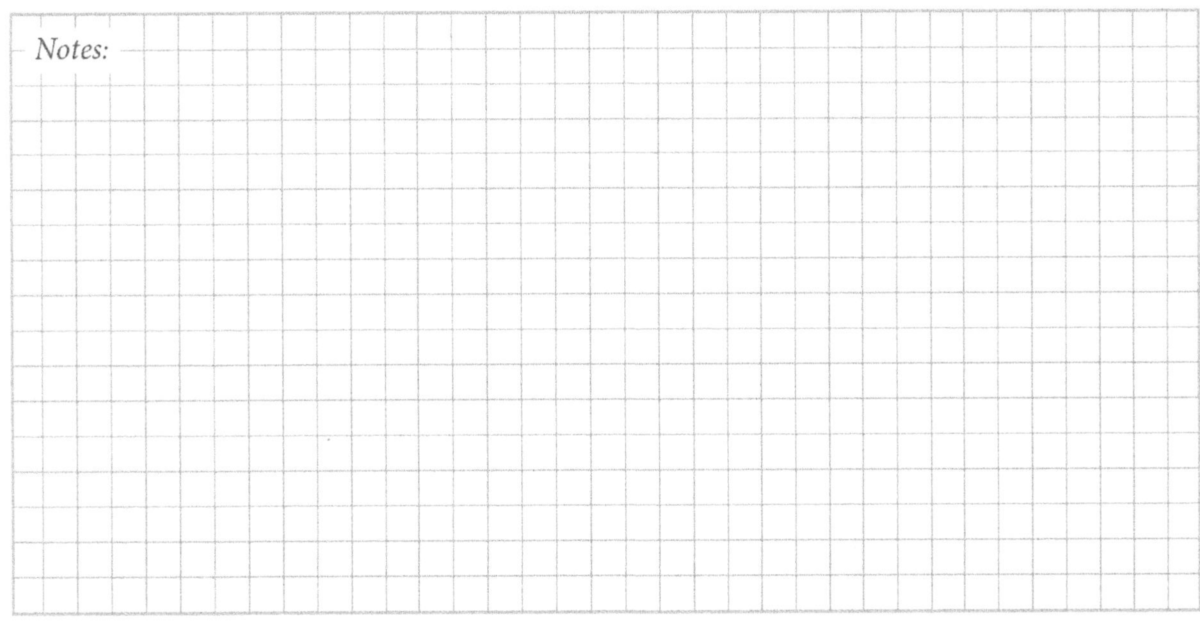

There are two calculation examples for extracting C_R from model test data; one for a semi-displacement hull and the other for a planing craft. Principal characteristics for the round-bilge hull, model 3-100, are given in Table 3-B. [*Note: Traditional displacement hull coefficients and parameters are given.*] Later as we discuss the design of planing hulls, you will see that different hull-form coefficients are significant.

TABLE 3-B: PRINCIPAL CHARACTERISTICS				
Example Hull: Model 3-100				
Hull Form: Round Bilge				
Dimensions	**Model**	**500 mt vessel**	**Coefficients**	
LOA ft (m)	6.38 (1.94)	182 (55.5)	C_B	0.381
LWL ft	5.97	170	C_P	0.648
BOA ft (m)	1.07 (0.326)	30.4 (9.27)	C_X	0.588
BWL ft	1.07	30.4	C_W	0.755
LCG fot* ft	2.76*	78.5*	L/B	5.58
Displacement	46.53 lbs	500 mt	B/T	3.48
Volume ft³ (m³)	0.747 (0.0212)	17,230 (487.9)	$L/\nabla^{1/3}$	6.58
Draft ft (m)	0.307 (0.0936)	8.74 (2.66)	LCG/L*	0.462*
Wetted Surface ft²	6.35	5,140		
Linear ratio	1.0	28.46		
*fot – forward of transom				

Table 3-C, model test data and calculation results are shown with the graph of C_R versus F_{nL} for the test condition in Figure 3-3. For this calculation example, the use of constant waterline length and wetted surface are appropriate and hydrodynamic phenomena are best represented by length Froude number F_{nL}.

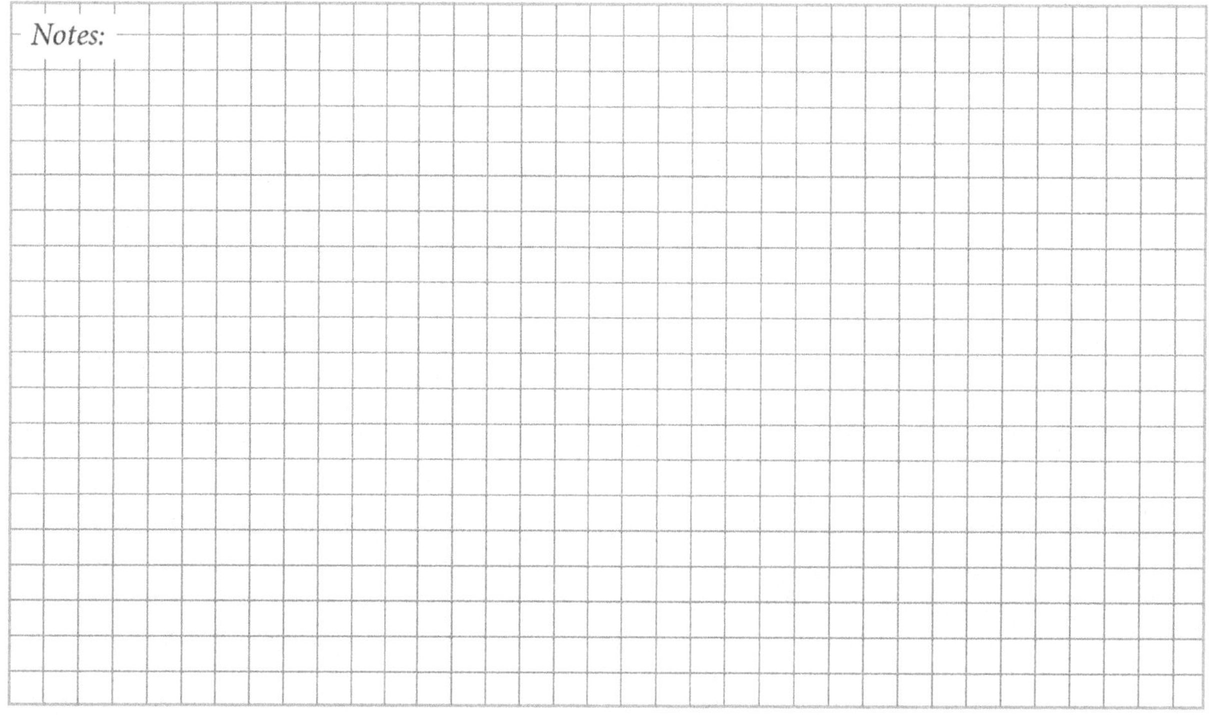

TABLE 3-C: MODEL TEST DATA AND CALCULATION OF C_R	
Example Hull: Model 3-100 Hull Form: Round Bilge Water: Fresh Temperature: 66°F Mass Density: 1.947 lb sec^2/ft^4 Kinematic Viscosity: 1.113 X 10^{-5} ft^2/sec	λ = 28.46:1.0 Method of Turbulence Stimulation: Trip wire Friction Formulation: ITTC 1957 L_m = 5.97 ft; Displacement = 46.53 lbs; Draft = 0.307 ft Wetted Surface, S_m = 6.35 ft^2

Test Data			Calculations				
A	B	C	D	E	F	G	H
Run No.	V – kt	R – lbs	$C_{T,m}$ X 10^3	$R_{n,m}$ X 10^{-6}	$C_{F,m}$ X 10^3	C_R X 10^3	F_{nL}
1	2.795	1.04	7.60	2.530	3.868	3.732	0.341
2	3.585	2.30	10.22	3.246	3.685	6.535	0.437
3	4.00	2.95	10.52	3.621	3.609	6.911	0.488
4	6.80	5.76	7.110	6.156	3.270	3.840	0.829
5	7.20	6.19	6.816	6.518	3.236	3.580	0.878
6	7.60	6.68	6.601	6.880	3.205	3.396	0.927
7	2.205	0.60	7.044	1.996	4.056	2.988	0.269
8	3.80	2.68	10.59	3.440	3.644	6.946	0.463
9	7.40	6.40	6.667	3.700	3.220	3.447	0.902
10	2.00	0.49	6.992	1.811	4.137	2.855	0.244
11	2.40	0.72	7.135	2.173	3.987	3.148	0.293
12	6.00	5.00	7.928	5.432	3.345	4.583	0.732
13	5.00	4.17	9.521	4.526	3.460	6.061	0.610
14	4.60	3.69	9.954	4.164	3.515	6.439	0.561
15	3.405	1.89	9.305	3.083	3.722	5.583	0.415
16	3.20	1.55	8.640	2.897	3.767	4.873	0.390

In Table 3-C, column inputs are briefly defined as follows:

Column	Source or Model Scale Calculations
A,B,C	Model data
D	$C_{T,m} = R/[(1/2)\rho_m S_m v^2]$
E	$R_{nm} = vL/\nu_m$
F	$C_{F,m} = 0.075/[(Log_{10} R_{n,m}) - 2]^2$
G	$C_R = C_T - C_{F,m}$
H	$F_{nL} = v/(gL)^{1/2}$

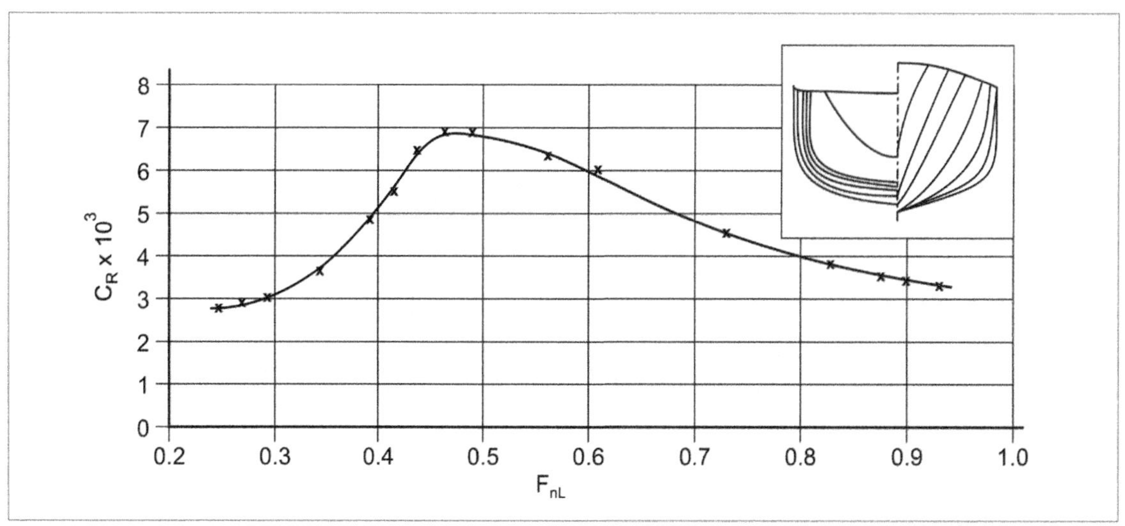

Figure 3-3: C_R versus F_{nL} for Example Hull 3-100

For the planing hull example, principal characteristics for hard-chine hull 3-200 are given in Table 3-D. Model test data and calculation results are in Table 3-E with the graph of C_R versus F_{nV} given in Figure 3-4. Note that speed-dependent hydrodynamic phenomena of craft supported substantially by dynamic lift are best represented by volume Froude number F_{nV}.

TABLE 3-D: PRINCIPAL CHARACTERISTICS			
Example Hull: Model 3-200			
Hull Form: Hard Chine			
Dimensions	**Model**	**500 mt Vessel**	**Coefficients**
LOA ft (m)	8.47 (2.58)	143 (43.7)	
L_P ft	8.00	135	
BOA ft (m)	2.44 (0.744)	41.3 (12.6)	$A_P/\nabla^{2/3} = 5.5$
B_{PX} ft	1.956	33.1	$(CA_P - LCG)100/L_P = 3.9\%$
LCG ft fot*	3.59*	60.7*	$L_P/B_{PX} = 4.09$
Displacement	221.1 lbs	500 mt	$LCG/L_P = 0.448$*
Volume ft³ (m³)	3.55 (0.101)	17,210 (487.4)	$B_{PX}/T = 5.0$
Draft ft (m)	0.392 (0.119)	6.63 (2.02)	$B_{PT}/B_{PX} = 0.64$
A_P ft²	12.80	3,660	$L_P/\nabla^{1/3} = 5.23$
CA_P ft fot*	3.90*	66.0*	
Linear Ratio	1.0	16.92	
*fot – forward of transom			

Table 3-E: Model Test Data and Calculations of C_R

Example Hull: Model 3-200 Hull Form: Hard Chine Water: Fresh Temperature: 74°F Mass Density: 1.9352 lbs sec^2/ft^4 Kinematic Viscosity: 1.0018 X 10^{-5} ft^2/sec	λ = 16.92:1.0 Method of Turbulence Stimulation: None Friction Formulation: ITTC 1957 L = Variable with Speed: ft; Displacement = 221.1 lbs; Draft = 0.392 ft Wetted Surface, S = Variable with Speed: ft^2

Test Data						Calculations					
A	B	C	D	E	F	G	H	I	J	K	L
Run No.	V-kt	R-lb	L_K-ft	L_C-ft	S_m-ft^2	$C_{T,m}$ X 10^3	$L_{M,m}$-ft	$R_{n,m}$ X 10^{-6}	$C_{F,m}$ X 10^3	C_R X 10^3	F_{nV}
1	1.00	0.35	7.90	6.70	14.74	8.615	7.30	1.230	4.484	4.131	0.24
2	1.51	0.75	7.90	6.70	14.67	8.135	7.30	1.857	4.116	4.019	0.36
3	2.07	1.70	7.85	6.70	14.55	9.893	7.28	2.539	3.866	6.027	0.50
4	2.50	2.87	7.90	6.70	14.60	11.411	7.30	3.075	3.724	7.687	0.60
5	3.07	5.58	8.00	6.60	14.58	14.732	7.30	3.776	3.580	11.152	0.74
6	3.50	8.67	8.00	7.30	15.16	16.937	7.65	4.511	3.462	13.475	0.84
7	4.12	13.96	8.00	8.00	15.71	18.992	8.00	5.553	3.332	15.660	0.99
8	5.18	23.93	8.00	8.00	16.43	19.693	8.00	6.982	3.196	16.497	1.25
9	6.20	26.49	8.00	6.60	14.91	16.768	7.30	7.625	3.146	13.622	1.49
10	8.24	30.31	7.40	6.00	13.18	12.288	6.70	9.301	3.038	9.250	1.99
11	10.36	31.69	7.10	5.00	10.90	9.847	6.05	10.560	2.972	6.875	2.50
12	12.58	33.03	6.40	4.50	9.68	7.822	5.45	11.551	2.926	4.896	3.03
13	14.44	34.88	6.10	4.20	9.06	6.698	5.15	12.529	2.886	3.812	3.48
14	16.64	36.38	6.00	3.90	8.68	5.492	4.95	13.877	2.836	2.656	4.01
15	20.73	43.40	6.10	3.40	8.24	4.447	4.75	16.589	2.753	1.694	5.00
16	24.88	52.55	6.40	3.10	8.24	3.738	4.75	19.910	2.671	1.067	6.00

In Table 3-E, column inputs are defined as follows:

Column	Source or Model Scale Calculations
A to F	Model data
G	$C_{T,m} = R/[(1/2)\ \rho_m S_m v^2]$
H	$L_{M,m} = (L_K + L_C)/2$
I	$R_{n,m} = vL/v_m$
J	$C_{F,m} = 0.075/[(\text{Log}_{10} R_{mn}) - 2]^2$
K	$C_R = C_{T,m} - C_{F,m}$
L	$F_{nV} = v/(g^{1/3})^{1/2}$

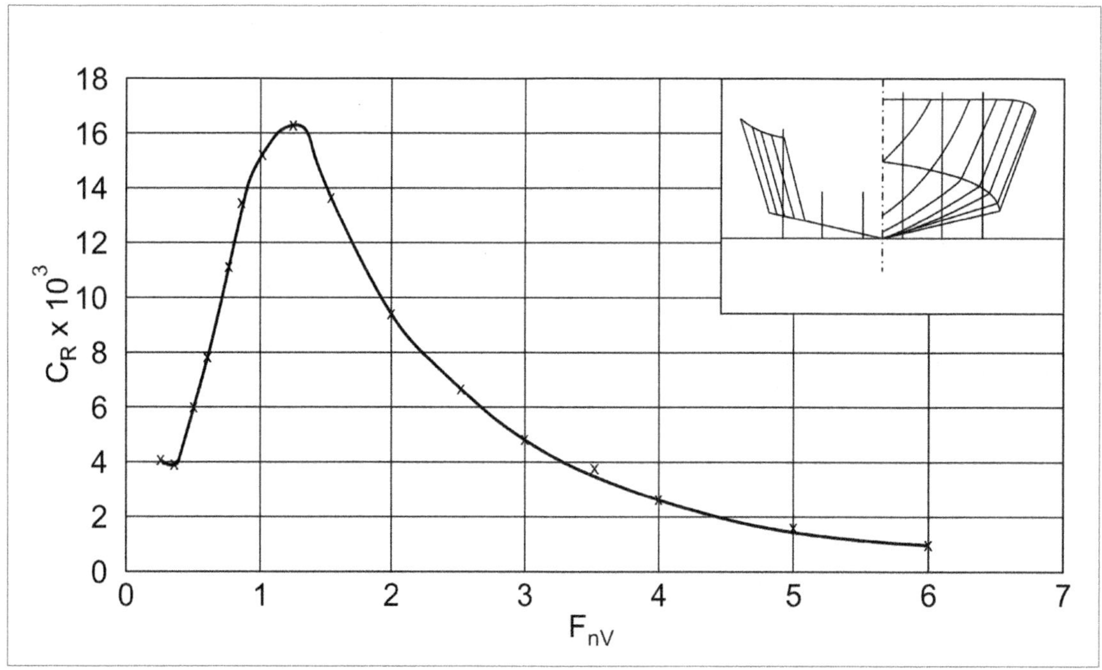

Figure 3-4: C_R versus F_{nV} for Example Hull 3-200

C_R versus F_{nL} and F_{nV}, respectively, from Figures 3-3 and 3-4 may be used to predict the hull resistance for any size vessel, larger or smaller, having the same dimensionless hull lines, loading, and speed range when mass, speed, and geometry are scaled by the Principle of Similitude.

The procedure for determining C_R versus F_{nL} and F_{nV} has been shown with examples when using ITTC 1957 Model-Ship Correlation Line. Follow the same procedure when using the ATTC 1947 Schoenherr Friction Formulation, remembering that the ATTC 1947 formulation is an implicit equation and estimates slightly lower values of C_F than ITTC 1957 for $R_n < 1 \times 10^7$. Tabulated values of C_F versus R_n for ATTC 1947 are given in Gertler (1947). An abbreviated table of C_F is also in the data tables in Appendix 1, Table A1-F. A numerical approximation for the ATTC 1947 friction formulation in an explicit equation format is:

$$C_F = 0.60804673\,[1.0353]^{\mathrm{Log}_{10} Rn}\,[\mathrm{Log}_{10} Rn]^{-2.8658} \qquad \text{Equation 3-10}$$

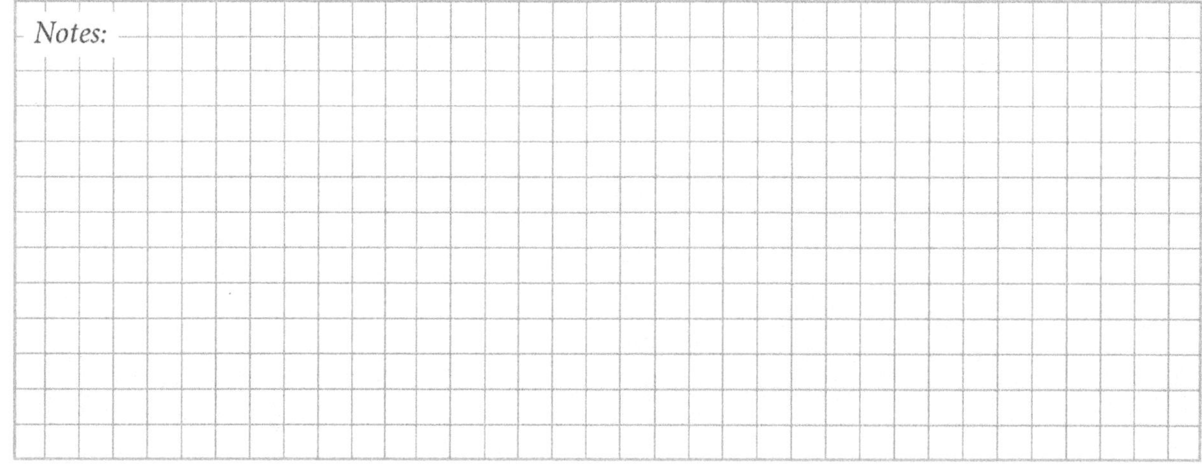

Correlation Allowance (C_A) and Margins

Correlation Allowance (C_A) is an approach to improve the quality and accuracy of model test data for prediction of speed and power performance for full-scale vessels. The use of C_A has wide application in the marine design profession for displacement and semi-displacement vessels and is addressed here relative to performance vessels. Also, C_A needs to be differentiated from "design margins." Based on either personal or corporate experiences, design margins account for assumptions and unexpected differences which might occur between completion of drawings and specifications and the as-built vessel delivered to the owner.

To establish correlation allowances for resistance prediction for a range of speeds, you will need both model test and full-scale trial data. The costs to acquire data for correlation factors are small relative to construction costs of commercial and military ships. As a result, most published references are for displacement vessels, and there is very little correlation information available in public sources for performance craft. C_A = 0.0004 has often been used for ship performance predictions. This displacement ship derived value of C_A = 0.0004 should not be arbitrarily used when making hull-resistance predictions for performance craft.

Some C_A data exist for hard-chine planing craft with various propulsors (Blount and Hankley, 1976; Blount et al.,1992; Blount et al.,1968.) This latter reference has quality thrust and power sea trial data; however, the model appendage resistance was accounted for in error which invalidated its usefulness for calculating C_A.

Note: The bit of correlation data of which I have is given in Table 3-F.

TABLE 3-F: CALM, DEEP WATER CORRELATION FOR HARD-CHINE CRAFT[2]

Craft Service	LOA ft	Approximate Full load $L/\nabla^{1/3}$	Propulsion	Correlation Speeds (F_{nV})	C_A
Military (prototype)	56	4.8	2 pod drives CRP propellers	0.8 to 1.7	0.0003
Military (patrol)	80	5.2	2 inclined shafts Fixed-pitch props	1.4 to 2.7	0[1]
Military (patrol)	95	6.7	3 inclined shafts Fixed pitch props	2.9 to 3.5	0.0001
GT/MY *Destriero* (yacht)	220	6.1	3 flush inlet waterjets	0.5 to 3.1	0

Footnotes:

1. C_A = 0 up to F_{nV} = 2.7 (approximately 34 knots). Trial data indicated propeller cavitation at higher speeds. Cavitation characteristics were not available to account for reduction of propeller efficiency for speeds above F_{nV} = 2.7 in order to establish C_A between F_{nV} from 2.7 to 3.4.
2. Model Reynolds numbers were sufficiently large such that C_A values may be used for scaling resistance to full-scale with either ATTC 1947 friction resistance coefficients or ITTC 1957 model-ship correlation line.

Builder's and acceptance sea trials of performance vessels with documented displacement and LCG afford, at a modest cost, an opportunity to measure power delivered to propulsors. Along with model data or analytical predictions, these trial data can be the basis for building a data base of C_A for various hull concepts.

C_A does not, however, achieve a constant percentage margin as it results in different percentages of resistance increase at each speed. This is demonstrated in Figure 3-5a, which shows changes relative to total resistance for a 500 mt vessel of Hull 3-200 using C_A = 0.0001 and 0.0004. C_A just incrementally elevates the fair curve of C_F versus R_n for either ATTC 1947 or ITTC 1957 friction formulations. However, for the simple case of $C_T = C_F + C_R$, C_F often becomes insignificant at hump speed: the speeds when the C_T of performance craft begins to rise from a displacement mode as the hull elevates toward the water surface.

Figure 3-5b gives the relative percentage proportions of C_F through the speed range and at F_{nV} = 1.2, C_F is only 10% of total resistance. Thus, C_A of 0.0001 or 0.0004 provide the hump resistance coefficient with margins of 0.5% and 2.2%, respectively, while providing margins of 3.9% and 15.6%, respectively at F_{nV} = 6.0. As this example indicates, C_A may provide false confidence in circumstances when it is inappropriately used as a design margin for resistance and power.

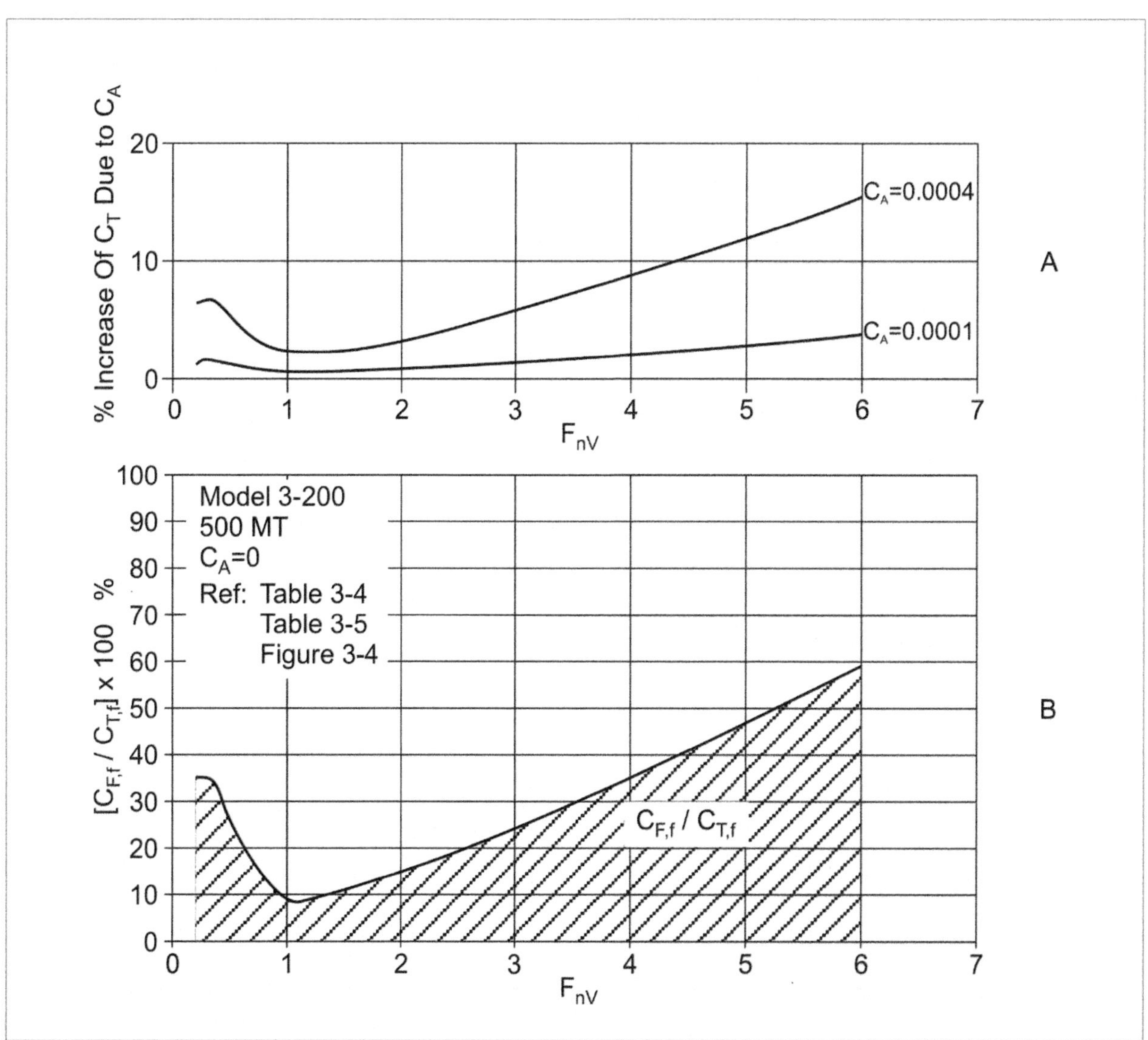

Figures 3-5a, 3-5b: Interaction of C_A with C_T when predicting hull resistance for performance craft

My experience suggests resistance and power are more difficult to predict at hump speeds than at top speed. Whenever it is determined that power margins are to be used for two critical design speeds—as often exhibited by performance craft—a larger percentage offset may be required at hump than for top speed.

In the existing circumstance of minimal C_A data for high-performance craft, I recommend that designers use a value of $C_A = 0.0$ for hull-resistance predictions. Should you choose to be cautious when uncertain about estimating power necessary for making contract speed, then their speed-power predictions should be increased by a percentage margin.

For example, you can obtain a five-percent design margin simply by multiplying predicted power by 1.05.

Note, however, that one resistance-prediction method for $0.5 \leq F_{nV} \leq 4.0$ for hard-chine craft may be used to develop a correlation factor which varies with speed, based on model test data. [*I am referring to Hubble's work from 1974 and 1982, summarized in Appendix 4.*]

Quantitative correlation is achieved by varying the standard deviation factor (SDF) for each F_{nV} until predicted and model test resistance are equated, or until a Hubble resistance prediction results in propulsive power for an achieved sea trial speed.

For me, it is counter-intuitive to consider only one C_A as a resistance or power design margin for the entire operational speed range for high-performance craft.

Assigning a C_A value for high speed may be appropriate when you are uncertain about your resistance prediction methods and there is a contract speed requirement.

I have found, however, that using one margin—resistance, thrust, or power—at hump near $F_{nV} \approx 1.5$ and a different margin at a required maximum speed often results in an agile craft when accelerating through the speed range.

Using a margin at hump is of great assistance when matching specifications of engines, gear-ratio, and propulsor-cavitation characteristics for dual design requirements at both hump and maximum speed.

Extrapolation to Full Scale

Here, I want to discuss several approaches for extrapolating resistance data to represent values for full-scale marine vessels. Although they provide the same predictions, the differences are reflected in the format in which designers collect and build their data base for air- and appendage-resistance coefficients.

The first approach is the ITTC guidelines, based on **all dimensionless coefficients normalized with static hull wetted surface**. Thus, total resistance coefficient and its components must be made dimensionless using hull-wetted surface.

In general, appendage-resistance coefficients are dependent on aspects of their individual geometries which have little or no relationship to hull lines, such as cylindrical shape of a propeller shaft has its resistance sensitive to diameter.

Therefore, I do not recommend it for performance vessels, but only for marine craft operating at displacement and semi-displacement speeds where hull-wetted surface and length are considered to be constant.

Following the guideline of the International Towing Tank Conference (ITTC 7.5-02-05-01, Rev. 01) the calm, deep-water total resistance of full-scale high-speed marine vehicles (HSMV) in non-dimensional form is the sum of several coefficients defined here:

$$C_{T,f} = C_R + C_{F,f} + C_{AA,f} + C_{APP,f} + C_A \qquad \text{Equation 3-11}$$

where the following definitions from ITTC 75-02-05-01, Rev. 01 apply only to Equations 3-11 and 3-12:

$C_{T,f}$ – Total resistance coefficient

C_R – Residual resistance coefficient (hull)

$C_{F,f}$ – Frictional resistance coefficient (hull)

$C_{AA,f}$ – Air resistance coefficient (superstructure and hull above water surface)

$C_{AAP,f}$ – Appendage resistance coefficient (underwater appendages)

C_A – Model to full-scale correlation allowance

Being reminded that $C_{AA,f}$ and $C_{AAP,f}$ are normalized with hull wetted surface, then the quantitative full-scale resistance is

$$R_{T,f} = (1/2)_f \rho_f V_f^2 S_f C_{T,f} \quad \text{Equation 3-12}$$

The bare hull component of total resistance is addressed in more detail later in this chapter. C_A was discussed in the previous section and aerodynamic and appendage resistance will be covered in Chapter 4.

My preference is to use the following approach, which is to first calculate bare-hull resistance with recommended $C_A = 0.0$.

$$R = [C_R + C_{F,f} + C_A]\,(1/2)\rho_f S_f V_f^2 \quad \text{Equation 3-13}$$

Then having R, you will find it convenient to evaluate total resistance for different possibilities by adding air-resistance plus only the appropriate appendage resistance components associated with propulsors: shafts, struts and rudders for propellers on inclined shafts; fixed course-keeping fins for flush inlet waterjets; and maybe no appendages for surface propellers, etc.

The approach I have just described here is what I am using throughout this book, rather than the ITTC guideline. Once model test data are analyzed to the form given in Tables 3-C and 3-E, C_R can be used to predict resistance for any size hull of the same dimensionless geometry and loading. The bare-hull resistance for full-scale 500 mt vessels is calculated for the two examples, Models 3-100 and 3-200. The principal full-scale characteristics for these two hulls are provided in Tables 3-B and 3-D, respectively.

For these examples, just a limited number of runs from Tables 3-C and 3-E have been analyzed. Analyzed data are provided in Table 3-G for Model 3-100 with Table 3-H furnishing calculations for Model 3-200.

Here is an additional teaching element: Calculated values for R/W and EHP_{BH} from Tables 3-G and 3-H are shown graphically in Figure 3-6.

Calm-water resistance is only one important element of design decisions. Figure 3-6 shows you that for the two vessels compared at 500 mt, Model 3-100 has the least resistance for the speeds presented. If data are slightly extrapolated, 3-100 is preferred up to approximately 45 knots. However, at high speeds, Model 3-200 has very little increase of resistance (about 6%) from 45 to 70 knots.

Thus, the increasing slopes of resistance curves at 40 knots show the round-bilge model increasing at 0.36%/knot while the hard chine model hardly changes. A hull-form decision for a single maximum-speed requirement would favor a round-bilge hull below 45 knots, and prefer a hard-chine hull above this speed. In the real world, design requirements are complex, and a naval architect's decisions are seldom as easy as developing a minimal-resistance hull at a single speed.

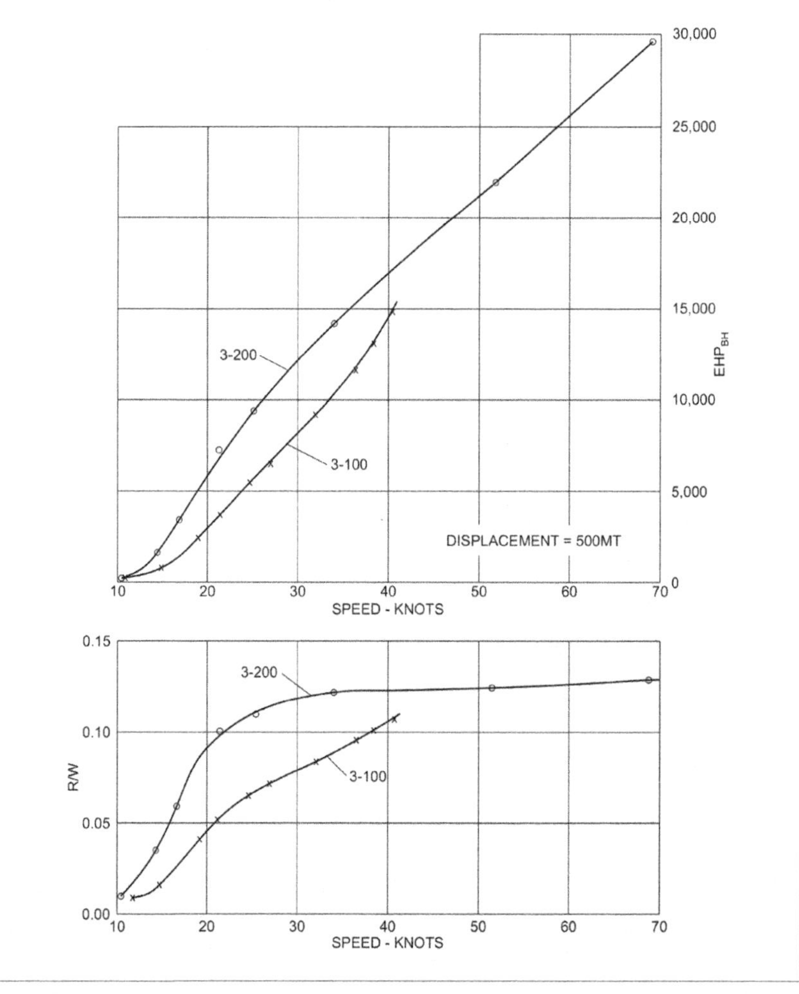

Figure 3-6: Bare-hull EHP and R/W graphs for calculation examples

TABLE 3-G: FULL-SCALE R/W AND EHP_{BH} FOR MODEL 3-100

Example Hull: Model 3-100
Hull Form: Round Bilge
Water: Salt Temperature: 59°F (15°C)
Mass Density: 1.9905 lb sec^2/ft^4
Kinematic Viscosity: 1.2260 X 10^{-5} ft^2/sec

λ = 28.46:1.0
Friction Formulation: ITTC 1957
L = 170 ft; Displacement = 500 mt; Draft = 8.74 ft
Wetted Surface, S_f = 5,140 ft^2
Correlation Allowance, C_A = 0

Test Data				Calculations						
A	E	G	H	I	J	K	L	M	N	O
Run No.	$R_{n,m}$ X 10^{-6}	C_R X 10^3	F_{nL}	$R_{n,f}$ X 10^{-8}	$C_{F,f}$ X 10^3	$C_{T,f}$ X 10^3	R lb	R/W	V_{Kt}	EHP_{BH} Hp
1	2.530	3.732	0.341	3.446	1.755	5.487	17,850	0.0162	14.94	818
2	3.246	6.535	0.437	4.474	1.696	8.231	43,940	0.0398	19.14	2,580
3	3.621	6.911	0.488	4.991	1.672	8.583	57,170	0.0518	21.38	3,750
4	6.156	3.840	0.829	8.486	1.562	5.402	103,800	0.0941	36.31	11,560
5	6.518	3.580	0.878	8.984	1.551	5.131	110,600	0.1003	38.46	13,050
6	6.880	3.396	0.927	9.483	1.541	4.937	118,600	0.1076	40.60	14,770
7	1.996	2.988	0.269	2.751	1.809	4.797	9,700	0.0088	11.78	351
12	5.432	4.583	0.732	7.487	1.587	6.170	92,410	0.0838	32.06	9,090
13	4.526	6.061	0.610	6.238	1.624	7.685	79,950	0.0725	26.72	6,550
14	4.164	6.439	0.561	5.739	1.642	8.081	71,090	0.0645	24.57	5,360

In Table 3-G, column inputs are briefly defined as follows:

Column	Source or Model Scale Calculations
A,E,G,H	Copied from Table 3-C
I	$R_{n'f}$ = (Column E) $\lambda^{3/2}$ (v_m/v_f)
J	$C_{F,f} = 0.075/[(Log_{10} R_n) - 2]^2$
K	$C_{T,f} = C_{F,f} + C_R + C_A$
L	$R = C_{T,f}[(1/2)\ \rho_f S_f v_f^2]$
M	R/W
N	V = (Column H) $(gL)^{1/2}/1.6878$
Reminder: V is in knots & v is in ft/sec	

TABLE 3-H: FULL-SCALE R/W AND EHP_{BH} FOR MODEL 3-200											
Example Hull: Model 3-200 Hull Form: Hard Chine Water: Salt Temperature: 59°F (15°C) Mass Density: 1.9905 lb sec^2/ft^4 Kinematic Viscosity: 1.2260 X 10^{-5} ft^2/sec					λ = 16.92:1.0 Friction Formulation: ITTC 1957 L_P = 135 ft; Displacement = 500 mt; Draft = 6.63 ft Wetted Surface, S_f = Variable with Speed: ft^2 Correlation Allowance, C_A = 0						
Test Data					Calculations						
A	F	I	K	L	M	N	O	P	Q	R	S
Run No.	S_m	$R_{n,m}$ X 10^{-6}	C_R X 10^3	F_{nV}	$R_{n,f}$ X 10^{-8}	$C_{F,f}$ X 10^3	$C_{T,f}$ X 10^3	R lb	R/W	V kt	EHP_{BH} Hp
4	14.60	3.075	7.687	0.60	1.749	1.924	9.611	11,940	0.0108	10.24	375
6	15.16	4.511	13.475	0.84	2.565	1.826	15.301	38,720	0.0351	14.34	1,700
7	15.71	5.553	15.660	0.99	3.158	1.775	17.435	63,500	0.0576	16.90	3,290
8	16.43	6.982	16.497	1.25	3.971	1.722	18.219	110,600	0.1004	21.34	7,240
9	14.91	7.625	13.622	1.49	4.336	1.703	15.325	120,000	0.1089	25.44	9,370
10	13.18	9.301	9.250	1.99	5.290	1.659	10.909	134,700	0.1221	33.97	14,040
12	9.68	11.551	4.896	3.03	6.569	1.614	6.510	136,900	0.1241	51.73	21,730
14	8.68	13.877	2.656	4.01	7.892	1.577	4.233	139,800	0.1268	68.46	29,370

In Table 3-H, column inputs are briefly defined as follows:

Column	Source or Model Scale Calculations
A,F,I,K,L	Copied from Table 3-E
M	$R_{n,f}$ = (Column I) $\lambda^{3/2}(\nu_m/\nu_f)$
N	$C_{F,f}$ = $0.075/[(\text{Log}_{10} R_n) -2]^2$
O	$C_{T,f} = C_{F,f} + C_R + C_A$
P	$R = C_{T,f}[(1/2)\rho_f S_m \lambda^2 v_f^2]$
Q	R/W
R	V = (Column L) $(g\nabla^{1/3})^{1/2}/1.6878$
S	EHP_{BH} = RV/325.9

Reminders: V is in knots & v is in ft/sec

For Δ = 500 mt, ∇= 17,226 ft^3 of salt water

As a calculation convenience for S_f varying with speed in Column P, $S_f = S_m\lambda^2$ is used.

Predicting Hull Resistance for Shallow-Water Operation

Of a variety of environmental conditions, water depth has significant influence on calm-water hull resistance for a wide range of operating speeds. At low boat speeds, shallow water substantially increases resistance while at planing speeds, resistance is reduced. This is depicted, in concept, in Figure 3-7.

Radojcic and Bowles covered this topic well in 2010. Their paper should be in the library of every performance vessel designer. The following material about shallow-water performance is condensed from applicable topics Radojcic and Bowles addressed in detail.

In short, designers of boats and craft intended to operate in rivers and near shores in littoral regions should give particular attention to influences of water depth. Radical change to both vessel resistance and propulsive factors occurs when vessels operate in shallow water; that is, in general terms, when vessels operate at critical speeds in water depths less than 0.8 (LOA).

The specific combination of speeds and water depths defining this critical region when resistance and propulsive factors must seriously be considered for estimating the "hole shot" accelerations which occurs when depth Froude numbers (F_{nh}) are between 0.70 and 1.20.

The critical region is defined for the following ratios of $V/h^{1/2}$, which are shown in Figure 3-8.

$4.27 \leq V/(h)^{1/2} \leq 7.32$ for h in meters Equation 3-14

$2.36 \leq V/(h)^{1/2} \leq 4.04$ for h in feet Equation 3-15

In shallow water, only the resistance component of wave making or residuary resistance (R_R) changes dramatically from its deep water characteristics (Radojcic and Bowles, 2010).

Ratios of $R_{Rh}/R_{R\infty}$ (where subscript "h" refers to the depth of the shallow water and ∞ refers to deep water) which are functions of L/h and F_{nh}. This ratio may be predicted for the subcritical range only for speeds up to $F_{nh} = 0.90$ with the numerical approximations given by Equations 3-16 to 3-20a (Radojcic and Bowles, 2010).

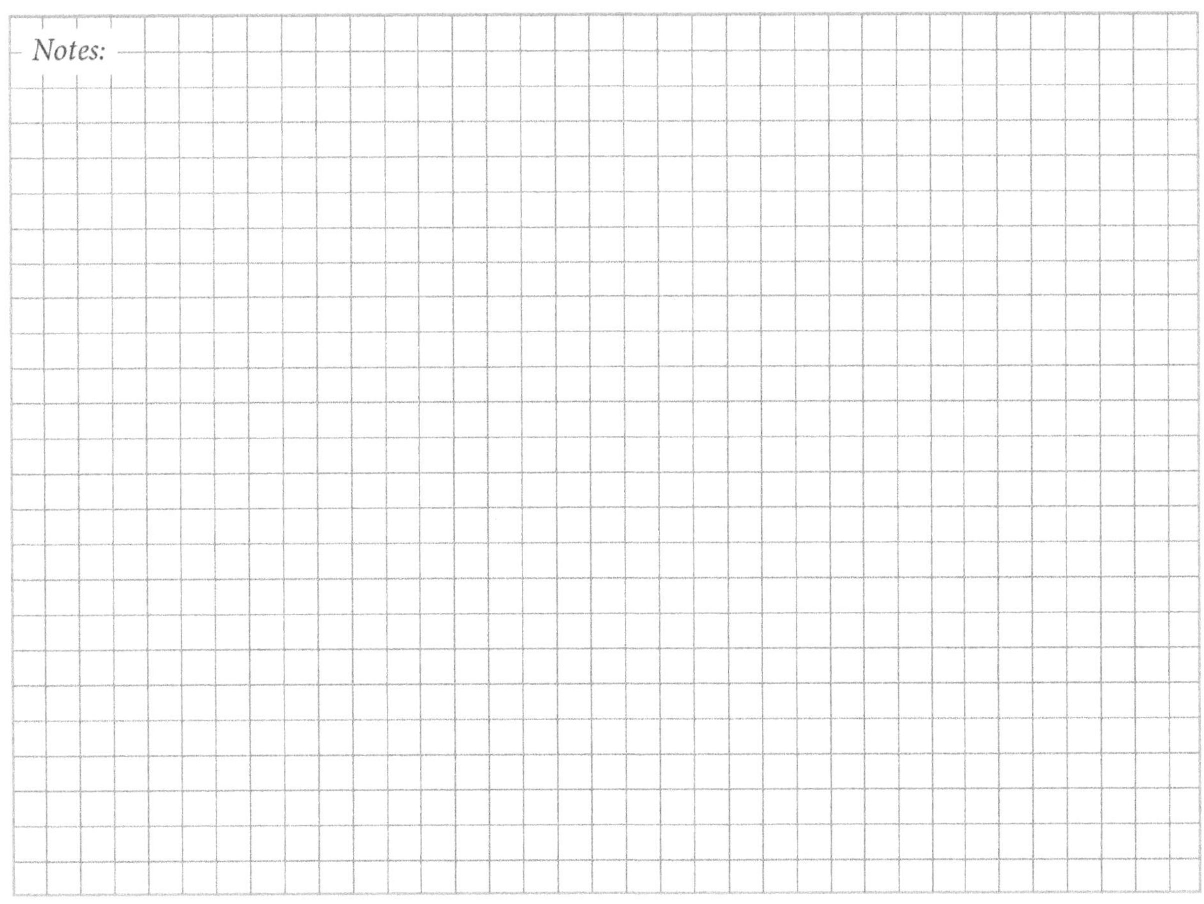

$R_{Rh}/R_{R\infty} = a + b(F_{nh})^c$ Equation 3-16

Where

$a = e^x$

$X = [-0.00370 + 0.00265(L/h)] / [(1 - 0.33444(L/h) + 0.03037(L/h)^2)]$

$b = 1/[-3.5057 + 0.0312(L/h)^2 + 14.7440/(L/h)]$

$c = 2.0306 + (10.1218/\pi)\{atan_{Rad}[((L/h) - 4.6903) / 0.7741] + \pi/2\}$

$(R_{Rh}/R_{R\infty})_{max} = 0.97476 + 0.01495(L/h)^3$ Equation 3-17

$(R_{Rh}/R_{R\infty})_{critical} = (R_{Rh}/R_R)_{max} - (L/h)/ 20$ Equation 3-18

$(F_{nh})_{max} = 0.92226 - 0.30827 / (L/h)^{1.5}$ Equation 3-19

$(F_{nh})_{critical} = 0.95$ (experimental) Equation 3-20A

$(F_{nh})_{critical} = 1.00$ (theoretical) Equation 3-20B

Having already predicted bare-hull resistance for calm, deep water, $R_T = R_R + R_F$, then for a constant shallow-water condition, bare hull resistance is R_{Th} and the total resistance coefficient is C_{Th}.

$R_{Th} = R_R (R_{Rh}/R_{R\infty}) + R_F$ Equation 3-21

$C_{Th} = C_R (R_{Rh}/R_{R\infty}) + C_F$ Equation 3-22

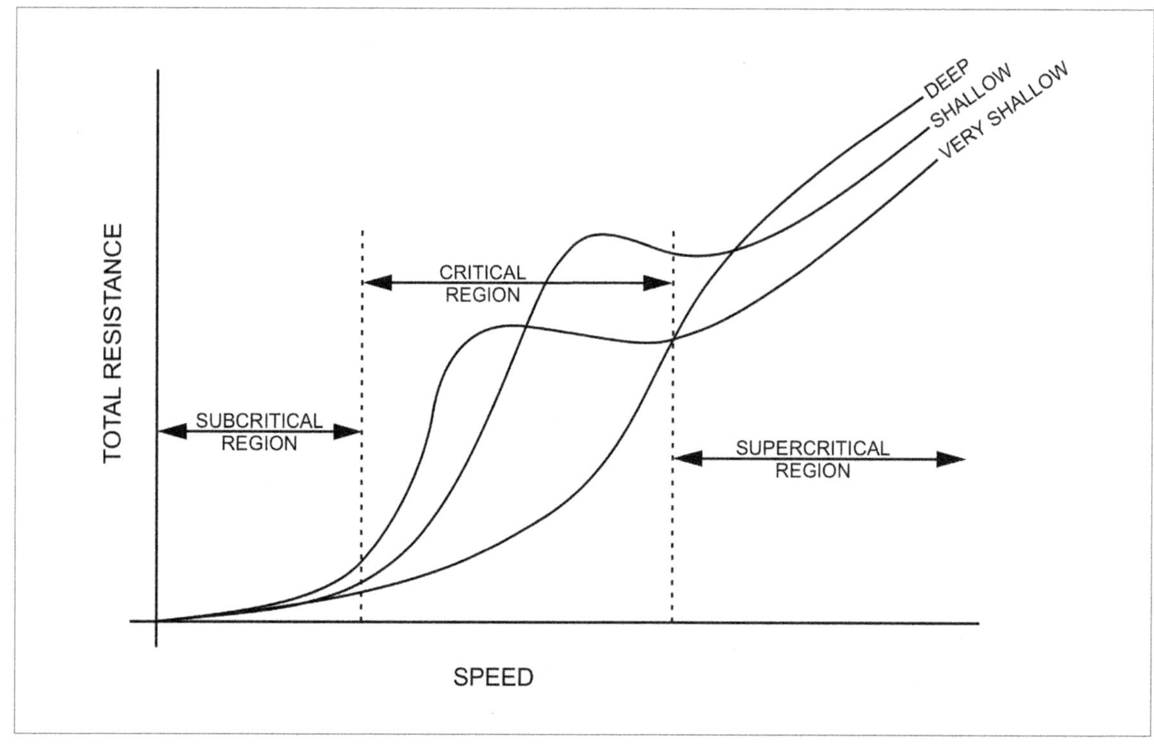

Figure 3-7: Conceptual resistance v. speed curves for deep and shallow water

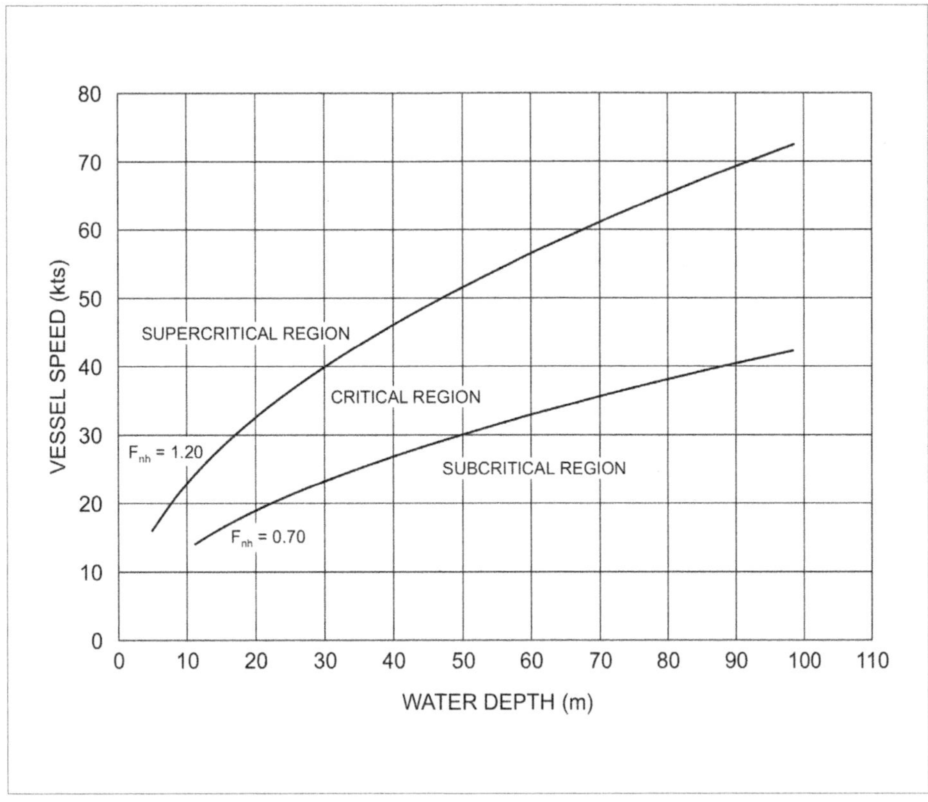

Figure 3-8: Relationship of speed v. water depth defining subcritical, critical, and supercritical regions

To develop a conservative resistance "envelope," you should determine the resistance curve based on shallowest water depth and critical peaks for various water depths. Resistance peaks are calculated from Equations 3-16 to 3-20A.

An example based on Model 3-200 demonstrates the prediction of bare-hull resistance for planing hulls in shallow water. The calculation of the shallow-water effect in the subcritical speed range is made using Equation 3-16 which only applies for speeds up to $F_{nh} = 0.90$.

The coefficients for this equation are variables of the ratio L/h. Fortunately, L is constant at displacement and semi-displacement speeds where, due to the effects of shallow water, the increase in resistance occurs. Also, for engineering purposes, the coefficients for Equation 3-16 may be made with an approximate constant length of about 0.95 L_P for planing craft at low speeds where shallow water causes increased resistance.

Besides making predictions for the assumed shallowest expected water depth for the operational region of the craft. You should also develop a conservative resistance envelope curve for estimating the least depth in which the boat can accelerate to planing speeds.

The calculated shallow water resistance addressed in this section can be combined with a procedure in Chapter 6 to include both engine and propulsive characteristics, along with gear ratio to estimate minimum water depth necessary for a boat to be capable of accelerating to planing speeds.

Table 3-I provides an example for this shallow-water resistance calculation procedure for subcritical and critical speeds where hull resistance is significantly increased. The results of these calculations are shown as R/W and EHP_{BH} versus speed in Figure 3-9, providing a comparison of predictions for deep water and shallow water (L/h = 4.0). While calculations are only given for critical speeds, Figure 3-9 does depict the reduced resistance benefit of planing craft operating in shallow water at supercritical speeds, $F_{nh} > 1.20$.

The ratio of $C_{Rh}/C_{R\infty}$ versus F_{nh} from subcritical to supercritical speeds is in Figure 3-10 (Sturtzel and Graff, 1963) which presents the ratio to be greater than one which documents increasing resistance due to shallow water at subcritical speeds. At supercritical speeds the ratio is less than one indicating a reduction in resistance.

Reduction in resistance when operating in shallow water at supercritical speeds is, at best, seldom more than 10 percent and reduces with increasing speed. I have not addressed a calculation procedure for shallow-water effects on resistance at supercritical speeds as it is not usually considered influential for design decisions.

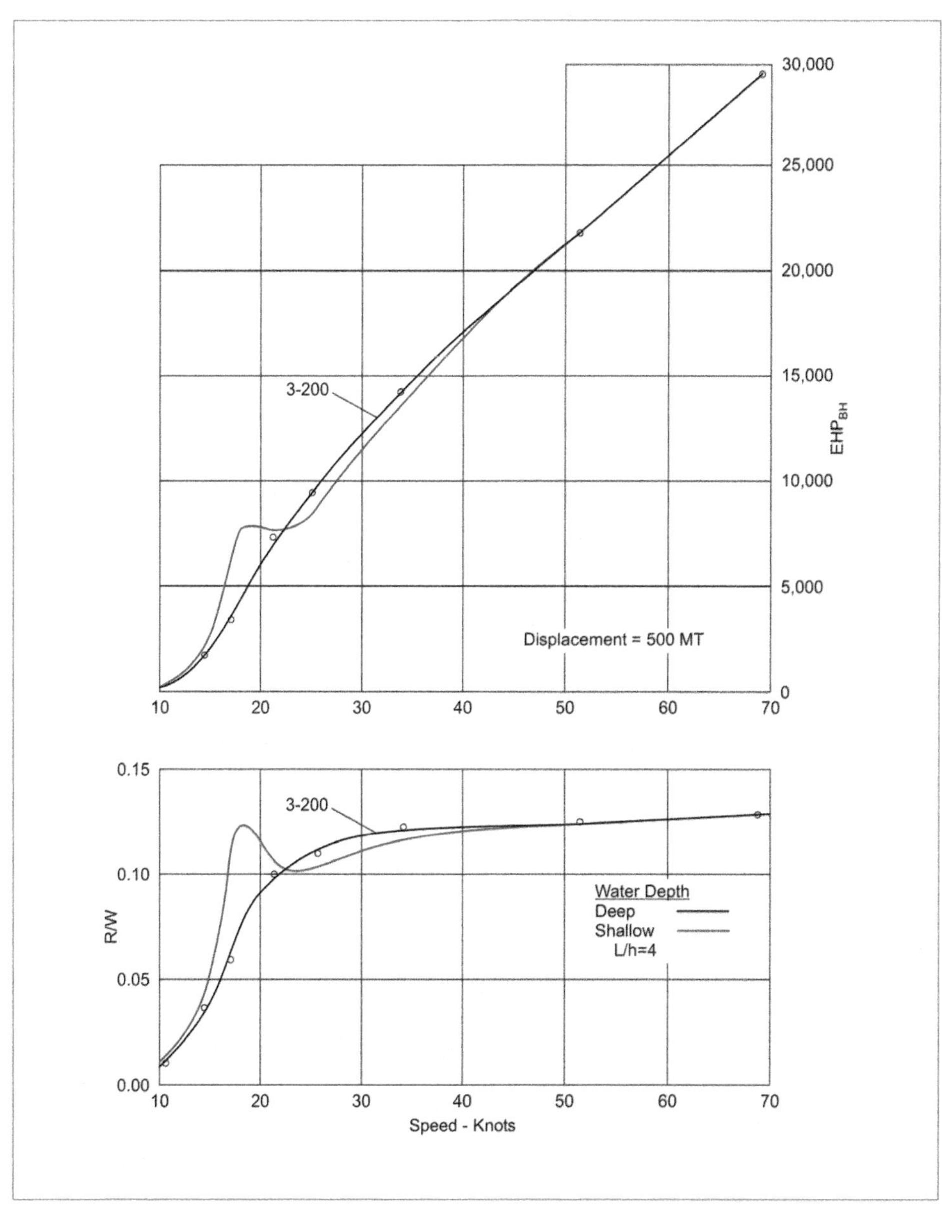

Figure 3-9: Deep and shallow curves for bare-hull EHP and R/W v. speed

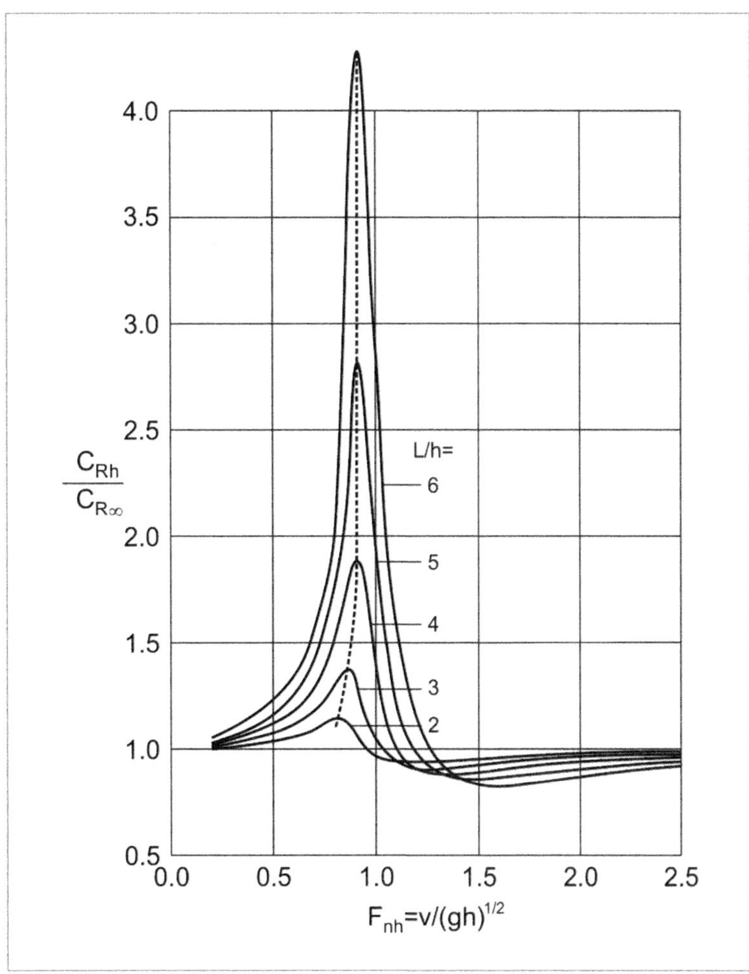

Figure 3-10: Ratio of $C_{Rh}/C_{R\infty}$ versus F_{nh}

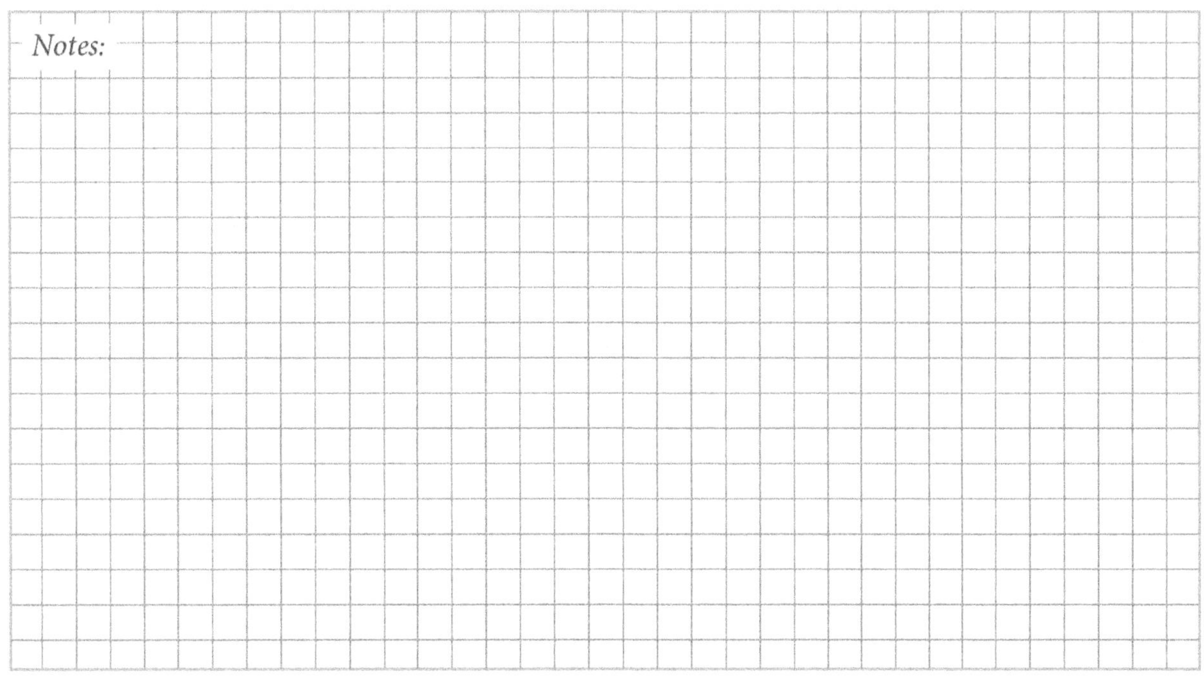

TABLE 3-I: SHALLOW WATER PREDICTION FOR FULL-SCALE R/W AND EHP_{BH} FOR MODEL 3-200

EXAMPLE HULL Model 3-200

Hull Form: Hard Chine

Water: Salt

Temperature: 59°F (15°C)

Mass Density: 1.9905 lb sec^2/ft^4

Kinematic Viscosity: 1.2260 X 10^{-5} ft^2/sec

λ = 16.92:1.0

Friction Formulation: ITTC 1957

L_P = 135 ft; Displacement = 500 mt

Draft = 6.63 ft

Wetted Surface, S = Variable with Speed: ft^2

Correlation Allowance, C_A = 0

Assumed: L = 0.95 (135) = 128 ft

h = 32 ft; L/h = 4

CALCULATION EXAMPLE

Source Table 3-E				Table 3-H + Calculations					Shallow Water Calculations						
A	K	L	F	N	O	Q	S	R	T	U	V	W	X	Y	Z
						Deep Water									
Run No.	C_R X10^3	F_{nV}	S_m ft^2	$C_{F,f}$ X10^3	$C_{T,f}$ X10^3	R/W	EHP_{BH}	V kt	F_{nh}	R_{Rh}/R_R L/h = 4.0	C_{Rh} X10^3	$C_{T,fh}$ X10^3	R_h/R	R_h/W	$EHP_{BH\infty}$
1	4.131	0.24	14.74	2.195	6.326	0.0012	17	4.10	0.216	1.049	4.332	6.527	1.032	0.0012	18
2	4.019	0.36	14.67	2.067	6.086	0.0025	52	6.15	0.324	1.055	4.239	6.306	1.036	0.0026	54
3	6.027	0.50	14.55	1.977	8.004	0.0062	179	8.54	0.449	1.081	6.513	8.490	1.061	0.0066	190
4	7.687	0.60	14.60	1.924	9.611	0.0108	375	10.24	0.539	1.126	8.656	10.580	1.101	0.0119	413
5	11.152	0.74	14.58	1.871	13.023	0.0223	556	12.63	0.665	1.260	14.051	15.922	1.223	0.0273	680
6	13.475	0.84	15.16	1.826	15.301	0.0351	1,700	14.34	0.754	1.433	19.311	21.137	1.381	0.0485	2,350
7	15.660	0.99	15.71	1.775	17.435	0.0576	3,290	16.90	0.889	1.890	29.593	31.368	1.799	0.1036	5,920
8	16.497	1.25	16.43	1.722	18.219	0.1004	7,240	21.34	1.123*	---	---	---	---	---	---

*Equation 3-16 is limited to $F_{nh} \leq 0.90$

Equation 3-16 coefficients for L/h = 4.0

X = 0.046571; a = 1.0477; b = 1.4717; c = 4.7452

$(R_{Rh}/R_{R\infty})_{max}$ = 1.932 Equation 3-17

$(R_{Rh}/R_{R\infty})_{critical}$ = 1.732 Equation 3-18

$(F_{nh})_{max}$ = 0.8837 = 16.8 knots Equation 3-19

$(F_{nh})_{critical}$ = 0.95 = 18.0 knots Equation 3-20a

In Table 3-I, column inputs are briefly defined as follows:

Column	Source or Model Scale Calculations
A,K,L,F	Copied from Table 3-E
N,O,Q,S,R	Copied from Table 3-H plus some calculations
T	$F_{nh} = V/(gh)^{1/2}$
U	Equation 3-16
V	$C_{Rh} = C_R(R_{Rh}/R_{R\infty})$
W	$C_{Th} = C_{F,f} + C_{Rh}$
X	$R_h/R_\infty = C_{T,fh}/C_{T,f}$
Y	$R_h/W = (R/W) X (R_h/R_\infty)$
Z	$EHP_{BH\infty} = (EHP_{BH}) X (R_h/R_\infty)$

Sources of bare-hull calm, deep-water resistance data

This chapter assumes that calm, deep-water resistance versus speed data are available to designers to use and manipulate for a variety of projects. Some sources of resistance data for $F_{nL} > 0.40$ listed here are followed by references reporting experimental data.

Sources of Resistance Data

- Model tests with known hull lines and test conditions
- Model tests of a series of geometrically-related hulls
- Regression analysis of experimental data from a number of hulls
- Analytical software developed using first principles (commercial or proprietary)
- Naval architects developing code or purchasing commercial software for predicting resistance versus speed are encouraged to personally compare computer output with experimental data to validate the quality and range of application of any analytical tools.

References for Calm, Deep-Water Resistance Data

Round Bilge Hulls

Taylor, D.W., "The Speed and Power of Ships," 3rd Edition, U.S. Government Printing Office, 1943.

Gertler, M., "A Reanalysis of the Original Test Data for the Taylor Standard Series," David Taylor Model Basin Report 806, March 1954.

Yeh, H.Y.H., "Series 64 Resistance Experiments on High-Speed Displacement Forms," *Marine Technology*, July 1965.

Ridgely-Nevitt, C., "The Resistance of a High Displacement-Length Ratio Trawler Series," SNAME Transactions, 1967.

Bailey, D., "The NPL High Speed Round Bilge Displacement Hull Series," RINA, Maritime Technology Monograph, No. 4, 1976.

Radojcic, D., Rodic, T., Kostic, N., "Resistance and Trim Predictions for the NPL High Speed Round Bilge Displacement Hull Series," Conference on Power, Performance and Operability of Small Craft, RINA 1997.

Beys, P.M., "Series 63 – Round Bottom Boats," Stevens Institute of Technology, Davidson Laboratory Report 949, 1993.

Compton, R.H., "Resistance of a Systematic Series of Semiplaning Transom-Stern Hulls," *Marine Technology*, October 1986.

Double-Chine Hulls

Anonymous, "Model Tests of a Coastal Patrol and Interdiction Craft (CPIC), Stevens Institute of Technology, Davidson Laboratory, Report 1553, Sept. 1971.

Anonymous, "Overload Tests in Smooth and Rough Water With Models of a Coastal Patrol and Interdiction Craft (CPIC)," Stevens Institute of Technology, Davidson Laboratory, SIT-DL-72-1605, May 1972.

Grigoropoulos, G.J. and Loukakis, T.A., "Resistance of Double-Chine, Large, High-Speed Craft," *Bulletin de L'Association Technique Maritime et Aeronautique*, ATMA Vol. 99, Paris, June 1999.

Radojcic, D., Grigoropoulos, G.J., Rodic, T., Tuvelic, T., and Damala, D.P., "The Resistance and Trim of Semi-Displacement, Double-Chine, Transom-Stern Hull Series," FAST 2001.

Hard-Chine Hulls in Deep Water

Clement, Eugene P. and Blount, Donald L., "Resistance Tests of a Systematic Series of Planing Hulls Forms," SNAME Transactions 1963.

Keuning, J.A. and Gerritsma, J., "Resistance Tests of a Series of Planing Hull Forms with 25 Degrees Dead-Rise Angle," *ISP*, Vol. 29, No. 337, September 1982.

Keuning, J.A., Gerritsma, J., and van Terwisga, P.F., "Resistance Tests of a Series of Planing Hull Forms with 30° Deadrise Angle and a Calculation Model Based on This and Similar Systematic Series," *ISP*, Vol. 40, No. 424, December 1993.

Kowalyshyn, D.H. and Metcalf, B., "A USCG Systematic Series of High Speed Planing Hulls," SNAME Transactions, 2006.

Morabito, M.G., "Re-Analysis of Series 50 Tests of V-Botom Motor Boats," SNAME Transactions, 2013.

Radojcic, D., Zgradic, A., Kalajdzic, M., and Simic, A., "Resistance Predictions for Hard Chine Hulls in Pre-Planing Regime," *Journal Polish Maritime Research*, Vol. 21, No. 2., 2014.

CHAPTER FOUR

Aerodynamics, Appendages, and Bulbous Bows

I addressed the essence of hydrodynamic relationships between appropriate hull forms and dimensionless speed in Chapter 2, and I covered calm-water resistance computation as a function of speed for basic hull forms in Chapter 3. Here, I'll discuss those necessary propulsion-related features and components both above and below the water's surface which influence control and maneuverability and contribute to a safe operational environment for personnel onboard performance vessels. I'll start by considering those horizontal forces which may either retard or enhance the potential speed of a vessel.

Aerodynamic Factors

Let's begin with drag, the combination of the effects of vessel speed and wind velocity. Figure 4-1 shows you the resultant direction of these two vectors (V_{AIR}).

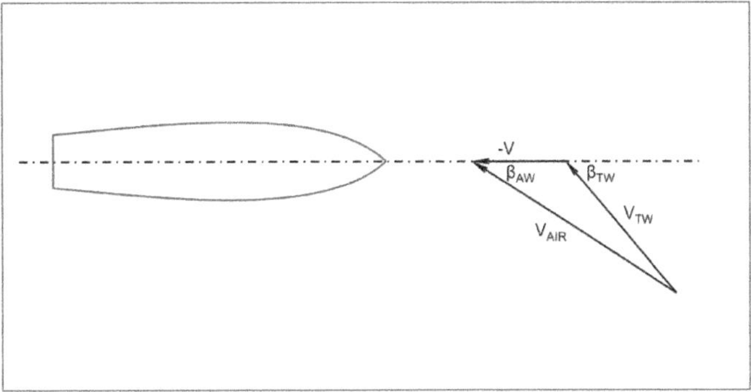

Figure 4-1: Relative wind (V_{AIR})

As a designer, you will need to estimate the forces that affect maneuvering and steering, taking aerodynamic side force and center of pressure into consideration. Then, to assure personnel safety at high speeds, you must also manage the air flow for open spaces. Remember: In the open spaces of performance vessels, both gale and hurricane wind velocities are possible!

Table 4-A is a list of threshold velocities for storm warnings. For designers of high-speed vessels, this is a reflective note to give appropriate attention to personnel safety. On a craft operating in calm water at 34-plus knots, exposed personnel feel gale-force winds.

TABLE 4-A: THRESHOLD WIND VELOCITY FOR STORM WARNINGS

Category	Velocity	
	MPH	Knots
Gale Warning	39	34
Storm Warning	55	48
Hurricane Category 1	74	64
Hurricane Category 2	96	83
Hurricane Category 3	111	97

Although they are more of an annoyance than a safety issue, odors in the air also need to be considered. Fumes from engine, galley, and sanitary-tank exhaust locations mixing with ambient-air movement can all result in an unsuitable environment for habitable spaces. Aerodynamics also affects spray and wetness patterns as well as the effectiveness and efficiency of inlets for ventilation and engine-combustion air flow.

Aerodynamic Resistance

At high speeds, the aerodynamic resistance of a high-performance vessel becomes a significant component of the total resistance. Full-scale trials of modern sportfishing boats, both with and without full towers, indicate that towers establish an aerodynamic resistance component on the order of nine percent at boat speeds of 32 to 35 knots. You can very easily visualize how the long pipes and tubes that make up the supporting structure create drag of a high magnitude.

Significant aerodynamic drag at high speed also results from the geometry of the hull and deckhouse. You may streamline the external shape of the superstructure to reduce drag only to discover conflicting requirements for needed usable internal volume, unusual styling features, and even stealth considerations. (Blount and Bartee, 1997)

Aerodynamic resistance for complex shapes is best predicted by wind-tunnel tests. From 1973 to 1983, three separate researchers—Isherwood, Van Berlekom, and Walshe—described testing procedures, analysis of data, and results for modern vessels. Additional test result analyses have been reported for wind directions from 0° to 360° relative heading, with 180° being headwinds.

Figure 4-2a shows typical dimensionless aerodynamic resistance coefficients normalized with projected frontal area above the water surface of a streamlined superstructure and hull, taking into consideration trim and heave at design speed, and compared with unfaired WWII DD 692 class destroyers.

The aerodynamic resistance coefficient is defined as:

$$C_X = R_{AIR}/(1/2\ \rho_{AIR} A_T v_{AIR}^2) \qquad \text{Equation 4-1}$$

For streamlined vessels, a resistance coefficient of $C_X \approx 0.50$ is possible at zero degree relative wind heading. As the area of the vessel exposed to the wind increases, however, peak values of C_X generally occur for a relative headwind of about 20 to 30 degrees from the centerline.

Further increases in relative wind angles then tend to reduce the component of wind-causing drag but increase the side force. Computing air resistance (R_{AIR}) is one component of total resistance necessary to make power predictions.

For vessels operating in head seas, where ambient wind direction can be the most variable, I use the highest C_X value for wind occurring in bow-quarter headings, not just C_X for relative wind direction of 180°. For the data in Figure 4-2a, I used $C_X = 0.73$ as representative of relative wind headings of 150° to 210°.

This approach can provide some small margin of predicted speed with R_{AIR} defined as:

$$R_{AIR} = C_X(1/2)\rho_{AIR} A_T v_{AIR}^2 \qquad \text{Equation 4-2}$$

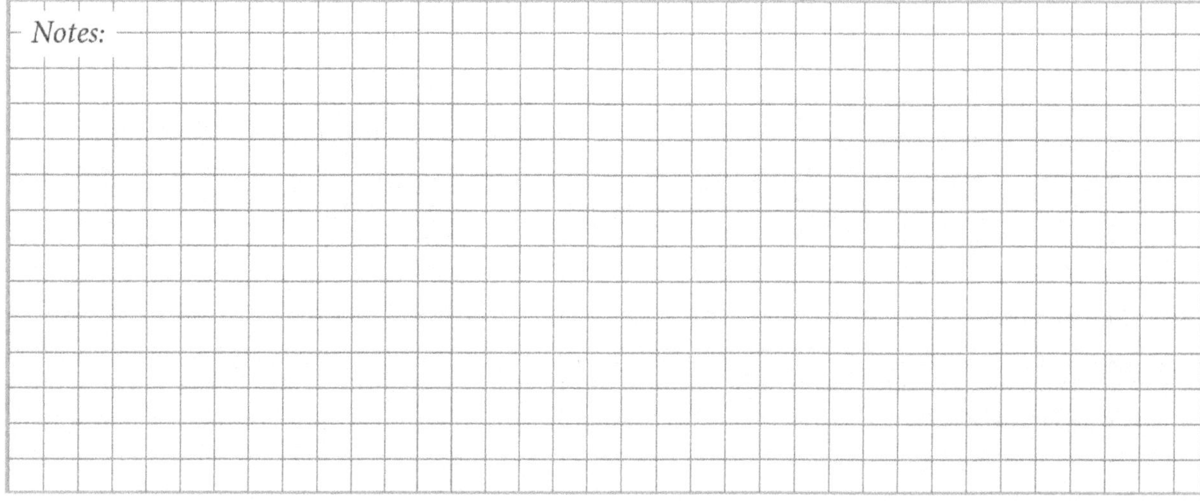

During concept design and feasibility studies, and where there is little or no deckhouse detail, I have found that A_T can be approximated by $(BOA)^2$ when first making total resistance predictions.

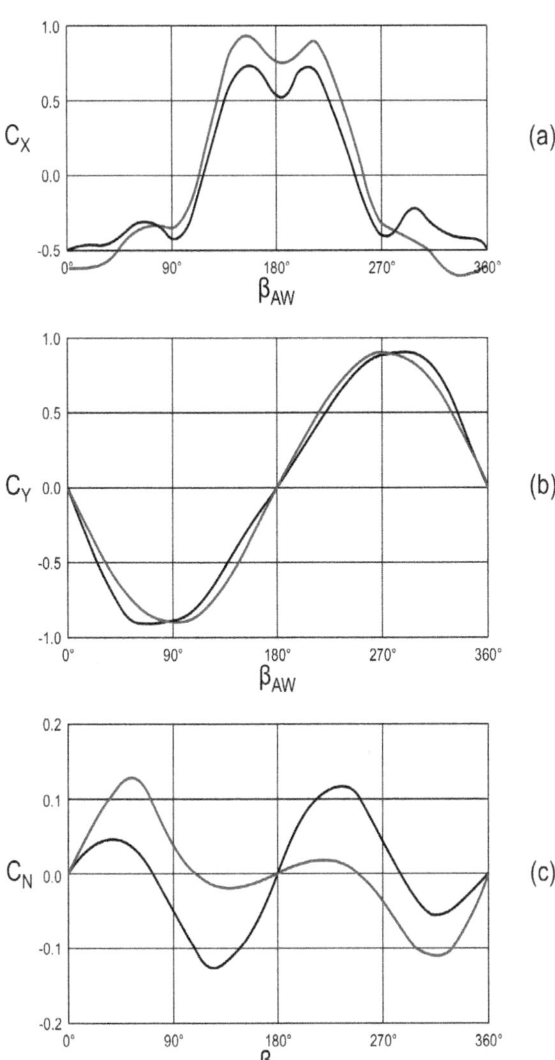

Figure 4-2 (a, b, and c): Aerodynamic coefficients for resistance (C_X), side force (C_Y), and yawing moment (C_N) v. relative to wind angle (β_{AW})

Figures 4-2a, 4-2b, and 4-2c provide more information than just aerodynamic-resistance (R_{AIR}),the force which acts in a direction to oppose craft speed. There are other forces and moments to consider, such as the lateral (side) force coefficient (C_Y) shown in Figure 4-2b.

The aerodynamic lateral coefficient (C_Y) and lateral force (F_Y) are defined as follows:

$$C_Y = F_Y/(1/2)\rho_{AIR}A_L v_{AIR}^2 \qquad \text{Equation 4-3}$$

$$F_Y = C_Y(1/2)\rho_{AIR}A_L v_{AIR}^2 \qquad \text{Equation 4-4}$$

Both lateral aerodynamic and hydrodynamic forces due to wind and currents define the mooring forces seen by dock lines and deck fittings. The aerodynamic yawing moment coefficients for the model data presented in Figure 4-2c highlight the influence of the longitudinal distribution of lateral projected area above water.

The yacht model tested had its centroid of lateral area aft of midships and when the resultant wind velocity direction was more than ±75° off the bow, the yawing moment turns the bow of the vessel toward the wind. Especially for high-performance craft, the design and sizing of the steering system needs to account for both aero- and hydrodynamic forces and moments.

The aerodynamic yaw moment coefficient (C_N) and yawing moment (YM) in Figure 4-2c are defined as follows:

$$C_N = YM/(1/2)\rho_{AIR}A_L(LOA)v_{AIR}^2 \qquad \text{Equation 4-5}$$

$$YM = C_N(1/2)\rho_{AIR}A_L(LOA)v_{AIR}^2 \qquad \text{Equation 4-6}$$

Now let's return to Figure 4-2, the aerodynamic comparison of the streamlined yacht with a WWII warship, the DD 692 class of destroyer. The deckhouse structure of DD 692 is essentially shaped like blocks stacked on the deck: aerodynamically unfaired geometry with the centroid of the lateral projected area forward of midships. Note that the general shapes of the curves for these designs are similar in 4-2a, 4-2b, and 4-2c. The magnitude of coefficients, however, are not the same.

The significant differences between the two superstructure shapes are demonstrated by the values of C_X in headwinds, and when wind is 30° off the bow. In headwinds, at the same relative velocity, the DD 692 has 43% greater aerodynamic resistance than the streamlined motoryacht. For wind 30° off the bow, the aerodynamic resistance of the DD 692 is 26% greater than for the yacht.

Virtually no difference exists between the two superstructure designs for the side-force coefficients (C_Y). The primary influence of C_Y is the projected lateral area of the hull and superstructure with little dependence of fairing for the direction of beam winds.

The yaw moment coefficient (C_N) shows the effect of the longitudinal distribution of the centroid of projected area. The yacht with its centroid of lateral area aft of midships reacts to relative wind direction off the bow by more than 75° to turn the bow into the wind.

While the DD 692 with its centroid of lateral area forward of midships, the aerodynamic yaw moment tends to destabilize yaw stability until the relative wind direction is greater than 110° off the bow.

Aerodynamic Effects for Safety and Comfort

Open areas of advanced craft may include sundecks, the quarterdeck, and external fore and aft passageways. When the craft is underway, these areas can be exposed to gale-force or greater wind speeds. This means that you will need to consider air-flow patterns carefully.

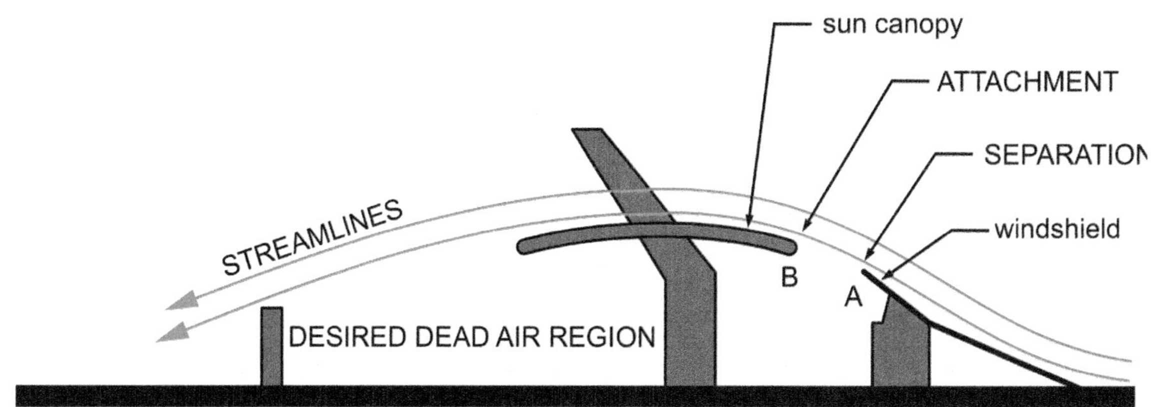

Figure 4-3: Desired relationship for flow alignment between windshield and sun canopy

For example, a flying bridge deck offers desirable space for sunning or social gatherings. Your thoughtful design of the windshield forward of the flying bridge helm and a sun canopy can result in a quiescent or dead-air space aft on this deck. Figure 4-3 provides guidance for achieving this atmosphere, even at very high speeds.

The objective is to have the streamlines separating from the top of the windshield (Point A), attaching at Point B to the top, and passing above the sun canopy. The principal parameters available to achieve this are the height and slope of the windshield, along with the curvature, tilt, and height of the sun canopy. If the streamlines go above the sun canopy, then the air flow in most parts of the flybridge is likely to be relatively slow.

You can confirm your degree of success at achieving dead-air space with computational fluid dynamics (CFD) or this simple wind-tunnel test: Place a smoke wand in the aft flybridge area of the model.

This will cause the entire flybridge area to fill with smoke, and the edges of the smoke will indicate the beginning of the high-speed air boundary.

If you withdraw the smoke wand quickly and observe the length of time required for the smoke to dissipate from the flybridge area, you'll have a good indication of the energy of the flow at model scale for the better configurations. A long residence time indicates a slower exchange rate with the exterior flow, confirming a quiescent atmosphere has been achieved.

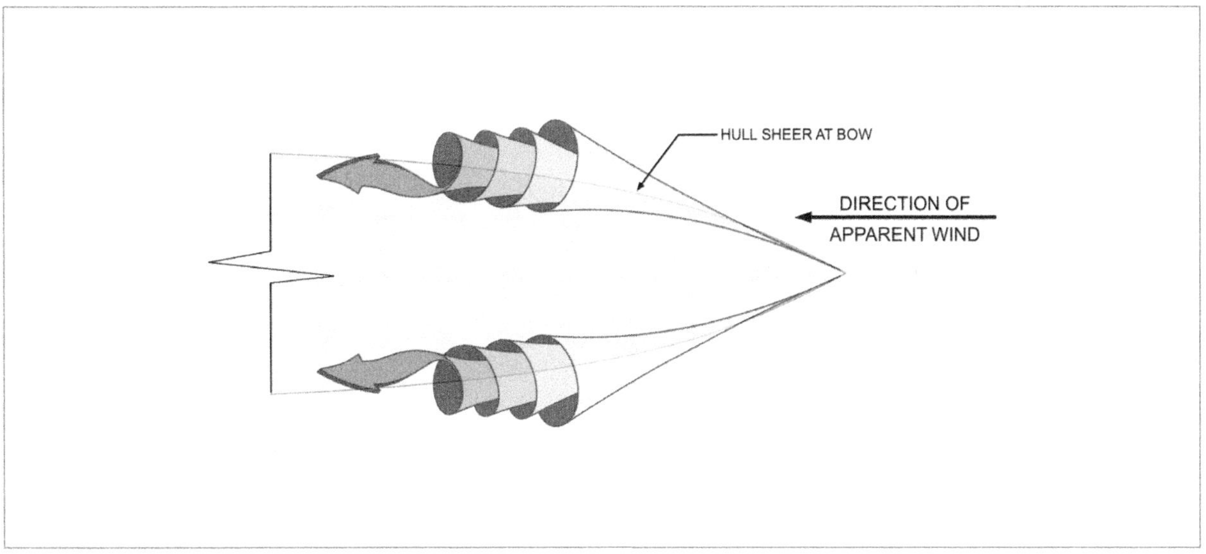

Figure 4-4: Combination of bow flare of hull sides and deck plan form can generate dual vortices, much like a delta wing aircraft

Another aerodynamic issue relates to the shape of the hull at the bow. The three-dimensional shape of the hull at the bow is the first contact to affect the flow of wind moving about a high-performance vessel. The plan-form shape of the bow sheerline in combination with flare of sides of the hull can generate vortices port and starboard, much like air flow about a delta wing aircraft. Figure 4-4 depicts a plan-view.

Should this vortex pattern occur and move downstream, passing the length of the hull, the vortices may create a problem if they follow the passageways on either side. I am not aware of any design guidance on how to avoid this flow phenomenon; however, the possibility of these vortices being generated can be evaluated with wind tunnel tests or CFD analysis.

Should flow studies be made, I suggest the bow geometry of the hull be varied by changing the plan form sheer shape defining locus of air-flow separation, as well as the hull side below this line to negate or control generation of these vortices. Should the vortices not be eliminated by redesigning the bow hull geometry, then personnel should not be permitted in locations exposed to wind at high speeds.

The Station-Wagon Effect

Sportfishing boat cockpits are susceptible to having a reputation for being either wet or dry. I find these are usually very subjective and/or opinionated descriptions. The size of the deckhouse, its proximity to the transom, and the height of any aft hull side wetting are influencing factors. For a dry cockpit, the desired design characteristics are a deckhouse with an air wake that does not extend past the inside of the transom in the cockpit, a transom with an independent or separate air wake, and cockpit hull sides that are dry as shown in Figure 4-5a.

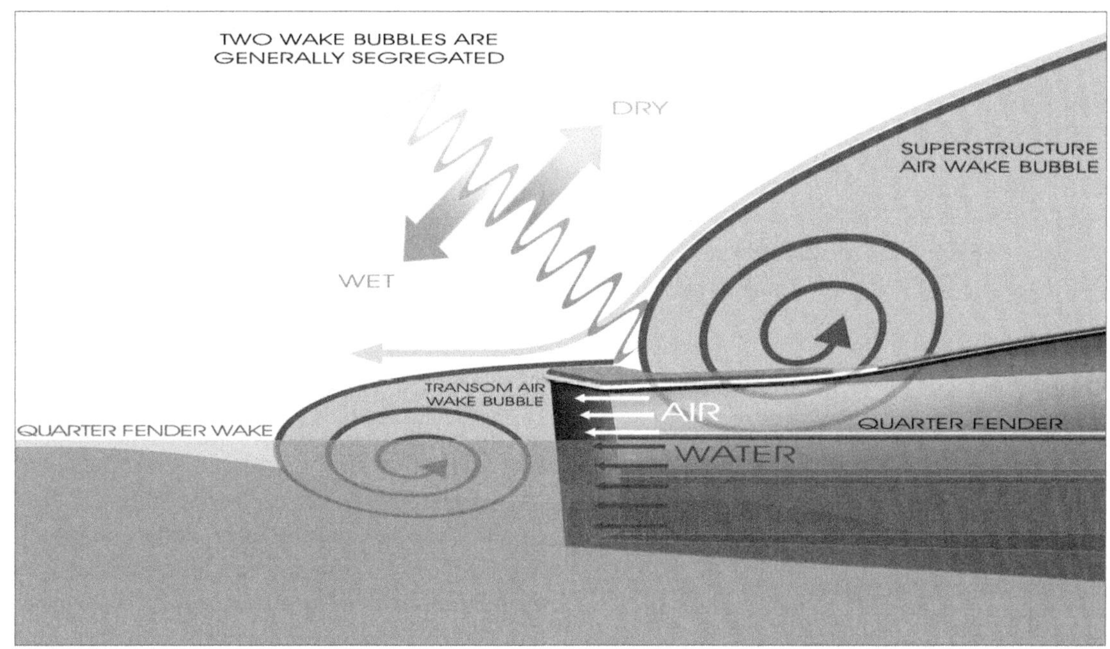

Figure 4-5a: Dry sportfishing boat cockpits usually segregate the air wake of the deckhouse from the air wake of the transom

Most sportfishing boat cockpits reputed to be wet have air flow as depicted in Figure 4-5b. The source of wetness or spray captured in air flow usually comes from aft of the transom.

Sportfishing boat cockpits are susceptible to having a reputation for being either wet or dry. I find these are usually very subjective and/or opinionated descriptions.

Spray is generated by turbulence of the hull's water wake and steam or spray originating from wet-exhaust systems exiting into the air wake of the transom.

With a sufficient number of examples of so-called wet and dry cockpit sportfishing boats, some form of quantitative design guidance may be established:

A ratio of height of top of deckhouse to transom covering board divided by the projected distance from the aft end of the top of the deckhouse to the back side of the transom may be graphed as a function of speed.

This ratio may possibly define the difference between wet or dry cockpit, when water has not attached to hull sides.

Take note that water reattachment to the aft hull side is very much dependent on an overloaded planing bottom for a too-low planing speed.

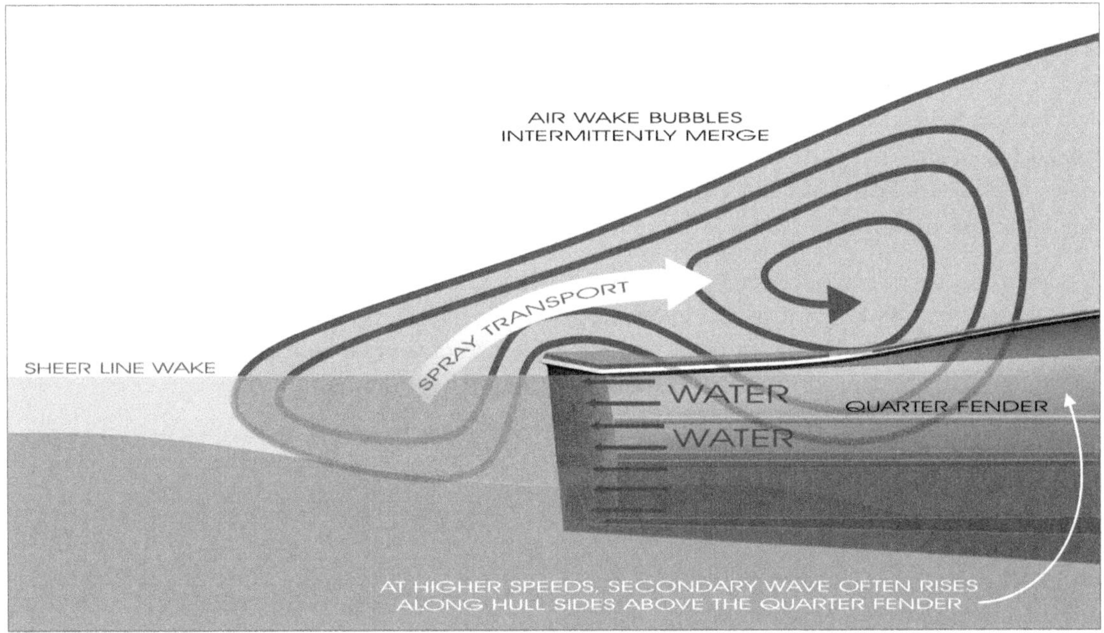

Figure 4-5b: When air wakes of the deckhouse and transom merge, a wet cockpit usually results

Underwater Appendage Drag

Raw-Water Inlets

Vessels have raw-water requirements for engines and mechanical systems such as gears, HVAC, water makers, etc. For performance vessels, seawater inlets supply the various systems inside the hull with raw water. After this water transfers heat away from the engines, generators, and gears, some of it may be sprayed into the exhaust system to cool the hot gas and lower the acoustic level; most of the remainder is drained overboard.

For the flow rate required by the raw-water system, a minimum pressure must be supplied. In some cases, there is also a maximum pressure not to be exceeded. To minimize clogging of the cooling system, the incoming raw water must be strained—to block sizable foreign objects—and then filtered.

A significant source of drag comes from hull-bottom mounted external water-inlet fittings, which should be minimized or eliminated. At moderately high speed, these fittings can initiate cavitation downstream, inducing noise into the hull and eroding or fatiguing metal, especially marine aluminum alloys.

Over time, owner requirements have steadily trended toward higher speeds. This has caused designers to rethink the configuration of raw-water inlets for management of drag and cavitation problems. To show you one effect of the trend of increasing design speed requirements, Figure 4-6 (a, b, and c) depicts the evolution of raw-water inlets, and Table 4-B provides guidance for application.

See Figure 4-6c for the current state of inlet design on operational craft having speeds greater than 50 knots.

My experience with openings of strainer plates is that they should have their long axis in the same direction as the water flow. (See Figures 4-6a and 4-6b.) Also, the open area of strainer plates and the transverse slot of inlet configuration (Figure 4-6c) should be three times the total areas of pipes and hoses connected for all systems, using raw water which is drawn from sea chests.

Underwater Exhaust Fairings

Underwater exhaust systems have benefits for high-performance craft, along with some risks, if not well-engineered. Benefits include both reduced weight and onboard noise level. Exhaust gas does not collect in the atmosphere around the vessel, and, best of all, very low engine exhaust back-pressure can be achieved.

On the downside, careful engineering of the exhaust system is necessary. The location of the underwater exhaust openings is critical. They must be aft and outboard of the transverse plane of propellers or waterjet inlets on single-chine and round-bilge hulls. On double-chine hulls, the exhaust opening can be farther forward near the machinery space if located between the upper and lower chines.

I'm introducing this topic here to emphasize that properly engineered underwater exhaust systems require external fairing ahead of the exhaust opening. These fairings must be shaped so as to develop low pressure at their base to educt or siphon engine exhaust gas into the water. The resistance of these fairing appendages can be negligible due to low base pressure. The fairings do generate lift, however, and this produces bow-down trim moments, especially when the underwater-exhaust fairings and outlets are near the transom.

Compared to craft with engines exhausting into the atmosphere, boats with underwater exhausts during both sea trials and model tests have had trim-angle reductions measured on the order of one degree at planing speeds.

For safety, dynamic stability, and performance considerations, both designers and builders need to offset this bow-down trim moment on high-speed craft whenever they want to install underwater exhaust systems. You can offset this bow-down moment with bow-up trimming moments produced through some aft shifting of light ship LCG, or during operations at full load by managing LCG shift of consumables.

You can also make changes to hull lines for increasing bow-up trim angle by reducing transom beam, increasing deadrise angle at the LCG, or incorporating some buttock rocker in the hull bottom near the transom.

High-speed craft with fuel tanks aft are at some greater risk than vessels with fuel tanks distributed longitudinally about the LCG when underwater exhaust systems are installed. When fuel is consumed for tanks aft, a bow-down moment develops as a result of removing aft weight. Boats also increase speed as weight is removed resulting in an additional reduction in trim angle as less hydrodynamic lift is necessary to maintain equilibrium.

For a high-performance craft with tanks aft, and underwater exhaust, operational time-and-fuel consumption result in trim angle becoming flat at the end of the day. It can be very tricky and dangerous for a boat captain to return to port in following seas with no means of increasing the trim.

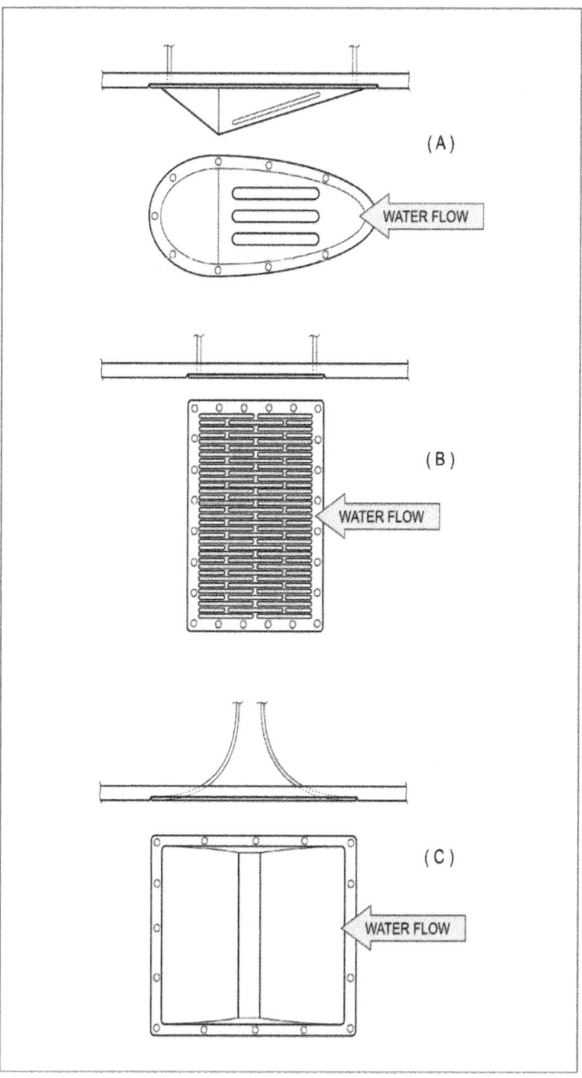

Figure 4-6: External water inlet fittings can be a significant source of drag and cavitation (4-6a: high drag; 4-6b: flush, low drag; 4-6c: recessed, can be zero drag and cavitation free with CFD study.)

TABLE 4-B: APPLICATION OF RAW-WATER INLET CONCEPTS		
Inlet	**Speed Range**	**Comments**
a	0 to 30 knots	Used on low planing-speed commercial, military, and recreational craft. Generally used on craft without internal sea chests. Drag coefficient, $C_D = 0.65$ using largest cross-section area (A) normal to flow direction. Drag in salt water = $0.65(1/2)1.9905(A)1.6782V_{KT}^2 = 1.84A\ V_{KT}^2$ (lb) With this type of seawater strainer, each square foot of cross-section area (A) requires 90 shaft horsepower on a craft at 20 knots, assuming a 0.5 propulsive coefficient. *Not recommended for performance craft.*
b	0 to 45 knots	Used on craft with internal sea chests. The external surface of the strainer plate is flush with the hull's exterior surface. To increase sea chest pressure, mount with a small angle to hull buttocks, the aft edge of the strainer plate may be mounted one inch from the hull's surface for $V_{KT} \geq 40$. Then, drag in salt water will increase from zero to $D_{1in} = 0.153W_{FT}V_{KT}^2$(lb) where W_{FT} is width of strainer plate. Internal filtering of raw water is expected.
c	0 to 70 knots	For very high speeds, use a long transverse, recessed inlet slot oriented at right angle to the flow. Used on craft having internal sea chests. Must be custom designed using CFD for each hull form at its design speed. Drag is zero and can be cavitation free. Internal filtering of raw water is expected.

Notes:

Other Appendage Considerations

Other underwater appendages not associated with propulsion systems, such as depth sounders, speed logs, and transducers for fish finders, all contribute to craft resistance and all must be located in non-aerated water. The shapes of transducer fairings must be carefully considered to minimize negative impact on speed. Hoerner (1965) is a good resource for experimental drag coefficients for a variety of shapes that might be used for fairings.

Calm-water, bare-hull resistance prediction versus speed has already been addressed in Chapter 3. Since the design process may include a task to study more than one propulsion concept, in that chapter I placed the emphasis on the bare hull, without propellers on inclined shafts, waterjets, pod drives, outboards, inboard out-drives, and surface propellers. By beginning with bare-hull resistance, it was only necessary for you to add the appendage drag of the propulsion systems that you are considering. Should a propulsion scheme require hull alteration that affects resistance, then you will need to adjust the resistance with the appropriate penalty or advantage as a change to the propulsion efficiency.

For example, a project with a navigational draft limitation might not require an alteration to the bare hull lines for waterjets or surface propeller drives, but altering the hull for propeller tunnels may be essential for it to meet the draft requirement. Therefore, any increase or decrease in bare-hull resistance due to propeller tunnels must be accounted for by adjusting propulsive efficiency. Following this approach will reduce your effort to arrive at a recommendation of a preferred hull-and-propulsion system combination. Hence, I will address appendage resistance in groups that are typical of each propulsion system.

Drag of Inclined Shafts, Struts, and Rudders

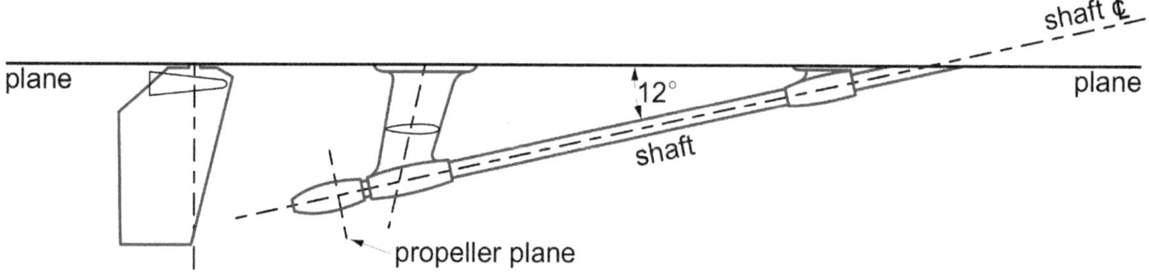

Figure 4-7: Inclined shaftline with rudder, single arm strut and shaft

Craft with inboard engines driving inclined shafts with fixed-pitch propellers, shown in Figure 4-7, tend to be the least costly due to minimum cost of the propulsion system. This prevalent shaftline arrangement includes a rudder aft of the propeller, a single arm strut, a shaft for transmitting power, occasionally an intermediate strut and the shaft entering the hull via a stuffing box (not shown) with connection to a gear box (not shown). A photo of a typical installation of a boat with an inclined-shaft propulsion system is shown in Figure 4-8.

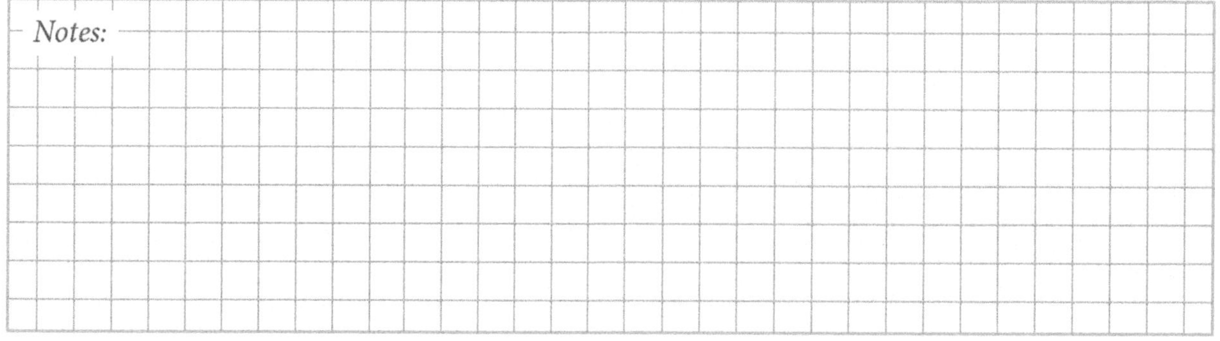

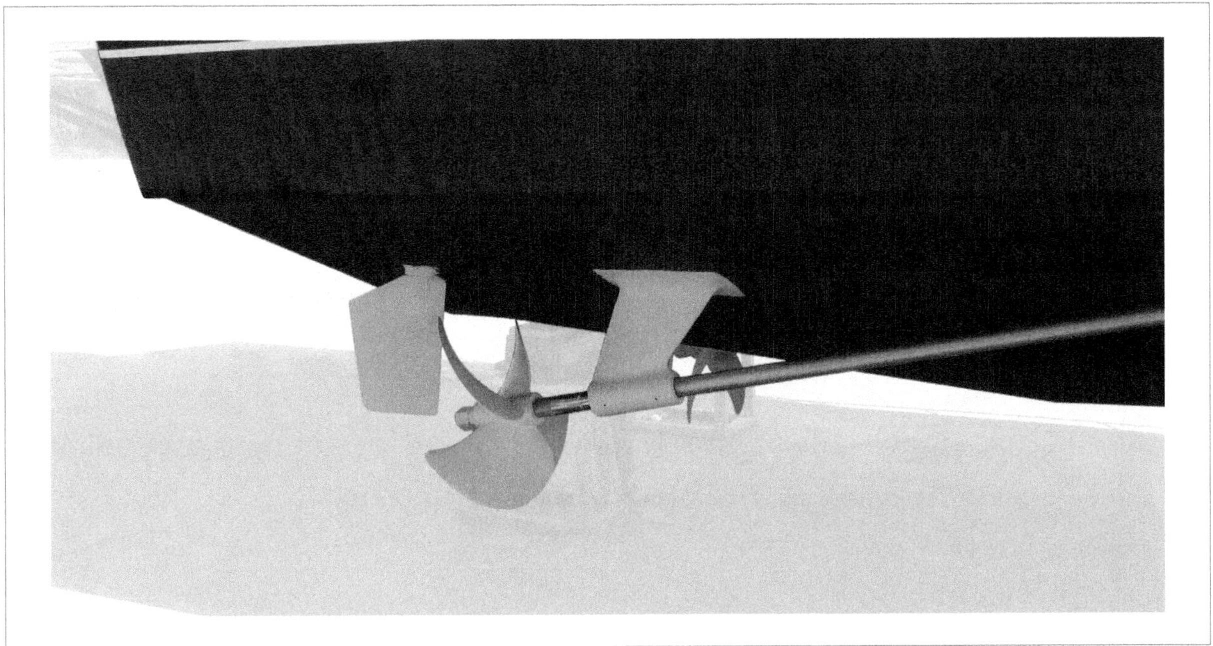

Figure 4-8: Photograph of inclined shaft installation

For very good reasons, performance craft have single arm struts, rather than "V" struts, just ahead of the propeller supporting the aft end of the shaft. The strut location is in the "shadow" or wake of the shaft. Thus, being in the shaft's wake, the strut arm experiences water velocity less than craft speed–a reduced drag location.

Structurally, the necessary lateral strength and stiffness of a single-arm strut at its connection to the hull is achieved with a thickness less than that of the shaft diameter. This is desirable to keep the strut arm within the shaft's wake shadow.

The strut palm, as a meaningful drag element, should be recessed to be fair with the outer surface of the hull.

You can predict the natural frequency of a single-arm strut and select the number of propeller blades to avoid resonant conditions excited by the propeller for the range of rpm.

Unless a "V" strut has been required by a client, I do not recommend nor do I use them. They are very difficult to manufacture accurately, hard to align, and have complex geometry with regard to flow. The "V" strut hub, with two arms attached, and the wake of the inclined shaft interact to develop early onset of cavitation which extends into the plane of the propeller.

A single-arm strut in the shaft shadow contributes to only one wake fault while a two-arm "V" strut plus the shaft shadow has three wake faults, which increases the probability of propeller vibration and cavitation. Improperly aligned with local flow patterns, "V" struts will also affect the running trim angles of a craft.

When dimensional details of a shaftline with single arm strut and rudder have not yet been established, I use the following empirical equation to estimate the appendage drag for each shaft line; doubling it for a twin screw boat and times three for a triple screw craft.

$$R_{APP} = R(0.003F_{nV}^2 + 0.025) \qquad \text{Equation 4-7}$$

The following equations (4-8 to 4-12) are useful in breaking out individual components of appendages for making drag estimates for actual dimensional sizes of components.

In addition, Appendix 2 provides experimental characteristic data for five rudders for high-performance craft including a rudder-sizing chart and a calculation procedure for determining the size of rudder stock along with steering gear.

Also, Table 4-C provides some experimentally determined rudder drag coefficients when in a propeller slipstream.

Appendage Resistance Equations As a Function of Speed

Inclined cylinder, that is, shaft and strut barrel:

$$D_{SH} = (\rho/2)\, l(d)v^2(1.1 \sin^3\epsilon + \pi C_F) \qquad \text{Equation 4-8}$$

Skeg:

$$D_K = (\rho/2)(2S_K)V_M^2C_F \qquad \text{Equation 4-9}$$

Strut palms:

$$D_P = 0.75\, C_{DP}\, (h_P/\delta)^{1/3}\, yh_P\, (\rho/2)V_M^2 \qquad \text{Equation 4-10}$$

where $C_{DP} \approx 0.65$ and $\delta \approx 0.016X_P$

Non-vented rudders and struts:

$$D_{R/S} = (\rho/2)\, sv^2 2C_F[1 + 2\, t/c + 60\, (t/c)^4] \qquad \text{Equation 4-11}$$

Interference drag: This component is small and may be ignored (Hadler 1966).

Non-flush seawater strainers:

$$D_O = (\rho/2)\, AV_M^2C_{DO} \qquad \text{Equation 4-12}$$

where $C_{DO} \approx 0.65$

Here are a few additional thoughts about appendages.

CENTERLINE SKEGS

Even though an equation is given to estimate resistance of skegs, I believe that centerline skegs have little use on high-performance craft except as a base (a shallow depth docking skeg) upon which to block a vessel while in dry dock. Directional stability is best achieved at high speed with a couple of small fixed fins on the aft end of the bottom of the hull.

Like feathers on an arrow, small aft fins will provide course restoring moments whenever a vessel deviates from heading. Unless there are overriding longitudinal structural necessities, the owner requires them, or there is some other compelling reason, you should avoid placing centerline skegs on high-performance craft.

RUDDERS WITH AIRFOIL SECTIONS

Rudders designed with airfoil sections have the potential to create problems that offset the advantages gained by the fuel saved due to their low drag coefficients. For positive steering control, performance vessels require rudders with section shapes which maintain positive control characteristics at high speeds.

TABLE 4-C: EXPERIMENTAL RUDDER DRAG COEFFICIENTS IN A PROPELLER SLIPSTREAM					
Rudder section	t/c	C_{DR}			
		σ = 4.0	σ = 2.0	σ = 1.5	σ = 1.0
NACA 0015	0.15	0.0015	0.0015	0.0015	0.0008
Parabolic (blunt base)	0.11	0.0417	0.0417	0.0433	0.0425
Flat plate	0.04	0.0278	0.0325	0.0371	0.0433
Wedge with six-degree angle	0.11	0.0495	0.0495	0.0495	0.0487

Notes:

Geometric aspect ratio = 1.5

$K_T/J_T^2 = 0.20$

Propeller 0.55D ahead of rudder stock

$D_R = (\rho/2)\, sv^2 C_{DR}$

In a 2002 article, Dudley Dawson and I compared characteristics of four rudders with different section shapes at a speed of 40 knots. Our results are summarized in Table 4-D.

TABLE 4-D: COMPARISON OF RUDDER SECTION SHAPES AT 40 KNOTS					
Section	SHP per sq. ft of rudder area	Maximum angle for positive side force	Rudder reaction to steering gear failure	Relative cost	Order of preferred technical choice
Airfoil (NACA0015)	10	18°	Does not trail at non-zero angle, uncontrol-lable turn of craft	high	---
Flat plate	40	35°	Rudder trails at 0°	low	3
Wedge	56	35°	Rudder trails at 0°	moderate	2
Parabolic	20	35°	Rudder trails at 0°	high	1

Notes:

All of these rudders had aspect ratio = 1.5.

Tests were conducted in a cavitation facility at the Navy's David Taylor Model Basin.

Tests were conducted in uniform flow.

Flush-Inlet Waterjets

Increasingly, flush-inlet waterjets are becoming the propulsor choice for large performance vessels. Their efficiencies above 25 knots are competitive with submerged propellers, they reduce machinery-related vibration, reduce navigational draft, and—except for small, needed course-keeping fins—they have no other underwater-appendage drag elements necessary for operation of vessels.

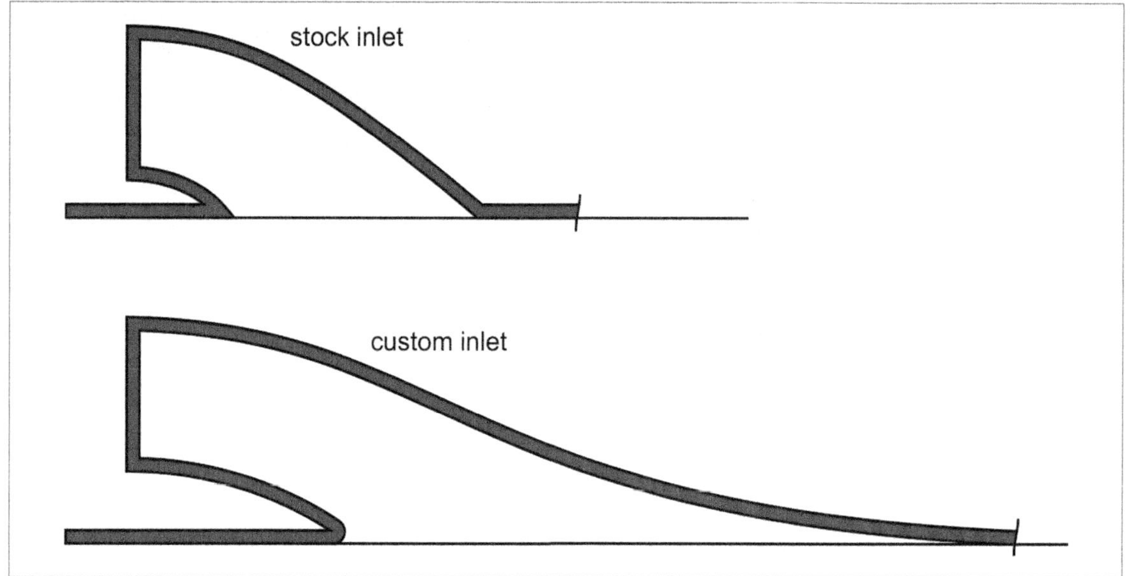

Figure 4-9: Representative profiles of two flush inlet waterjets

Essentially, production waterjets are offered commercially in two forms.

- First, complete, including stock inlet, pump, and exit nozzle with steering and reversing mechanisms. These waterjets tend to be selected for craft having design speeds of less than 45 knots.
- Second, similar to the first form, except that a custom inlet is designed by the waterjet manufacturer for the vessel in lieu of providing inlet hardware: The shipyard is responsible for fabricating inlets to the waterjet supplier's design. These waterjets tend to be selected for large vessels having design speeds from 25 to 70 knots.

Figure 4-9 depicts representative inlet profiles for these two available forms of waterjets.

To recommend waterjet models and performance predictions, designers must provide suppliers with the following: hull geometry and resistance versus speed predictions for lightship and full-load displacements for calm and rough water.

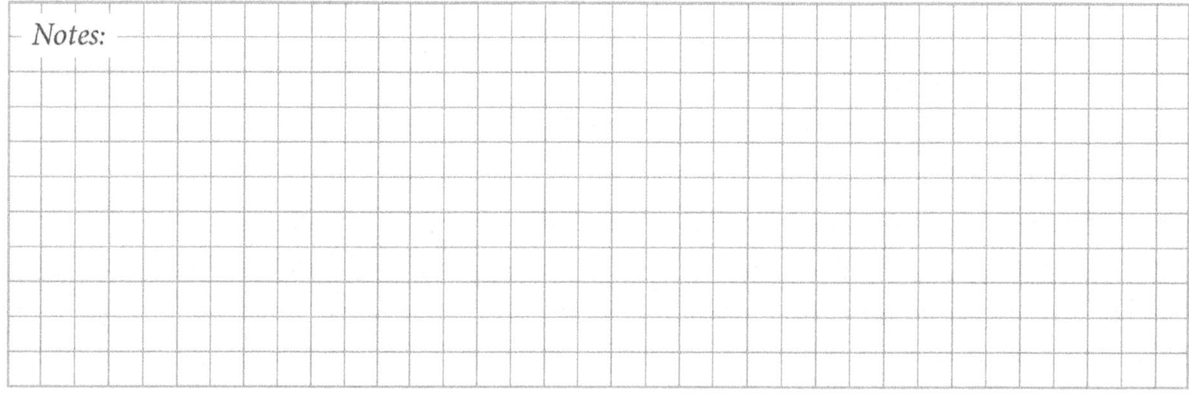

Fixed Fins for Course Stability

With very few exceptions, high-performance waterjet-propelled craft will have several small, fixed fins on the aft hull bottom designed by a naval architect. These provide course-keeping stability and minimize spinout during high-speed turns.

In the remainder of this section, I'll discuss hydrodynamic fin characteristics obtained experimentally, representing speeds from 30 to 70 knots. Course-keeping stability fins are small in total area. Most effectively used in pairs, they are conveniently located slightly aft of and between waterjet inlets. Here I'll give recommendations for wedge-section fins and provide hydrodynamic characteristics.

The blunt base of these fins should be located about half of their mean chord forward of the transom to minimize reduction of side force due to ventilation should yaw angles exceed six degrees at very high speeds. The water-flow direction at this location is usually not parallel to the centerline of a craft so fins must be attached to the hull at some small angle on the order of one to four degrees to minimize drag. This flow angle may be established by model flow tests, CFD, or sea trials to measure flow angles before fins are attached.

Recommended total projected fin area may be approximated by either of the following:

$$\text{Total fin area} \approx 0.0175\nabla^{2/3} \qquad \text{Equation 4-13}$$

$$\text{Total fin area} \approx 0.01L_P \qquad \text{Equation 4-14}$$

Hydrodynamic characteristics of the wedge-section fin in Figure 4-10 are provided for a range of cavitation numbers for vessel speeds from 30 to 70 knots.

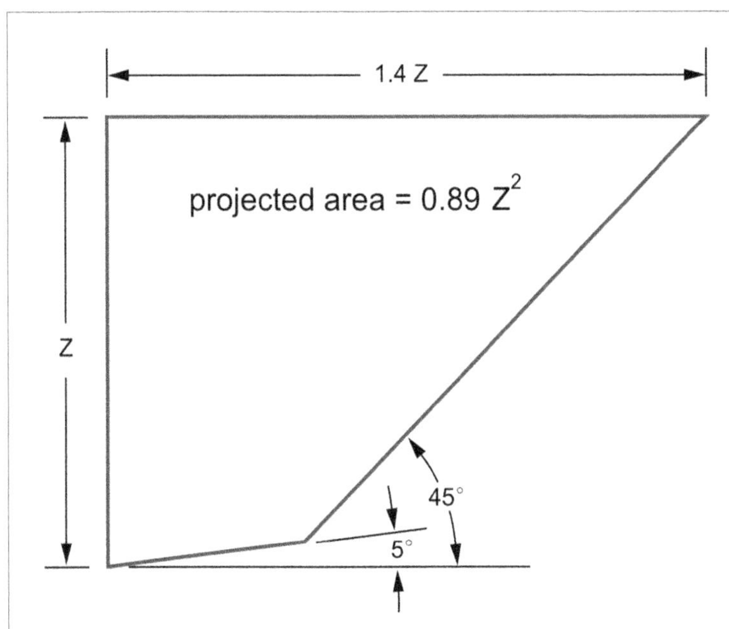

Photographs from tests in the cavitation facility appear in Figure 4-11. In Figures 4-12 and 4-13, drag and side-force coefficients are given respectively for non-ventilating flow and forced base ventilation. During tests, forced base ventilation was obtained by air pressure slightly above atmospheric supplied at the top of the fin base.

Figure 4-10:
Profile of wedge section fin for course-keeping stability for high-speed craft propelled by flush inlet waterjets. The leading edge is sloped at 45 degrees.

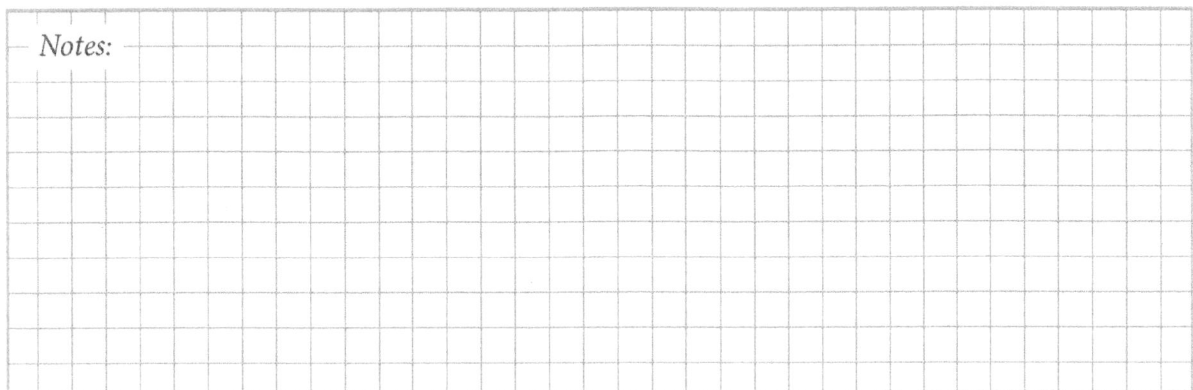

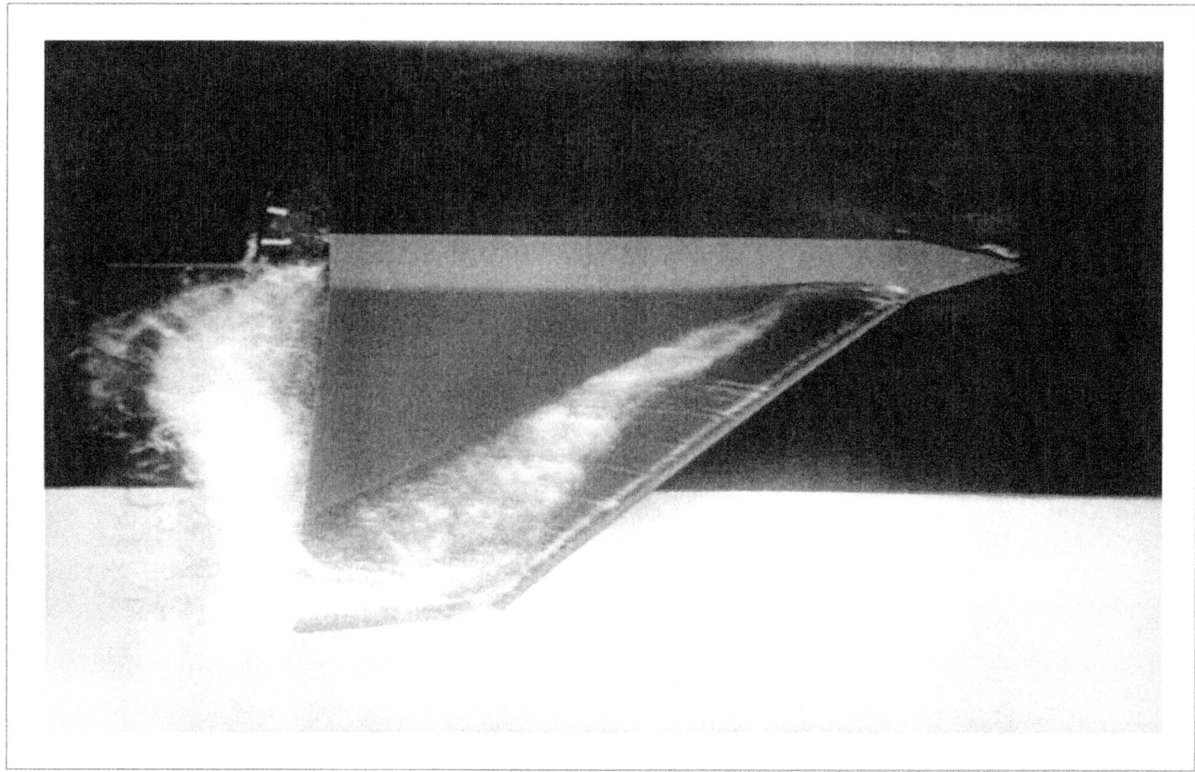

Figure 4-11a: Profile

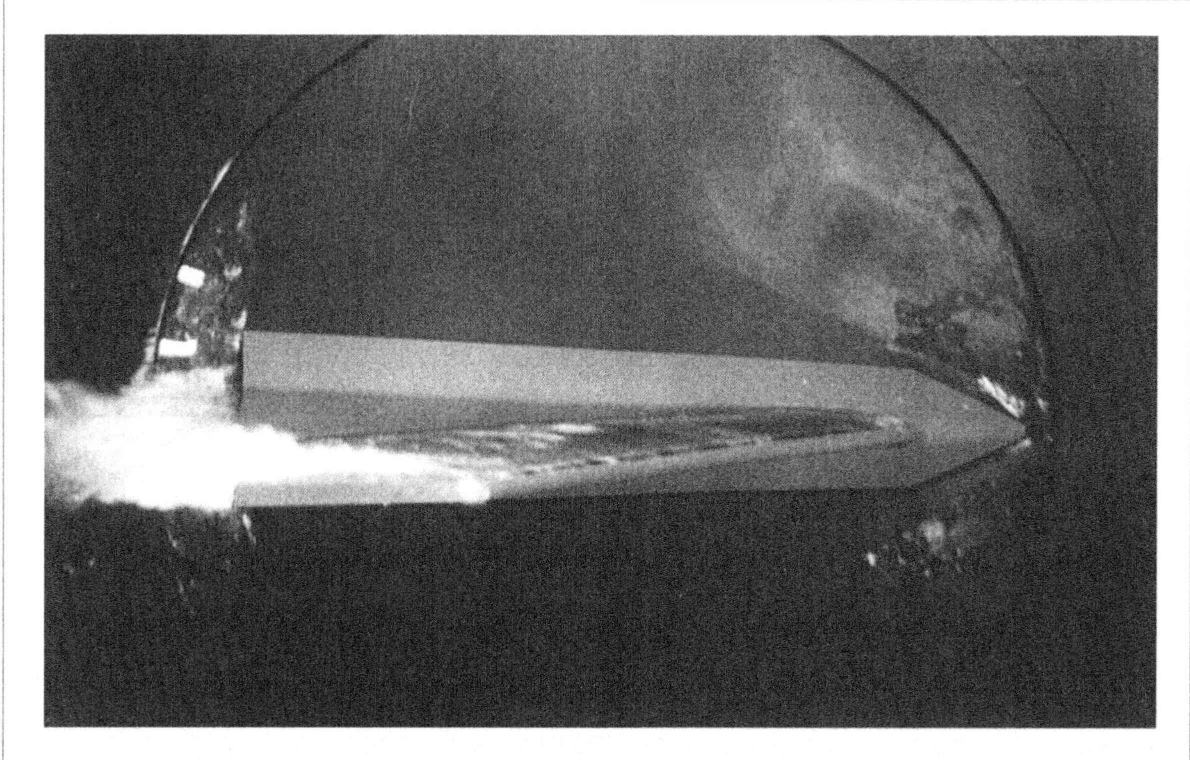

Figure 4-11b: Bottom view, looking up

Figure 4-11: Cavitation of fin for a $\sigma = 0.319$ (50 knots) at angle of 3° for non-ventilated condition

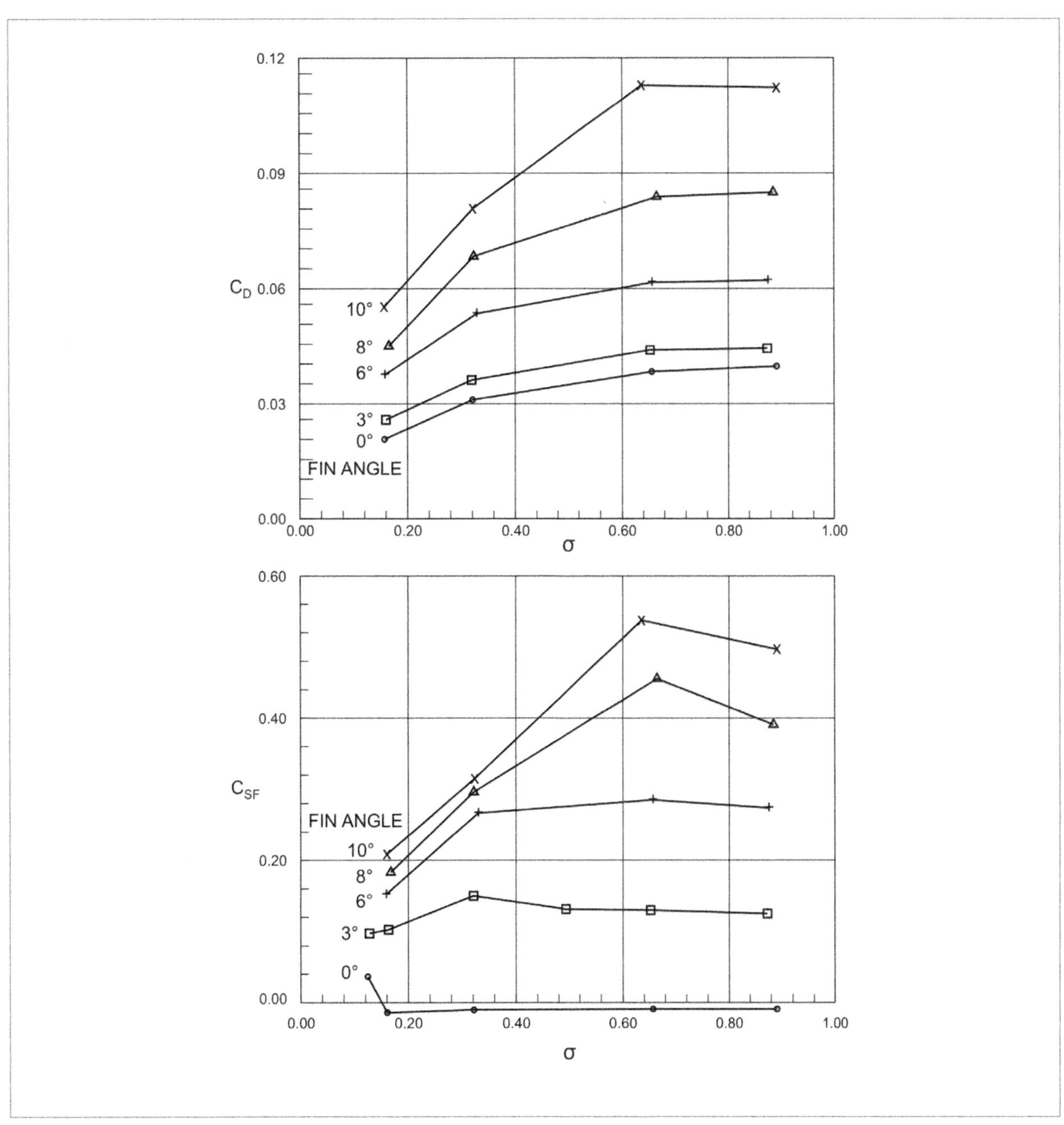

Figure 4-12: Drag and side-force coefficients for fin in Figure 4-10 for non-ventilating flow

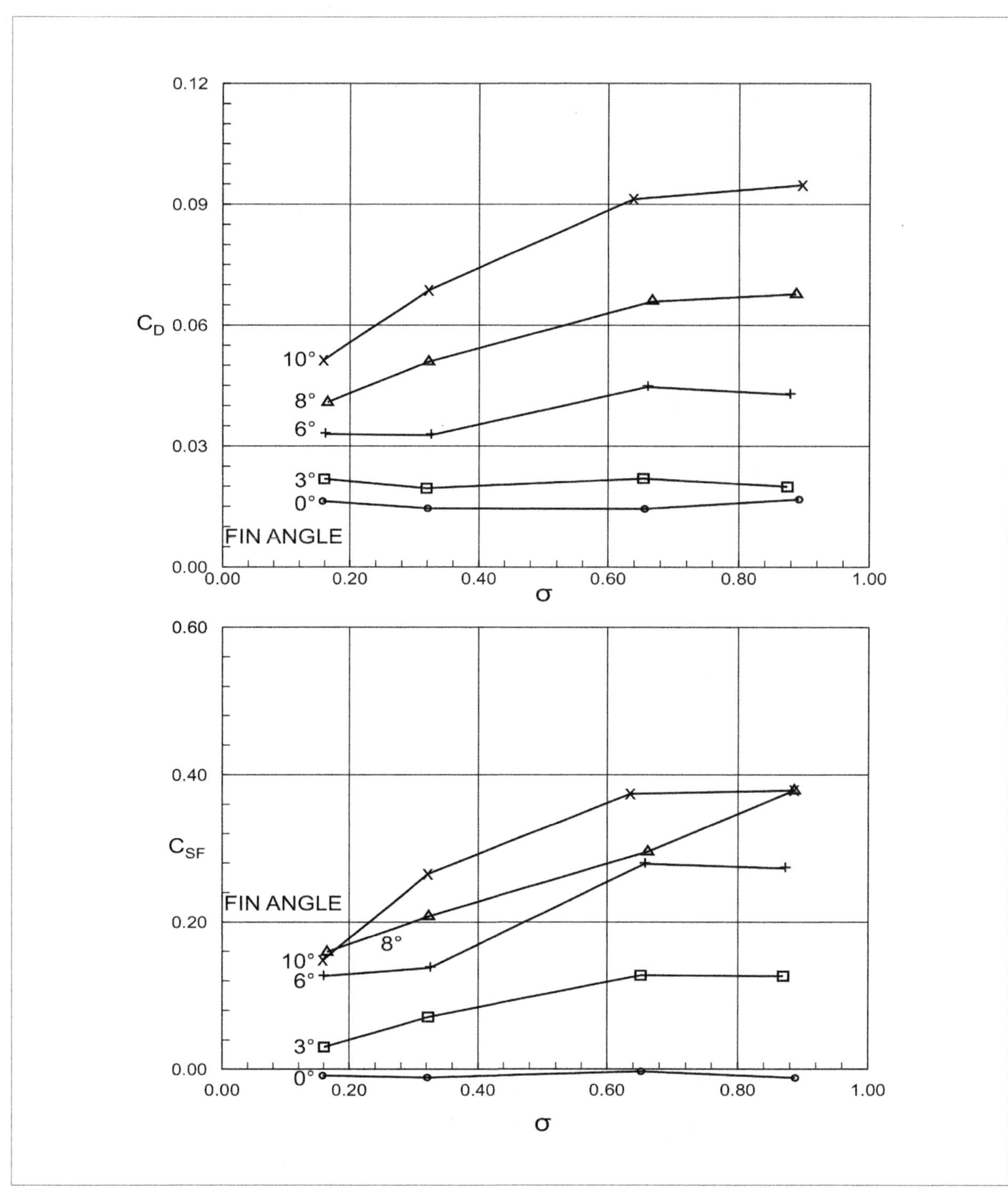

Figure 4-13: Drag and side-force coefficients for fin in Figure 4-10 for forced ventilation

Pod Drives

Pod drives are propulsion systems where inboard engines transmit power through the hull bottom via two right-angle gears to drive submerged propellers which are steerable and provide thrust to propel vessels. For high-performance craft, pod drives with counter-rotating propellers are commercially available in both pusher and tractor configurations.

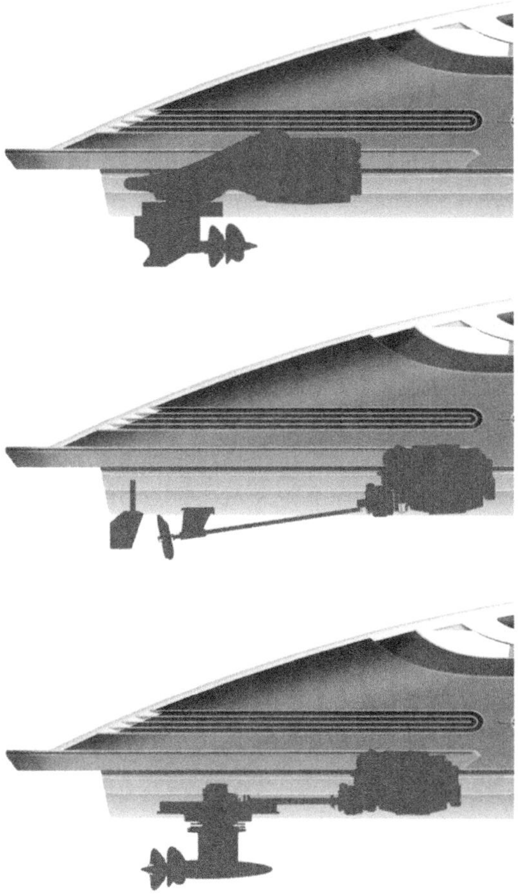

Figure 4-14 depicts a silhouetted comparison of propulsion systems of an inclined shaft system with tractor and pusher pod drives. Take note that compact pod drives make it possible to increase interior hull volume for commercial, military mission, and recreational purposes.

At the same craft speed, pod drives with counter -rotating, fixed-pitch propellers have noticeably higher propeller efficiency than a fixed-pitch propeller on an inclined shaft.

Propulsor thrust and hull resistance vectors are shown in Figure 4-15. The lower unit of a pod drive is the source of appendage drag for this propulsion concept. However, dimensional information of lower unit is generally not provided by suppliers.

Designers must provide sufficient hull geometry, weight, resistance versus speed, and LCG information to suppliers to get assistance for performance prediction as is done by waterjet manufacturers.

Figure 4-14: Comparison of propulsion systems of tractor, inclined shaft, and pusher pod drives

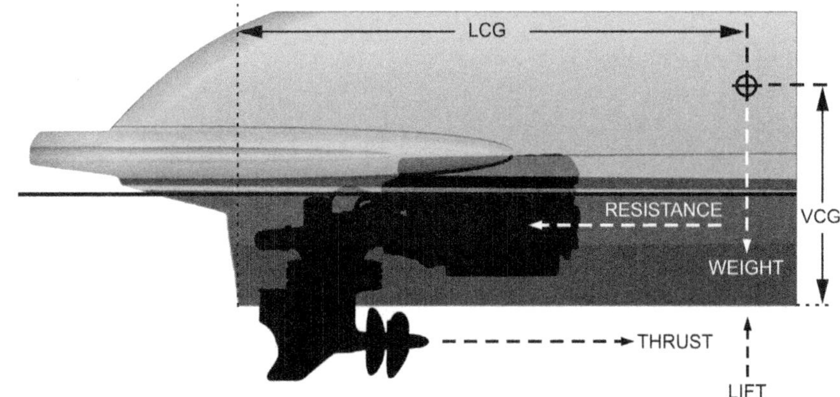

Figure 4-15: Simple force vectors for pod drives

Surface Propeller Drives

Surface propeller drives have a long history. They began as a simple convenience to allow the shaft to pass through the transom to connect to the propeller. The bottom of the propeller hub was at, or slightly above, the bottom of the hull with about 40 percent of the propeller diameter extended below the hull to develop thrust.

In the early 1900s, Hickman's sea sled hulls (inverted "V" hull having β = -10°) won many boat races, a feat which he attributed to the hull form. Leaping forward to the 1950s, we learn through experimental analysis that his surface propeller efficiencies, combined with the absence of appendage drag of shafts and struts, is the real source of the winning ways of Hickman's boats.

In addition to Hickman's arrangement of a solid shaft through the transom with rudder steering, surface propeller drives are commercially available in several other forms. For example, Figure 4-16 depicts the shaft aft of the transom. Not shown is a flexible-shaft joint near the transom allowing both transverse steering and vertical shaft angle movement. Other configurations have a rudder cantilevered aft of the propeller for steering.

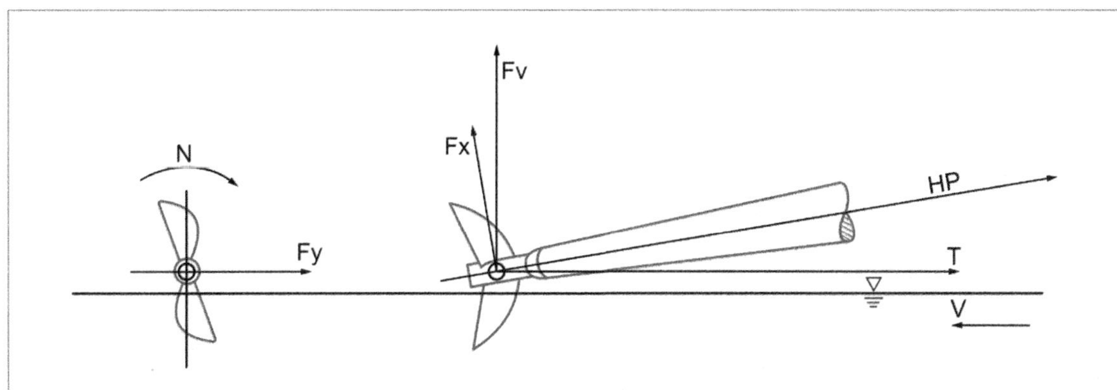

Figure 4-16: Surface propeller drive

The advantages of surface-propeller drives are that there is little or no appendage drag, and thrust is developed only by blades in the lower half of the diameter. Also, in addition to thrust, both lift and side forces are generated as blades enter and exit the water's surface. [Best craft control is achieved when surface drives are installed in pairs to counterbalance propeller side forces when propellers rotate in opposite directions.] This application is best for extremely high-speed operation on craft that become airborne in sea conditions.

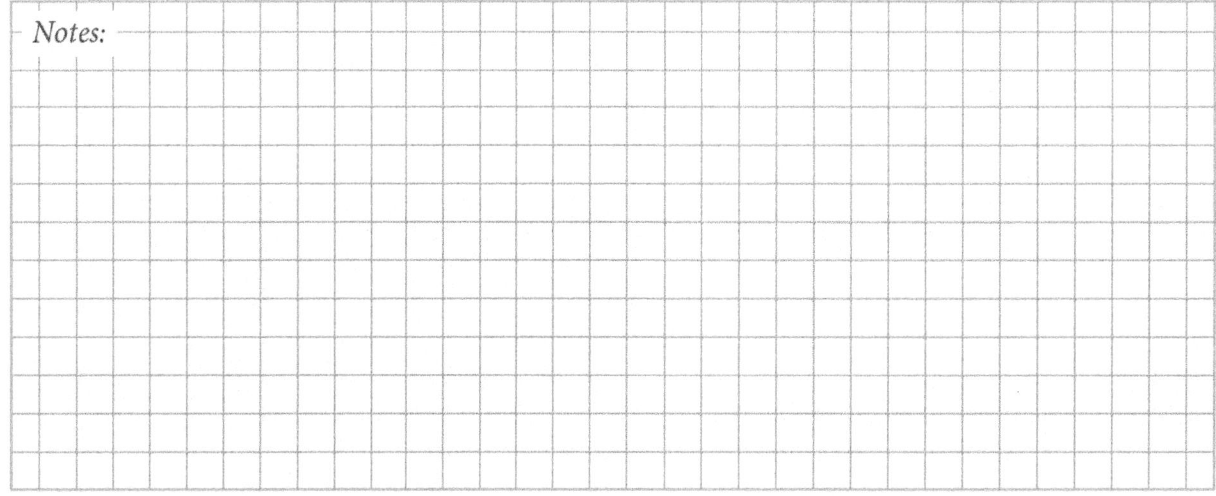

Disadvantages include poor acceleration from low to planing speeds, the craft can be difficult to maneuver at low speeds, and there is relatively high propeller blade rate vibration. Appendage resistance prediction is zero, unless a small fin or skeg is ahead of the propeller or separate rudders are used for steering.

Note: I have already addressed prediction of rudder resistance for inclined-shaft propulsion arrangements.

Commercially available surface propeller drives are primarily used on high-speed craft of lengths up to 82 feet (25 m) in recreational racing and military service.

Bulbous Bows

A bulbous bow is an elementary device to reduce wave-making resistance in the approximate speed range $0.18 \leq F_{nL} \leq 0.60$. The protruding bulb form affects the velocity field to attenuate the bow wave system.

I think of bulbous bows as "good" appendages as they generally reduce bare-hull resistance in the indicated speed range and can be added to some existing vessels to increase speed without adding power. Outside of this speed range, however, bulbous bows can increase bare-hull resistance.

Design guidance for bulbous bows is based on three basic transverse section shapes and nine profiles. Some of the bulbous bows studied by Kracht (1978) and Hoyle et al. (1986) are shown in Figure 4-17.

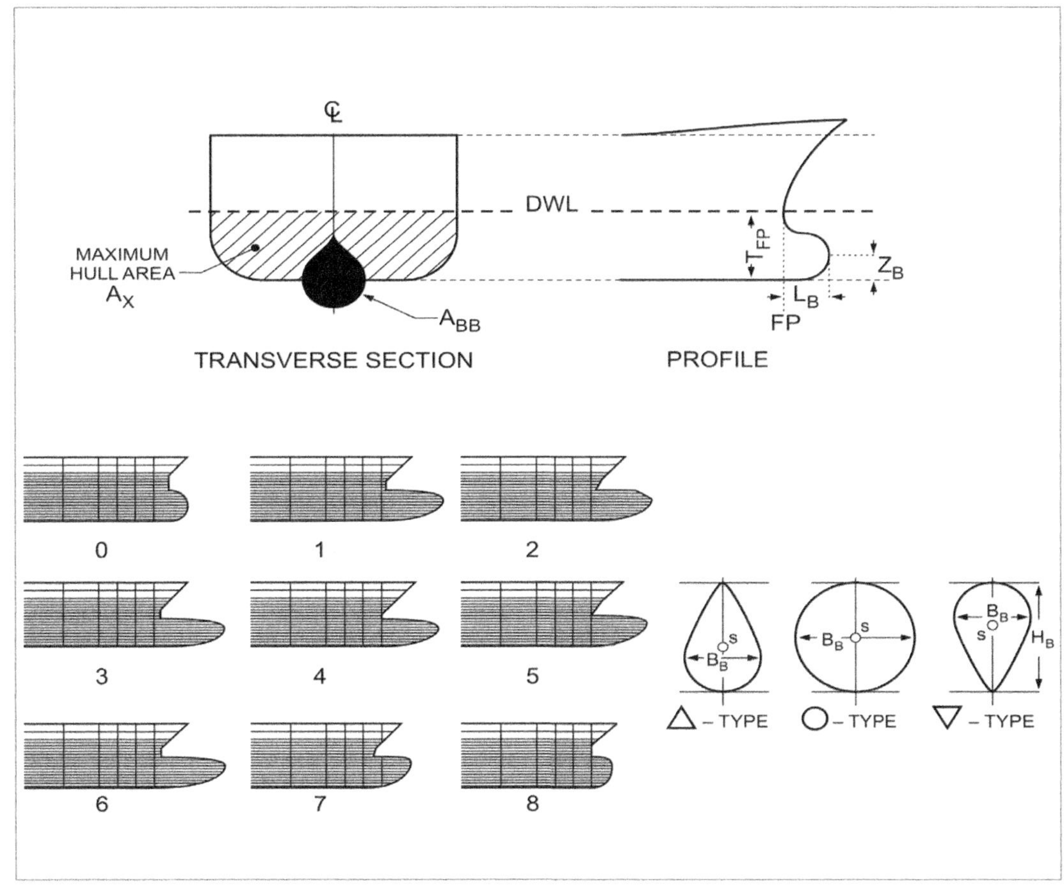

Figure 4-17: Definition of significant bulbous bow geometry, profiles and section shapes (Kracht 1978 and Hoyle et al. 1986)

The ratio of hull resistance with a bulbous bow to a hull without one, suggested by these references is that maximum improvement occurs between $F_{nL} \approx 0.30$ and 0.40. Resistance reduction for vessels with bulbous bows relative to bare hulls may be attained for $0.18 < F_{nL} < 0.60$. Potential results of resistance reduction versus speeds reported by Kracht, Hoyle, and van Oossanen (2009) for their best bulb designs is provided in Figure 4-18, with bulb geometry summarized in Table 4-E.

TABLE 4-E: RANGE OF BULB GEOMETRY COEFFICIENTS

Parameter	Value	Comment
L_B/L	0.020 to 0.055 (typical – 0.04)	Ratio of length of bulb forward of FP to L
A_{BB}/A_X	0.07 to 0.15 (typical – 0.11)	Ratio of transverse bulb area at FP to maximum hull area
Z_B/T_{FP}	0.33 to 0.58 (typical – 0.46)	Ratio of height of most forward profile point of bulb to draft of vessel at FP

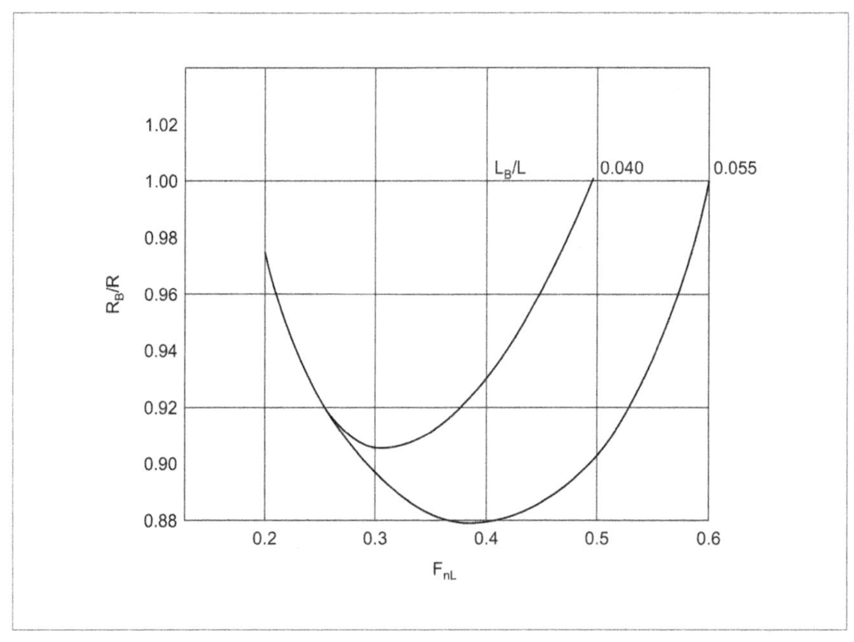

Figure 4-18: Possible resistance reduction with a bulbous bow relative to bare hull for the speed range of effective benefit

There are several additional factors which you must consider. Bulbous bows do have effects on motions in a seaway.

Also, from a practical perspective, when designing bulbs, you must be aware of anchoring, docking, and dry docking, especially when $L_B/L > 0.04$.

A dropping anchor must not strike the bulb, and your docking drawings and procedures must be clear in defining appropriate blocking placement.

Should vessels having bulbous bows be powered for speeds of $F_{nL} \geq 0.5$, bulbs having ∇-type sections should be used with regard to ride quality in waves.

Bow Thruster Fairing

Bow thrusters provide zero- and low-speed maneuvering capability for advanced craft when supported by buoyancy.

The distinguishing design aspect of bow thrusters for performance vessels is that their tunnel openings can be both in and out of the water as the bow of craft lifts with increasing speed and pitches in response to waves.

Tunnel openings can be a source of spray coming onboard in windy conditions or—when partially submerged—induce air under the hull, which can interfere with engine-cooling water flow and functioning of transducers.

Results from model tests show that when a hull has transverse cylindrical, carefully faired thrusters completely submerged, resistance can be estimated by Equation 4-15 (Stuntz, 1964).

$$R_O = 0.07\,(\rho/2)\,v^2\,(\pi\,D_T^{\,2}/4) \qquad \text{Equation 4-15}$$

where v is in ft/sec, and D_T is the diameter of the thruster in feet and R_O is resistance in pounds.

With regard to "carefully faired" thrusters, I do not recommend external wedges or "eyebrows" forward of the cylindrical openings. These can be a significant source of spray when a hull is pitching in a seaway.

I do recommend scalloped openings, or a recess in the hull side aft of the thruster opening aligned with mean local flow direction which, in profile, is usually 30° to 45° above a static water plane through the transverse axis of the cylinder.

CHAPTER FIVE
Propulsion

Power Prediction

Up to this point, I have presented data sources and logic leading to the assumptions necessary to establish the hull-resistance data for making a power prediction for craft. Now is the time to put it together.

Keep these thoughts in mind when working with propulsors: The function of propulsors is to produce thrust to overcome hull resistance to move the craft relative to the water. Also, while engine power is converted to thrust by propulsors, selecting propulsors to absorb power at a particular rpm and speed does not necessarily yield the maximum speed potential of a craft.

The propulsors that I will address include submerged propellers, surface propellers, and waterjets. Coupled with engine power, these expand further into subsets of drive lines to become inboard craft with inclined shafts and tunnel drives, outboards, inboard-outboards, pod drives, and flush-inlet waterjets. I will, however, only describe methods to effect a speed-power prediction for submerged propellers, beginning with speed versus resistance.

Propulsive Data

After you have identified a means of predicting the speed-resistance relationship for a craft, next you must describe the interrelation of the hull and propulsor in order to properly include their characteristics. Propulsive data are the transfer functions that describe this interrelation, and there is little published information about this area of technology for high-performance vessels.

In 1966, Hadler described propulsive effects to be considered which interact with high-speed hull forms. In 1971, Hadler and Hubble together presented a very complete synthesis of the planing-craft propulsion problem for single, twin, and quadruple-screw configurations. This work reports computed values of $(1-W_T)$ and $(1-t)$ for various shaft angles and speeds.

In 1968, my colleagues and I reported experimental propulsive data obtained from full-scale trials of twin-screw craft. In addition to these sources, other model and full-scale experimental data for twin-screw craft have been collected and found to fall consistently within reasonable bounds while having some variation with speed.

In Figure 5-1, graphs of mean values of data reported in 1976 are provided with bandwidths of experimental scatter outlined as dashed lines.

The function of propulsors is to produce thrust to overcome hull resistance to move the craft relative to the water.

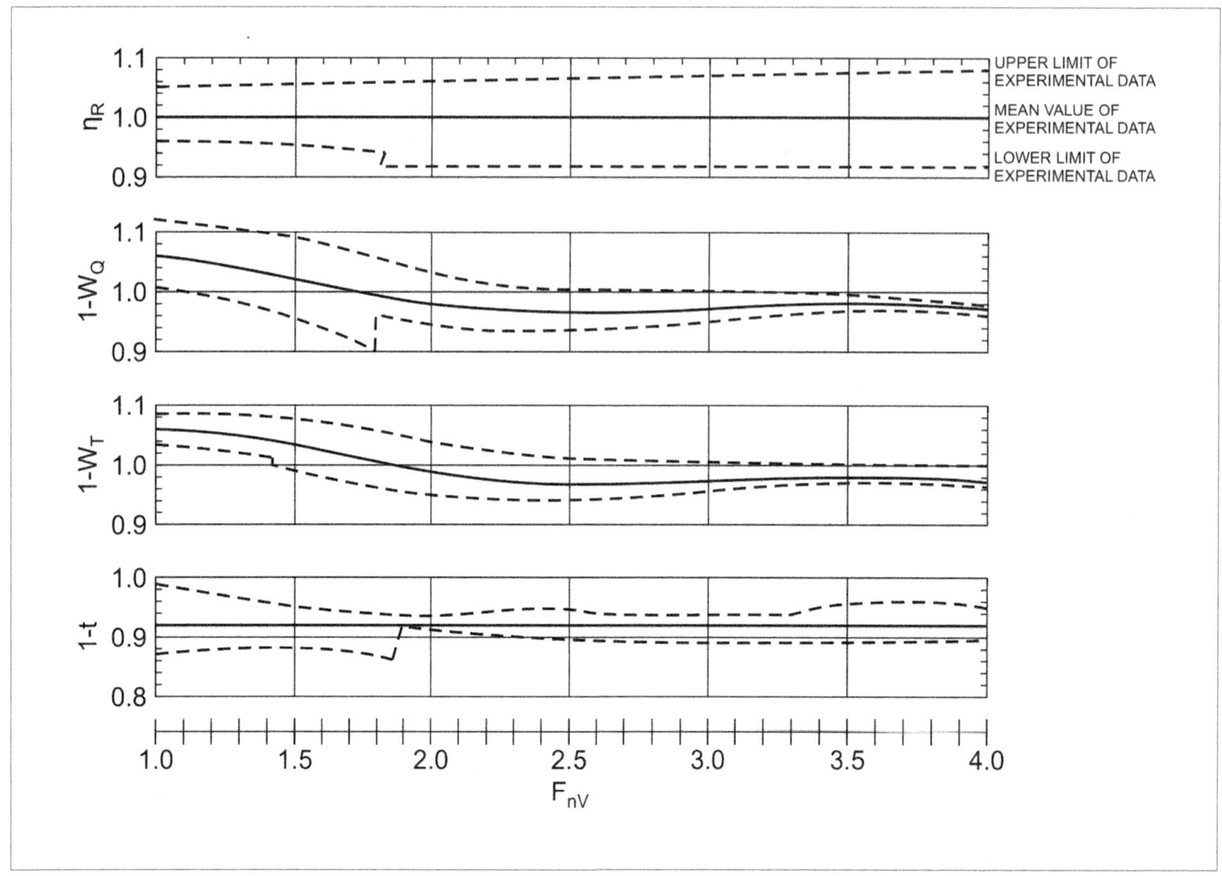

Figure 5-1: Twin-screw propulsive coefficients versus speed (F_{nV})

For these craft the range of shaft angles were from 10 to 16 degrees, measured relative to buttocks in the plane of the shaft line, which may be partially responsible for the bandwidth of data.

Data collected for various propulsors are reported in Table 5-A with mean values or observed variations. Three years earlier, I reported limited propulsive data for a single-screw small craft with a skeg.

To properly apply the thrust-deduction factor (1—t) reported in Table 5-A, and in Figure 5-1, you must understand the significance of the definition used here.

The thrust-deduction factor (1—t) reported here was experimentally obtained and computed as the ratio of appendaged resistance (i.e. the horizontal component of resistance force when towed in the shaft line) to total shaftline thrust when propelled at full-scale self-propulsion point.

Thus, this modified definition of (1—t) includes the effects *for the angle difference between the resistance and thrust vectors, the trimming effects, and resulting hull resistance change, due to the propeller pressure field acting on the hull, as well as similar trimming effects for propeller lift resulting from operation in inclined flow.*

This is different from using (1—t) in the classical sense, that is, which describes a resistance augmentation where the pressure field alone influences just hull-flow patterns.

TABLE 5-A: TYPICAL PROPULSOR HULL INTERACTIVE FACTORS FOR HIGH-SPEED CRAFT

Propulsion Concept		Speed Range								
		$F_{nL} \le 0.4$			$F_{nL} > 0.4$ to $F_{nv} \le 2.5$			$F_{nv} > 2.5$		
		W_T	t	η_R	W_T	t	η_R	W_T	t	η_R
Propeller on Inclined Shaft	6 deg shaft	0.01 To -0.02	0.01	0.97 To 1.01	0 To 0.04	0.01 To 0.02	0.97 To 1.01	0 To -0.10	0.03	0.97 To 1.01
	12 deg shaft	0.02 To -0.02	0.05	0.97 To 1.01	0.04 To -0.05	0.05 To 0.07	0.97 To 1.01	0.03 To -0.05	0.07 To 0.11	0.97 To 1.01
Propeller in Tunnel	40% d	-0.03	0.10	0.92	0.02 To -0.03	0.07 To 0.10	0.93 To 0.90	0.03	0.03 To 0.07	0.88 To 0.90
	65% d	-0.03	0.12	0.92	0 To 0.05	0.10 To 0.12	0.93 To 0.90	0.04 To 0.05	0.08 To 0.10	0.88 To 0.90
Outboard & Inboard/ Outdrive Propeller		0.03	0	0.97 To 1.01	0.03	0	0.97 To 1.01	0.03	0	0.97 To 1.01
Partially Submerged Propeller		0	0	0.97 To 1.01	0	0	0.97 To 0.98	0	0	0.97 To 1.01
Flush Inlet Waterjet		0 To 0.02	0.05 To 0.08	0.99	0.02 To 0.04	0.05	0.99	0.05	-0.02 To -0.07	0.99
Pod Drive - Pusher Propeller (Under Hull)		0	0 To 0.05	1.00	0	0 To 0.05	1.00	0	0 To 0.05	1.00
POD DRIVE - PUSHER PROPELLER (UNDER HULL)		0.05 To 0.07	0.05 To 0.07	0.97 To 1.01	0.05 To 0.07	0.05 To 0.07	0.97 To 1.01	0.05 To 0.07	0.05	0.97 To 1.01

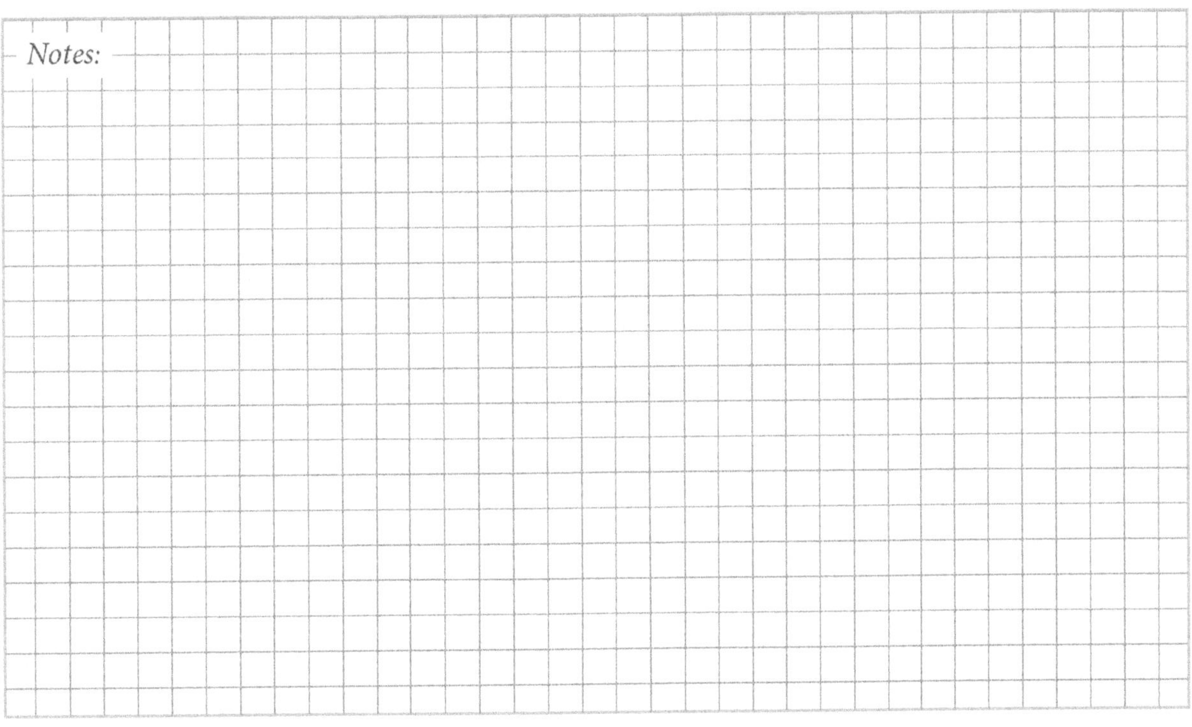

Propeller Characteristics with Cavitation

Many working craft operating at a fairly high speed and loading result in propellers likely to have some degree of cavitation which adds another dimension to their characteristics.

Operationally, this variable is most often reflected in a nonlinear speed-to-rpm relationship near top speed, generally detected as vibration or heard as a "gravel-passing-through-the-propeller" sound in the lazarette above the propellers. It may also appear as erosion of the propeller-blade material.

Cavitation must be accounted for because of changing propeller characteristics due to craft speed and thrust load, as reported by Gawn and Burrill (1957). Cavitation occurs when local pressure becomes low enough that the liquid transitions into a vapor phase. Figure 5-2 shows propeller thrust and torque coefficients changing with cavitation number (σ). The data represent flat-face propeller sections similar to most commercial propellers manufactured for small craft.

Gawn-Burrill propeller characteristics may be replicated by Blount and Hubble (1981) for open-water characteristics and with Blount and Fox (1978) for partially and fully cavitating conditions. Radojcic (1988) is an additional source for calculating Gawn-Burrill propeller characteristics for open-water and cavitating conditions.

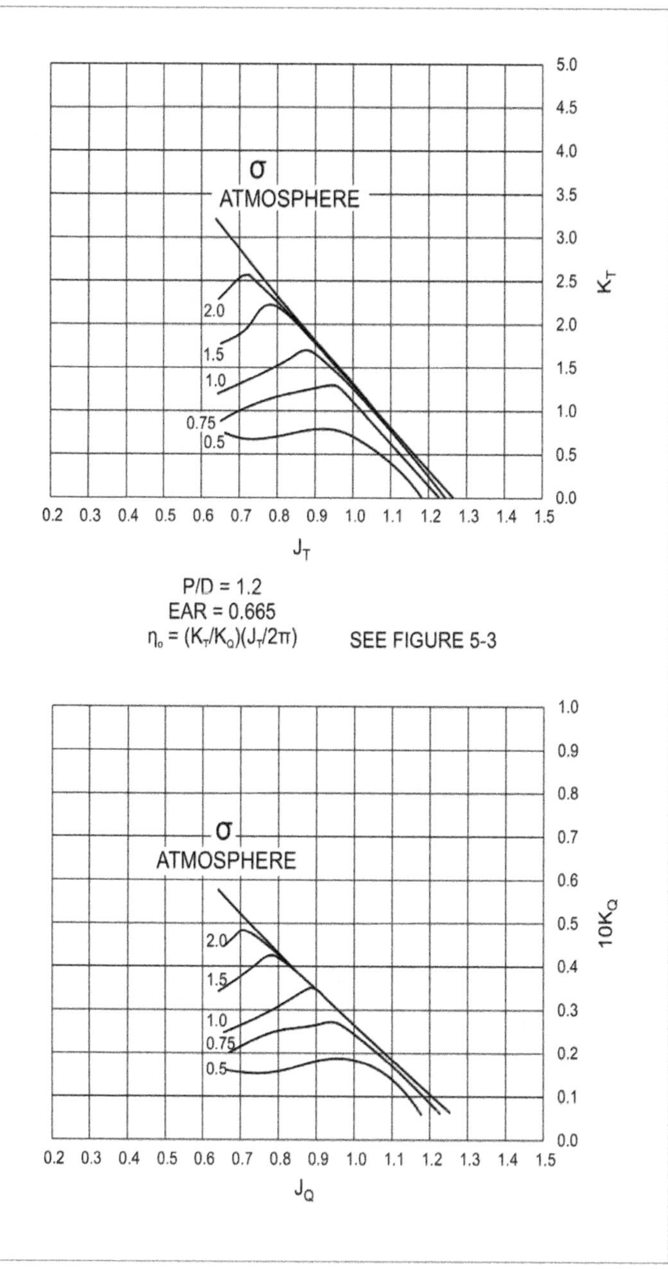

Figure 5-2: Effect of cavitation on propeller characteristics of K_T and K_Q

In 1981, Hubble and I compared cavitation data for four-bladed commercial propellers with the equivalent blade area as the three-bladed Gawn-Burrill propellers.

Our examination of these two sets of data indicates that the Gawn-Burrill criteria can be used to make engineering estimations of this type of commercial propeller performance when operating in a cavitating condition.

This determination is consistent with other researcher's experiences where they measured speed, power, and rpm on new craft, with one important exception, which is that the Gawn-Burrill data are optimistic when commercial propeller blade sections are thick and the manufactured leading edge is blunt with a poor quality of surface finish.

The Gawn-Burrill propeller characteristics are in the form of K_T, K_Q, η_O, versus J_T for various pitch ratios;

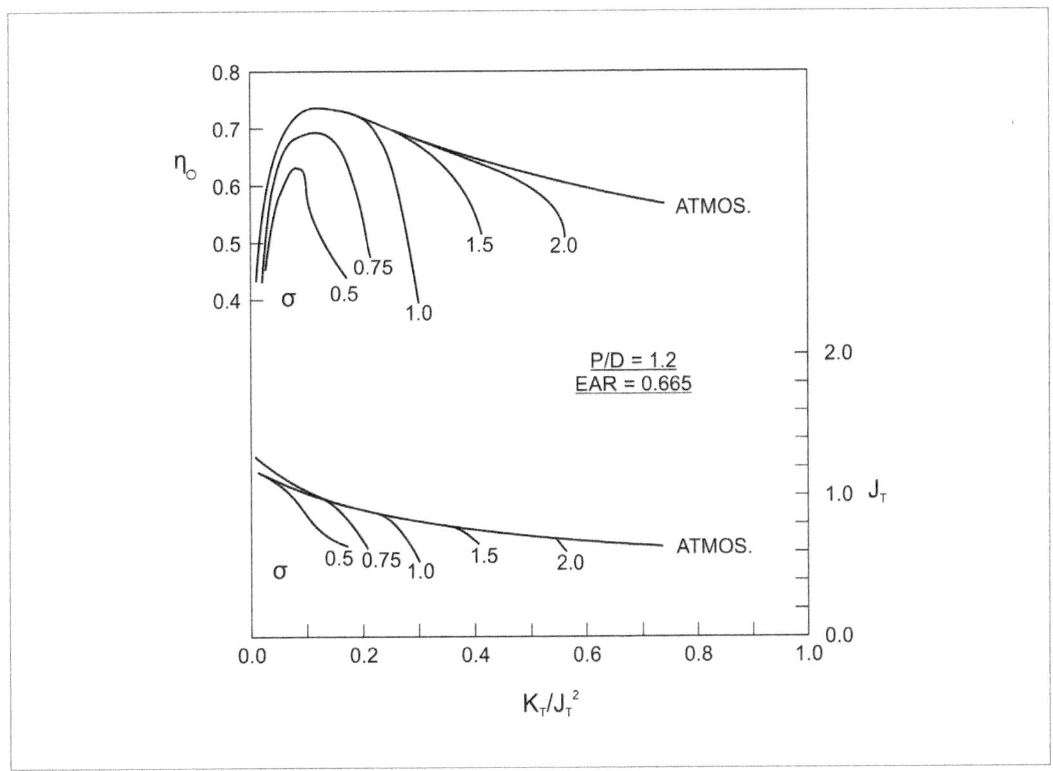

Figure 5-3: Effect of cavitation on propeller efficiency and J_T versus K_T/J_T^2

blade-area ratios, and the cavitation numbers (σ). An example of this format for one Gawn-Burrill propeller is shown in Figure 5-2 without η_O but which is included in Figure 5-3.

The familiar format of K_T, K_Q, η_O, versus J_T can be replaced with another graphic composition which reduces the effort required to optimize the propeller, select the gear ratio, and make the speed-power-rpm predictions.

This alternate format is η_O and J_T versus K_T/J_T^2. The characteristics for the propeller in Figure 5-2 are presented in this alternate format in Figure 5-3.

The entire Gawn-Burrill propeller series for a range of P/D, EAR (expanded area ratio) and σ is presented in this format in Appendix 5.

The effort required to re-compute these characteristics is compensated for by the time savings in preparing performance predictions and optimizing propulsion systems.

There is much to be learned from yet another format of propeller-characteristic data which includes cavitation effects. This format provides the relationship of dimensionless thrust-load coefficient (τ_C) versus cavitation number based on the resultant velocity at the 0.7 radius of the propeller ($\sigma_{0.7R}$) [Equations 5-19 and 5-20 are defined on Page 148.]

Seen in Figure 5-4, the format is consistent with various references defining the limits of propeller thrust load (τ_C). Contours of advance coefficient (J_T) and cavitation number at constant vessel speed are also included.

You will also find this format to be helpful when analyzing sea-trial data from under-performing boats. Referring back to Chapter 1, ask yourself: "If Charles Parsons had had propeller cavitation characteristics in this format in 1894, would the under performance due to thrust breakdown of steam turbine-powered *Turbinia* at 19.7 knots have been initially avoided?" The answer is "Yes."

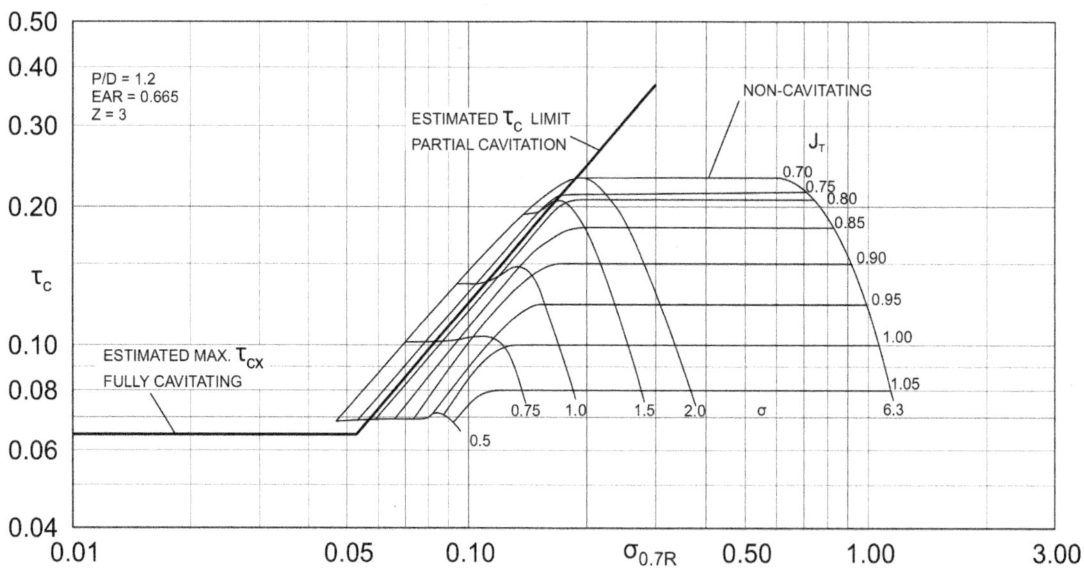

Figure 5-4: Changes of propeller characteristics during different phases of cavitation development

In fact, *Turbinia* did attain 34.5 knots in 1897 when additional propeller blade area was fitted to correct the previous thrust breakdown problem. For designing advanced craft, your understanding of this third format is essential in establishing minimum blade area to efficiently use propulsion power. Propellers having different blade section shapes can accommodate different thrust loads (τ_C).

Table 5-B gives empirical equations derived from experimental data to define both partial and maximum thrust load coefficients (τ_C) and maximum torque load coefficients (Q_C). See Equation 5-24 for definition of Q_C.

Propeller equations represent different blade-section shapes: flat-face, airfoil, transcavitating, and supercavitating. Figures 5-5a and 5-5b provide comparisons of τ_C and Q_C versus $\sigma_{0.7R}$ of the equations in Table 5-B for P/D = 1.2 when propeller cavitation is present.

Flat-face and airfoil blade section propellers are essentially identical regarding partial loading until the back has a full cavity, then the flat-face sections continue to develop thrust as τ_C becomes constant with further increase of power and rpm (reducing $\sigma_{0.7R}$). Transcavitation and supercavitating blade sections generate the greater part of thrust and absorb torque with the positive pressure of the contoured surface shape of the blade's face. Thus, these blade sections develop the greatest thrust load at very high speeds.

TABLE 5-B: EQUATIONS FOR PARTIAL AND MAXIMUM THRUST AND TORQUE LOAD COEFFICIENTS DUE TO CAVITATION		
Propeller Series	**Equations**	**Range of Application**
Gawn-Burrill (1957) Flat Face	$\tau_C = 1.2\ \sigma_{0.7R}$ $\tau_{CX} = 0.0725(P/D) - 0.034\ EAR$ $Q_C = 0.2(P/D)\sigma_{0.7R}^{X}$ where $X = 0.7 + 0.31(EAR)^{0.9}$ $Q_{CX} = [1/(EAR)^{1/3}]\ [0.0185(P/D)^2 - 0.0166(P/D) + 0.005941]$	$\tau_C = \tau_{CX}$ and $\sigma_{0.7R} \leq 0.4$ $Q_C \geq Q_{CX}$ and $\sigma_{0.7R} \leq 0.4$
Wageningen B (1969) Van Lammeren et al. Airfoil	$\tau_C = 1.2\ \sigma_{0.7R}$ $Q_C = [-0.0167 + 0.247(P/D)]\sigma_{0.7R}^{1.04}$	$0.03 \leq \sigma_{0.7R} \leq 0.2$ $0.03 \leq \sigma_{0.7R} \leq 0.2$
Newton-Rader (1961) Transcavitating	$\tau_C = [0.703 + 0.25(P/D)]\ \sigma_{0.7R}^{\{0.65 + 0.1(P/D)\}}$ $\tau_{CX} = -0.0142\ EAR + 0.0833(P/D)$ $Q_C = [-0.12 + 0.24(P/D)]\sigma_{0.7R}^{\{0.5 + 0.165(P/D)\}}$ $Q_{CX} = -0.024(EAR)^{1/2} + 0.0335(P/D)$	$\tau_C \geq \tau_{CX}$ and $\sigma_{0.7R} \leq 0.4$ $Q_C \geq Q_{CX}$ and $\sigma_{0.7R} \leq 0.4$
Supercavitating CRP (1962) Hecker et al. Supercavitating	2 blades: $\tau_C = 0.43(P/D)\ \sigma_{0.7R}^{0.26}$ $Q_C = 0.067(P/D)^2\ \sigma_{0.7R}^{0.26}$ 3 blades: $\tau_C = 0.44(P/D)\ \sigma_{0.7R}^{0.35}$ $Q_C = 0.072(P/D)^2\ \sigma_{0.7R}^{0.36}$ 4 blades: $\tau_C = 0.39(P/D)\ \sigma_{0.7R}^{0.39}$ $Q_C = 0.063(P/D)^2\ \sigma_{0.7R}^{0.40}$	$0.02 \leq \sigma_{0.7R} \leq 0.5$ $0.03 \leq \sigma_{0.7R} \leq 0.5$ $0.04 \leq \sigma_{0.7R} \leq 0.5$

Notes:

Reality of Propeller Loading

The maximum thrust limit is a tenuous design criterion as demonstrated performance under laboratory conditions is not always achieved during craft operations.

It may seem paradoxical to employ data obtained under controlled testing conditions to establish limits and then immediately question their validity. Model propeller testing is in no way questioned, as most successful marine vehicles with propellers designed to minimize cavitation perform as predicted. The point to be made is that propellers operating with significant cavitation seldom produce the thrust or absorb the power predicted at the expected advance coefficient (J). In general, higher propeller RPMs than predicted are required to develop thrust or to absorb the desired power, or both, when significant cavitation is present. The reasoning for this variance may be traced to essentially air-free water used in laboratory test facilities to achieve consistent results while saturated-air content in seawater near the surface results in premature cavitation as well as a change in propulsive coefficients (1–t), (1–W_T), and (η_R).

The maximum propeller load limits given in Table 5-B are not generally attained in full-scale operation (as seen in Figures 5-5a and 5-5b for flat-face Gawn-Burrill type propellers).

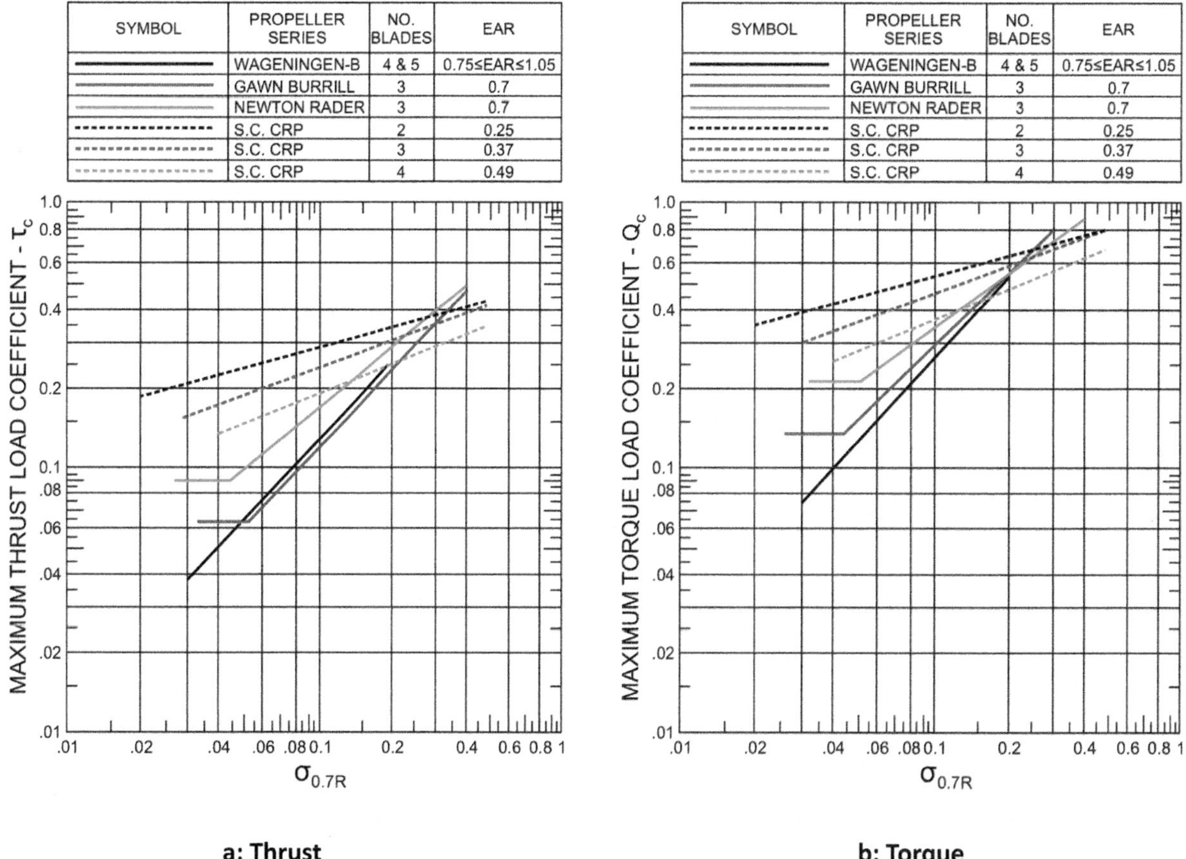

Figures 5-5a & 5-5b: Comparisons of maximum thrust and torque-load coefficients for P/D 1.2

Similar speed-power-rpm data for Newton-Rader propellers (cambered face) support this trend. For this reason, propeller dimensions should be selected for no more than 80% of the maximum values given for thrust and torque load coefficients (τ_C and Q_C) when significant cavitation is likely. This recommended design criterion is based only on possible attainment of predicted performance (modest risk). Erosion-free criteria in this transcavitation region will be established by full-scale experience.

Logic for Propeller Characteristics Format Change

The development of K_T/J_T^2 as the common variable between hull thrust requirements and the propeller characteristics has the distinct advantage of eliminating propeller rpm from the early prediction calculations.

For each propeller

$$K_T = T_P/(\rho n^2 D^4) \quad \text{Equation 5-1}$$

and $$J_T = [v(1 - W_T)/nD] \quad \text{Equation 5-2}$$

Therefore propulsive thrust for each propeller

$$K_T/J_T^2 = T_p/[\rho D^2 v^2 (1 - W_T)^2] \quad \text{Equation 5-3}$$

K_T/J_T^2 and the thrust load coefficient (τ_C) are related by

$$\tau_C = 2.55\ (K_T/J_T^2)/(J_T^2 + 4.84)EAR(1/J_T^2)[1.067 - 0.229(P/D)] \quad \text{Equation 5-4}$$

Hull resistance and total propulsion thrust are related by

$$T = R_T/(1 - t) \quad \text{Equation 5-5}$$

Assuming that each propeller produces the same amount of thrust for a multiscrew craft, the thrust required by each propeller would be:

$$T_P = R_T/(1 - t)\ N_{PR} \quad \text{Equation 5-6a}$$

and

$$T = T_P(N_{PR}) \quad \text{Equation 5-6b}$$

The thrust requirement each propeller must produce to satisfy total hull resistance leads to:

$$K_T/J_T^2 = R_T/[\rho D^2 v^2 (1 - W_T)^2 (1 - t)(N_{PR})] \quad \text{Equation 5-7}$$

If unequal thrust is produced by each of several propellers of a multi-screw craft (N_{PR}) should be omitted from equations 5-6 and 5-7. Then revised equation 5-7 should be multiplied by the percentage of thrust each propeller is expected to produce to obtain the K_T/J_T^2 value for designing that propeller. The total percentage of thrust for multi-screw craft must equal 100%. An example for a triple-screw craft is that it might have 30% of thrust by each of two outboard propellers and 40% by the centerline propeller.

Once the number of propellers has been established for a craft design, the only significant variables that can influence K_T/J_T^2 are the propeller diameter D with EAR being the second most important factor. Equation 5-7 equates hull resistance with required propulsor thrust, which is the basis for the format change in propeller characteristics as presented in Appendix 5. This format can be used for any type of propeller such as the Gawn-Burrill series, Wageningen B-Screw series, and supercavitating CRP series.

Once the design thrust loading (K_T/J_T^2) has been established, the equilibrium condition between hull requirements and propeller capacity leads, in general, to a unique value of open-water propeller efficiency (η_O) and advance coefficient (J_T) for EAR, craft speed and corresponding cavitation number σ:

$$\sigma = [P_A + P_H - P_V]/(1/2)\rho v^2 \quad \text{Equation 5-8}$$

The appendaged propulsive coefficient is computed as

$$\eta = \eta_O \, \eta_H \, \eta_R = EHP/SHP \quad \text{Equation 5-9}$$

where

$$\eta_H = (1 - t)/(1 - W_T) \quad \text{Equation 5-10}$$

The total shaft horsepower (SHP) is computed from total EHP where capital "V" is speed in knots vice lower case "v" which is ft/sec.

$$EHP = R_T V/325.9 \quad \text{Equation 5-11}$$

By the following equation

$$SHP = EHP/\eta \text{ (SAE power definition)} \quad \text{Equation 5-12}$$

The corresponding value of J_T defines the propeller rpm (N) as

$$N = V(1 - W_T)\,(101.3)/J_T D \quad \text{Equation 5-13}$$

Most designers of small craft are familiar with bare-hull propulsive coefficient which is commonly referred to as overall propulsive coefficient (η_D, or OPC). The magnitude of 0.5 for η_D has been used for years for preliminary power estimates. For current design practice a value of $\eta_D = 0.55$ is readily attainable for twin- and triple-screw craft with propellers on inclined shafts. I mention this to emphasize that the propulsive coefficient (η) in Equation 5-9 and η_D are different.

Bare hull EHP is

$$EHP_{BH} = R_{BH} V/325.9 \quad \text{Equation 5-14}$$

used to compute bare hull or overall propulsive coefficient as follows:

$$\eta_D = EHP_{BH}/SHP \quad \text{Equation 5-15}$$

The difference between equations 5-9 and 5-15 for smooth-water conditions is mostly a result of the appendage drag factor (η_A) with minor effect due to the tendency of propeller efficiency to be reduced with increasing thrust loading (K_T/J_T^2) which becomes very important should the propeller begin to cavitate. Thus, for moderate speeds and thrust loading in smooth water.

$$\eta \approx \eta_D/\eta_A \quad \text{Equation 5-16}$$

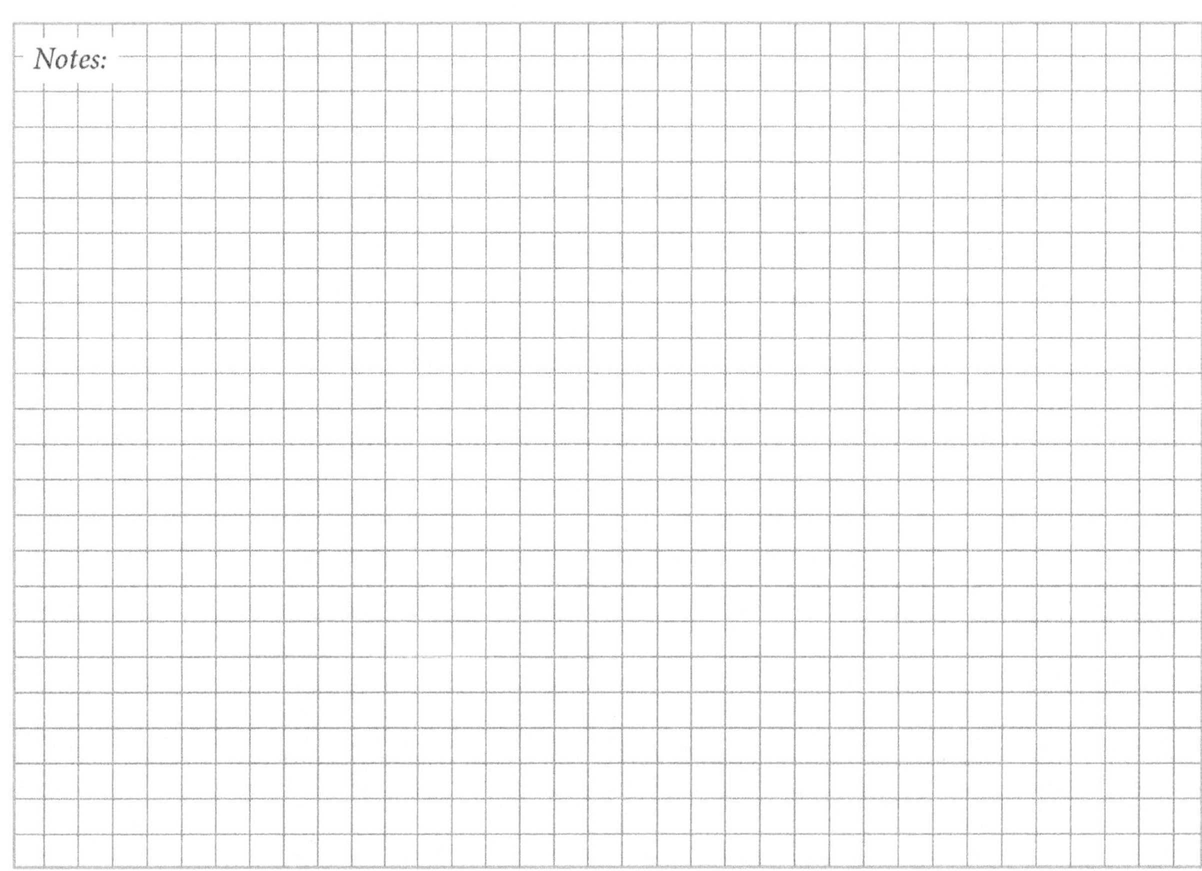

Speed-Power Prediction

The speed-power calculation procedure, applying the approach discussed, is best demonstrated by following through a data calculation form. The sample form with column-by-column calculation procedures as given in Tables 5-C and 5-D will show the interrelationships of hull resistance propulsive data, and propeller characteristics.

The numerical example depicts a 65-foot, 100,000-lb craft operating in rough water. For this example, the following assumptions are used. The boat will have two 1,600 BHP (1,193 kW) diesel engines rated at 2,300 rpm, driving fixed-pitch propellers on inclined shafts. The stock propellers have D = 36 inches, P/D = 1.20, Z = 3, and EAR = 0.82.

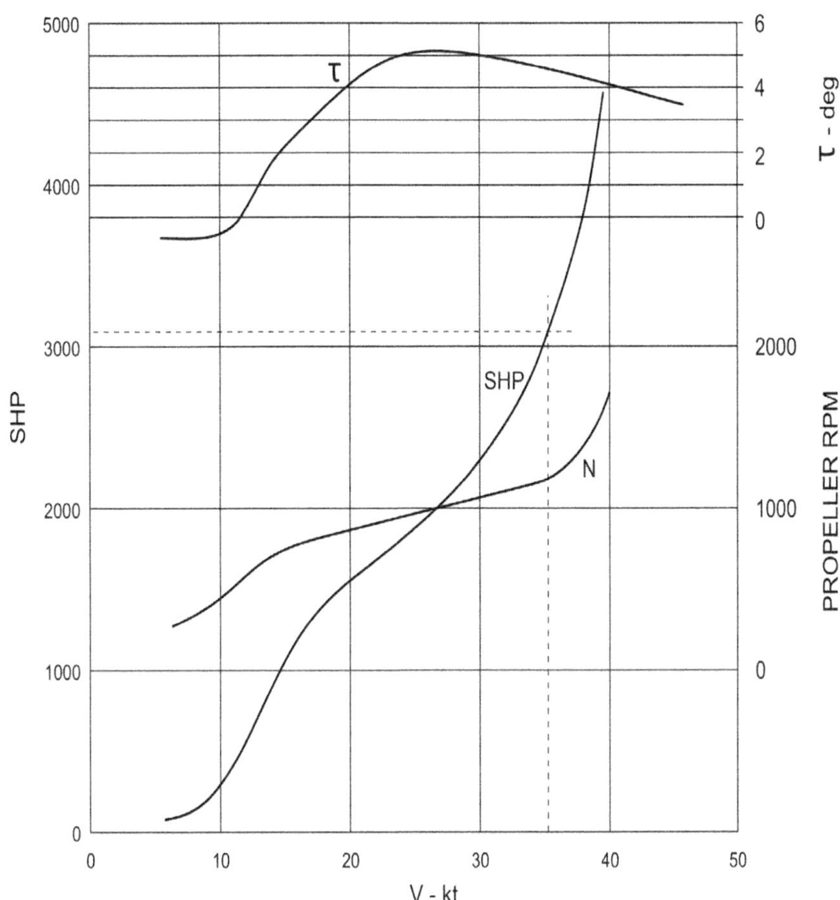

Figure 5-6: Speed-power prediction of total SHP, propeller rpm, and trim for the example from Table 5-C

It is important to note that engine characteristics play no part in the speed-power requirements—other than impact of machinery and fuel weight on total displacement—once hull geometry and loading, LCG, appendages, and propeller geometry are fixed. Since the propeller controls the engine power output at a given rpm up to the maximum power capability of the engine, you must select an engine-and-gear ratio with characteristics compatible with predicted power and propeller rpm versus speed for fixed-pitch propellers. I will discuss the selection of propeller dimensions and gear ratio later in this chapter.

TABLE 5-C: SPEED-POWER CALCULATIONS

Craft: 3-200 Date: 08/13 Calculated By: DLB

Displacement: 100,000 lb LCG: 27.3 ft fot B_{PX}: 14.9 ft β: 12°

Water Characteristics: ρ: 1.9905 lb sec²/ft⁴ ν: 1.226 x 10^5 ft²/sec h: 100 ft (deep) C_A: 0

Sea State: 2 Wind Speed: 12 kt $H_{1/3}$: 2.2 ft T_W: 4.8 sec

Propeller Style: flat face D: 36 in P: 43.2 in. P/D: 1.2 EAR: 0.82 Z: 3 No. of Shafts: 2 Propeller Centerline Depth: 4.5 ft

Line	1	2	3	4	5	6	7	8	9	10	11	12	13	14
	V	R	R_{APP}	R_A	R_{AIR}	R_T	F_{nV}	τ	1 - t	1- W_T	η_R	σ	D	K_T/J_T^2
1	5.72	660	34	396	172	1,260	0.50	-0.45	0.92	1.06	1.0	25.53	3.0	0.365
2	6.87	1,130	59	475	195	1,860	0.60	-0.49	0.92	1.06	1.0	17.70	3.0	0.374
3	8.47	2,300	122	586	230	3,240	0.74	-0.50	0.92	1.06	1.0	11.64	3.0	0.428
4	9.61	3,600	195	665	256	4,720	0.84	-0.55	0.92	1.06	1.0	9.04	3.0	0.484
5	11.33	5,880	328	783	299	7,290	0.99	0	0.92	1.06	1.0	6.51	3.0	0.538
6	14.30	10,230	607	989	380	12,210	1.25	2.35	0.92	1.05	1.0	4.08	3.0	0.577
7	17.16	11,270	716	1,187	466	13,640	1.50	3.02	0.92	1.04	1.0	2.84	3.0	0.456
8	22.77	12,580	928	1,574	663	15,740	1.99	4.90	0.92	0.98	1.0	1.61	3.0	0.337
9	28.61	12,890	1,138	1,978	905	16,910	2.50	5.10	0.92	0.96	1.0	1.02	3.0	0.239
10	34.79	13,070	1,378	2,405	1,201	16,680	3.04	4.75	0.92	0.97	1.0	0.69	3.0	0.156
11	39.82	13,410	1,645	2,753	1,473	19,280	3.48	4.30	0.92	0.98	1.0	0.53	3.0	0.135
12	45.89	13,520	1,980	3,173	1,838	20,510	4.01	3.67	0.92	0.97	1.0	0.40	3.0	0.110

Line	15	16	17	18	19	20	21	22	23	24	25	26
	η_O	J_T	P/D	η	EHP	SHP	N	V	EAR	P	EHP_{BH}	η_D
1	0.660	0.81	1.20	0.573	22	39	253	5.72				
2	0.655	0.80	1.20	0.568	39	69	307	6.87				
3	0.635	0.78	1.20	0.551	84	153	389	8.47				
4	0.625	0.75	1.20	0.542	139	256	459	9.61				
5	0.610	0.73	1.20	0.529	253	479	556	11.33				
6	0.600	0.71	1.20	0.526	536	1,019	714	14.30				
7	0.635	0.75	1.20	0.531	718	1,353	803	17.16				
8	0.660	0.81	1.20	0.620	1,100	1,775	930	22.77				
9	0.705	0.87	1.20	0.676	1,484	2,197	1,066	28.61				
10	0.710	0.96	1.20	0.673	1,780	2,644	1,187	34.79				
11	0.540	0.78	1.20	0.507	2,356	4,647	1,689	39.82				
12	---	---	1.20	---	---	---	---	45.89				

Comments:

For this example, I used the following:

V – Calculated from F_{nv}

R – Calculated from C_R for F_{nv} from Table 3-E

R_{APP} – Equation 4-7

R_A – Hoggard (1979)

R_{AIR} – Equation 4-2 for A_T = 222 ft², and β_{TW} = 0°

C_X = 0.73

τ – from model tests

1 - t, 1 - W_T & η_R from Figure 5-1

Sea water and air temperatures assumed to be 59°F (15°C)

TABLE 5-D: DEFINITIONS FOR CALCULATION PROCEDURE		
Col.	Data	Source
1.	Speed	Model tests, assume or calculate from σ or F_{nV}
2.	Bare-hull resistance	Model tests or predictions
3.	Appendage resistance	Computed from resistance sum of shafts, struts, rudders, etc. or estimate with Eq. 4-7
4.	Added resistance in waves (0 for calm water)	Model tests or Eq. 9-2, Hoggard (1979)
5.	Aerodynamic resistance	Computed with Eq. 4-2
6.	Total resistance	Sum of Col. 2, 3, 4, and 5
7.	Volume Froude No.	Computed with Equation A7-1 of General Notes and Definitions in Appendix 7
8.	Trim relative to buttock at $B_{PX}/4$	Model tests or analytical prediction
9.	Thrust deduction factor	Table 5-A or Figure 5-1
10.	Thrust wake factor	Table 5-A or Figure 5-1
11.	Relative rotative efficiency	Table 5-A or Figure 5-1
12.	Cavitation No.	Computed with Eq. 5-8
13.	Propeller diameter	Assumed values or design diameter
14.	Thrust loading	Computed with Eq. 5-7
15.	Propeller efficiency	Obtained from propeller characteristics for cavitation no. and K_T/J_T^2 at proper P/D and EAR (Appendix 5 for Gawn-Burrill props.)
16.	Advance coefficient based on thrust	Obtained from propeller characteristics for cavitation no. and K_T/J_T^2 at proper P/D and EAR (Appendix 5 for Gawn-Burrill props.)
17.	P/D	Assumed P/D for speed-power calculations or P/D for maximum propeller efficiency for K_T/J_T^2 when selecting best propeller
18.	Appendaged propulsive coefficient	Approximation computed with Eq. 5-9
19.	Total EHP	Computed with Eq. 5-11
20.	Shaft horsepower	Computed with Eq. 5-12
21.	Propeller rpm	Computed with Eq. 5-13
22.	Speed	Repeat of Col. 1
23.	EAR	Assumed, design or as necessary due to τ_C
24	Propeller pitch	Computed with Col. 17 X D
25	Bare-hull EHP	Computed with Eq. 5-14
26	Overall propulsive coefficient	Computed with Eq. 5-15

Total installed BHP = 3,200. Accounting for transmission losses, (η_T - 0.965), total SHP delivered to the propellers will be 3,088 SHP.

SHP = 0.965 (BHP) Equation 5-17

In Figure 5-6, the predicted speed is 35 knots for this condition with 1,200 propeller rpm. The reduction gear ratio will be (2,300/1,200) 1.917:1 for P/D = 1.20.

Should this performance equal or exceed your client's requirements, you might not consider further refinements except to better match P/D to the nearest available stock gear ratio.

Assuming the nearest stock gear ratio was 2.0:1, then P/D must change slightly in order to match the engine rating of 1,600 BHP at 2,300 rpm. The approximate P/D match would be 1.2(2.0/1.917) = 1.25.

Figure 5-6, however, does raise a warning about potential propeller cavitation as indicated above 35 knots by the increasing slopes of both the rpm-speed and SHP-speed predictions. The assumed propeller diameter of 36 inches with EAR = 0.82 appears to be marginal with regard to cavitation avoidance.

Based on results of this prediction, I would continue further design study to find a better propeller, providing a greater design margin, and for reducing shaft angle.

Also, if diameter cannot be increased, it would be worth considering partial tunnels recessed about 0.25D to have a reduced shaft angle appropriate for a speed of 35 knots.

Selecting Best Propeller and Gear Ratio

The "best propeller" for a craft is that one which satisfies the craft thrust requirements within geometric, power and financial limitations. If there were no design constraints, an "optimum propeller" could be designed for maximum efficiency and speed.

Because of this slight distinction of terminology, optimum propeller and best propeller, are not considered to be equivalent descriptors. We will work with the term "best propeller" here.

The geometric constraints to be considered may well preclude the selection of an operationally suitable propeller. Thus, it is important for you to establish the maximum propeller dimensions allowable for the shaft angle, tip clearance, and draft limitations.

Most craft have shaft angles in the range of 7 to 15 degrees measured relative to the hull buttock of the shaft line, and propeller tip clearance of 15 to 25 percent of diameter.

The smallest shaft angles and tip clearances are generally employed on craft with highest design speeds. However, tip clearances are controlled to a large extent by propeller-induced blade rate pressures, which are often traced to cavitation.

The key to selecting a best propeller to satisfy craft propulsion depends on equating required craft thrust with propeller developed thrust. Equation 5-7 defines the thrust-speed-propeller diameter relation required for equilibrium conditions, and the speed-power calculation procedure outlined in Tables 5-C and 5-D should be followed.

- First, assume three or more values for propeller diameter, one of which may exceed geometric constraints requiring tunnels.
- Next, as defined by Equation 5-8 assume three values of cavitation numbers corresponding to speeds above and below the design speed.
- Perform the calculations according to the procedure described in the numerical example in Tables 5-D and 5-E for each combination of speed and diameter.
- In Column 13, record the assumed propeller diameters used for the calculations.

Based on previous experience, a designer will usually have an estimate of the expanded area ratio of propellers on similar craft. If so, choose an equal or larger value of EAR available in the Gawn-Burrill series (Appendix 5) with the propeller characteristics for the remaining calculations.

As some propeller cavitation is likely for performance craft, I suggest that EAR $\geq$ 0.82 be the minimum value for calculations. For this example of a 65-ft (20 m) planing boat having a displacement of 100,000 lb (45.4 mt), I estimated hull resistance using C_R and other necessary data from Table 3-H Model for 2-300 data scaled with a linear ratio, λ = 7.613.

- For each line, use the values of K_T/J_T^2 and σ from Columns 14 and 12 to enter the appropriate propeller characteristics curves in Appendix 5, to locate the maximum value of efficiency (η_O) for that thrust loading and record the maximum η_O and corresponding (J_T) and (P/D) in Columns 15, 16, and 17 respectively.
- Then complete calculations and prepare a graphical solution in the format of Figures 5-7 (a & b).

TABLE 5-E: PROPELLER SELECTION

Craft: 3-200 Date: 08/13 Calculated By: DLB
Displacement: 100,000 lb LCG: 27.3 ft fot B_{PX}: 14.9 ft β: 12°
Water Characteristics ρ: 1.9905 lb sec²/ft⁴ ν: 1.226 X 10^{-5} ft²/sec h: 100 ft (deep) C_A: 0
Sea State: 2 Wind Speed: 12 kt $H_{1/3}$: 2.2 ft T_W: 4.8 sec
Propeller Style: flat face D: assumed P: --- P/D: --- EAR: 0.82 Z: 3
No. of Shafts: 2 Propeller Centerline Depth: 4.5 ft

	1	2	3	4	5	6	7	8	9	10	11	12	13	14
	V	R	R_{APP}	R_A	R_{AIR}	R_T	F_{nV}	τ	1-t	$1\text{-}W_T$	η_R	σ	D	K_T/J_T^2
1	28.9	12,900	1,136	1,991	918	16,940	2.52	5.7	0.92	0.97	1.0	1.00	2.67	0.290
2													3.00	0.230
3													3.33	0.186
4	33.4	13,000	1,315	2,307	1,131	17,750	2.92	5.4	0.92	0.98	1.0	0.75	2.67	0.223
5													3.00	0.176
6													3.33	0.143
7	40.9	13,300	1,682	2,820	1,535	19,340	3.57	4.9	0.92	0.97	1.0	0.50	2.67	0.165
8													3.00	0.131
9													3.33	0.106

	15	16	17	18	19	20	21	22	23	24	25	26
	η_O	J_T	P/D	η	EHP	SHP	N	V	EAR	P	EHP_{BH}	η_D
1	0.65	0.83	1.2	0.62	1,502	2,423	1,281	28.9	0.82	3.2		
2	0.71	0.88	1.3	0.67		2,242	1,076			3.9		
3	0.73	0.94	1.4	0.69		2,177	907			4.7		
4	0.62	0.84	1.4	0.58	1,819	3,136	1,478	33.4	0.82	3.7		
5	0.73	1.11	1.5	0.68		2,654	996			4.5		
6	0.74	1.22	1.6	0.69		2,618	816			5.3		
7	0.48	0.77	1.4	0.46	2,427	5,276	1,955	40.9	0.82	3.7		
8	0.56	0.97	1.4	0.53		4,569	1,381			4.2		
9	0.68	1.17	1.5	0.64		3,792	1,032			5.0		

Comments:

For this example, I used the following:

R_{APP} — Equation 4-7

R_A — Hoggard (1979)

R_{AIR} — Equation 4-2 for A_T = 222 ft²

β_{TW} = 0°

C_X = 0.73

These data are plotted as shp, rpm, and pitch versus speed for curves of constant propeller diameter.

- Construct a horizontal line at the installed power level that intersects the predicted speed-power curves.
- Construct vertical lines passing through each speed-power intersection point up to the speed-rpm and speed-pitch curves for the corresponding propeller diameter.

These intersecting points are then plotted on a base of propeller diameter. That is, (1) speed versus diameter at design power; (2) rpm versus diameter at predicted speed for design power; and (3) pitch versus diameter at predicted speed for design power. This process is illustrated in Figures 5-7a & 5-7b with intersection points identified in both graphs. The ratio of propeller rpm to engine rpm yields the desired reduction ratio. Slight adjustments in propeller pitch are usually required to match stock gear ratios, as previously shown with the results of speed-power prediction.

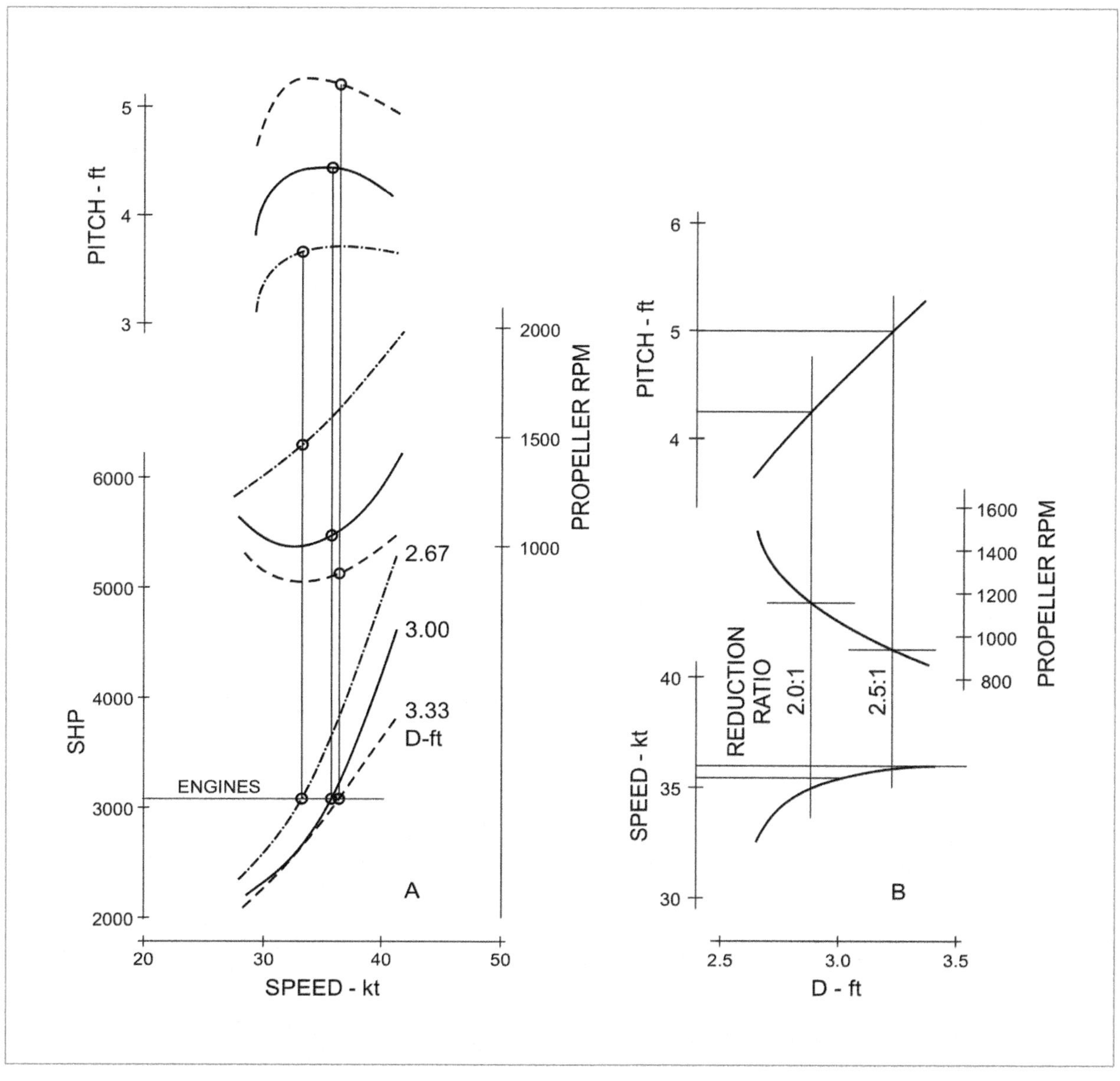

Figures 5-7a & 5-7b: Plot of calculation example for propeller and reduction ratio selection

The results of this best propeller calculation for EAR = 0.82 only show a speed gain of 0.5 knot with D = 38.4 in. This is a speed improvement with an ungainly impact on shaft angle and overall craft design. To show the influence of EAR on the best propeller, calculations in Table 5-E were repeated (but not included in the book) for EAR = 1.00. Without adjusting pitch for stock gear ratios, the results are compared in the following table:

EAR	Speed (knots)	D (inches)	P (inches)	P/D	Reduction Ratio
0.82	35.5	38.4	58	1.53	2.37:1.0
1.00	37.0	36.0	49	1.36	2.00:1.0

Tables 5-C and 5-E are defined for a spreadsheet format with interim manual input from propeller cavitation characteristics obtained from experimental sources, regression analysis of propeller series tests, or complex analytical computer code. (This calculation procedure is defined in Table 5-D.)

A design office with limited resources may wish to use a fourth alternative suggested in 1981 by Blount and Hubble to supplant the manual input of propeller cavitation characteristics. This approach combines the equations from Table 5-B with a regression analysis of propeller open-water series, such as that by Gawn-Burrill or the Wageningen B series.

Figure 5-8 provides a comparison of test data for one Gawn-Burrill propeller with the partially and fully cavitating characteristics derived from Table 5-B equations. A comparison for most of this series of propellers is provided in the 1981 Blount and Hubble reference.

This method is suitable for feasibility studies and initial propeller sizing for $0.8 \leq P/D \leq 1.4$ and $\sigma \leq 2.0$. Equations 5-4, 5-8, 5-20, and 5-27, with the assumptions that $(1\text{-}W_T) = 1.0$ and $J_T = J_A = J_Q$ are used to transform the equations from Table 5-B into K_T, K_Q, and η_o versus J_T format.

While the procedure for selecting the best propeller and gear ratio establishes the combination of propeller diameter and pitch for the assumed EAR, it does not confirm that blade area is adequate relative to cavitation effects other than from a performance point of view.

Blade rate-induced hull pressures on the order of 4 to 5 psi can be generated by a badly cavitating propeller, and can fatigue and crack metal hull plating after short periods of operation. In addition, these blade cavities can be destructive to the propeller, eroding blade material to the point of requiring frequent propeller replacement.

These effects can be minimized by selecting EAR such that τ_C (a thrust loading coefficient related to blade pressure) does not exceed the 10 percent back cavitation relationship defined by Gawn-Burrill (1957) so that the propeller loading is in agreement with the following approximation:

$$\tau_C \leq 0.494\,(\sigma_{0.7R})^{0.88} \qquad \text{Equation 5-18}$$

τ_C and $\sigma_{0.7R}$ take into account the resultant of both rotational and axial velocities and are computed as follows:

$$\tau_C = T/[1/2(\rho A_B)(v_{0.7R})^2] \qquad \text{Equation 5-19}$$

$$\sigma_{0.7R} = [\sigma/(1-W_T)^2]\,[J_T^{\,2}/(J_T^{\,2} + 4.84) \qquad \text{Equation 5-20}$$

Full-scale tests of planing craft have confirmed that advances in propeller design and manufacturing technology indicate that Equation 5-18, thrust load criteria (τ_C) can be increased. I now use Equation 5-21 vice 5-18 as the criteria to establish diameter and EAR for fully submerged propellers.

$$\tau_C \leq 0.60\,(\sigma_{0.7R})^{0.88} \qquad \text{Equation 5-21}$$

Both EAR and diameter can be changed to increase projected blade area (A_B) in order to reduce τ_C to acceptable thrust loading defined by Equation 5-21 by using the following equation.

$$A_B = EAR(\pi D^2/4)[1.067 - 0.229(P/D)] \qquad \text{Equation 5-22}$$

When EAR exceeds 0.90, the number of sources of available stock propellers decrease. For alternatives, you have the option of preparing or ordering a custom propeller design from a manufacturer or obtaining relief from geometric constraints by using propeller tunnels for advanced craft. Increasing shaft angles to permit use of a larger-diameter propeller to reduce τ_C to acceptable loading often ends with disappointment.

Limitations Due to Propeller Loading

The ability of a given propeller to develop thrust and absorb power has finite limits under specific operating conditions. This limit requires quantification to establish minimum propeller diameter and blade area.

Gawn-Burrill provided, very nicely, limits based on dimensionless thrust (τ_C) and speed ($\sigma_{0.7R}$) coefficients. Few other sources report these limits. Using this familiar format, you will need to establish maximum and recommended thrust limits for the other propeller series data when cavitation becomes significant. As noted previously, the maximum thrust limit is a tenuous design criterion, a realistic criterion needs to be established for each propeller type, much as the 10 percent back cavitation curve is often used as an erosion-free design goal for the Gawn-Burrill Series.

Under some circumstances, such as analyzing full-scale trial data, it is necessary to make assumptions to go from measured torque and rpm data to thrust. To ascertain if thrust limits or thrust breakdown are being approached, here is an alternative to estimating performance limits, dimensionless torque, and speed relations for a propeller: If thrust limits are definable, you can deduce torque limits, neglecting scale effects. Using a form analogous to τ_C to derive a torque load coefficient (Q_C).

$$\tau_C = T/(1/2)\rho A_B v_{0.7R}^{\,2}] \qquad \text{Equation 5-23}$$

So in a similar manner let

$$Q_C = Q/(1/2)\rho D A_B(v_{0.7R})^2] \qquad \text{Equation 5-24}$$

By definition

$$A_B = EAR(\pi D^2/4)(1.067 - 0.229\,P/D) \qquad \text{Equation 5-25}$$

$$(v_{0.7R})^2 = v_A^{\,2} + (0.7\pi n D)^2 \qquad \text{Equation 5-26}$$

Substituting into equation 5-24 results in

$$Q_C = 2.55K_Q/[(J_Q^{\,2}+4.84)EAR(1.067 - 0.229P/D)] \qquad \text{Equation 5-27}$$

The 0.7 radius cavitation number ($\sigma_{0.7R}$) was defined earlier, in Equation 5-20. Maximum torque load coefficients (Q_C) are provided in Table 5-B and Figure 5-5b for various propeller series.

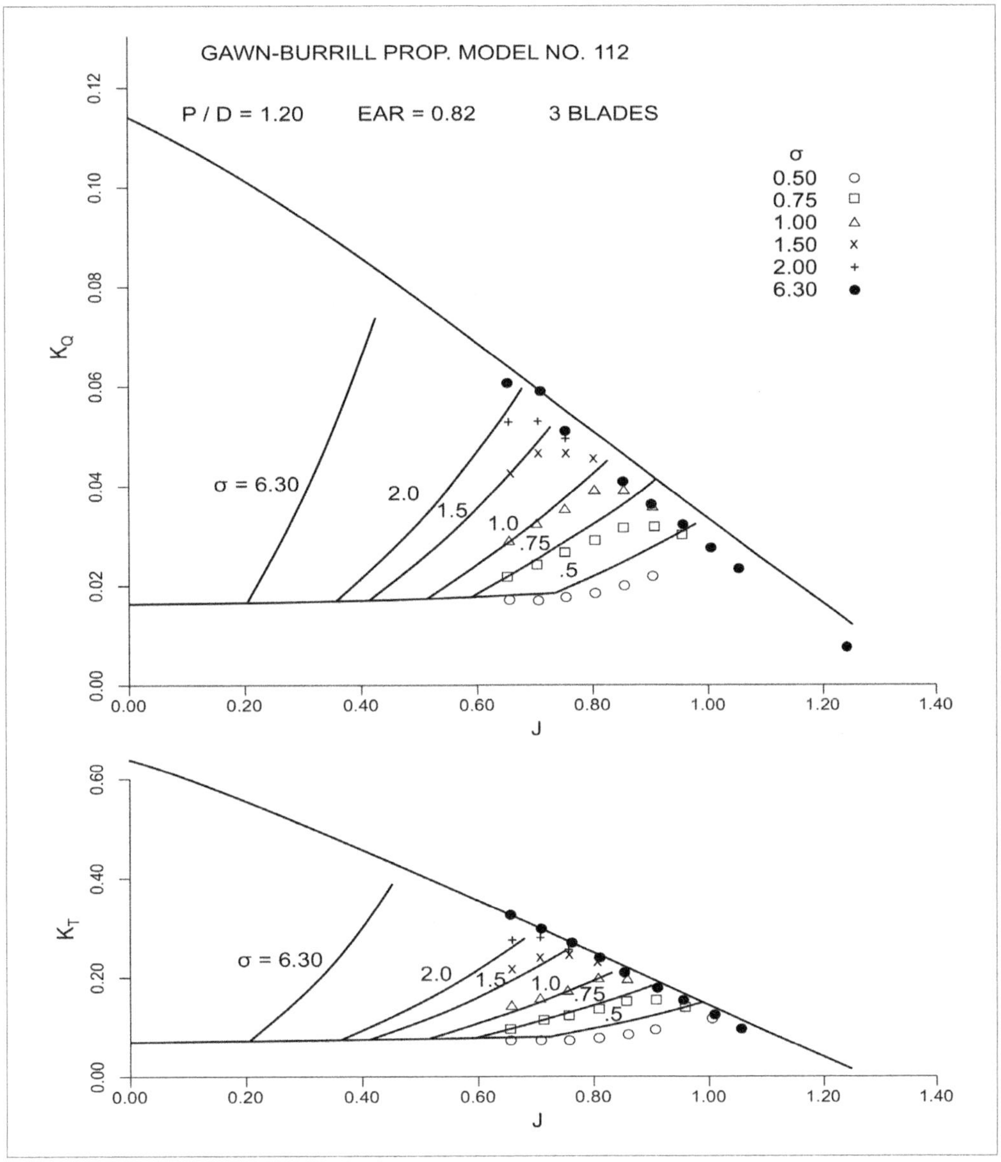

Figure 5-8: A comparison of Gawn-Burrill data with cavitation characteristics calculated for propeller used in Table 5-C with Table 5-B equations

Notes:

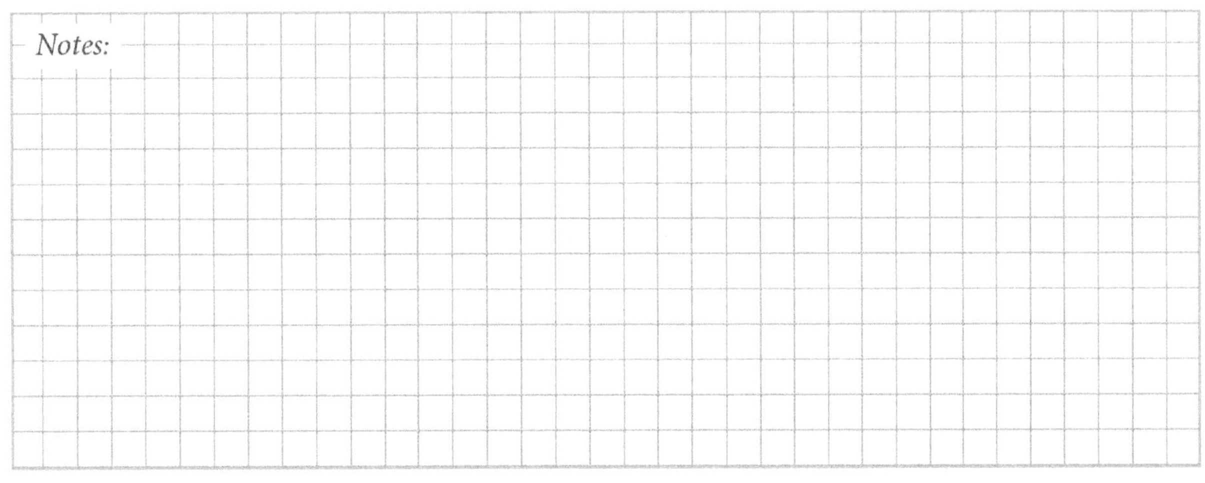

Number of Propulsors

The number of propulsors for a craft may be fixed by client requirements; however, there are technical considerations which may strongly suggest the best combination is from one to six powered shafts, and "best" can have more than one meaning. "Best" can mean fastest speed for total installed power or greatest range for quantity of fuel used at full power, but there may be operating speed gaps which are not possible due to engine characteristic or propulsor limitations, and increasing the number of engines can result in increasing craft availability.

For the same total power, the cost of purchasing multiple low-horsepower engines can be less than a few high-horsepower engines. Conflicting client requirements or classification rules can also be detailed to the degree that one may prohibit the attainment of the other. Therefore, you need to be prepared to educate and guide your clients about how multiple numbers of engines may affect powering requirements.

Figure 5-9 provides some technical guidance for studying and discussing the effects of the number of powered propulsors for craft. Using prediction of total resistance (R_T) versus speed for a craft Equation 5-7, K_T/J_T^2 can be computed for each speed. Then, for the total craft thrust, the following approximation is useful for a trend analysis of the effect of the number of powered propulsive shafts (N_{PR}).

$$K_T/J_T^2 \propto T/D^2V^2 \qquad \text{Equation 5-28}$$

Since ρ is constant, $(1\text{-}W_T)^2$ only changes slightly with speed and let $N_{PR} = 1$ so Equation 5-28 represents the baseline condition of a single shaft. Simplifying this example further, we may assume D = 1 ft since the only interest is of the change in T/V^2 with speed, which then results in

$$K_T/J_T^2 \propto T/V^2 \qquad \text{Equation 5-29}$$

For the purpose of discussion, three speeds are indicated: hump speed, design speed A, and design speed B. The number of shaft lines are noted at hump speed as 1 (single propulsor) through 5 (indicating five engines and propulsors). Advanced craft have two significant design points in their operational speed range: a low speed, high-thrust loading condition at hump as support of craft weight transitions from buoyant to dynamic lift, and the other at high speed when weight is primarily supported by dynamic lift. The critical hump speed thrust loading condition must be met and exceeded or there will be no advanced-performance craft. The design solution at hump speed may be different

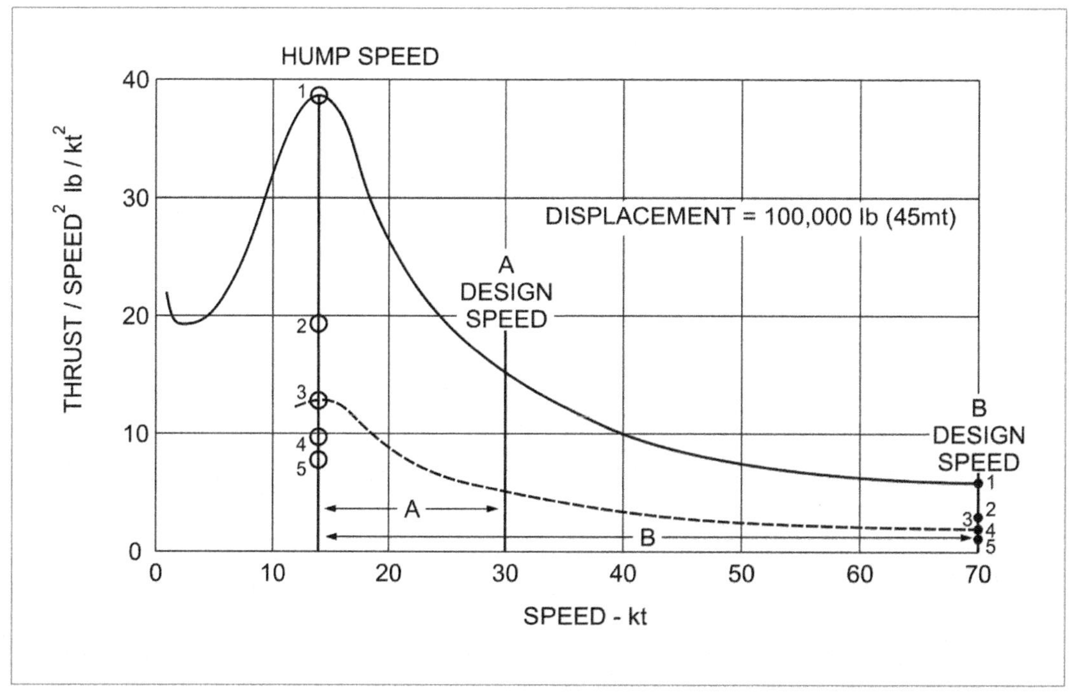

Figure 5-9: Example of the effects of reducing thrust loading by increasing the number of propulsors

considering whether the high-speed requirement is close to hump as design speed A or very much higher as design speed B.

The difficulty of designing for the hump speed has two parts: low rpm torque characteristics of engine and gear ratio and cavitation characteristics of propulsors at low vessel speed could result in limited capability due to thrust loading.

As seen in Figure 5-9, at hump speed, required thrust loading quickly reduces with increasing number of propulsors reaching a condition of diminishing return at about a four-shaft machinery arrangement. Often manageable thrust solutions may be achieved more easily with three and four shaftlines than with twin-screw arrangements.

The speed differential between hump speed and design speed A is relatively small and may well be suited for twin propulsor craft, most of which have fixed-pitch propulsors with a fixed reduction gear ratio.

The design circumstance for the differential between hump speed and design speed B is very difficult. Besides needing three or four shaft lines, two-speed gears, variable-pitch propellers, or waterjets with variable inlet geometry could potentially be required.

Propellers on Inclined Shafts—With and Without Tunnels

Earlier in this chapter, I discussed selecting the best propeller diameter and gear ratio. Having an established propeller diameter essentially fixes the shaft angle relative to the hull buttock line. The typical range of shaft line angle is from about 7° to 15°.

You may use large angles for slow vessels, but small angles are necessary for high-speed craft. Shaft angle has a negative effect on propeller performance. Characteristic effects include reduced efficiency and lower speed at which cavitation inception occurs and—during propeller rotation—blade hydrodynamic angles are different for downward and upward movement.

For fixed-pitch propellers, shaft angle causes hydrodynamic blade angles to have radial variation from hub to tip of the propeller which increases the probability of cavitation damage at the root. Propeller-induced blade rate hull vibration is often traced to large shaft angles and clearance of the tip of the propeller blade from the hull.

Minimizing large shaft-angle-induced blade rate problems for high-performance craft having fixed-pitch propellers is possible with at least two options. The first is to increase the number of propellers so that the best propeller diameter will be smaller as thrust loading of each is reduced. The second option is to have propellers recessed into tunnels.

Propellers recessed into tunnels is worthy of consideration as a machinery-design arrangement alternative to just propellers on inclined shafts. The enhancements achieved by using a partial tunnel include reducing the shaft angle, decreasing navigational draft and allowing the propulsion machinery to move aft for an appropriate longitudinal center-of-gravity location for ride quality and speed as well as improved internal machinery arrangements. A partial tunnel allows you to fit large-diameter propellers which may reduce cavitation or reduce shaft angle to minimize the variation in hydrodynamic blade angle.

There is an important relationship between the propellers and the geometry of the tunnel. They must be designed together as a propulsion system. Design guidelines for partial propeller tunnels and relative placement of propellers are detailed in Blount (1997).

Key Propulsion Design Decisions

A detailed example of integrating the propulsion system for a propeller-driven craft with its speed-resistance requirements has now been completed.

Significant lessons learned? Essential factors must be accounted for, regardless of the propulsor concept selected to convert engine power to thrust.

These essential factors are:

- The best number of propulsors is a function of the magnitude of the T/V^2 at hump and the difference between hump and maximum speed.
- Having cavitation (or ventilated) characteristics of propulsors available to designers for the entire operational thrust loading and speed range of the craft.

- The low rpm engine power and torque characteristics in combination with definition of heavily thrust loaded capability of propulsors dominate craft agility; that is to have sufficient longitudinal acceleration for attaining planing speeds.

The remainder of this chapter provides brief information for other propulsor systems presently found on performance marine craft. Often suppliers of these systems must be provided with information about craft being designed before they can recommend a propulsor model number, size, and give data about performance characteristics. Where technically possible, the following paragraphs provide some guidance for sizing other propulsor systems for use in concept design studies.

Surface Propeller Loading

Partially submerged or surface propellers, seen in Figure 4-16, generally have applications for performance marine craft up to about 25m (82 ft) which may operate in rough water. Steerable and fixed variations of these propulsion devices offer the advantage of minimal shafting, struts, and control surfaces arrayed underwater, while requiring less interior volume. Conventional propulsion system appendages can account for substantial craft resistance.

Partially submerged propellers have had many successes, but there have also been instances in which they did not produce adequate thrust to accelerate through hump to planing speeds due to thrust breakdown. In addition, the lack of ability of a craft having surface propellers to accelerate through hump can also result from a mismatch of inadequate engine torque being less than the torque demand of the propeller.

This prompted Denny (1988) to investigate and develop a criterion for establishing practical thrust limits for surface propellers. Denny's development followed the same general concept for thrust load limits used for submerged propellers, but assumed only the submerged, wetted disk area developed propulsive thrust.

Correlation with trial data indicated that most successful craft achieved the best high-speed performance with approximately 40 to 50 percent propeller diameter submergence. For maximum thrust loading, Denny's equation 5-30 is for 50 percent propeller diameter immersion.

$$K_T/J_T^2 = 585/v_a^2(EAR)[1.067-0.229(P/D)]$$

Equation 5-30

In τ_C format, this maximum thrust load becomes:

$$\tau_C = 0.7\ \sigma_{0.7R}$$

Equation 5-31

It should be noted that W_T (Table 5-A) for partially submerged propellers is usually zero. For this condition (v_a) is equal to boat speed. Thus $v_a = v$ and $J_T = J_A$.

This maximum-thrust load limit—Equations 5-30 and 5-31—for sizing a surface propeller and selecting a gear ratio leaves little margin for error.

Thus, if the total resistance is too optimistic because of the prediction method utilized or the as-built boat weight and/or LCG exceed design expectations, then performance is likely to be very much different than anticipated. For this reason, the recommended design thrust load criteria should be no more than 80 percent of the maximum values.

This approach will reduce the technical risk for surface-propeller applications. Thus, the recommended thrust load-design criteria for surface propellers defined in equations 5-30 and 5-31 become:

$$K_T/J_T^2 = 468/v_a^2(EAR)[1.067 - 0.229(P/D)]$$

Equation 5-32

and:

$$\tau_C = 0.56\ \sigma_{0.7R}$$

Equation 5-33

Surface propellers generate significant vertical and horizontal forces in addition to propulsive thrust.

Two definitive papers (Rose et al. [1991 & 1993]) provide experimental characteristics for a family of four-bladed surface propellers. These two sources provide sufficient detail to account for vertical and horizontal forces generated by surface propellers.

Additional references about surface propellers are worthy of study by naval architects who frequently design craft having this propulsion concept. They are the PhD thesis by Olofsson (1996) and the experimental work of Misra (2012).

Flush-Inlet Waterjet Loading

Flush-inlet waterjets have made significant inroads as the preferred propulsors for craft requiring operational speeds of greater than 25 knots. Considering overall propulsive efficiency, they offer a distinct advantage at higher speeds.

In addition, requirements for reduced vibration and navigational draft often support the decision to select waterjets in favor of other propulsors. The combination of gas turbines driving waterjets is particularly well matched for larger vessels requiring high power levels. Both are free-turbine fluid machines with power characteristics proportional to rpm cubed. Thus, overload/offload boat conditions are readily accommodated by this combination of propulsion machinery.

With regard to sizing waterjets for a vessel and its performance requirements, some geometric differences need to be acknowledged. Figure 5-10 depicts a flush-inlet waterjet with some geometric aspects indicated.

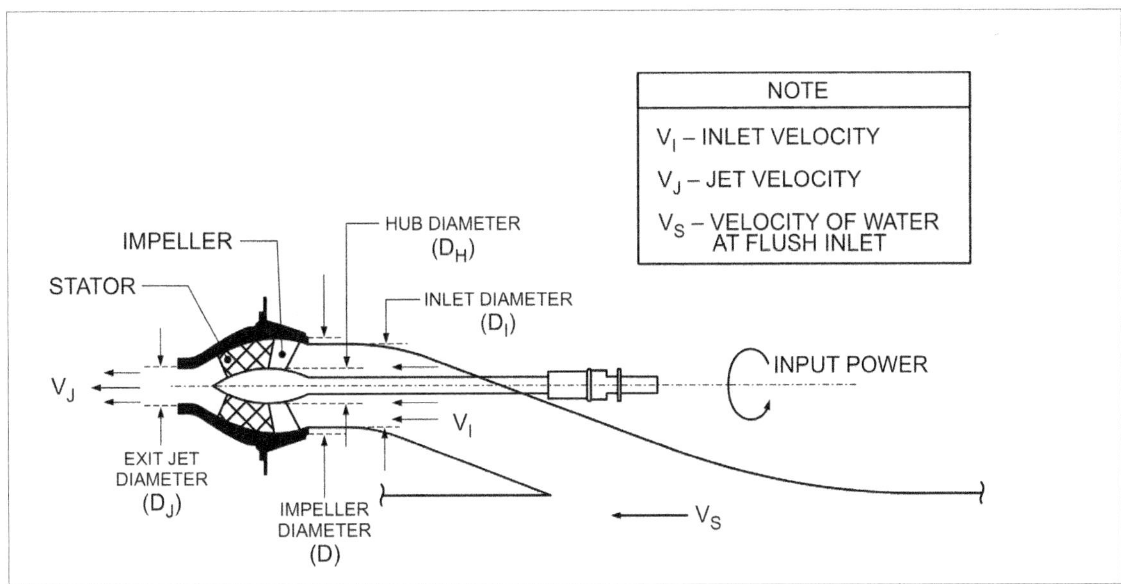

Figure 5-10: Cut-away centerline section of a mixed-flow flush inlet waterjet

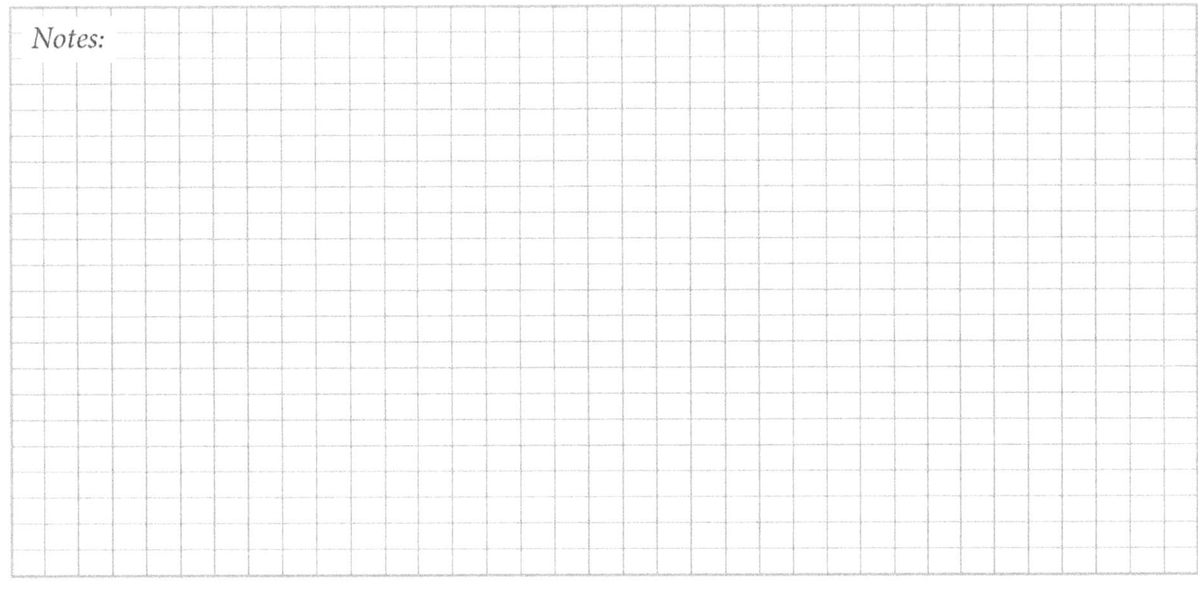

I can recommend an excellent compilation of waterjet technology presented by Allison (1993) for additional information regarding these propulsors.

The ultimate waterjet size must be confirmed by providing hull-speed and resistance data, engine power, rpm, and other appropriate information to a vendor; however, the naval architect/marine engineer has the ultimate responsibility to establish the preferred model and size for integration within the envelope of hull geometry and/or owner requirements.

See Figure 5-10: The waterjet inlet diameter (D_I) has often been taken as the significant propulsor sizing dimension for the model size designation by manufacturers of large mixed-flow waterjets. Manufacturers of axial flow waterjets tend to size their equipment by impeller diameter.

Denny (1979) developed an engineering approximation for the propulsive thrust force which could be generated by a waterjet with known engine power. This relationship utilized exit jet-nozzle geometry. Assuming that the exit nozzle diameter is approximately 70 percent of the inlet diameter for mixed-flow jets, then the thrust (in pounds) approximation becomes

$$T = \rho(0.385)D_I^2V_{JB}^2[1 + (v/V_{JB})x^{-1.737}\, v/V_{JB}] \qquad \text{Equation 5-34}$$

for: $v > 0$ and $x = (V_{JB}/v) + 1$

where:

$$V_{JB} = ((1611.7(SHP)^{1.0556})/(\rho D_I^2))^{1/3} \qquad \text{Equation 5-35}$$

These equations provide an estimate of the maximum propulsive thrust without considering pump or inlet cavitation and hull-waterjet interactive factors as mentioned in Table 5-A. It is important to take these interactive factors properly into account in the characteristics provided by the waterjet supplier when predicting speed for known hull resistance.

The naval architect/marine engineer has the ultimate responsibility to establish the preferred model and size for integration within the envelope of hull geometry and/or owner requirements.

Low boat speed/high-horsepower situations expose the waterjet to flow rates which can result in reduction in thrust production due to cavitation. This pump-cavitation limit can be defined by suction specific speed. Current experience indicates that for mixed-flow waterjets, suction specific speeds are recommended to be below 10,400 for continuous operation and not to exceed 16,000 for transient conditions.

The use of suction specific speed as a cavitation limit usually requires information about pump characteristics which may be difficult to obtain from waterjet manufacturers. For preliminary decisions, the following equation provides guidance which should be used for a half-load displacement condition for establishing a minimum inlet diameter (feet) for maximum boat speed (v)

$$D_I = \{SHP[1.241 - (0.1267)v^{1/2}]\}^{1/2}/12 \qquad \text{Equation 5-36}$$

for

$10 \le v \le 85$ ft/sec

After the minimum inlet diameter is established, speed, thrust, and power characteristics should be calculated following Equations 5-34 and 5-35 to compare with hull resistance and to evaluate the margin at hump speed. It is likely that slightly larger inlet diameters than that indicated by Equation 5-36 may result in improved overall propulsive coefficients.

As with propellers, the design rpm must also be established simultaneously with the waterjet size. There will be some rpm variation between waterjets supplied by different manufacturers. However, the following equation serves as an approximation for rotational speed:

$$N = u(SHP)^{1/3} \qquad \text{Equation 5-37}$$

where:

$$u = -0.01341 + 4.04\,[D_I] - 1.628 \qquad \text{Equation 5-38}$$

As a final reminder, a water jet's thrust, speed, power, and rpm relationship must eventually be obtained from the supplier.

Natural Aeration of Flat Planing Surfaces

Zero deadrise hulls have been observed in calm water to develop substantial aeration on the hull bottom. Size and density of bubbles increase with trim and speed.

This phenomenon has resulted in flat-bottom boats propelled by flush inlet waterjets having erratic propulsive performance; exhibiting very sluggish response or failure to accelerate through hump speed or not attaining predicted speed.

This is primarily due to diminished waterjet thrust from an air-water mixture flowing into the inlet. Experiments in model basins show that planing surfaces aerate very readily at certain conditions both when the stagnation line is on the straight surfaces and the curved buttocks in the bow region. This aerated condition has been observed both on models and full-scale hulls with section shapes having flat surface of at least a width of 0.40 times maximum chine beam and straight afterbody buttocks for the primary planing bottom (Ward-Brown, 1980).

Criteria for minimizing or avoiding this hull-bottom aeration condition is provided in Table 5-F when CV is calculated with the maximum chine beam of the planing surface and trim angle is measured relative to the slope of the buttock at stagnation (initial longitudinal intersection of still water surface with the hull).

TABLE 5-F: HULL BOTTOM AERATION AVOIDANCE	
1.	Do not use $\beta = 0°$ for planing hull bottoms.
2.	For planing hulls, use $\beta \geq 5°$.
3.	Avoid β between 0° and 5° for planing hulls with propulsors which can exhibit erratic performance in aerated water.
4.	Conservative recommendation: Operational speed and running trim criteria C_V Tan $\tau \leq 0.13$.
5.	Less conservative, but with some operational risk: Do not exceed C_V Tan $\tau \leq 0.35$.

Notes:

Pod Drives

A complete pod drive propulsion system is made up of an inboard engine, drive train consisting of a reduction gear, two right-angle gears, and a nacelle housing shafting for single or counter-rotating propellers. Inboard engine power is transmitted through the aft hull bottom, by means of a steerable lower unit having an attached nacelle. Pod drives are available in both tractor and pusher configurations, propellers forward and aft, respectively, of the lower unit. Figure 5-11, repeated from Chapter 4, shows for comparison a tractor configuration at the top and a pusher at the bottom, separated by a familiar inclined shaft and propeller arrangement.

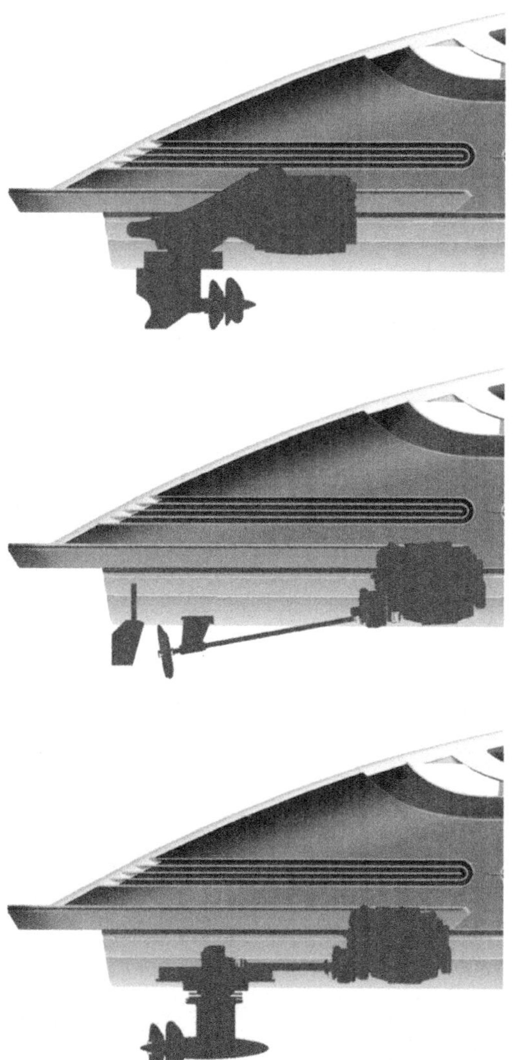

For hulls of the same length, pod drives free up interior hull volume for additional accommodations when compared with inclined shaft propulsion arrangements. Pod drives with their counter-rotating propellers aligned with water flow, offer measurably improved propeller efficiency on the order of five to eight percent when compared to a single propeller on an inclined shaft.

Steering forces are a combination of the surface area of the lower unit and vectored propeller thrust which results in "sports car" agility; a maneuvering capability which should be used so as to maintain human factor comfort. Cost of installation of pod drives during new construction is reduced and can be accomplished with less skilled workers as compared to inclined shaft installations.

Presently, known installations of one-to-four pod drive units on marine craft are in service. Commercially available tractor drives are installed with their steering axis normal to the hull bottom, except on the centerline of a boat. In this latter case, you can usually obtain design guidance for a hull recess from the pod-drive supplier.

Pusher drives are installed with their steering axis vertical. Small tunnels or recesses are necessary for these drives so that propellers do not contact the hull when steering. For this type of pod drive, the suppliers can provide design guidance for the recess.

Installation of both tractor and pusher pod-drive systems can employ a jack shaft to have longitudinal separation of engine and gear of the through-hull lower unit. This allows some flexibility of machinery arrangement for an appropriate LCG with some loss of interior hull volume. Pod-drive and waterjet suppliers are both resources who can provide data for naval architects who are making speed-power predictions.

Figure 5-11: Comparison of propulsion systems of tractor, inclined shaft, and pusher pod drives

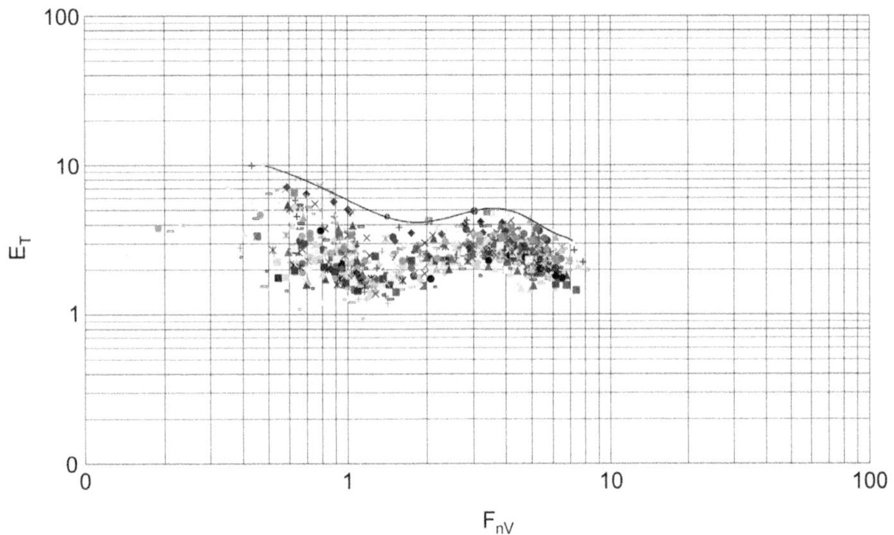

Figure 5-12: Transport efficiency versus F_{nV} for a collection of outboard-powered boats

Outboard and Inboard/Outboard Engines

Outboard engines are complete, portable propulsion units from the power source to propellers, either single or contra-rotating. They are portable in the sense that they are mounted on the transom of planing boats without penetrating the watertight hull structure. They are offered in a wide range of powers. Outboard-powered boats are generally designed for one-to-three engines, while four or more are occasionally used for specialized purposes.

Inboard/outboard propulsion units are fixed installations, with the engine mounted inboard of the hull and the power driveline through the transom to a retractable, steerable lower units with propellers. As inboard/outboard lower units are hydrodynamically similar to that of outboard engines, speed prediction of both may be treated with the same approach.

Prediction of speed has been less rigorous for outboard-powered boats as most approaches are based on the manufacturer's boat testing with combinations of engines and propellers. Engine manufacturers may provide experienced speed predictions based on trial data to designers and boatbuilders for new hulls.

Boatbuilders who maintain performance records of previous craft may also make their own performance predictions. Finally, boat-test articles in marine magazines can be an excellent source for speed estimates for installed power, with the caveat that a most important piece of data—actual craft-test weight—may need to be estimated.

You can analyze the outboard-powered boat test data provided in magazine articles to estimate maximum calm, deep water speed. Equation 5-39 provides an empirical prediction of probable speed in knots when newly designed boat weight has been estimated, and the total rated outboard engine power has been established.

$$V = 350(W/HP)^{-0.75} \text{ for } 10 \le W/HP \le 30 \qquad \text{Equation 5-39}$$

For this equation, W is in pounds and HP is total rated horsepower. Speed of outboard-powered boats is often reported in statute miles per hour (MPH) rather than knots; conversion is given in the following:

$$V_{MPH} = 1.15V \qquad \text{Equation 5-40}$$

Test data collected from a number of different outboard-powered boats was available to prepare a graph of transport efficiency (E_T) versus dimensionless speed, F_{nV}, in Figure 5-12. A line at the highest boundary of this experimental information represents the state of the art (SOA) of outboard-powered recreational boats in 2002.

Conducting sea trials of new boats to obtain E_T versus F_{nV} benefits both designers and builders for measuring advances as well as being a tool assisting the design process.

In order to estimate outboard engine characteristics for the feasibility studies, I first assume that engine thermal efficiency is constant from 50-to-100 percent of rated power. This latter assumption then implies the power delivered to the propeller(s) is proportional to the measured fuel rate at maximum speed. Engines designed by different manufacturers have different thermal efficiencies.

I use an average thermal efficiency which means that the variation of speed prediction for power less than 100 percent for different engine manufacturers could be +/- 10 percent.

Usually during preliminary discussions with clients, a few controlling specifications are established, such as the maximum overall length and the craft's stated full load speed. With these key bits of information, you can begin to assess the reality of the project requirements.

You will start with the following approximations: The calm-water speed of outboard-powered boats is primarily a function of weight to power ratio. Planing boats are in service which meet or slightly exceed Equation 5-39. An experienced naval architect or designer makes an approximate weight estimate for the boat length discussed, then power for the required speed can be established.

Then fuel rate for this power can be approximated by

$$\text{fuel rate (U.S. gal/hr)} \approx HP/10.4 \qquad \text{Equation 5-41}$$

Another useful approximation for outboards is the relation between percent engine RPM_E and fuel rate.

$$\%\ \text{maximum fuel rate} \approx 0.0055(\%RPM_E)^{2.13} \qquad \text{Equation 5-42}$$

for $60\% \leq RPM_E \leq 100\%$

Transport Efficiency

Transport efficiency (E_T) is becoming a significant marketing factor as it provides a quantified method of comparative performance of craft in calm water. E_T provides a measure of power delivered to the propulsor required to move weight (W) at a steady speed (V).

$$E_T = WV/SHP(325.9) \qquad \text{Equation 5-43}$$

Manufacturer's rated horsepower, HP is used for outboard engines.

Therefore

$$E_T = WV/HP(325.9) \qquad \text{Equation 5-44}$$

It is important to note that E_T can be rewritten in dimensionless form as

$$E_T = \eta_D/(R/W) \qquad \text{Equation 5-45}$$

By collecting a sufficient amount of calm, deep-water test data for various marine craft, you can prepare a graph of E_T versus F_{nV} for similar hull forms such as hard-chine boats.

Within the group scatter of E_T of marine craft performance being compared, you can show the highest E_T for each speed, F_{nV} is the best performer and the lower, dense cluster of data represents average or below average vessels or propulsion concepts.

When a large collection of comparative data is available, think of the upper boundary of this graph as representing the state-of-the-art (SOA) of the most recent date of available data.

As design and production technology progress, the SOA (ET) line, should rise as future boats and craft will be more efficient compared to those from previous generations.

CHAPTER SIX

PROPULSION DURING TRANSIENT CONDITIONS AND IN SHALLOW WATER

TRANSIENT CONDITIONS: CALM, DEEP WATER

Marine-craft performance is measured by more than just meeting steady-state operational requirements. A design may well attain both cruise and maximum performance, but still be perceived as underperforming if it doesn't have impressive acceleration from zero to planing speeds (hole shot). To avoid under performance, you must take into account transient or non-steady operations during the design process (Peach, 1963). A mismatch of propulsor-power demand at craft speeds that occur at low-engine rpm can even result in the inability to accelerate to planing speeds, even though maximum power is sufficient to meet higher speed requirements.

When longitudinal acceleration (a) is greater than 0.05 g (either 1.6 ft/s^2 or 0.49 m/s^2), it can result in a total propulsive force (accelerating plus steady-state thrust) which dominates selection of engines having the full range of appropriate BHP versus rpm characteristics. This occurs when the design decision is controlled by a transient/acceleration requirement, which might also be stated as "time to plane." When requirements include both maximum speed and acceleration, the combination of design reduction gear ratio and propulsor characteristics may also be in conflict.

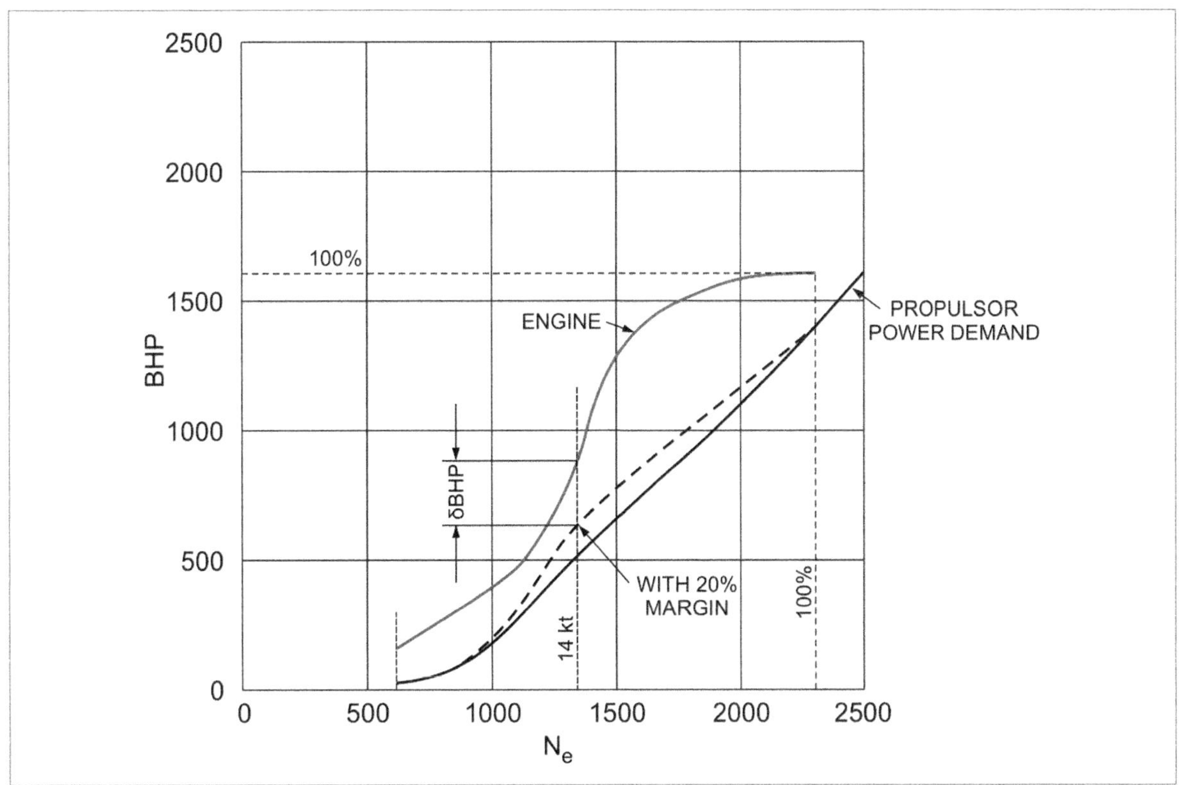

Figure 6-1: Excess engine power (δBHP)available—assuming 20% thrust margin—for accelerating boat speed

Full-scale trial data indicate that longitudinal acceleration is likely to be lowest near the thrust hump speed and will be greater at both higher and lower speeds, rather than remaining constant through the speed range.

Two methods to be considered for predicting transient performance are

(1) designing for specific longitudinal acceleration, or

(2) using a thrust margin approach.

Designing a craft for longitudinal acceleration is based on the difference between the propulsor power demand for steady speed and maximum engine power available at each corresponding propulsor rpm. The concept is depicted graphically in Figure 6-1, which indicates the differential power available for acceleration (δBHP) which must be converted to incremental propulsor thrust for craft acceleration.

Should predicted acceleration be less than desired, then you must increase δBHP. You can achieve this by selecting an engine having higher power within the operating range of rpm or reducing the propulsive-power demand in some way. For improving acceleration, there is a particular advantage in selecting engines having high torque characteristics at low rpm.

Propulsor-power demand as shown in Figure 6-1 is a further development of the calculated results of the example from Table 5-C using Equation 5-17 to calculate BHP from SHP. Engine characteristics shown in Figure 6-1 are representative of a typical marine diesel.

The acceleration distribution observed on full-scale trials can be simulated during design by applying a thrust margin approach. That is, to add a thrust margin at the speed for maximum steady thrust loading. With an assumed design thrust margin of 20 percent at the thrust loading hump, a peak acceleration of about 0.03 g can generally be attained; a design thrust margin of 50 percent can yield an 0.08 g acceleration at hump speed.

The thrust-margin approach provides realistic design guidance with respect to engine selection having suitable characteristics throughout the entire operational rpm demand of the propulsors. Incorporating this thrust margin fortifies design decisions which provide the resultant vessel with transient performance to accelerate, to tow a device in or under water, as well as overcome additional resistance from waves, overloading, hull fouling, degraded engine performance and problems such as minor propulsor damage. Small δBHP at low rpm of propulsor power demands relative to available engine power indicates risk of underperformance or lack of hole shot agility which will become apparent later as acceleration is measured during full-scale trials.

For the design of propulsion systems, therefore, I recommend using a thrust margin of at least 20 percent added at the maximum steady-state thrust loading speed. For design purposes, the thrust margin percentage should increase in a linear manner from zero at zero speed to the 20 percent at maximum thrust loading speed and then decrease in a linear manner to zero at design speed as depicted in Figure 6-2.

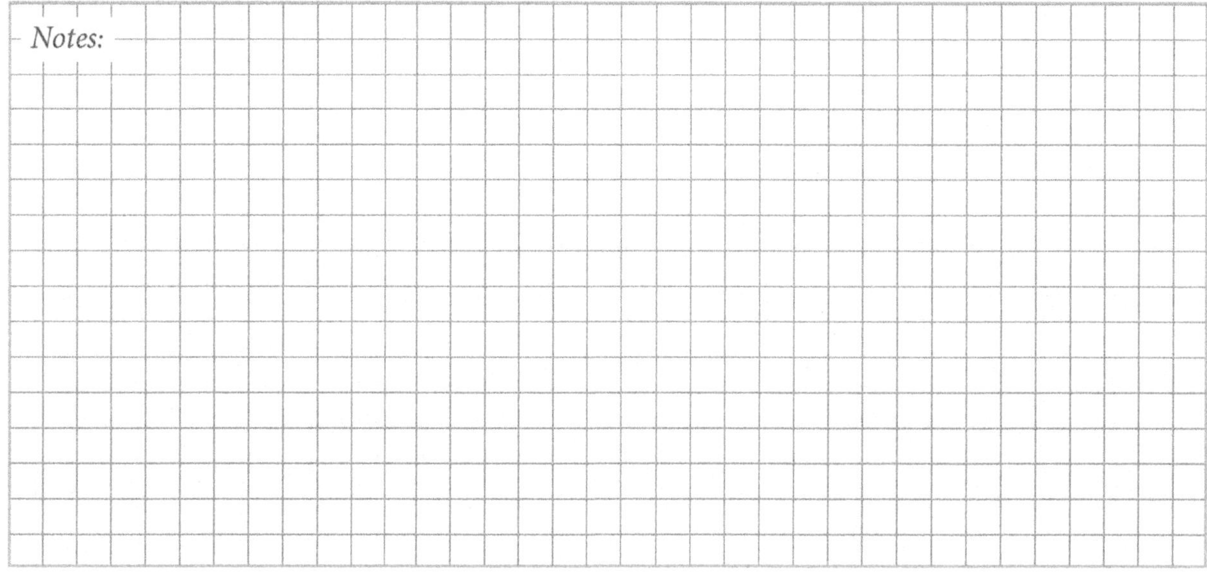

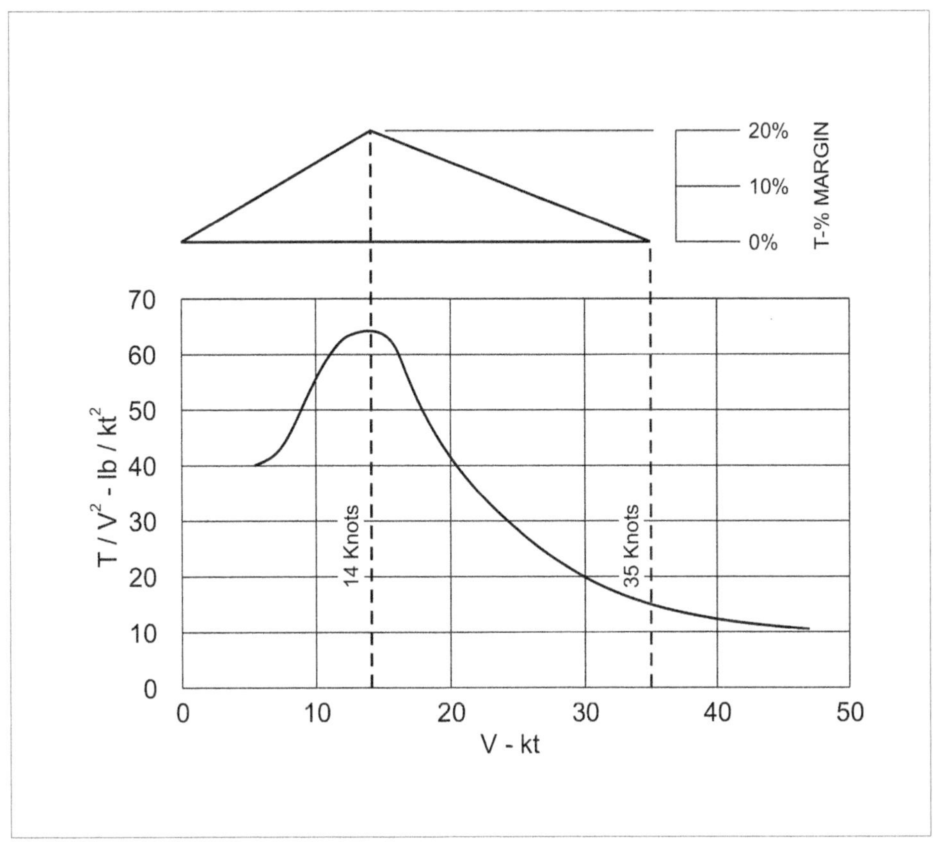

Figure 6-2: Maximum steady state thrust loading versus speed

Applications

- **New Boat Design**

  You must specify the proper propulsion drive line machinery so that the vessel meets client requirements for maximum and cruise speeds in both calm and rough water. Whether stated by the client or not, the vessel needs to be relatively agile and responsive to the throttle when accelerating from displacement to planing speeds. Thus, the selection of size and characteristics of propulsors, gear ratio, and full range engine characteristics (BHP versus engine rpm) must reinforce agile acceleration for all expected displacements and LCGs. In your design practice, consider using the recommended thrust margin at the hump speed where T/V^2 is maximum.

- **Evaluation of Propulsors, Gear Ratio, and BHP versus Engine RPM**

  This approach allows you to evaluate the relative significance and influence each of these three factors as they all effect agile acceleration, i.e. the hole shot. This is especially useful when analyzing sea trial data of craft which exhibit poor acceleration characteristics to define problems and recommend corrections.

- **Advising Engine Manufacturers**

  With this analysis approach you may also advise engine manufacturers of possible changes to engine characteristics, BHP versus rpm, of some models to enhance their marketability, especially for marine propulsion of high-performance vessels.

- **Shallow-Water Performance**

  This analysis approach will, to a limited extent, allow naval architects to evaluate shallow water effects on craft with regard to attainable speeds. There is some risk in shallow water that performance craft may not be able to accelerate from displacement to planing speeds; that is, they may not get over the hump when water depth is lacking.

Predicting Longitudinal Acceleration in Deep Water

The accelerating propulsive thrust is determined using the following approach for the accelerating force (thrust).

$$F = ma \quad \text{Equation 6-1}$$

where

m = The mass of the marine craft including the added mass of longitudinally entrained water

a = Longitudinal acceleration of a craft

I. Establishing Thrust Margin

A. Predict steady state propulsive thrust, SHP and rpm versus vessel speed, following an acceptable procedure for estimating total resistance to calculate thrust required for propeller or waterjet propulsors. (Refer to Chapters 3, 4 and 5.)

B. Next, calculate total steady state thrust using Equation 5-5 and divide by speed in knots squared (T/V^2) remembering that

$$T = R_T/(1-t) \quad \text{Equation 5-5}$$

C. Plot T/V^2 versus speed in knots to establish the speed for maximum steady state thrust loading of the propulsors as shown in Figure 6-2. The example in Figure 6-2 uses results from Table 5-C.

D. Add percent thrust margin distribution as shown in Figure 6-2.

II. Procedure for Relating Engine Characteristics to Propulsor Power Demands

A. For propulsion engines, obtain characteristics for brake horsepower (BHP) from idle to maximum rated rpm from the manufacturer or distributor.

B. Divide the total predicted thrust and BHP including margin for propellers or waterjets by the number of shaft lines rpm (N_{PR}) to obtain propulsor power demand on each engine.

C. Predicted shaft, N, is converted to engine rpm (N_e) by multiplying by gear ratio, GR.

$$N_e = N(GR) \quad \text{Equation 6-2}$$

D. Plot both engine BHP characteristics and predicted propulsor BHP demand versus engine rpm for the full range of steady vessel speeds, as shown in Figure 6-1. The propulsor power demand curve should be below the engine characteristics curve except when, or if, it intersects at maximum rated rpm. Should the propulsor demand curve cross above the engine BHP characteristic curve below maximum rated rpm, the maximum craft speed will be that predicted for the engine crossing rpm.

III. Thrust with Margin Available for Vessel Acceleration

At each engine rpm, there must be separation between the engine BHP characteristics and the propulsor BHP demand (δBHP) to produce thrust (F) to accelerate the vessel to a higher speed. Limiting factors affecting the magnitude of accelerating thrust could be that the maximum engine BHP is less than propulsor demand at any rpm, possible propulsor cavitation, or even thrust breakdown.

Torque (Q_e=5252BHP/N_e) versus N_e characteristics available from an engine to drive the propulsor must be converted to maximum thrust in order to calculate the differential force which accelerates a craft to higher speed. Transition from engine Q_e to propulsive T must involve the characteristics of the propulsor.

This transition is possible with some assumptions which, fortunately, have little effect on the ultimate outcome of this prediction procedure. For each steady vessel speed at its engine rpm, assume that $J_Q \equiv J_T$ and 1-t, 1-W_t and η_R remain unchanged when accelerating in deep water, but not in shallow water. *Note: I'll discuss this later.*

$$J_Q \equiv J_T \quad \text{Equation 6-3}$$

It is necessary to calculate maximum K_Q available from an engine's maximum characteristics to a propulsor connected through a gear. Shaft horsepower (SHP) delivered to the propulsor by BHP is reduced by transmission efficiency (η_T). A typical value of η_T = 0.965. Output of engine power and rpm will be changed by a gear ratio (GR) to be compatible with characteristics of propulsors.

For example, a reduction gear with GR = 2.0:1.0 will deliver to the propulsor almost double engine torque at half of the corresponding engine rpm. A step-up gear having GR = 0.833:1.0 will deliver to the propulsor reduced engine torque ($0.833Q_e$) at higher than corresponding engine speed with rpm $N = N_e/0.833$.

The maximum K_Q available to a propulsor begins with the output BHP of the engine, which is then reduced by combined losses (η_T) from all driveline losses including gear, shaftline bearings, and seals to deliver SHP to the propulsor.

$$SHP = 0.965\ BHP \qquad \text{Equation 5-17}$$

By definitions

$$SHP = 2\pi QN/33{,}000 = QN/5252 \qquad \text{Equation 6-4}$$

and

$$n = N/60 \qquad \text{Equation 6-5}$$

Torque (Q) in lb–ft and n in rps at input to propulsors is

$$Q = (GR)SHP(5252)/N_e \qquad \text{Equation 6-6a}$$

or

$$Q = (GR)Q_e(\eta_T) \qquad \text{Equation 6-6b}$$

and

$$n = N_e/(GR)(60) \qquad \text{Equation 6-7}$$

For propellers K_Q is defined as

$$K_Q = Q/\rho n^2D^5 \qquad \text{Equation 6-8}$$

Maximum K_Q available at each N_e from the engine which may be converted to propeller thrust becomes

$$K_Q = (GR)^3\ SHP\ (1.8907X10^7)/N_e^{\ 3}\ \rho D^5 \qquad \text{Equation 6-9a}$$

or

$$K_Q = (GR)^3\ Q_e\ \eta_T(3600)/N_e^{\ 2}\ \rho D^5 \qquad \text{Equation 6-9b}$$

Equations 6-9a or 6-9b provides the means to convert engine characteristics to K_Q, knowing gear ratio and propeller diameter. Then craft speed can be introduced through propeller characteristics when in the format of K_Q versus J_Q, defined as

$$J_Q = v/nD = V(1.6878)(1\text{-}W_Q)/nD \qquad \text{Equation 6-10}$$

Solving Equation 6-10 for n and substituting into Equation 6-8, K_Q becomes

$$K_Q = Q\,J_Q^{\ 2}/\rho D^3[V^2(2.849)(1\text{-}W_Q)^2] \qquad \text{Equation 6-11}$$

Since it is assumed that $J_Q \equiv J_T$, it follows that $(1\text{-}W_Q) \equiv 1\text{-}W_T$).

It is now possible to estimate transient craft longitudinal acceleration with known engine characteristics in either BHP versus N_e or Q_e versus N_e format, propeller dimensions, gear ratio, and a steady-state thrust prediction for a hull.

In the next section, this process is outlined, followed with a numerical example.

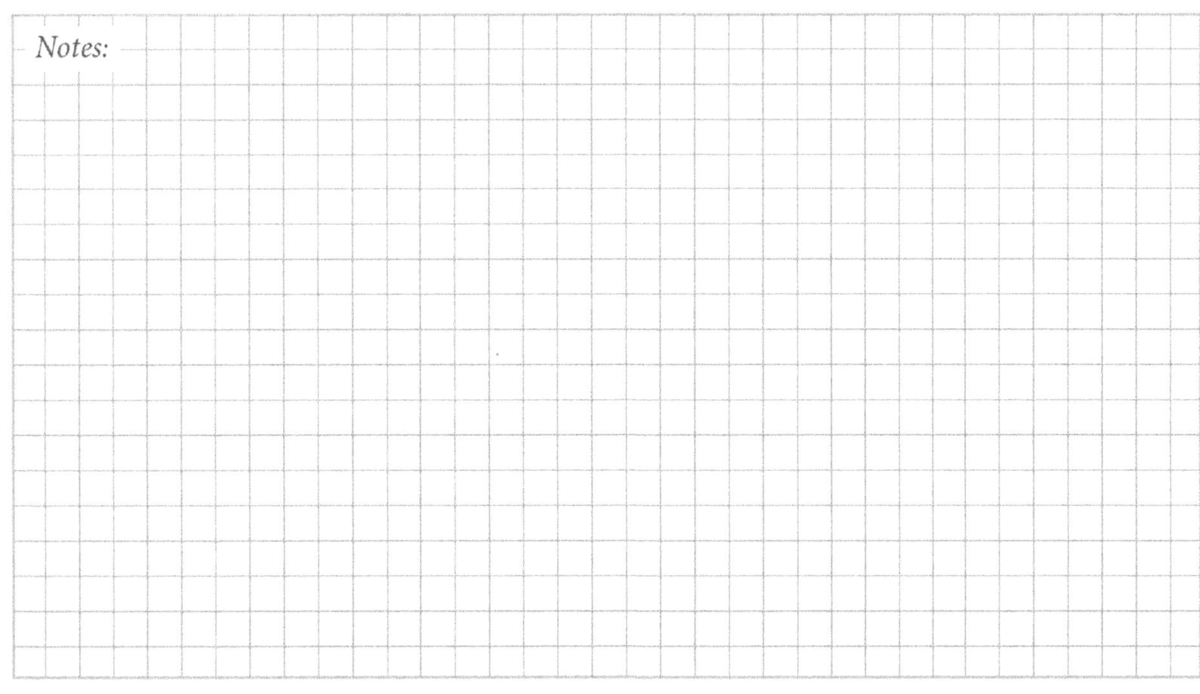

IV. Longitudinal Acceleration Prediction

A. At each vessel speed from engine idle up to maximum speed (condition of zero acceleration), the total accelerating force (F) available from all propulsion shafts is the difference between the thrust available at total engine power gear and propulsors minus steady state thrust required by the hull.

B. Divide this total available accelerating force by the mass of the vessel including an increase for mass of entrained water. For displacement and semi-displacement vessels, I suggest a value of 20 percent of weight of displacement be used for longitudinal added mass of entrained water as their hulls remain submerged in the water. For hard chine planing craft, I suggest for longitudinal added mass of entrained water, five percent of vessel mass. As simplifying assumptions constant values of added mass are provided, in reality added mass varies as a function of speed decreasing from 20% at slow speed to 5% when planing. Mass of planing craft plus entrained water = m = (1.05) displacement (lb)/32.15 ft/sec^2. (*Note: If hole-shot trial data is available for a comparable craft, you may also approximate an estimate of added mass of entrained water.*)

C. Since F = ma, then a = F/m = dv/dt_i

Then

$$1/a = m/F \qquad \text{Equation 6-12}$$

Plot 1/a which is mass (m) divided by accelerating thrust (F) versus vessel speed in knots.

D. By integrating the area under the 1/a versus speed curve shown in Figure 6-3 from "A" to "B" (for example) for increments between two speeds, a speed versus time curve (hole shot curve) is generated. And the elapsed time from "A" to "B" would be plotted at point "B" in figure 6-4. Thus,

$$dt_i = (1/a)dv \qquad \text{Equation 6-13}$$

and

$$t_i = \int(1/a)dv \qquad \text{Equation 6-14}$$

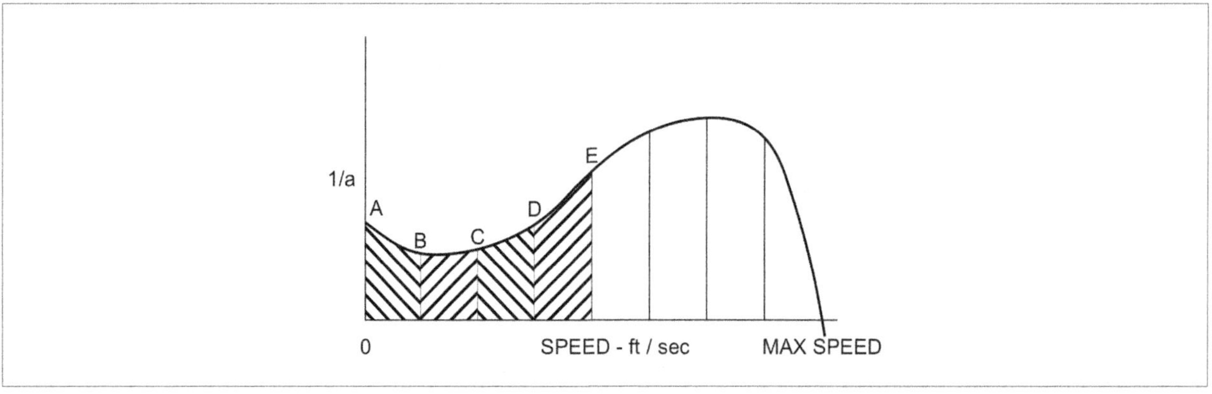

Figure 6-3: Reciprocal of longitudinal acceleration versus craft speed

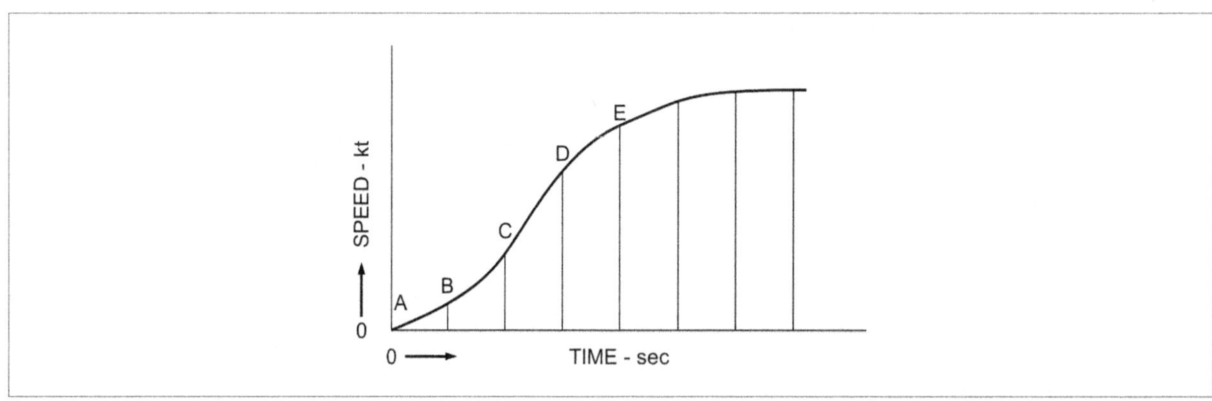

Figure 6-4: Time history of craft speed when accelerating

Within the context of the calculation of the speed-versus-time relationship, it is also possible to determine the time to increase from one operational speed to another at a higher speed.

For example, referring to Figures 6-3 and 6-4, when a vessel operating at a steady speed "C" and the throttle is increased, propulsive acceleration results in speed increasing until a faster steady speed is attained at the new engine rpm. By integrating the area under the 1/a versus speed curve, beginning at speed "C" for example will determine the time to accelerate from that steady speed to that of the new steady speed "E" at a higher engine rpm.

Distance Transited During Acceleration: Having developed the speed-versus-time curve, or the "hole shot" curve, then you can calculate the distance in feet (S) the vessel transits during acceleration.

This is determined by the following equation and executed by integrating the area under the speed-time curve, Figure 6-4 where V_{av} is the average speed, in knots, taken over a short interval, not the numerical average.

$$S = 1.688\, V_{av}(t_i) \qquad \text{Equation 6-15}$$

This calculation procedure is best demonstrated in the following numerical example.

V. Numerical Example

This calculation example for longitudinal acceleration is a continuation of the performance prediction for the 65-ft, 100,000 lb craft from Table 5-C which is shown graphically in Figure 5-6. This is a twin-screw boat having fixed pitch propellers with D = 36 inches, P/D = 1.20, EAR = 0.82 and Z = 3. The gear ratio is 1.917:1.0.

To begin, you need to have a graphical representation or an analytical expression for the output power, BHP or SHP or torque, Q versus rpm (N_e) for the entire engine operational range from idle to rated, N_e. The procedure being presented is based on engine characteristics which ultimately can be defined as Q_e being a function of N_e.

For this example, the clutches of the gears are engaged and at engine idle rpm the boat has a speed of 5 knots. the 65-foot craft is powered by two 1,600 BHP diesel engines rated at 2,300 rpm. The following approximation represents Q_e versus N_e characteristics for each engine as shown in Figure 6-5.

$$Q_e = -0.002385(N_e^{2}) + 8.677(N_e) - 3{,}692 \qquad \text{Equation 6-16}$$

for

$$600 \le N_e \le 2{,}300$$

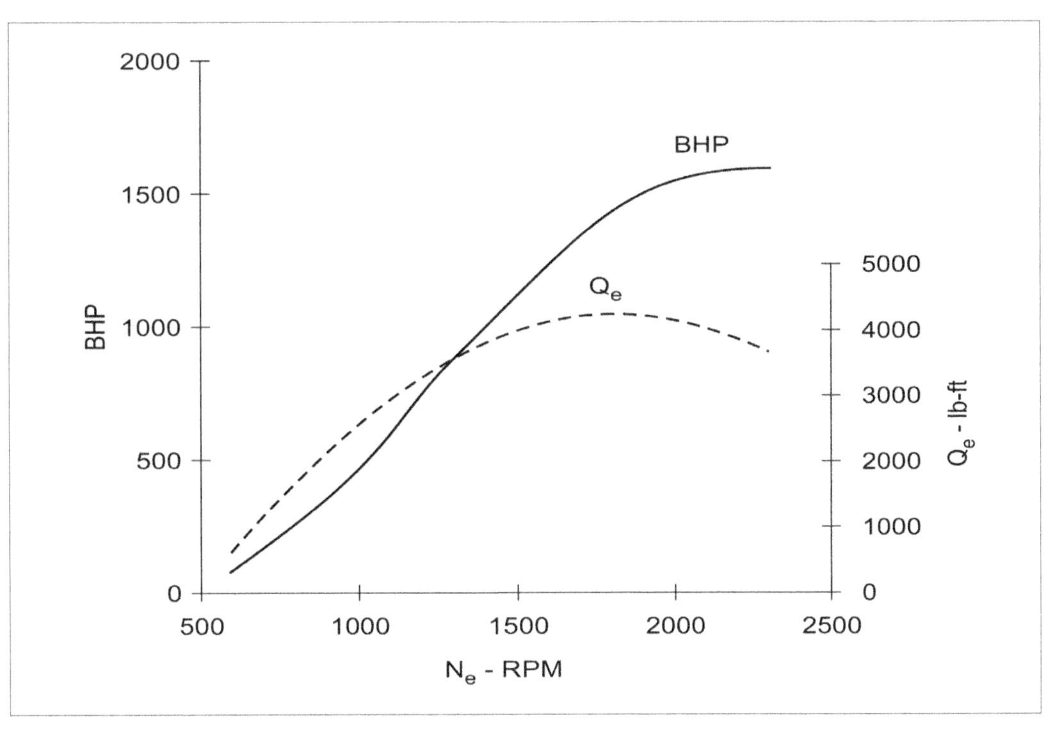

Figure 6-5: Engine characteristics

Each engine can deliver K_Q to a propeller over its range of N_e according to Equations 6-9a or b. Engine rpm (N_e) is related to propeller rps,n by Equation 6-7 and engine torque (Q_e) related to propeller torque by Equations 6-6a or 6-6b.

For D = 36 inches, GR = 1.917:1.0, $\eta_T = 0.965$ and $\rho = 1.9905$ lb sec²/ft⁴ for salt water, Equation 6-11 becomes

$$K_Q = Q\,J_Q^2/(153.1)[V^2\,(1-W_Q)^2] \qquad \text{Equation 6-17}$$

With propeller rps (n) and torque (Q) Equations 6-7 and 6-6b, respectively, combined with engine characteristics, Equation 6-16 and dimensional data for this example defines the maximum Q versus n available to each propeller.

$$Q = -58.35n^2 + 1{,}846\,n - 6830 \qquad \text{Equation 6-18}$$

Solving the advance coefficient, Equation 6-10, for n, craft speed (V) can be introduced into Equation 6-18 by substitution

$$n = V(1.6878)(1-W_Q)/(J_Q D) \qquad \text{Equation 6-19a}$$

With numerical data for this example

$$n = V(1-W_Q)0.5626/J_Q \qquad \text{Equation 6-19b}$$

and 6-18 becomes

$$Q = -18.47\,[V^2(1-W_Q)^2/J_Q^2]\ 1038.6\,[V(1-W_Q)/J_Q] - 6830 \qquad \text{Equation 6-20}$$

The calculation procedure to follow begins with the format of Table 6-A. Divide the speed range over which longitudinal acceleration is to be predicted into a suitable number of equal increments for using Simpson's or some other method for integrating areas under a graph. $(1-W_Q)$ is assumed for each speed to be equal to $(1-W_T)$, which is taken from Table 5-C.

For each speed, K_Q is calculated only for the range of J_Q values needed and presented in the format suggested by Table 6-A. K_Q is calculated with Equation 6-17 using Q from Equation 6-20.

Figure 6-6 gives the open water characteristics for the propeller in this example using the Gawn-Burrill regression by Blount and Hubble (1981), along with partially and fully cavitating characteristics from Blount and Fox (1978).

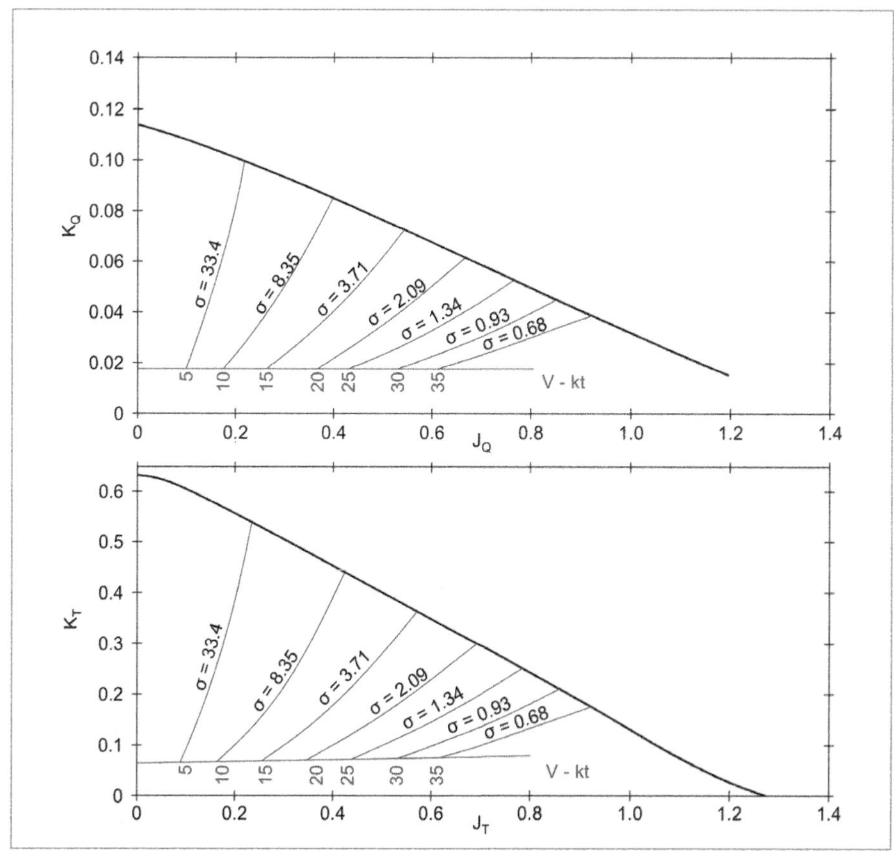

Figure 6-6: Propeller characteristics including the effects of cavitation

TABLE 6-A: MAXIMUM K_Q CALCULATED FROM ENGINE CHARACTERISTICS FOR THE CRAFT PROPELLER							
K_Q							
V-knots	5	10	15	20	25	30	35
$1\text{-}W_Q$	1.06	1.06	1.05	1.01	0.97	0.97	0.97
σ	33.4	8.35	3.71	2.09	1.34	0.93	0.68
J_Q							
0.20	0.0718						
0.25	0.1001						
0.30	0.1204	0.0356					
0.40		0.0718	0.0229				
0.50		0.1001	0.0498	0.0199			
0.60			0.0730	0.0415	0.0199		
0.70			0.0927	0.0609	0.0380	0.0213	
0.80				0.0780	0.0546	0.0331	
0.90					0.0697	0.0465	0.0278
1.00						0.0598	0.0405
1.10							0.0523
1.20							

Notes:

Continuing toward development of accelerating thrust available for the machinery arrangement of this example, the K_Q versus J_Q data from Table 6-A is overlaid onto the characteristics seen in Figure 6-6 to locate intersecting J_Q values for each craft speed. In Figure 6-7 these intersecting J_Q values are projected down to the K_T versus J_T characteristic since $J_Q \equiv J_T$ and taking care to note where the intersection takes place. It is either on the open water or the cavitation characteristics lines, as shown in Figure 6-7. K_T for the intersecting J_T is used to calculate the total thrust available for both steady speed and craft longitudinal acceleration.

$$K_T = T/\rho n^2 D^4 \qquad \text{Equation 6-21}$$

It may be difficult to define the J_Q at which K_Q from Table 6-A intersects the propeller cavitation characteristics in Figure 6-6. This possibility occurs since there can be similar slopes for the intersecting curves. Figure 6-8 is provided for definition of this problem using an example with the 20 knot K_T versus J_T cavitation characteristics. K_T / J_T^2 used in this figure is not the same as K_Q versus J_Q data calculated in Table 6-A. However, they both have similar slopes, which mimics the condition for defining the intersection of two curves having similar slopes.

The shaded area of the characteristics at 20 knots defines where this propeller can develop thrust. This area is bounded by three lines, the open water characteristics from $J_T \geq 0.70$ until $K_T = 0$, the fully cavitating propeller condition for $0 \leq J_T \leq 0.32$ and the partially cavitating condition where K_T connects these two boundaries between $0.32 \leq J_T \leq 0.70$. Overlaid on the cavitation characteristics are three curves of K_T / J_T^2 calculated for the same thrust required at 20 knots for different diameter propellers having the same P/D and EAR.

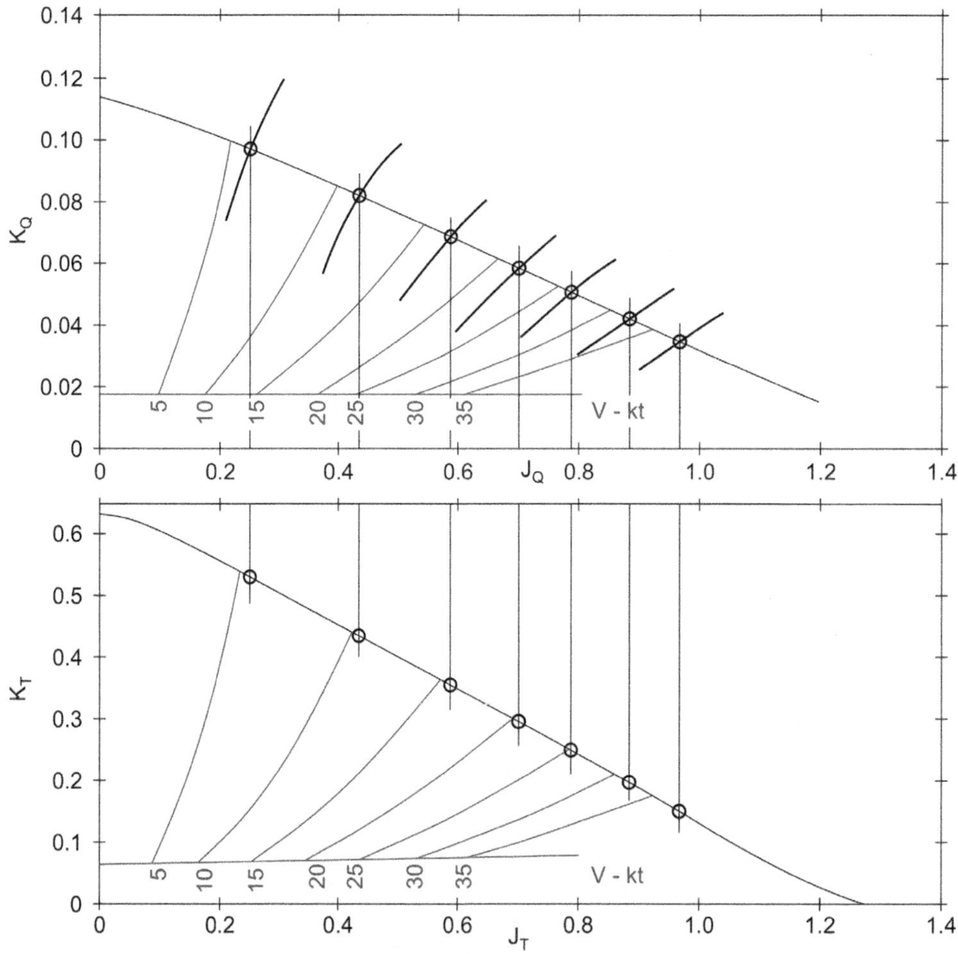

Figure 6-7: Overlay of K_Q from engine onto propeller characteristics to define available propeller K_T for calculating propulsion thrust.

The propeller with D = 42 inches is lightly loaded and intersects the open water characteristic boundary at Point No. 1.

The smallest propeller, D = 30 inches, is heavily loaded and intersects the full-cavitating boundary at Point No. 4. The middle propeller, D = 36 inches, almost lays on the partially cavitating boundary and intersection could be anywhere between Points No. 2 and No. 3. The water environment in which the boat operates would control the intersection point.

This latter condition should be avoided by the designer. Avoidance options include, at least:

- Select D > 36″ with same EAR.
- Increase propeller EAR, which moves the partial cavitation characteristic boundary left.
- Cup propeller blades' trailing edges, which changes both open-water and cavitation characteristics.
- Use different propeller-section shapes, which introduce new, open-water and cavitation characteristics.
- Change gear ratio.

The propeller selected for the prediction of a transient condition can also foretell of a potential propeller cavitation problem as seen by studying Figure 6-7 and the discussion related to Figure 6-8.

The calculation procedure for determining the total thrust available with engine power and propeller characteristics is outlined in Table 6-B.

The values for K_T and J_T are taken from the intersecting points from Figure 6-7 and entered into Table 6-B for each boat speed. The resulting calculation gives the total available thrust for the twin screw propulsion arrangement.

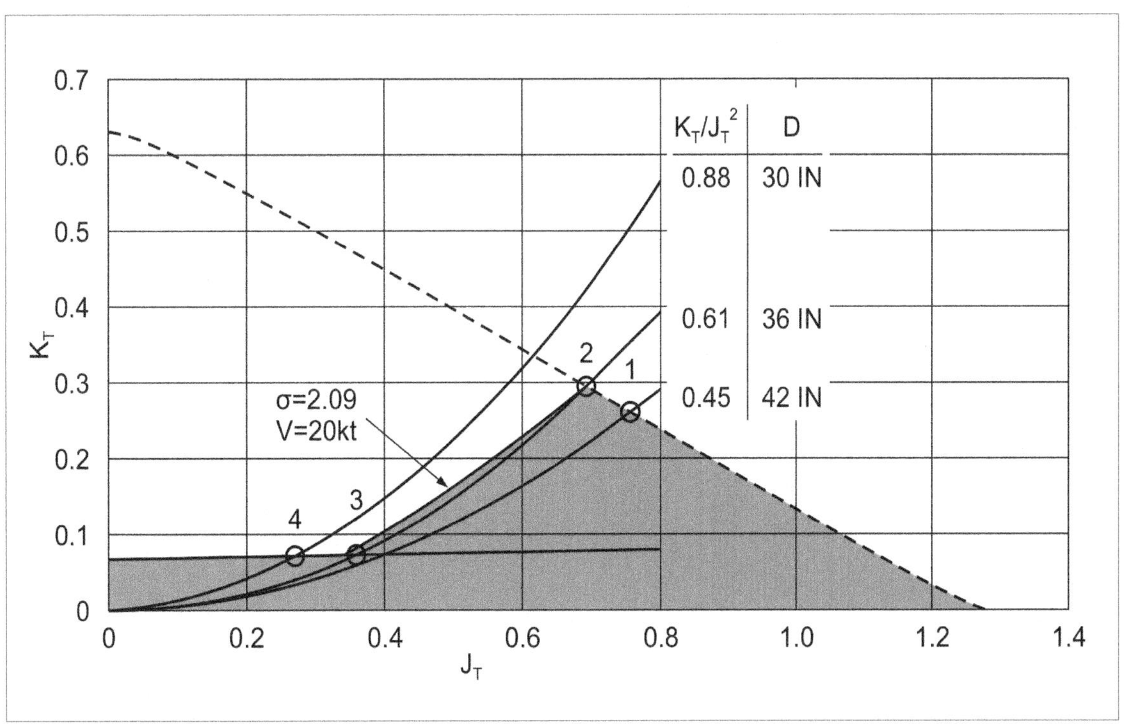

Figure 6-8: Decision guidance regarding propeller thrust loading relative to cavitation characteristics

TABLE 6-B: PREDICTION OF TOTAL THRUST AVAILABLE WITH POWER AND PROPELLER CHARACTERISTICS						
V kt	K_T	J_T	K_T/J_T^2	$1-W_T$ from Table 5-C	Per Shaft T lb	Total T lb
	from Figure 6-7					
5	0.535	0.24	9.288	1.06	13,315	26,630
10	0.440	0.43	2.380	1.06	13,645	27,290
15	0.355	0.59	1.020	1.05	12,910	25,820
20	0.300	0.70	0.612	1.01	12,750	25,500
25	0.250	0.79	0.400	0.97	12,020	24,040
30	0.200	0.88	0.258	0.97	11,160	22,320
35	0.160	0.96	0.174	0.97	10,210	20,420
Comments: K_T at intersecting points for $J_T \equiv J_Q$ K_T/J_T^2 – calculation				$T = (K_T/J_T^2)\ [\rho D^2(1.688)2\ V^2\ (1-W_T)^2]$ $T = (K_T/J_T^2)\ (51.04)\ V^2 (1-W_T)^2$ for this example Total T to be input into Table 6-C		

Table 6-C gives the procedure for total boat thrust with the thrust margin for each speed.

Then, with the total available thrust from Table 6-B the accelerating force (F) is calculated for each speed.

$F = T_P - T$ Equation 6-22

With the mass of the vessel plus mass of entrained water longitudinal acceleration, "a" may be calculated.

m = W(1+ entrained water allowance)/g Equation 6-23

Using an added mass allowance of five percent for this example:

$m = 100{,}000\ (1.05)/32.15 = 3{,}266\ \text{lb} \cdot \text{sec}^2/\text{ft}$

Recalling that F = ma, for this calculation procedure, it is convenient to use the reciprocal of a, that is 1/a which is:

$m/F = 1/a$ Equation 6-24

The results from Table 6-C provide 1/a versus speed (V) for this prediction example are seen in Figure 6-9.

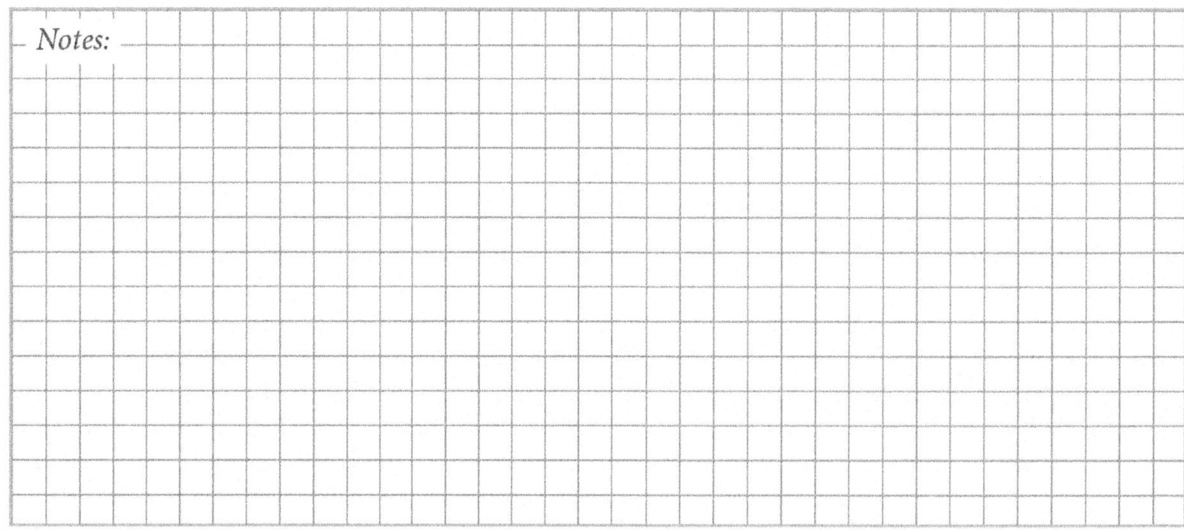

TABLE 6-C: PREDICTION OF BOTH 1/a AND LONGITUDINAL ACCELERATION FOR HOLE SHOT								
V kt	R_T lb	1-t	% margin distribu-tion	T lb	T_P lb	F lb	1/a sec^2/ft	Longitudinal acceleration g's
	Faired data from Table 5-C		From Fig. 6-2	Total steady state thrust	Total thrust avail. w/ total power	Accelerating force		
5	1,000	0.92	7	1,160	26,630	25,470	0.128	0.24
10	5,800	0.92	14	7,190	27,290	20,100	0.162	0.19
15	12,500	0.92	19	16,170	25,820	9,650	0.338	0.092
20	14,800	0.92	14	18,340	25,500	7,160	0.456	0.068
25	16,800	0.92	10	20,090	24,040	3,950	0.827	0.038
30	17,000	0.92	5	19,400	22,320	2,920	1.118	0.028
35	16,700	0.92	0	18,150	20,420	2,270	1.439	0.022

Comments:

20% maximum thrust margin applied at 14 knots – thrust hump speed

$T = R_T (1 + margin)/(1\text{-}t)$; example for 20 knots: $T = 14{,}800 (1.14)/(0.92)$

m – mass of craft plus 5% added mass of entrained water

$m = 100{,}000 \text{ lb } (1.05)/32.15 \text{ ft/sec}^2 = 3{,}266 \text{ lb sec}^2/\text{ft}$

$1/a = m/F$ Equation 6-12

$g\text{'s} = 1/[(1/a)(32.15)]$

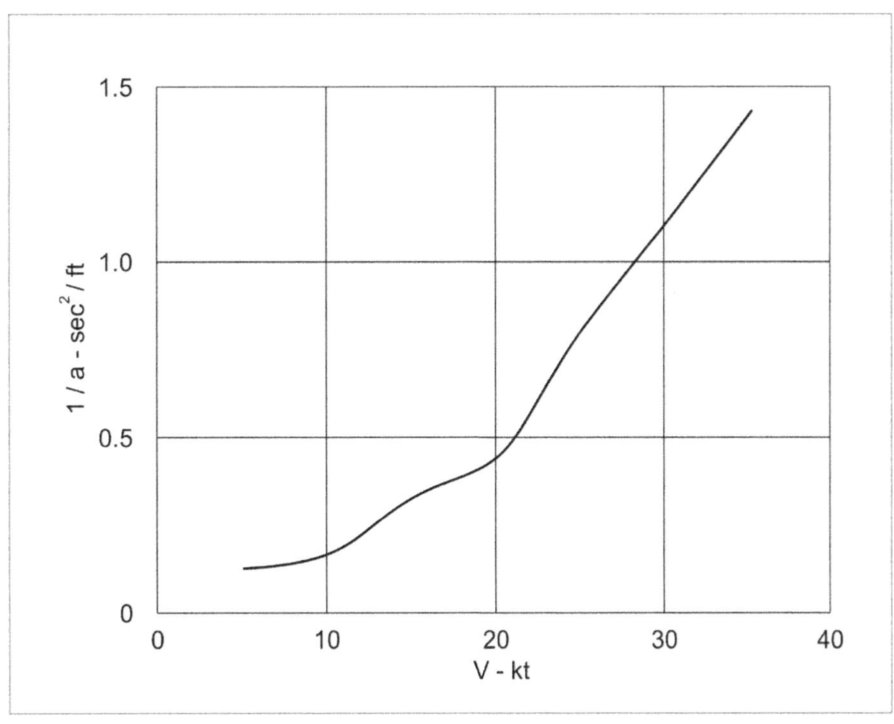

Figure 6-9: Prediction of the reciprocal of longitudinal acceleration versus speed

Time Versus Speed

The time history of speed change is found by incrementally integrating the area under the graph of 1/a versus V in Figure 6-9. There are various software tools for integrating the area under a curve. For this example, I used Simpson's first (1.4.1) rule in Tables 6-D and 6-E. The graph in Figure 6-10 of the results from Table 6-D gives the time history of velocity during acceleration.

The calculation approach for Table 6-D is divided into three 10-knot intervals, but it could have been a greater number in order to refine the time history of speed change.

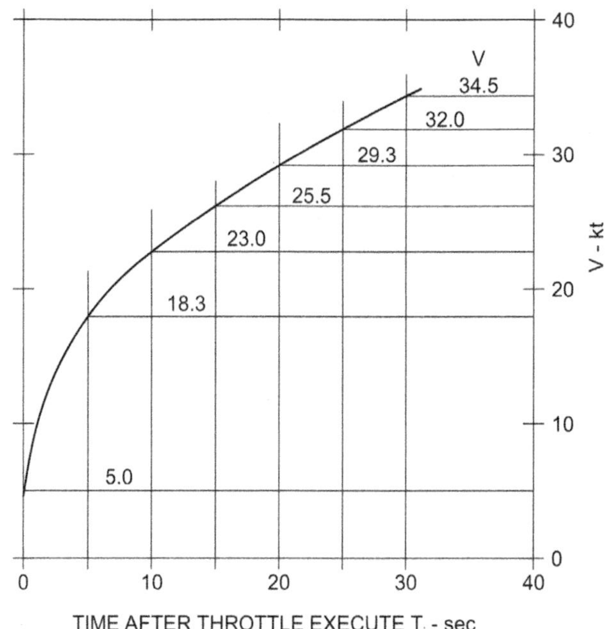

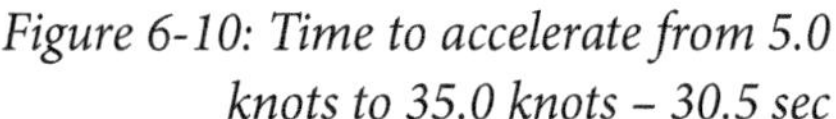
Figure 6-10: Time to accelerate from 5.0 knots to 35.0 knots – 30.5 sec

TABLE 6-D: TIME VERSUS SPEED INTEGRATION

V kt	Simpson's multiplier	1/a sec²/ft	f(1/a)		t_i sec
			incremental	cumulative	
5	1	0.128	0.128		
10	4	0.162	0.648		
15	1	0.338	0.338		
	Σ = 6		Σ = 1.114	Σ = 1.114	3.13
15	1	0.338	0.338		
20	4	0.456	1.824		
25	1	0.827	0.827		
	Σ = 12		Σ = 2.989	Σ = 4.103	11.54
25	1	0.827	0.827		
30	4	1.118	4.472		
35	1	1.439	1.439		
	Σ = 18		Σ = 6.738	Σ = 10.841	30.50

Comments:

V increment = 5.0 knots
t_i (time in seconds) = [1.688 (5)/3] Σf(1/a) [cumulative]
t_i = 2.813 Σf(1/a) [cumulative]
Simpson's first rule (1.4.1) used

Some designers, however, may only be interested in the elapsed time to accelerate to a particular speed, not time history. In this latter case, the incremental sums are not needed and the time to accelerate from five to 35 knots remains unchanged at 30.5 seconds unless the increment spacing is increased, which reduces accuracy.

Distance Versus Time and Velocity

Continuing this approach leads to prediction of the time history of distance as well as the relationship between speed and distance transited during acceleration which has utility for developing algorithms for control systems. Distance transited, X, is calculated by

$$X = 1.688\ V_{av}(t_i) \qquad \text{Equation 6-25}$$

The calculation example is in Table 6-E with a graph during acceleration of the time history of X is shown in Figure 6-11 and the velocity (V) relationship with Distance (X) is in Figure 6-12.

TABLE 6-E: PREDICTION OF DISTANCE CRAFT TRANSITS DURING TIME OF ACCELERATION

t_i – Time Sec	Simpson's multiplier	V kt	f(v) incremental	f(v) cumulative	V_{av} kt	X-Distance ft
0	1	5.0	5.0			
5	4	18.3	73.2			
10	1	23.0	23.0			
	Σ = 6		Σ = 101.2	Σ = 101.2	16.9	285
10	1	23.0	23.0			
15	4	25.5	102.0			
20	1	29.3	29.3			
	Σ = 12		Σ = 154.3	Σ = 255.5	21.3	719
20	1	29.3	29.3			
25	4	32.0	128.0			
30	1	34.5	34.5			
	Σ = 18		Σ = 191.8	Σ = 447.3	24.8	1,256

Comments:

Actual time to transit to 35 knots is 30.5 seconds while this integration is only for 30.0 seconds. The additional distance traveled for 0.5 seconds is

$x = 1.688\ V_{av}t_i = 1.688\ [34.5 + 35.0/2]\ 0.5\ \text{sec} = 29$ ft.

Total X = 1,256 + 29 = 1,285 ft.

t_i increment = 5.0 seconds

x (distance in feet) = $1.688\ V_{av}\ t_i$

Simpson's first rule (1.4.1) used

$V_{av} = \Sigma f(v)/\Sigma$ (Simpson's multiplier) [cumulative]

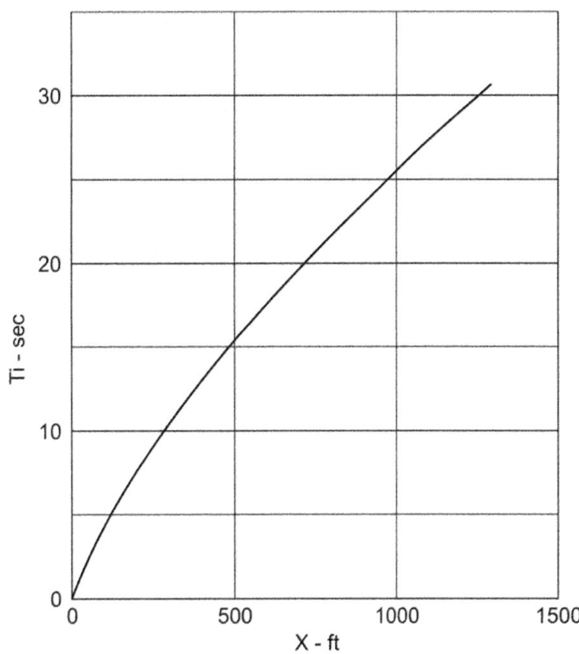

Figure 6-11: Prediction of distance transited during the time of acceleration from 5.0 to 35.0 knots

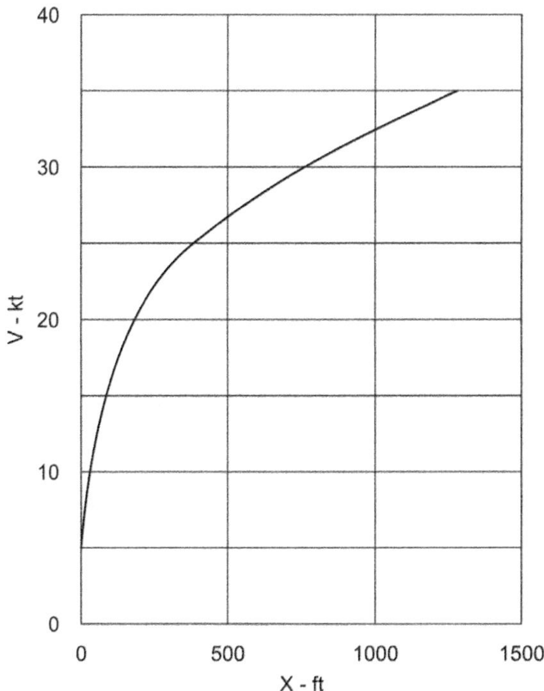

Figure 6-12: Prediction of speed for distance transited during acceleration from 5.0 to 35.0 knots

Shallow-Water Effects on Propulsion

Near the critical water depth speed, both hull resistance and propulsive factors are also dramatically influenced during shallow-water operation. Shallow water resistance is addressed in Chapter 3.

Here we'll discuss the components making up the propulsive efficiency (η).

Propulsive efficiency is reduced by the increase in thrust loading such as K_T / J_T^2 for propellers or waterjets. Also hull efficiency $\eta_H = (1\text{-}t)/(1\text{-}W_t)$ changes as both thrust-deduction and wake factors are effected by propulsor thrust loading.

Experimental data defining the trends for changing values of thrust deduction and wake factors due to shallow water are not readily available.

In 1968, I reported some limited trial data, which provides some guidance as to changes of wake factors due to shallow water. With sea trial data, you may also approximate a ratio of shallow-to-deep water OPC values as a function of F_{nh} and h/L.

There are some deep- and shallow-water full-scale sea trial data (Blount et al. 1968) for a propeller-driven, hard-chine craft for a range of speeds from displacement up to planing conditions $0.8 \leq F_{nV} \leq 3.5$).

Analysis of these data provides some trend information regarding the influence that water depth has on propulsive factors near and above the theoretical critical depth Froude number ($F_{nh} = 1.0$).

The shallowest depth in which this craft was sea-trialed was $h/L_p = 0.156$ as operation in less water depth resulted in the propellers touching the sandy bottom at hump speed.

Propulsive efficiency (η) is defined as:

$$\eta = \eta_O \, \eta_H \, \eta_R \qquad \text{Equation 6-26}$$

The following can be inferred from anecdotal information for the effective change of propulsive factors relative to deep water when in shallow water of $h/L_P \approx 0.20$ and $F_{nh} = 1.0$.

Propulsive factors	Rough order of percent change due to shallow water
η	-15 to -20
η_R	0 to 4
η_H	-2 to +4
η_O	-15 to -20 (due to increased thrust loading)

The major increase in power necessary to accelerate this boat through the critical depth Froude number (F_{nh} = 1.0) speed is the reduced propeller efficiency caused by the increased propeller loading (K_T/J_T^2).

Thus, an engineering approximation for calculating the hole shot response of propeller-driven boats is likely to be adequately represented when accounting for the increase in shallow water resistance, thereby increasing K_T/J_T^2 causing propeller efficiency η_O to decrease.

When the combination of propulsion machinery does not produce agile operation, study your sea trial results for specific boats, engines, gears, and propellers. You could need to make changes in components or consider options such as two-speed gears or controllable-pitch propellers.

Evaluating Shallow Water Depth Performance

To a limited extent, the methodology discussed allows a designer to evaluate the combined effects of shallow-water and propulsion machinery. These combined effects can impede a high-performance craft from accelerating from displacement to planing speeds; in other words, a boat may not get over the planing hump.

It becomes important to isolate the primary problem. Is it a shallow water hull resistance problem, or is the difficulty a result of selecting borderline propulsion components and the design displacement with its LCG?

Therefore, first conduct acceleration sea trials in deep water, $\eta \geq 0.8$ (LOA). Should acceleration not be satisfactory, analyze trial results and make necessary design changes. Conduct additional trials to verify that acceptable acceleration is achieved before testing in shallow water.

CHAPTER SEVEN

SEAKEEPING: SIZE DOES MATTER

There is much yet to learn about what designers can control early on with their choices in hull geometry and mass distribution. How do these decisions influence ride quality for onboard personal comfort, crew proficiency, and the functionality of equipment and safety? What about vessel seakeeping characteristics, considering dynamic stability and structural integrity? If you are able to quantify sensitivity to sea surface of hull design details, you can also make the best choices to resolve conflicts between calm- and rough-water requirements.

In a general sense, the factors which influence ride quality in a seaway include hull-transverse section shape (deadrise, bow shape, etc.), dynamic or running trim angle, hull-bottom loading, and length-to-beam ratio. These elements also affect calm-water trim angle, hull resistance, and forward-looking sight line from the helm. Thus, to design high-performance vessels to be operated in a seaway, you must come to understand the interaction of these variables and incorporate the best features for ride quality while maintaining acceptable and efficient calm-water performance, but operational characteristics in a seaway must take precedence overall.

In your design, you must also describe the environment in which the vessel will operate. The form and function of an effectively designed vessel is strongly tied to its operating environment. For example, the North Sea, Arabian Sea, and Caribbean offer very different design challenges which you can obtain from the wave statistics for ocean regions shown in Figure 7-1.

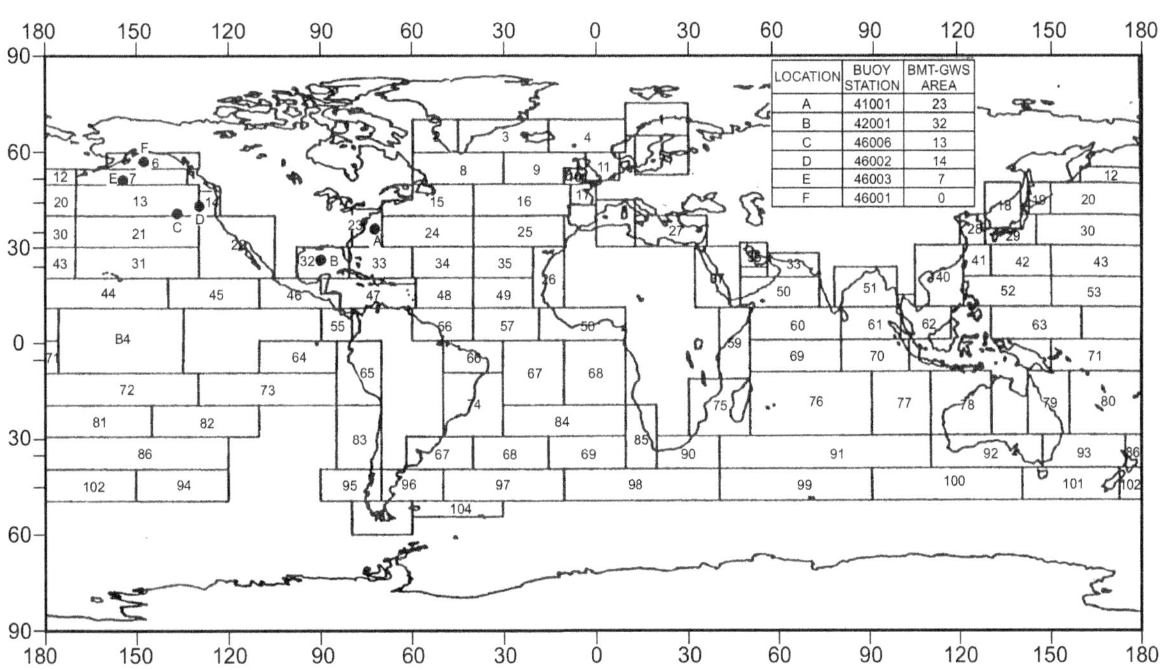

Figure 7-1: Designated regions of the oceans of the world where wave statistics are available

As you develop the hull and mechanical and electrical systems, keep the environmental characteristics of the vessel's operational area in mind for proper functioning of equipments.

You must specify materials that are compatible with temperature, humidity, salt, sand, dust, ice, wind, and waves as appropriate.

Wind and wave specifications are generally summarized by a sea-state definition. Wave conditions are commonly typified as "irregular, confused seas."

"Irregular" means that the waves are not uniform in height or length.

"Regular seas" are generally encountered when the center of the weather disturbance generating the waves is a great distance from an observation point.

"Confused" means that the waves are arriving from different directions.

"Short crested waves" is synonymous with "confused waves."

"Long crested" or "unidirectional" describes waves running in the same direction.

Finally, there are local conditions which have a significant effect on sea state. These include water depth, currents, fetch, wind, and the presence of crushed, broken ice. For our purposes, however, we'll assume that all operations will occur in an ice-free environment.

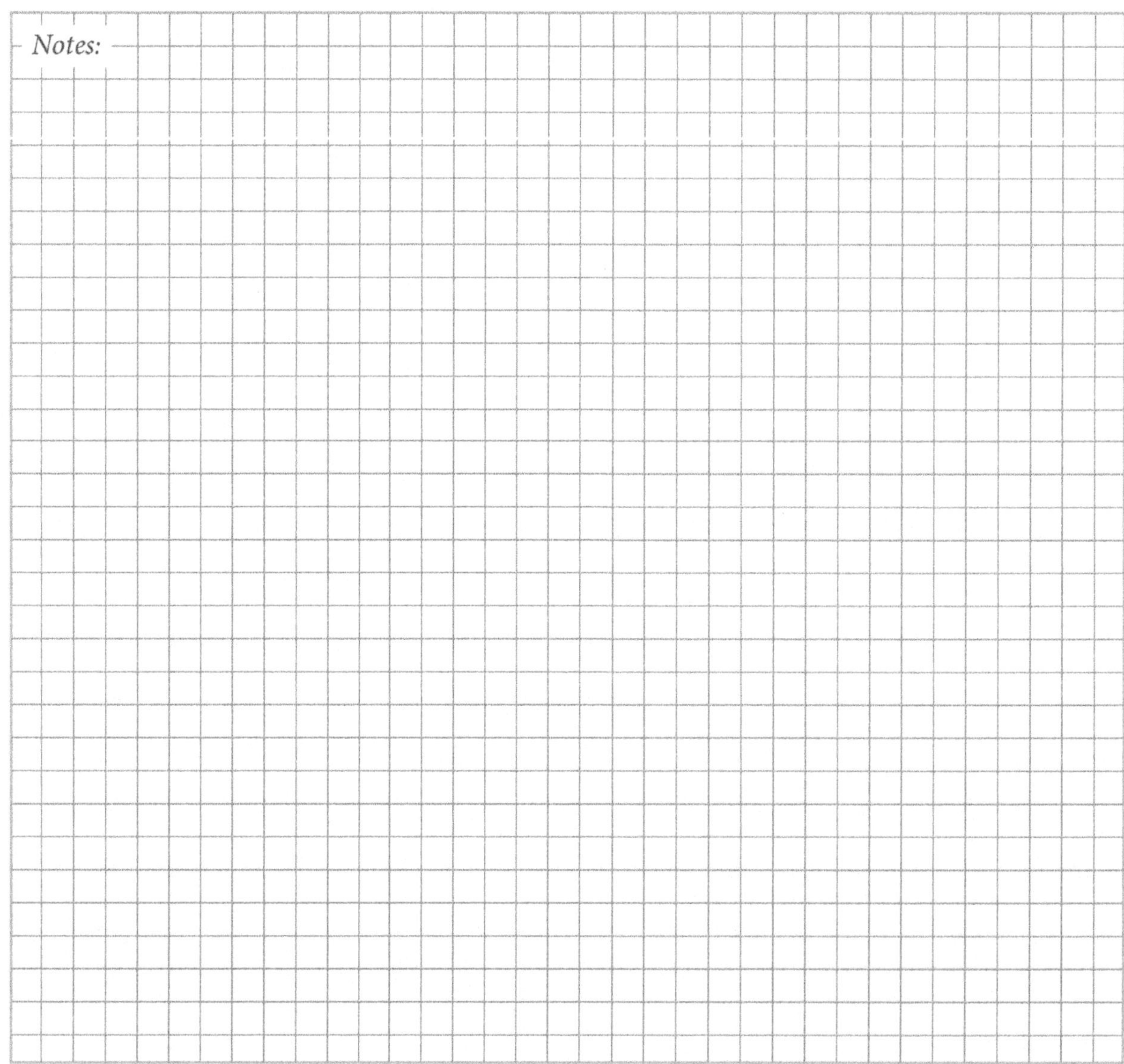

Planing Craft Operating in Irregular Waves

From 1971 through the 1980s, several researchers organized and presented irregular wave-test data in a format for making design predictions.

These included Fridsma (1971), Savitsky and Brown (1976), and Hoggard and Jones (1980) for experimental data. Both Fridsma and the Savitsky and Brown data are from prismatic models having rather bluff bows.

The experimental program for each model was conducted at two calm-water trim angles achieved by varying LCG for each speed. At that time, the researchers obtained their experimental acceleration data by manually reading time-history oscillograph records.

Hoggard and Jones collected seakeeping data for a number of planing-boat designs and performed a regression analysis to explore significant variables. Since data were available for only a small number of hulls with no systematic relationship, the regression results are limited to the geometric and hull-loading variables existing for the study.

To obtain seakeeping qualities for concept and feasibility studies when there were no expectations of subsequent model tests, I relied on the Hoggard and Jones vertical acceleration prediction method. This approach, within engineering expectations, has been validated whenever full-scale seakeeping data became available.

In 1995, Joseph Koelbel revisited the subject of seakeeping in a paper which had two objectives:

- to quantify the uncertainties in design procedures for hull structural loads resulting when craft operate in irregular seas, and
- to recommend an alternative design approach which could reduce design uncertainty.

The Koelbel paper elicited prepared comments from a broad spectrum of knowledgeable professionals who had contributed much to the technology and design approaches for high-performance craft in irregular seas.

These discussions greatly enhanced the value of the Koelbel paper, and from these discussions, a consensus grew that establishing vertical acceleration at the longitudinal center of gravity (LCG) is the essential starting data necessary for proceeding with a rational hull structural design process.

Response motions and accelerations of a craft operating in irregular seas, however, have many significant implications beyond structural integrity of a hull.

The nature of hull motions, velocities, and accelerations—both magnitude and frequency—have important human-factor influences on the safety of crew members and their ability to perform their duties for the duration of their assignments. These same hull responses in a seaway affect equipment/systems operability as well as the vessel's capability to interface with other manned and unmanned craft.

Koelbel's paper suggests that some uncertainty in most structural-design approaches can be mitigated by establishing, with some degree of confidence, comparison of vertical LCG acceleration with similar hulls of known geometry, mass, and its distribution in response to defined sea spectra.

While the prediction techniques of Fridsma, Savitsky and Brown, and Hoggard and Jones have been meaningful for design purposes, there is a need for better technical knowledge between LCG acceleration, hull geometry, and loading for making valued trade-off studies.

Some early analytical approaches for predicting accelerations and motions of hard-chine planing craft in irregular seas include Martin (1978) and Zarnick (1979). These studies provided technical direction but required further development into a user-friendly format before wide use by designers.

In 1999, Richard Akers developed commercially available software extended from Zarnick's work to offer a rudimentary design tool for predicting, in response to irregular waves, planing-craft motions and accelerations.

Akers' POWERSEA® [Source http://www.shipmotion.com/pwrsea.htm] software offers improved motion and acceleration predictions when experimental data are available to correlate some software constants for defined hull geometry.

Seakeeping Database

I have planned and managed a number of seakeeping model-test programs for a rather wide range of vessels which are now in operation. I prefer model tests of high-speed planing craft be conducted in irregular waves having characteristics of the expected region of operation.

My goal is threefold:

- Obtain motion and vertical acceleration data to evaluate the vessel's design for suitability of its intended purpose,
- Validate the design assumptions/classification authority requirements for structural integrity, and
- Expand seakeeping technology through organized test programs of disparate designs taking place over a period of years.

Within some of these seakeeping programs, I have made micro R&D studies with systematic variations in hull geometry. For example, using a monohull with fixed forebody shape, I evaluated seakeeping response due to changes in the afterbody of the hull with hook, straight, and rocker buttock shapes.

Most models were instrumented with three or four accelerometers to define its longitudinal distribution along the hull's length for locating personnel and equipments for the best ride-quality environment. These tests are often conducted in two sea states. Also, I have conducted some model tests in different modal wave periods while maintaining constant significant wave height to obtain knowledge of the influence of encounter frequency and wave slope on hull form.

Figure 7-2: Geometry definitions for the round-bilge and hard-chine monohulls regarding seakeeping

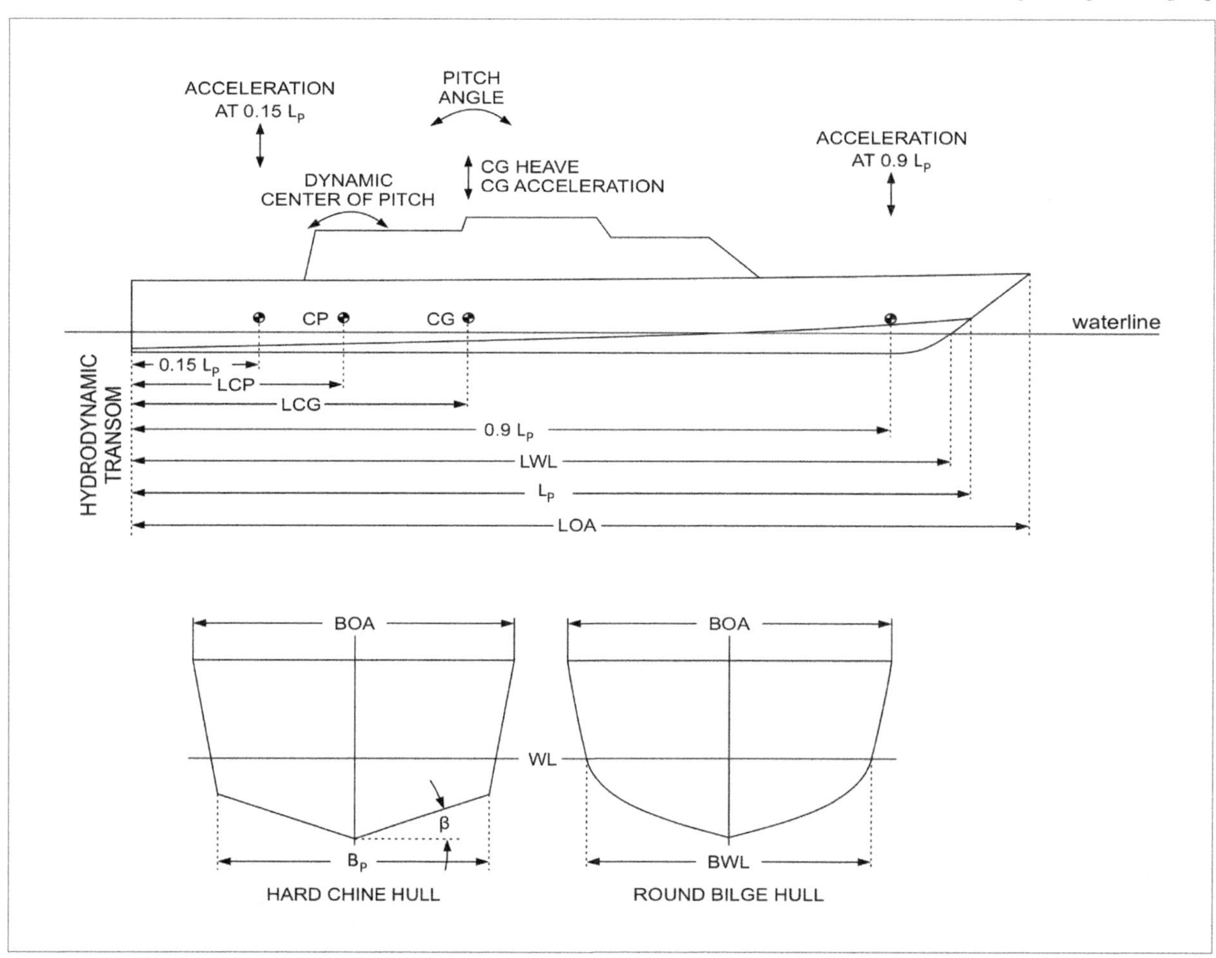

Thus, the long-term approach I have taken is somewhat similar to that used by Hoggard and Jones, but with more detail in order to collect experimental data at the same dimensionless speeds for different designs which I could then analyze for technology trends.

This database has grown to be a significantly larger size than either held by Fridsma or Hoggard and Jones. As of 2014, I have more than 800 lines of data available for analysis and developing design tools for making predictions for planing craft in irregular waves from displacement up to very high planing speeds of $F_{nV} \approx 7.1$.

The preponderance of information making up this seakeeping database of monohull models has been obtained since 1990. These model-test data are, for the most part, from four different international hydrodynamic research facilities with model sizes reasonably large, but in keeping with the avoidance of tow-tank blockage effects. Hull geometry used for analysis of motion and acceleration relationships are defined in Figures 7-2 and 7-3.

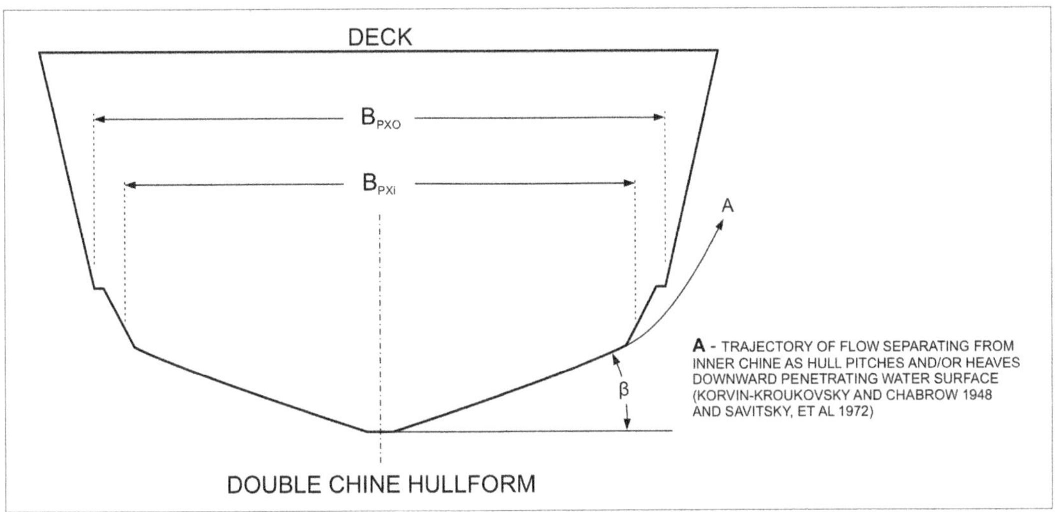

Figure 7-3: Chine beam definitions for double-chine hulls regarding seakeeping

The range of model hull geometry and loading, relative head-sea conditions and speeds are given in Table 7-A. The entire database is established in model-scale metric dimensions. To date, this data collection has utilized seakeeping results from more than 20 vessel designs which have been built. The database remains open for expansion as additional model-test results become available.

TABLE 7-A: RANGE OF CONDITIONS FOR THE DLBA SEAKEEPING DATABASE

Maximum Sea					Maximum Speed	
$H_{1/3}/\nabla^{1/3}$	$H_{1/3}/B$	$H_{1/3}/B_{pxo}$	$H_{1/3}/B_{pxi}$	Full-Scale Modal Wave Periods	F_{nV}	F_{nL}
0.781	0.648	0.603	0.625	2.9 to 9.0 sec	7.10	2.85

Hull Loading Range			Model Geometry Range					
$L_p/\nabla^{1/3}$	$L/\nabla^{1/3}$	$\nabla/B_{pxo}{}^3 = C_\Delta$	LOA/BOA	L_p/B_{pxo}	L/B	β_{mid}(deg)	β_T (deg)	Pitch Gyradius/L_p
2.92 to 8.14	2.61 to 8.03	0.182 to 1.326	2.55 to 6.06	1.65 to 6.82	2.20 to 6.29	-5 to 24	0 to 22	0.196 to 0.302

The format in which experimental seakeeping data have been reported is not the same for all research facilities.

In some cases, it may be as specified by customers. For example, I have found vertical accelerations to be reported in various references in one or more of the various forms: RMS, minimum, mean, maximum, significant, and average of one-tenth of the highest; some using total signal and others analyzing only positive peaks.

Within the broad collection of references available to develop this database of experimental information, the RMS format of the total signal was most commonly available.

The majority of these tests were performed to provide acceleration data at the LCG for hull structural design. Some programs also reported analysis format for response for ride quality/human factors requirements, such as one third octave RMS analysis or evaluating the vessel as a suitable platform for specialized mission equipments.

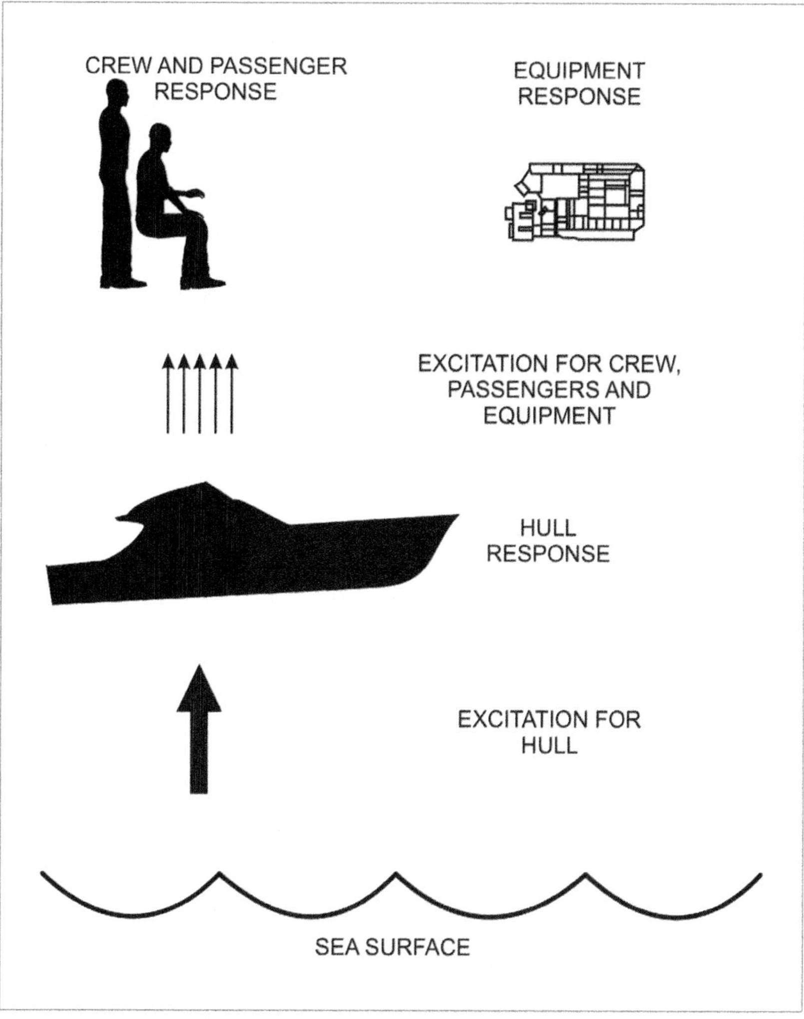

Figure 7-4: Reciprocal of longitudinal acceleration versus craft speed

The ultimate goal in designing surface vessels to operate in irregular waves is to produce a craft with minimal motions and accelerations in response to the waves in order to provide a stable platform for crews and operation of equipments.

Figure 7-4, in a simplified way, depicts relationships between high-speed vessels and the sea surface. Take note that my focus here is only on the head-sea condition and experimental response of the hull having analysis of vertical acceleration containing statistical parameters of at least RMS format.

An elementary depiction of planing craft in relationship to the sea surface is provided in Figure 7-5.

The through-wave concept ("A" in Figure 7-5) has been mentioned in literature since the mid 1980s. It approaches the ultimate goal of minimal motions and acceleration. The VSV (very slender vessel) is an example of a through-wave concept which might achieve the desired minimal motion ride quality.

Designers and operators need to better understand the motion and acceleration response of a high-speed vessel to the sea surface ("B" in Figure 7-5).

This condition of free-to-trim-and-heave hull response without dynamic motion control operating in irregular head seas, represents the major content of the database. The application of dynamic motion control, especially dynamic pitch control, tends to move a vessel from "B" toward "A."

The third circumstance ("C" in Figure 7-5) relates to a planing boat which has become airborne due to the combination of speed, wave height, and slope. This is shown as it is about to re-enter the water. This is analogous to a seaplane landing. This latter technology has been studied by NACA and NASA both experimentally and analytically.

This NACA/NASA technology relates to small craft operating in a relatively large sea state at high speed when it comes off the crest of a wave and re-enters the water. This action serves as a technology-bounding condition for very high dimensionless speeds of surface craft.

Fridsma, Savitsky, and Brown considered this seaplane landing-impact technology as an analogy to define significant variables when they were formulating equations to predict vessel accelerations in waves. Pitch oscillations of the hull relative to the wave surface, hull bottom loading, and—to a lesser extent—heave velocity are factors affecting motions and vertical accelerations.

Cross-section shapes of the hull bottom (concave, vee, ox bow, convex, round) are significant with regard to rate of change of vertical accelerations as craft impacts waves.

The goal through research is to improve the quality of motion and acceleration prediction of craft operating in a seaway. Then there will be an opportunity to minimize rigid hull motions and accelerations, within constraints of requirements, by selecting hull geometry, bottom loading, LCG, and wetted bottom shape to influence trim. Anticipating irregular seaways, it is the responsibility of the naval architect or design manager to minimize pitch oscillations and control the center of pitch which is necessary to restrain vertical velocity and thus, vertical acceleration of a vessel.

Dynamic pitch control becomes vital when sufficient motion and acceleration reduction cannot be achieved within available design technology for hull geometry and bottom loading.

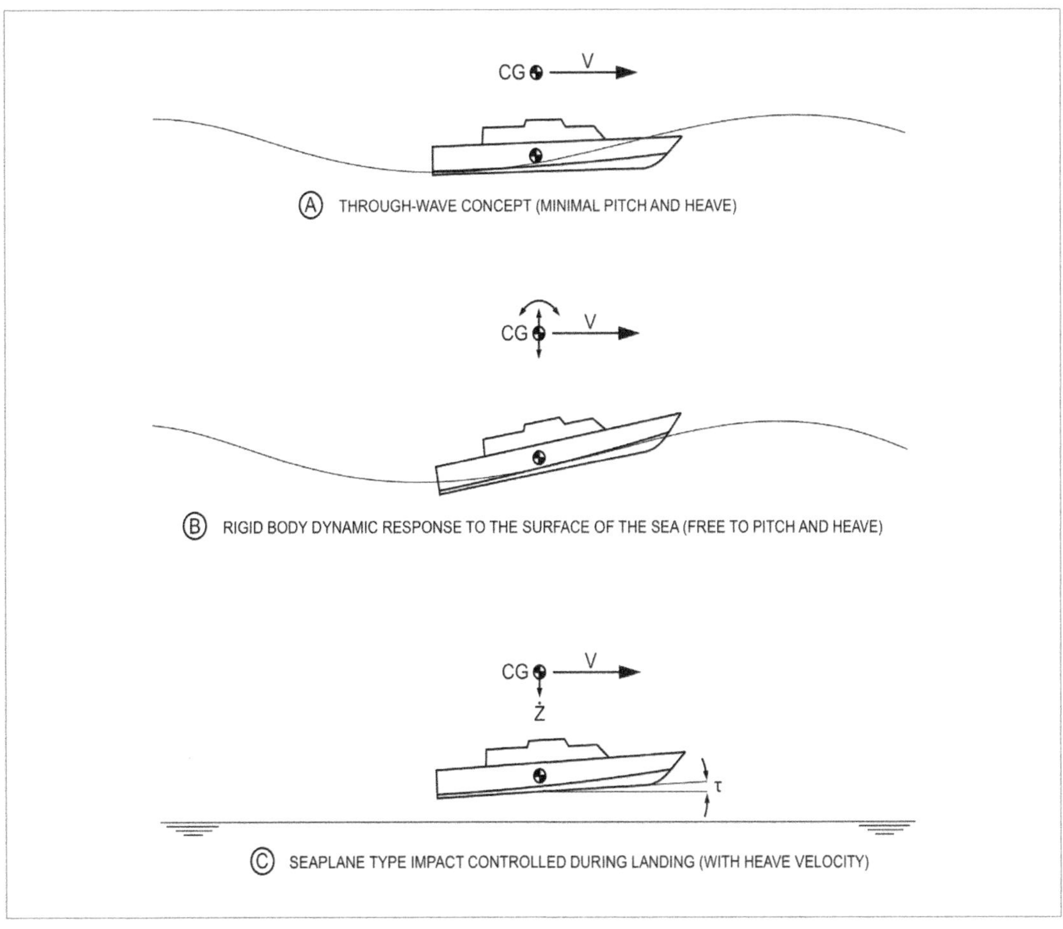

Figure 7-5: Relationship of high-speed vessels with respect to the surface of the sea

Influence of Wave Period

With the arrival of dockships onto the yachting scene, and the military capability of airlifting combatant craft to worldwide operational areas, vessels become open to irregular seas of all oceans and large bodies of water. Thus, conducting model tests or making predictions for single-sea characteristics may not expose undesirable hull responses.

Vessels having an operational profile with seasonal activities in different oceans still must undergo model tests or predictions in irregular head seas for likely significant wave heights, with modal wave periods representative of diverse bodies of water. You must determine a hull's response for sea conditions in all of its expected operational areas.

Sea conditions may be obtained for worldwide locations seen in Figure 7-1, for a fee, for observed data from the Global Wave Statics database maintained by BMT Fluid Mechanics (Teddington, United Kingdom).

A hard-chine, deep-vee craft was model-tested to provide a good example of the influence of wave period on ride quality. This example is representative of military and recreational hulls.

These model tests were conducted at a loading condition for a slenderness ratio ($L_P/\nabla^{1/3} = 6.9$) with its equivalent planing hull area coefficient ($A_P/\nabla^{2/3} = 7.1$). The irregular head sea condition for the three modal periods was at a constant significant wave height ($H_{1/3}/\nabla^{1/3} = 0.44$).

Vertical accelerations were measured at four longitudinal positions with LCG data presented in Figure 7-6. RMS accelerations are shown for the craft versus F_{nV} for three wave periods.

There are three distinct curves, one for each modal wave period. The highest accelerations were recorded for the shortest wave period, the sea condition which has the greatest wave slope. Trends seen in Figure 7-6 for RMS vertical acceleration are replicated when maximum vertical acceleration (slam peaks) are graphed in this format.

Note that at $F_{nV} = 4.5$, a relatively high planing speed, the RMS vertical acceleration at the LCG is fifty percent higher for a 4.6 second wave period than for a 9.0 second period. By approximating the period of irregular head seas with the modal wave period (T_W) the craft frequency of encounter (ω_e) becomes

$$\omega_e = \omega\,[1 + (\omega v)/g] \qquad \text{Equation 7-1}$$

where wave circular frequency is approximately

$$\omega = 2\pi/T_W \qquad \text{Equation 7-2}$$

When the RMS acceleration data shown in Figure 7-6 is graphically presented in Figure 7-7 as RMS acceleration versus frequency of encounter, it has a slightly different look. These data for different wave periods, however, continued to be three distinct curves.

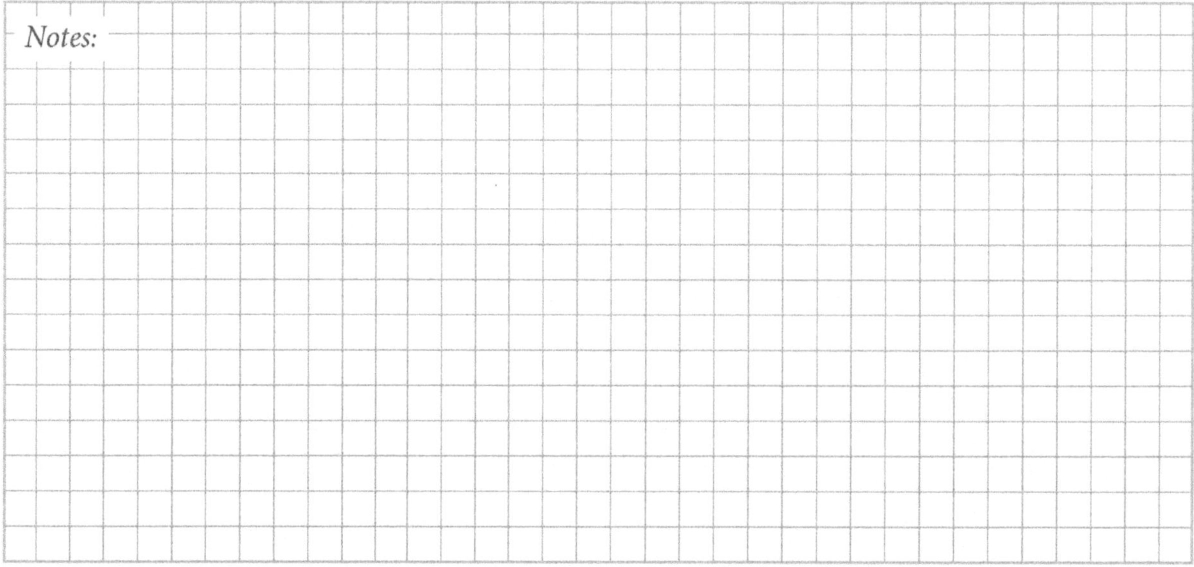

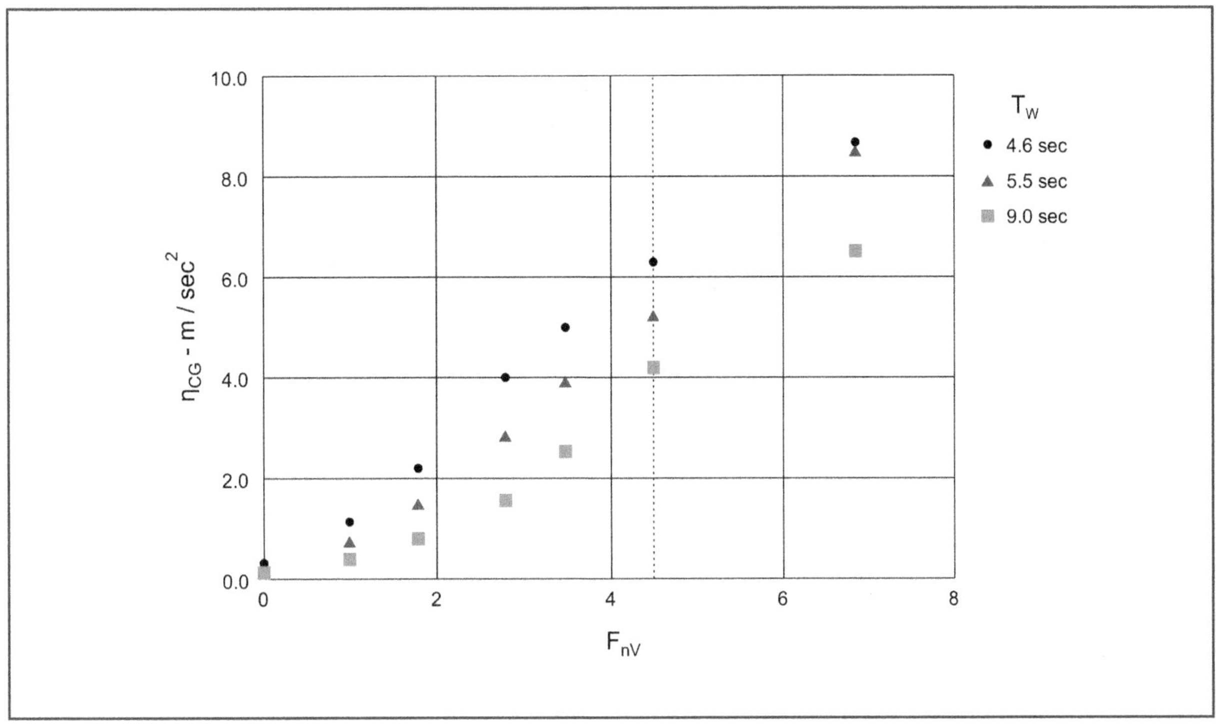

Figure 7-6: RMS vertical acceleration in head seas at LCG versus F_{nV} for three wave periods

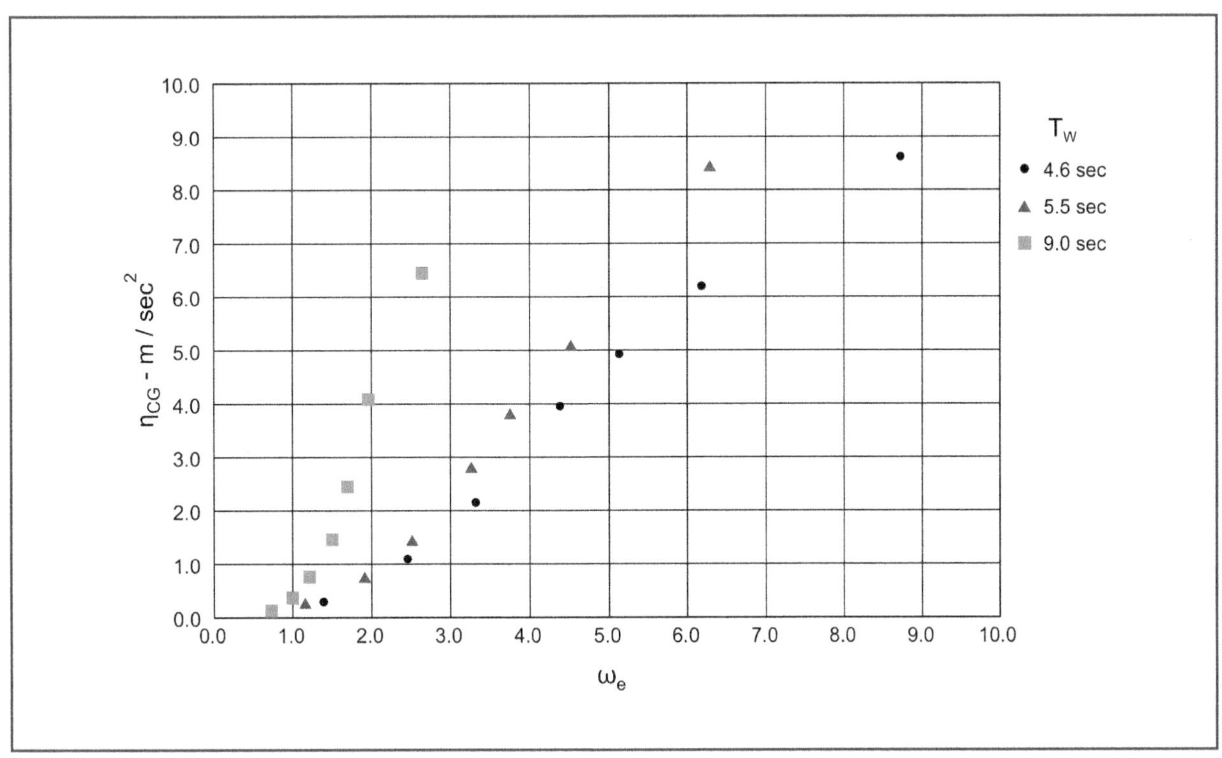

Figure 7-7: RMS vertical acceleration in head seas at LCG versus ω_e for three wave periods

I have found that, for engineering purposes, CG acceleration data for the three wave periods tend to come together to define a single trend line when plotted versus wave encounter number seen in Figure 7-8. This dimensionless wave encounter number is defined as

$$\Omega = \omega_e T_W \qquad \text{Equation 7-3}$$

Finally, RMS vertical acceleration versus mean running trim angle is provided in Figure 7-9a. Again, the acceleration data form three clearly defined patterns, one for each wave period.

Beginning with displacement speeds, RMS acceleration increases with both speed and mean trim angle, increasing at a rate more than twice when operating in $T_w = 4.6$ second waves than when $T_w = 9.0$ seconds.

At and above maximum hump trim speed ($F_{nV} \approx 2.8$ for this model) RMS acceleration, when planing, increases dramatically with speed and reducing wave period even though trim angle is also reduced with speed increase.

However, it cannot be determined from Figure 7-9a how much, if any, the acceleration is reduced as the mean trim angle continues to go down from 4+° at $F_{nV} = 2.8$ to less than 2° as the craft increases to high planing speeds.

Note that the same experimental acceleration data are seen graphically, in different formats, in Figures 7-6 to 7-9a and 7-10.

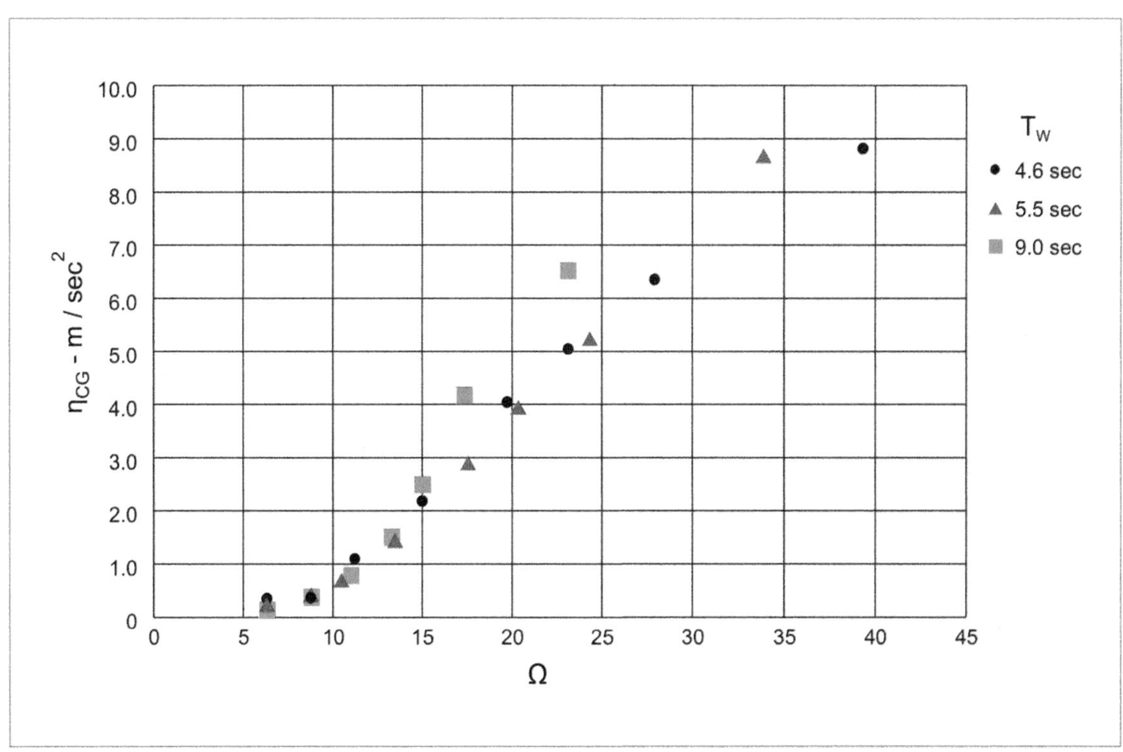

Figure 7-8: RMS vertical acceleration in head seas versus encounter number for three wave period

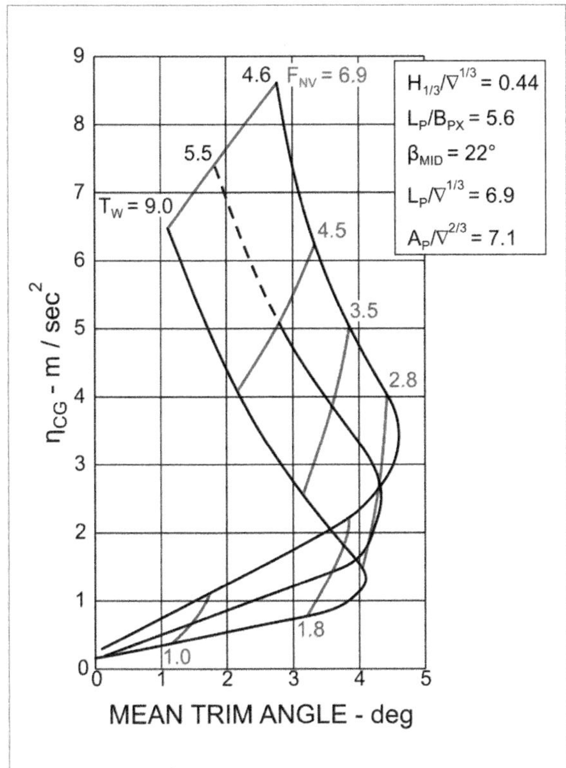

Figure 7-9a: RMS vertical acceleration in head seas having constant $H_{1/3}/\nabla^{1/3}$ = 0.44 for three wave periods versus mean trim angles

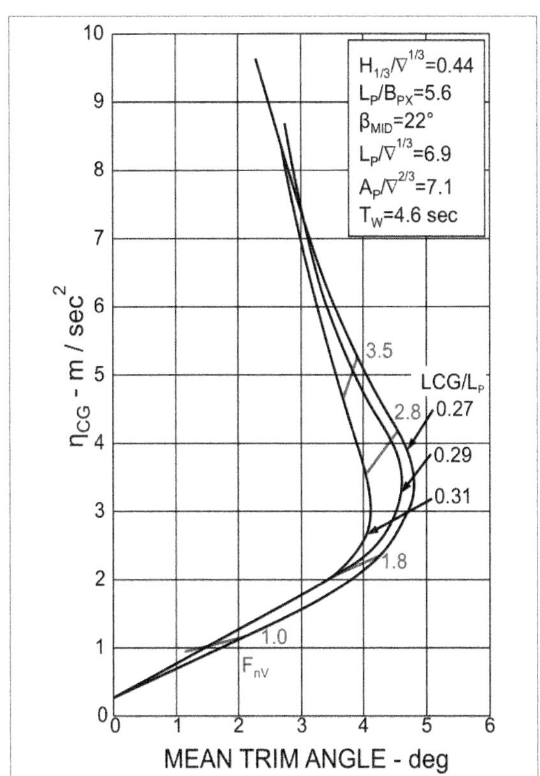

Figure 7-9b: RMS vertical acceleration in head seas having constant $H_{1/3}/\nabla^{1/3}$ = 0.44 with T_w = 4.6 seconds versus mean trim angles varied by three LCGs

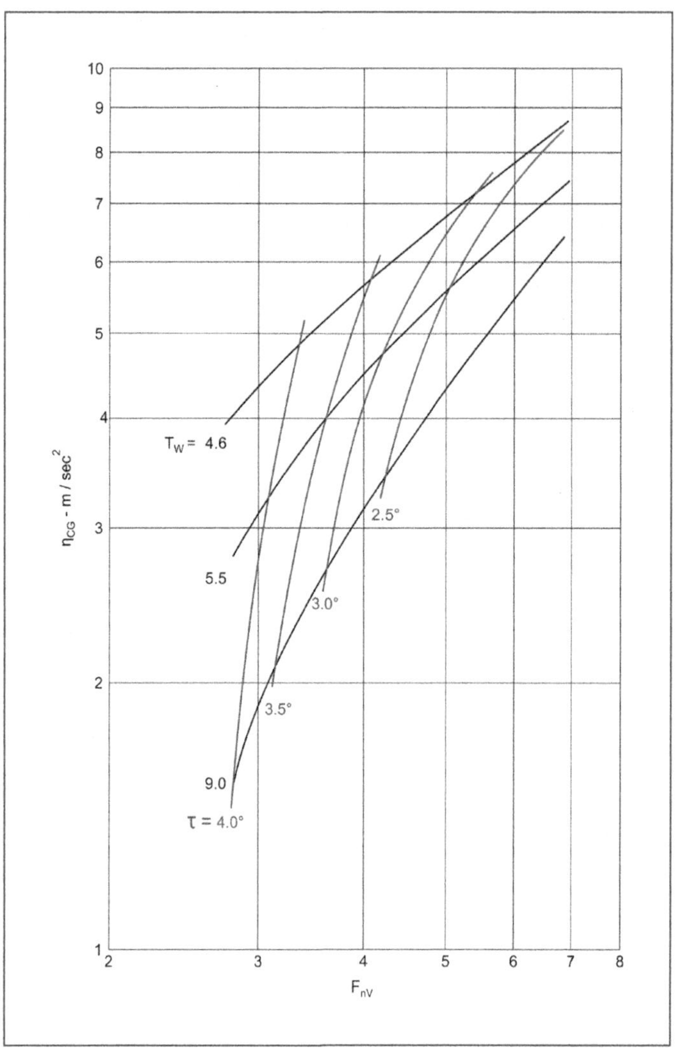

Figure 7-10: RMS vertical acceleration showing contours of mean trim angles in head seas having constant $H_{1/3}/\nabla^{1/3} = 0.44$ for three wave periods versus $F_{nV} \geq 2.8$

As seen in Figures 7-9a & 7-9b, the relationship of vertical acceleration at the LCG of hard-chine craft is meaningfully influenced by speed, sea conditions, and to a lesser extent by trim angle of the craft, especially since significant wave height and modal period combine to define wave slope. This influence is very different when vessel weight is supported mostly by buoyancy rather than hydrodynamic lift.

Vertical accelerations for speeds up to $F_{nV} \approx 1.8$ are relatively minor compared to accelerations anticipated at high speeds, $F_{nV} \geq 2.8$. For speeds, F_{nV} between 1.8 and 2.8 vertical accelerations at the LCG show the effect of transitioning from being primarily supported by buoyancy to hydrodynamic support.

Figure 7-10 is reformatted data having constant $H_{1/3}/\nabla^{1/3}$ from Figure 7-9a to explore the relative significance of vertical acceleration at the LCG of a planing craft for various speeds and trim in response to irregular head seas of different modal wave periods. At constant trim angles of three and four degrees, which are somewhat vertical near to constant planing speeds, vertical acceleration at the CG more than doubles when modal wave period is halved.

Clearly, with this matrix of variables, there can be large differences in experimentally obtained values of vertical acceleration for each hull design that is dominated by characteristics of irregular sea conditions.

Designers control planing hull geometry, mass and mass distribution, dynamic motion-control systems, and propulsion power and, therefore, they must incorporate the combinations of these aspects which minimize the response of a craft to the sea surface.

To obtain minimum vertical acceleration of planing craft, I believe it is more important to eliminate or minimize pitch oscillations in response to the sea surface—rather than to design for the lowest mean trim angle in a seaway at which dynamic stability can be maintained. Also, if requirements place a craft in a variety of sea conditions, then the design sea condition should be for the lowest modal wave period and seasonal wave heights of the expected operational areas. The lesson here is that experimental or analytical methods for predicting vertical acceleration for hard-chine craft must account for *both* wave period and some measure of wave height.

These two components—significant wave height and modal period—are the necessary minimum information for defining the operational sea conditions in order to obtain valid magnitude of vertical accelerations to be used for both design of hull geometry and structure in addition to predicting speed-wave height operational envelopes based on human-factor criteria.

Airborne Craft

While collecting seakeeping information for the database, I observed that measured vertical acceleration aft of the LCG was occasionally greater than at the LCG. Initially, I speculated that this resulted from random or spurious measurements. Over time, as additional seakeeping data became available from tests of more hull designs, vertical accelerations aft of, rather than at, the LCG were noted to be higher for more than five percent of the recorded conditions.

I require video recordings of model tests and carefully review these at reduced frame rates. This visual review documented that higher vertical acceleration aft of the LCG frequently occurred when the hulls became entirely airborne and re-entered the water with first contact being at the stern. I have model-tested nine different hull designs up to sufficiently high speeds whereby wave height and modal periods combined to result in the craft becoming airborne.

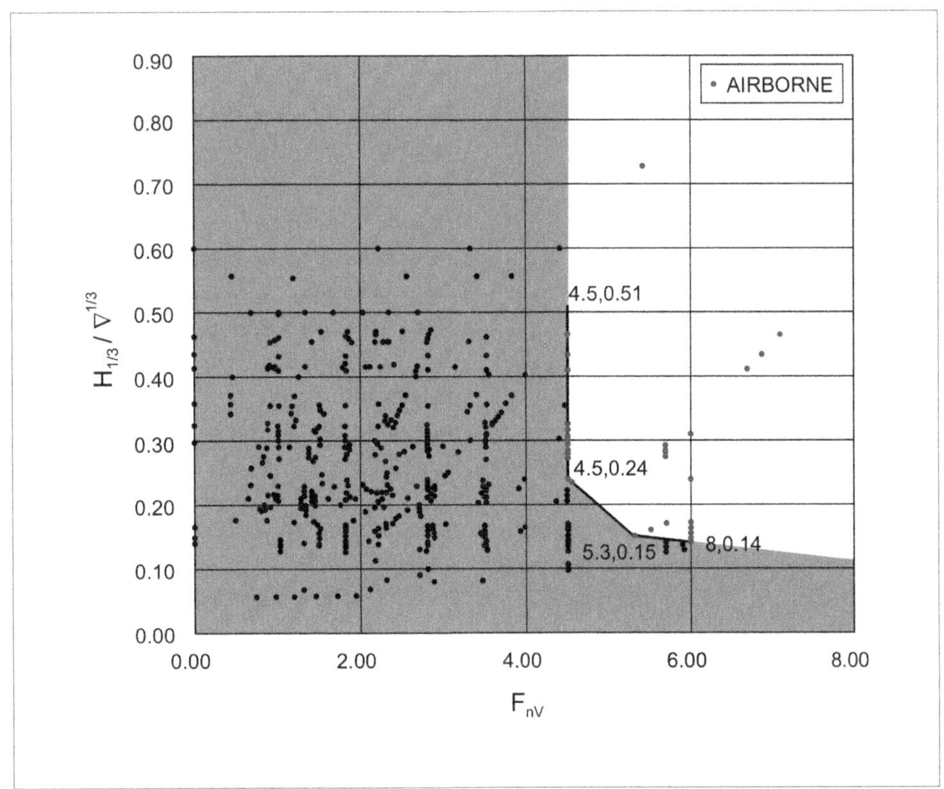

Figure 7-11: Relative sea conditions when craft may become airborne in head seas

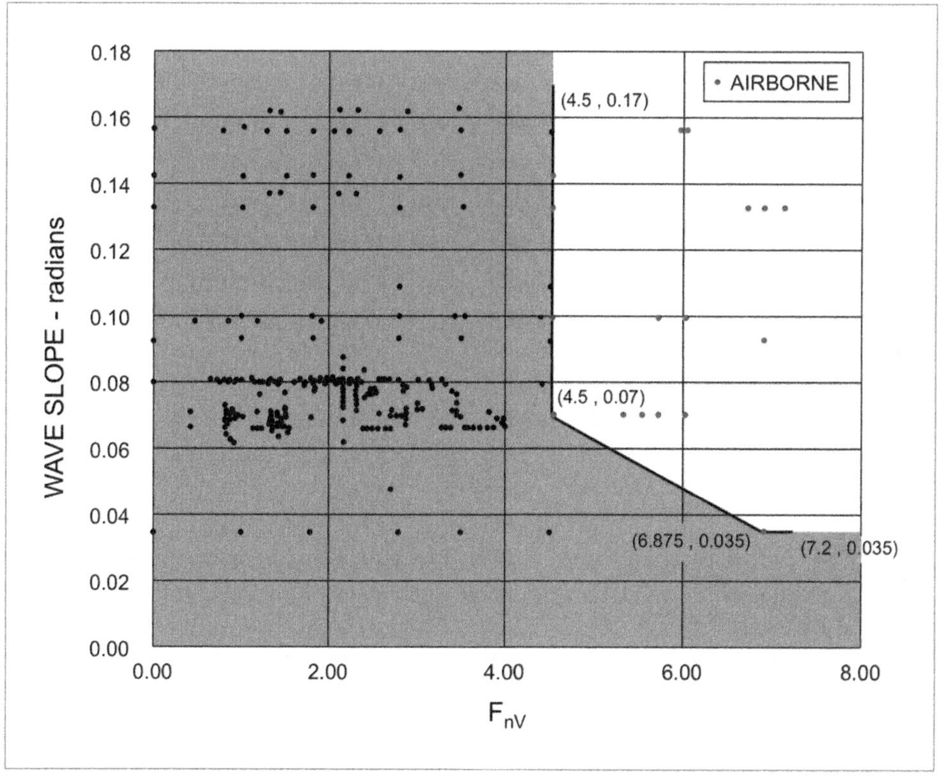

Figure 7-12: Wave slope conditions when craft may become airborne in head seas

Data for craft which became airborne are presented in Figures 7-11 and 7-12, and are based on speed (F_{nV}). The red data points are for airborne events and the black data points represent conditions where hulls did not leave the surface of the water.

The boundary-separating conditions whereby hulls maintain contact with the sea or become airborne based on nine designs are defined by lines with coordinates of connecting points noted in both figures. Future tests of additional hull designs could change these boundaries.

For simplicity, I used a definition for wave slope for irregular seas patterned after that of regular waves. The numerical values for wave slope in Figure 7-12 were calculated by the following approximation for wave slope in radians:

$H_{1/3}(0.6140)/T_W^2$ (FPS) Equation 7-4A

$H_{1/3}(2.014)/T_W^2$ (metric) Equation 7-4B

The sea conditions relative to vessel size in Figure 7-11 are defined by $H_{1/3}/\nabla^{1/3}$. Figure 7-12 brings in the influence of modal period to define a nominal random wave slope.

The ratios of maximum (slam) vertical acceleration near the transom to that at the LCG in some cases are below 1.0. However, in the extreme, these data show that stern vertical accelerations (9-to-15 percent L_P forward of the transom) can be as much as 1.4 times the acceleration measured at the LCG. For planing boats which became airborne during testing at F_{nV} = 6.0, Figure 7-13 shows a simplified longitudinal distribution of the ratio of local vertical acceleration relative to that at the LCG. Acceleration ratios at 0.15 L_P, LCG and 0.90 L_P are connected by straight lines.

Accelerations aft of the LCG when extrapolated to the transom can be very severe structurally. Thus, if you are designing a high-performance craft which is likely to become airborne during operational service, I caution you to provide reinforced structural and shock mitigation for personnel stationed near the transom—and for the structural integrity of the hull's afterbody, especially when outboard engines are mounted.

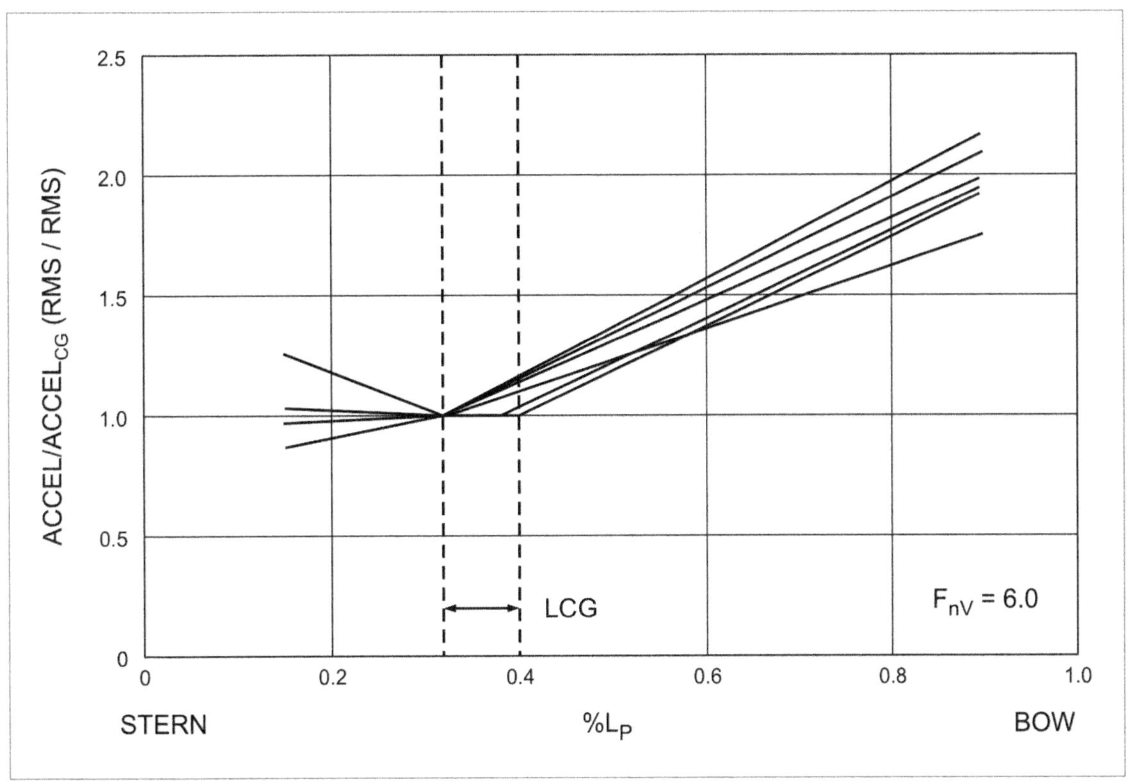

Figure 7-13: Simplified longitudinal distribution of vertical acceleration relative to that at the LCG

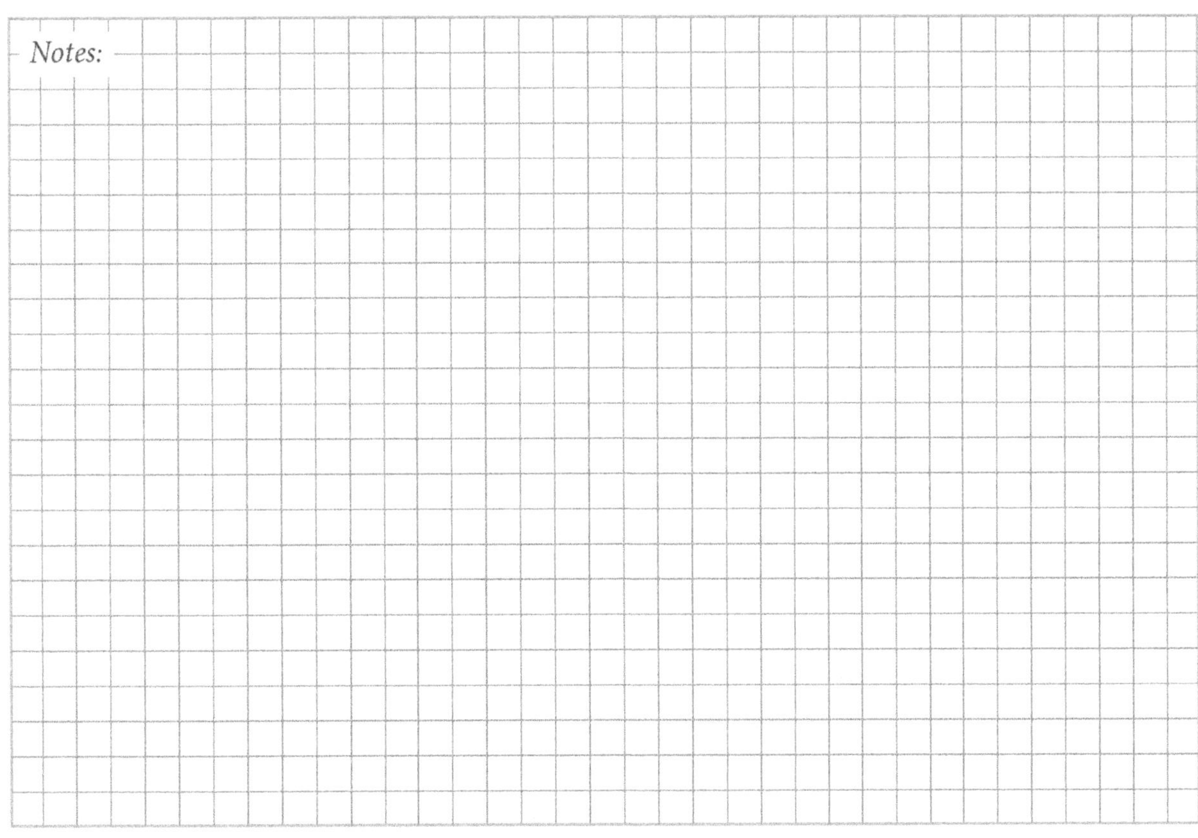

General Observations Regarding Hull Geometry

During the design stage, you must establish a range of hull dimensions for length, beam, deadrise, loading, LCG, and so on, to meet craft requirements. Vertical acceleration of the hull in response to high planing speeds in irregular seas is influenced by the decisions you make concerning the geometry, mass, and mass distribution of the vessel; however, it is not clearly understood which of the following has the primary influence on the resulting vertical accelerations in a defined sea condition: speed, hull loading, running trim, or deadrise angle.

The question for you then is, "What is the effectiveness of each designer-controlled element relative to mitigating the magnitude of vertical acceleration?" Over time, I am coming to believe that mean running trim and deadrise only have secondary influence on vertical acceleration at the LCG.

Restating my earlier opinion: *To attain the best ride quality result for a high-speed, rough-water craft, you must achieve minimum pitch oscillations about as low a mean running trim angle within the boundaries that dynamic stability can be maintained.* I consider heave oscillations to be only a minor contributor to the magnitude of vertical acceleration.

Unfortunately, hull dimensions, loading, and LCG all interact in ways which change the mean running trim while hull beam loading, pitch gyradius, wave height, and slope (modal period) influence pitch oscillations. Within limitations of requirements and technology the designer and builder have control of almost everything but sea conditions.

Many believe that increasing deadrise angle alone will reduce vertical accelerations in rough water. Thus, $\beta_{mid} \geq 20°$ is commonly found in hull lines drawings of high-speed constant deadrise planing craft. These designers ignore the fact that just increasing deadrise (β) also requires trim (τ) to increase to achieve the necessary hydrodynamic lift to support the weight of the craft at planing speeds. Impact tests of seaplane models which were analyzed by Milwitzky (1952) indicate that for the same lift coefficient and L/B, the relationship of deadrise and mean trim angle must change together as seen in Table 7-B.

When studying Table 7-B, note that *in all cases* large deadrise angles may not always reduce vertical acceleration in rough waters as the incremental increase in mean trim occurs at a higher rate than that of β.

TABLE 7-B: TO MAINTAIN DYNAMIC LIFT AS DEADRISE ANGLE INCREASES, TRIM MUST INCREASE

β Deg	τ Deg	Increments		Rate of deg. τ increase per deg. β increase
		β	τ	
0	3.00			
		10°	0.32°	0.032
10	3.32			
		10°	0.57°	0.057
20	3.89			
		10°	0.75°	0.075
30	4.64			
		10°	0.89°	0.089
40	5.53			

Acceptance for constant, deep-deadrise hulls for high speeds in irregular seas is only justified for craft which become airborne as then the stern typically reenters the water first. For craft which are not expected to become airborne, hulls having a deep deadrise forward are best in rough water, especially when they are also provided with capability to have some reduced running trim angles at initial hull-water contact.

My recommendation is to have reduced deadrise, but more than 10° at the transom (warp) and operator-controlled trim. I prefer some warp with the possibility of dynamic pitch control rather than to add the responsibility for manually adjusting trim, as the helmsman's full attention is essential during high-speed operation in irregular seas.

If a test model has been "fixed in roll," be wary of seakeeping experimental study results for complete hulls as the relationship of vertical acceleration clearly trends as a function of deadrise angle. Fixed in roll is not representative of operation in irregular seas in the real world. All boats and craft respond with six degrees of freedom relative to irregular waves.

It is common for both pitch and roll to occur simultaneously resulting in an off-axis wave hit. For example, a boat having $\beta = 20°$ may come off a wave with 10° roll, which effectively results in an impact on one side of a craft with $\beta = 10°$. If you increase the deadrise a few degrees on a drawing to obtain a calculated vertical acceleration reduction to meet a requirement, you may be very disappointed in the results when you conduct full-scale trials in irregular seas.

For design decisions, therefore, you need to know how much the vertical acceleration increases as a result of greater mean trim angle relative to the reduction of vertical acceleration from higher deadrise angle. Or, vice versa: Is vertical acceleration increased by the net effective deadrise reduction along with the expected mean trim angle reduction in multi-directional, irregular seas?

Roll oscillations of boats operating in irregular seas effectively diminish the significance of deadrise angle. Some empirical prediction procedures even omit β as a factor, and trim angle to a limited degree. Hull loading, however, remains as a factor in most analytical and empirical acceleration prediction approaches. Nevertheless, it is generally understood that technology trends studied with model experiments may require attachment fixtures to towing carriages having less degrees of freedom than are experienced by full-scale craft having unconstrained freedom of motion when operating at sea.

I believe that even in head sea testing in towing tanks, models must be free in pitch, heave, and roll, but they may be restrained in yaw and sway.

Wave Impact: Size Does Matter

What does a wave slam—the big hit—look like? Operators of small performance craft in irregular seas certainly know what it *feels* like. Large vertical acceleration spikes hurt, visual focus of craft instrumentation becomes intermittent, and bodily injury is not uncommon. As vessel size increases relative to significant wave height ($H_{1/3}/\nabla^{1/3}$) reduces, then the magnitude of vertical acceleration tends to be mitigated. Size does matter. *Destriero* never experienced CG vertical acceleration as high as 1.0 g in 6.0 ft (1.8 m) significant waves when transiting at an average speed of 58 knots during the last 24 hours of the Atlantic crossing record.

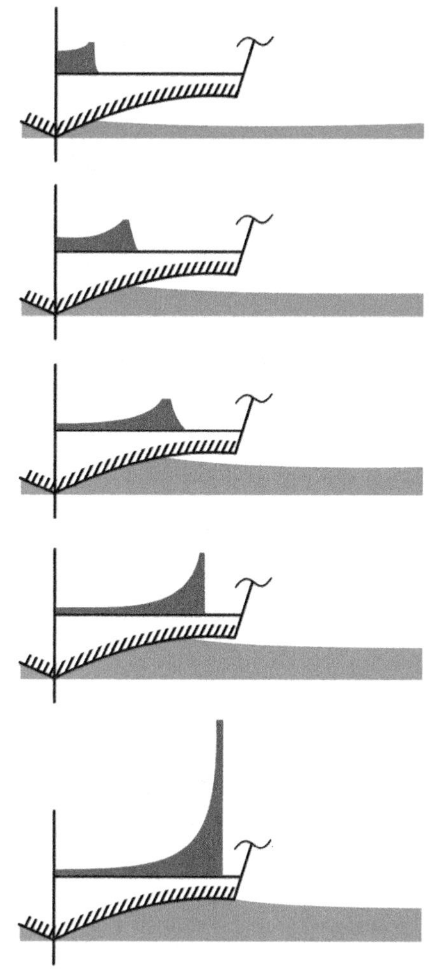

Figure 7-14: Graphic representation at water entry over time increments of bottom wetting with associated transverse pressure distribution (red color)

Looking at a single slam of a small craft in waves helps to understand the impact phenomenon: Imagine a stop-action camera taking photographs of a bow section of a planing boat entering a wave while pressure data is superimposed on the pictures.

The results are shown in Figure 7-14, which is a graphic from Wagner (1948). Starting at the top of the figure, you can see this progression: As relative water immersion increases, the transverse pressure spike at the stagnation line (red color) moves outward toward the edge of the wetted bottom. The reaction of the hull to pressure build-up is vertical acceleration. If you are a structural designer, take note of the peak pressure spike. It is ultimately slightly inboard of the chine, highlighting where careful attention to detail is needed at the chine where the hull bottom and side connect.

While Figure 7-14 shows time-varying pressure of a transverse section of the hull, Figure 7-15 shows typical contours of pressure distribution at the instant of impact of a symmetrical wave hit.

Contours of constant pressure are outlined by black lines. This figure is a colored version from Allen and Jones (1978). As with Figure 7-14, take note that the highest pressures are essentially parallel to and slightly inboard of the chines. Since a primary responsibility of performance craft designs is to control flow separation, you can alter these pressure contours somewhat by carefully locating longitudinal strakes to the smooth bottom of planing hulls.

I have been addressing this topic from a hydrodynamic perspective, understanding pressure-load distributions is also important for structural designers. When operating in irregular seas, the reaction of the hull to hydrodynamic loads results in an acceleration and motion environment. This most detrimental reaction of vertical acceleration must be safely endured by personnel, mission equipment, cargo, and all things on board.

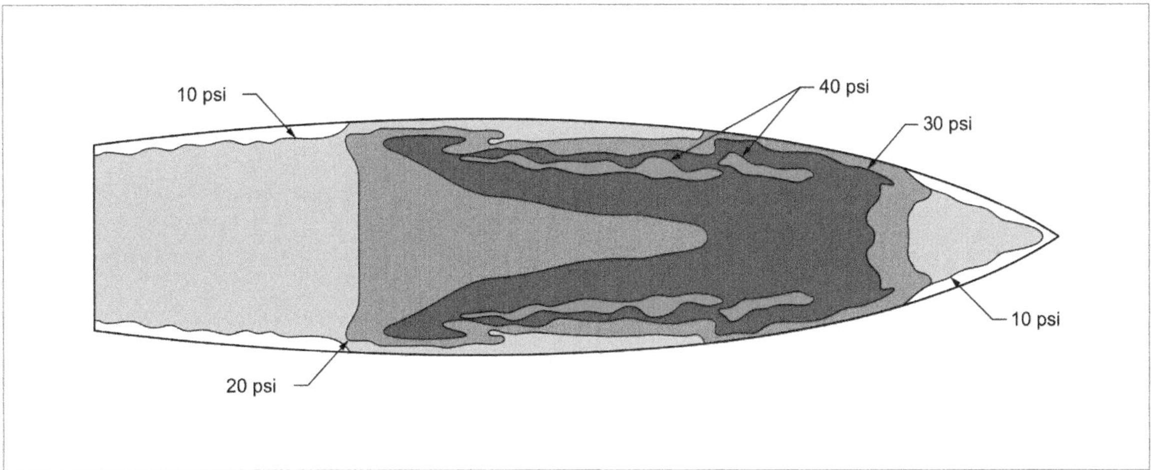

Figure 7-15: Typical contours of pressure distribution at the instant of impact (Allen and Jones – 1978)

In 2010, researchers dissected the time history of the rapid change in vertical acceleration caused by a single wave impact,. Figure 7-16 illustrates the sequence of events in a typical wave-slam event for planing speeds.

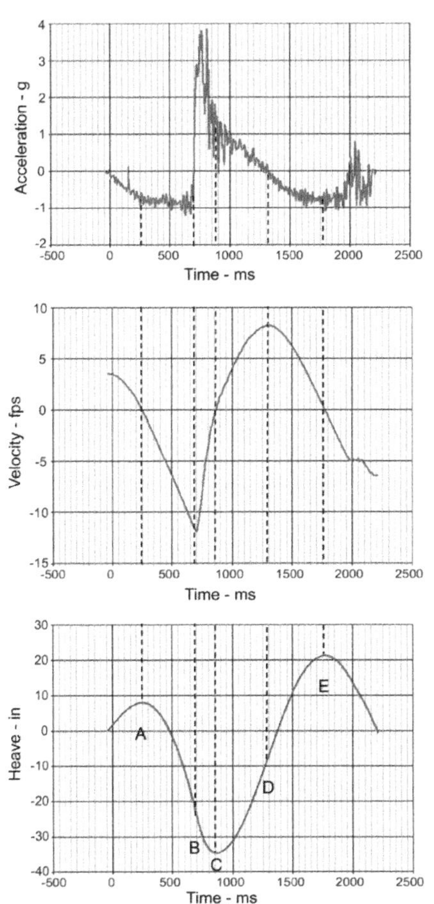

Figure 7-16: Wave slam sequence of events. (Riley, Haupt, and Jacobson. 2010.)

At the location of an accelerometer, the upper curve is the individual unfiltered vertical acceleration time history extracted from a longer sea trial recording. The middle curve is the vertical velocity time history obtained by integrating the acceleration curve, and the lower curve is the integral of the velocity to show the absolute vertical heave at the accelerometer.

When the craft impacts a wave at Time B, the downward velocity reaches a minimum and changes rapidly to an upward value. The force of the impact is seen as an almost instantaneous jump to a peak vertical acceleration. The duration of the wave-slam event from Time B to C is roughly 0.16 seconds. From Time B, the heave continues to move down in the water and approaches zero at Time C. Acceleration decreases rapidly toward a value of approximately 1.0 g at Time C.

At Time C, the heave at the accelerometer location reaches a minimum, and the momentary velocity is zero. However, forces due to buoyancy, hydrodynamic lift, and components of thrust and drag combine to produce a net positive force upward. The impact event is complete at Time C and the craft motion is now dominated by forces of buoyancy and hydrodynamic lift effects. This brief summary here is described in technical detail in Riley, Haupt and Jacobson (2010). For craft weights between 14,000 and 105,000 lbs (16.3649 gmt) Riley, Haupt, and Murphy (2014) found that impact durations may vary from 0.10 to about 0.45 seconds.

I recommend that you study these and other related references, including Timothy Coats, if you intend to design performance craft to be operated in irregular seas.

Hard-Chine Monohulls in Head Seas

Currently the most reliable predictions of vertical acceleration in irregular seas can be obtained by conducting model tests in representative seas having scaled $H_{1/3}$ and T_W. In 2014, Akers' commercially available software POWERSEA® is, in my opinion, the preferred analytical prediction method while Hoggard and Jones (1980) regression analysis technique provides a realistic engineering order of magnitude estimate of CG vertical acceleration.

Depending on client funding level for high-speed planing craft design projects, my own preference for predicting vertical acceleration based on descending order of cost is model tests, POWERSEA®, analysis with DLBA's seakeeping database, and the Hoggard and Jones technique.

With regard to the seakeeping characteristics of high-speed planing monohulls there are two distinct variants to be considered: single-chine hulls and double-chine hulls. Transverse hull sections for these two hull forms are shown in Figures 7-2 and 7-3.

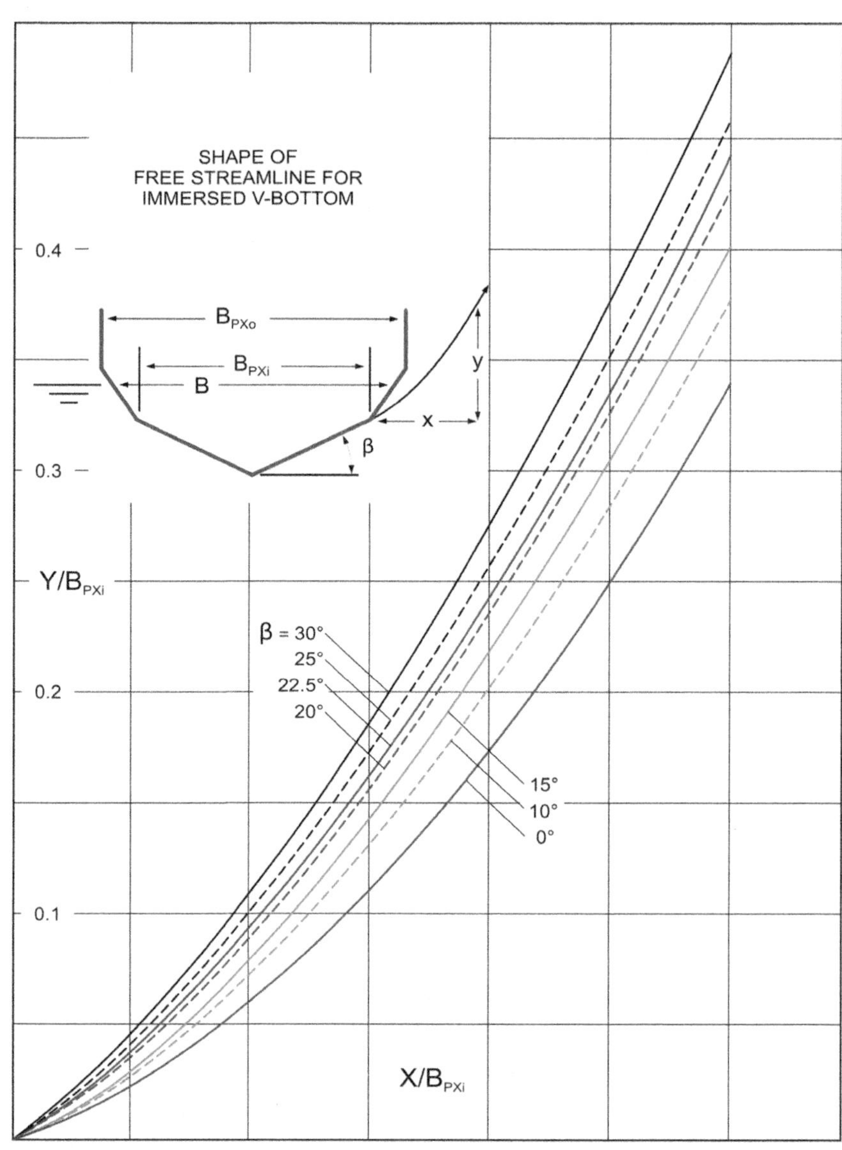

Figure 7-17: Shape of free streamline for immersed V-bottom

The design of a planing monohull can be reconfigured to a double-chine hull which can have measurably better ride quality than a single-chine hull. If, however, the distinguishing criteria defined in the following paragraph is not taken into account for the design of a double-chine hull, then the same seakeeping characteristics as a single-chine planing hull would result at a higher cost of construction.

To achieve the advantage of a double-chine hull in irregular seas, you must begin by designing the inner chine projected beam (B_{PXi}) so that $C_\Delta \geq 0.50$. Savitsky et al. (1972) used a slightly larger $C_\Delta = 0.85$ in their design example of a double-chine concept. Then, locate the elevation of the projected outer chine beam (B_{PXO}) such that when, in irregular waves, the vertical component of the streamline separating from the lower inner chine beam does not contact the hull between the inner and outer chines.

Or, stated another way, "...the upper chine must not extend into the separated flow cavity formed by the steamline separating from the lower chine." (Savitsky et al., 1972). And remember the waterline beam (B) must, as a minimum, be sufficient for the craft to have adequate static stability.

Figure 7-17 provides for various deadrise angles, the two-dimensional streamline from vertical motion of the hull as it separates from the inner chine (Korvin-Kroukovsky and Chabrow, 1948; Savitsky et al., 1972).

Summary Thoughts about Trim-Control

Through technology improvements, vessels operating in calm water have made meaningful speed increases for efficient transport of goods and personnel. The ocean environment, however, continually changes with variable winds. Naval architects and designers have come to realize that hull-design optimization for calm water is not likely to be in agreement with one optimized for rough seas. Calm-water design is focused on efficiency, but for rough seas, you should be primarily interested in ride quality and personnel safety, consistent with its mission requirements.

Pitch oscillations in irregular seas are a major source of the extremely disorienting motions and harmful accelerations that affect personnel, equipment, and structures of marine vessels. At high speed, trim control then becomes an added capability as well as a responsibility for the operator who already is very busy controlling speed and direction through the throttle, and steering. Either manual or, better yet, dynamic pitch control capability becomes essential for operating offshore.

Designers must accommodate interfacing systems requirements for dynamic pitch and trim-control equipment, such as trim tabs and interceptors, by allowing for space, weight, and power, as well as incremental main-engine propulsive power.

CHAPTER EIGHT

IMPORTANT DESIGN ASPECTS FROM THE HYDRODYNAMIC VIEWPOINT

Designers must be visionaries. They are conceiving maritime vessels to be constructed and operated in the future. Designers must be ever vigilant and open to adopting emerging technology—without introducing excessive risk. These are its essential guiding principles if a design office is to expand and attract clients with challenging opportunities,

A military vessel, small boat, or a world-class motoryacht design grows from some fundamental process whether performed by an individual or a team of specialists. This undertaking integrates the knowledge of hydrodynamics, aerodynamics, propulsion machinery, structures, control systems, vessel dynamics, performance, and weights. Clearly the design process blends many disciplines under the leadership of a chief designer, the decision maker who understands and appreciates all of the disciplines required to move the project toward the best combination of technology and aesthetics to meet project goals.

Forged by experiences gained while working on small teams, these talented people understand the pressures of time and budget, with all team members being responsible for their own contributions to the diverse elements of marine-craft design. Potential chief designers hone their experience by first developing solutions to specific requirements, and then increasingly become generalists with a broad level of knowledge of a wide range of technologies and a deep appreciation of their interactions.

In the context of this book, I refer to the design of a marine craft as "the process of activities that span the creation on paper or in computer files, details of a vessel which can be constructed to perform its intended mission." A design begins with an articulated need, including a definite mission, or use requirements, established by a client who has the resources to afford the cost of construction and operation of the desired vessel.

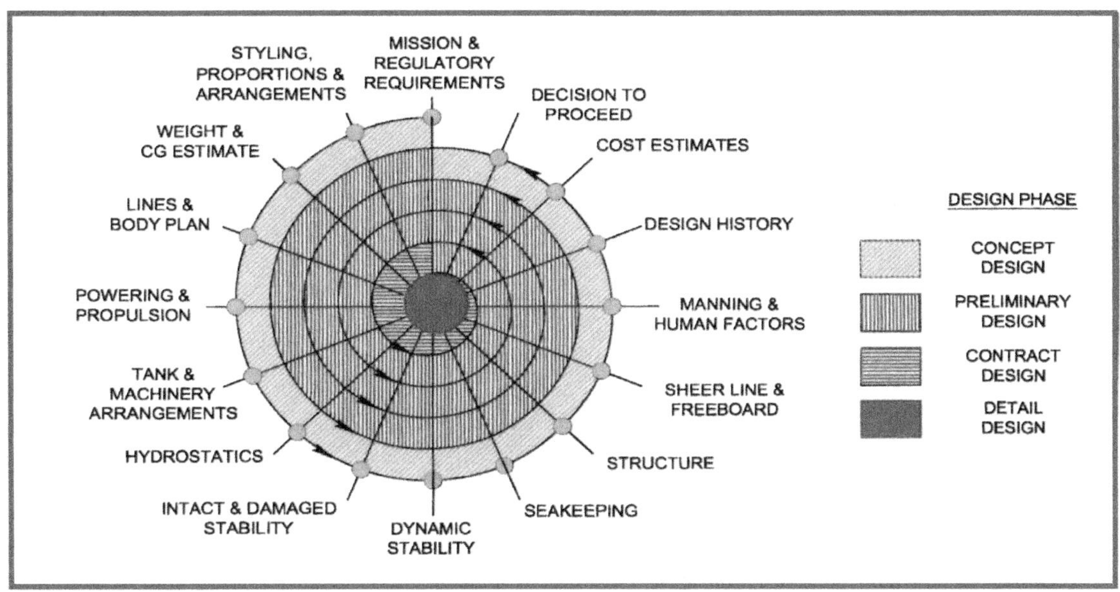

Figure 8-1: Design spiral for high-performance marine vessels

The objectives of design projects will also be affected by specific situations, such as:

- The design of a new marine craft
- The modification of an existing marine craft affecting hydrostatic, hydrodynamic, and/or propulsive characteristics while maintaining the original requirements
- Extending or augmenting performance of an existing marine craft which will be different from the original requirements

Is there a need to spend much more time discussing design? Yes. There is a need to address design and the design process for high-performance marine craft to characterize the boundaries between artistic design such as renderings or picturesque depictions and the technical delineation of boats, yachts, or vessels intended to be built.

This discussion will help you understand the similarities and contrasts between the design process for displacement ships and high-performance craft which are, at least if not almost entirely, affected by dynamic forces from the shape of the hull and deckhouse reacting to high velocity of water and air at these fluid boundaries.

There is also a need to interpret the similarities of the design process for high-performance craft and aircraft. They are more akin than the often-used approach by some authors who modify displacement-ship design techniques to represent the design process for performance craft. In general, displacement-ship design practice does not account early enough in the design process for the fact that at some increasing speed, hydrodynamic and aerodynamic flows over the surface of the vessel begin to dominate. Dynamic pressures begin to change sensitivity for the balance of stability at high speeds.

Let's consider the broad aspects of design: To design is to be creative, so don't expect wide agreement. The design process begins as mental evolution, and it ends when the designer has stopped revising the work and is satisfied with the end product, which may be a work commissioned to meet specific client requirements or something created for personal fulfillment.

The end product might be a painting, a statue, the design of an automobile, a great building or park, a bridge, or a yacht. Anything! If the project is executed by an individual who is "The Designer," then that person is the decision maker.

Or if a design team is carrying out the work, as is common with complex projects, team leadership is an additional responsibility of the Chief Designer. The decision maker already described as understanding and appreciating all the disciplines needed to move toward the best combination of technology and aesthetics to realize the goals of the end product.

Without a single authority or strong design czar, a complex-project's team will have a difficult time in reaching a good outcome.

Design Process

For some time, the marine design process has been graphically depicted as a "design spiral" with focus on traditional aspects generally important to operation, and showing hydrostatics/buoyancy as providing principal support for the displaced weight of the vessel. This thinking has roots going back to the earliest days of sailing ships when naval architects first developed practices and technology to reduce design risks.

The vintage displacement-vessel design spiral was subsequently modernized by replacing sail-power with hydro-mechanical systems such as paddle wheels, propellers, and waterjets, while adding hydrodynamic and structural-dynamic loads which impact speed-related stability.

The design spiral is a readily accessible concept for explaining an orderly process beginning with requirements for developing an initial concept and continuing around the spiral to develop increasing levels of design detail. The end goal, in the shortest reasonable time, is to reduce risk, increase confidence, and define the most cost-effective way that a performance vessel can be constructed from the detailed design.

As the project proceeds around each design spiral phase—concept, preliminary, detail, and contract—to develop increased definition, the designer must continually confirm that the requirements have been met. If they have not, the project options are to "go around" that phase of the design spiral again, request waivers or changes to requirements from the client, or determine that it is not technically possible to satisfy the requirements. Fortunately, when a decision is made to go around the spiral again, it is only necessary to revisit those deficient technologies requiring further definition of detail to meet requirements before moving on to the next phase.

High-performance marine vessels, however, bring additional decision factors to the design spiral process, and these can be as important as any issues related to displacement operation. If these high-performance related factors are ignored early in the design stage, the resulting vessel can ultimately have operating limitations at maximum power.

I've discussed the threshold for defining high speed in terms of dimensionless speeds rather than dimensional speed due to its dependence on hydrodynamic vice being limited to hydrostatic technology. I've also provided a discussion of various definitions of dimensionless speed in *Nomenclature and Definitions* in Appendix 7 and in Chapter 2.

For the moment, however, I want you to simply consider 22 knots as the dimensional speed that becomes the potential threshold for full-scale vessels whereby hydrodynamic forces are likely to become significant with regard to dynamic stability. Hydrodynamic vice hydrostatic forces tend to dominate, and dynamic stability is not guaranteed, even though traditional hydrostatic criteria are met.

All marine vessels, including high-performance craft need to be watertight and float upright when at rest in an intact condition, but also they must be dynamically stable at high speed. Thus, the process for high-performance marine vessels is more akin to aircraft rather than ship-design practices.

For an aircraft, the designer must evaluate the controlling dynamic requirements from the very beginning:

- Will it operate at supersonic or subsonic speeds for lengthy periods?
- Will there be take-off limitations due to weight of mission equipment or length of runways?
- Will range dominate the size of the aircraft due to fuel load?
- Will type of operation—fighter, bomber, cargo, passenger, etc.—result in dominant factors for internal appointments and volume, agility, and ride quality?

Some parallel examples of questions for performance marine-vessel designers might be:

- Will the vessel operate for extended periods above 50 knots or just below 35 knots?
- Must it be transported in a cargo aircraft, operate from a wet-well ship, or in water less than three meters in depth?
- Will fuel load for extended range dominate vessel size; or will type of operation—patrol boat, passenger ferry, littoral combatant, motoryacht, etc.—result in dominant factors for internal appointments and volume, agility, and ride quality?

As designer, you must always be alert to the fact that it may not be feasible to satisfy requirements developed by every client. *Learning when to back away from an infeasible project is just as important as knowing how to efficiently and professionally execute a technically-challenging one.*

After going around the design spiral several times, you will develop ever-increasing levels of detail built on previous progress. For example, preliminary design refines and expands on the concept level of detail for any number of technical areas such as weights and center of gravity, structures, and intact and damaged stability. This design spiral approach is efficient and instructive in a simple sense. There can, however, be aspects of each design phase which are not expanded in further detail in subsequent cycles.

The design process for high-performance marine vessels must have expanded components for their design spiral. You may need to develop and/or refine components of this design sequence (see Table 8-A and The Design Spiral in Figure 8-1) in each of the stages.

TABLE 8-A: COMPONENTS OF DESIGN SPIRAL FOR HIGH-PERFORMANCE MARINE VESSELS		
Activity		Comments
1	Mission & Regulatory Requirements	Necessary information provided by client to begin a design
2	Styling, Proportions & Arrangements	Sketches and drawings to confirm form, fit and functions for space of mission requirements
3	Weight & Center of Gravity (CG) Estimate	Estimate of weight and center of gravity with margins commensurate with level of detail of each design phase
4	Lines & Body Plan	3-D geometry of the appropriate or dictated hull form
5	Powering & Propulsion	Speed and power predictions for various displacements and LCGs as well as appropriately sized and number of propulsors
6	Tank & Machinery Arrangements	Fuel tank capacities sized for range requirements and appropriately located about CG to achieve best performance as fuel is consumed, establish all other tank sizes and location, develop propulsion machinery arrangements
7	Hydrostatics	For hull lines, establish draft, center of buoyancy relationship versus displacement, and necessary hydrostatic calculations
8	Intact & Damaged Stability	Defines necessary longitudinal locations of watertight bulkheads for defined margin line as well as down-flooding locations
9	Dynamic Stability	Determines probability of dynamic instabilities in roll, yaw, pitch/ heave (porpoising) and chine walking
10	Seakeeping	Predicts the rigid body motions and accelerations for the vessel in the required sea state with defined significant wave heights and periods, establish human factor ride quality characteristics, accelerations for structural and equipment design loads
11	Structure	Determines scantlings of structural members based on regulatory rules and/or based on first principles
12	Sheer Line & Freeboard	Establishes aesthetic and technical weather deck heights
13	Manning & Human Factors	Establishes crew size for vessel and/or mission equipment operation, owner and guest services, passageway patterns for personnel movement, etc.
14	Design History	Documents the technical procedures, criteria and logic applied and decisions made while developing the details of vessel design
15	Cost Estimating	Estimates material and labor costs for constructing the vessel

Note: The sequence of components listed in Table 8-A has significance. For example, damaged stability calculations locate longitudinal positions of watertight bulkheads and seakeeping predictions provide CG accelerations which define hull bottom design pressures. Both of these pieces of information are essential before executing the structural design.

The design process is often divided into phases with increasing levels of detail definition. These design and engineering phases are:

- Concept Design
- Preliminary Design
- Detail or Contract Design
- Production Engineering

Following the establishment of requirements, you will cycle your new marine vessel design through these phases although a conversion, modification, or refit of an existing boat may begin at an interim phase.

To satisfy the needs of your client, you should provide some clarity of services to include the activities and decisions that you made during the various phases outlined here.

Concept Design Phase

The concept design phase includes the necessary trade-off studies to establish the hull form, general size, and macro arrangements which, if feasible, define the best combination of technology, aesthetics, and other features which meet all performance and regulatory requirements.

During the concept design phase, you will establish the hull form—round bilge, double chine, hard chine, monohull, multi-hull—along with its general size, exterior styling, profile, and initial arrangements, as well as its plan form, allocation of deck areas, and internal volumes. You will also conduct trade-off studies so that construction materials, type and size of propulsion machinery, weights, and longitudinal center of gravity (LCG) converge, and you can confirm that the best hull-loading, static- and dynamic-stability, performance, and seakeeping qualities are possible. Otherwise, you will make the decision to go around the design spiral one more time to improve the fit of the conceptual design with your client's requirements.

If it is not feasible, the project requirement may be cancelled, revised to suit known or anticipated emerging marine technology, or redirected, if for commercial or military purposes, changed from being a marine product toward other techniques such as trucks, trains, or aircraft. However, these trade-off studies to consider other approaches to meet requirements are not usually repeated or expanded as a subsequent marine design.

Preliminary Design Phase

During the preliminary design phase, you will realize the form, fit, and functional aspects of the design as dimensional quantities of components, systems, and arrangements to be brought together within the hull and stylized profile. The generalizations that were depicted during the conceptual design phase will now become scaled arrangements with respect to internal and external space, including passageways for movement of personnel about the vessel, deck areas and volume, as well as longitudinal, vertical, and lateral distribution of weight.

This will bring together either an aesthetically pleasing design—or a reduced degree of signature in the case of military craft. It is essential to include, for service-life maintenance, an allowance for space and/or removal paths for major machinery components. You can achieve this by designing and engineering space allowances necessary to make craft safe, reliable, accessible, and functional. Additional details may indicate a substantial change from the resulting weight and center of gravity from the concept design phase. This change could make it necessary for you to relocate watertight bulkheads and refine the hull lines. Thus, your project may go around the preliminary design cycle several times to overcome this problem with an alternative but acceptable rearrangement.

You may need to conduct model tests to fine tune the design and make performance predictions during this phase. You will also need to make significant decisions concerning hull lines, hull and superstructure scantlings, number of engines, horsepower, gear concept and ratios, propulsor size, size and location of tanks, generator capacity, heating/ventilation/air conditioning capacity, electronic systems, galley layout, and all other necessary systems.

Upon completion of these and other engineering decisions, you will make the hydrostatic and damaged stability calculations necessary for the hull lines and arrangements. With the resulting overall weight and center-of-gravity location, you can make calculations to confirm that adequate static stability exists, then make performance predictions to confirm that cruise range and dynamically stable speeds can be achieved with acceptable operations expected in a seaway.

Detail Design and Contract Design

At the completion of this phase, the configuration is frozen until the client makes a decision to construct the craft. You must clearly specify—in writing and with drawings—any requirements which are essential for the intended use of the craft, for the performance, or for quality of construction.

This drawings-and-specifications package should consist essentially of the preliminary design drawings with added detail to control the vessel's essential features, along with structural details for the hull, deck, and deckhouse. You must also prepare a detailed design weight and center-of-gravity estimate; a detailed building specification for hull, mechanical, electrical, and all necessary systems; and also outfit and furnishings.

Note: In general, Detail Design is for a builder client and Contract Design is for an owner client.

Production Engineering

You will need to establish manufacturing considerations and provide an approach to minimize construction time and cost within resources and facilities available to the shipyard. This will include establishing tooling details, jigs, and production schedule, and developing a detailed Bill of Materials for the shipyard to place orders to assure the components are received in time to meet the production schedule.

Additional preparation includes generating detailed shop drawings for joints, fittings, attachments, equipment mounts, cable runs, and wiring bundles; fabricating mock-ups where necessary; establishing production quality-assurance checks; meeting industry standards; and producing boat operating, maintenance and repair manuals and videos.

Comparison of Various Hull Forms

The first topic that I am going to address in comparing various hull forms is the concept of transport efficiency: The relationship between vessel weight, speed, and power, based on instrumented sea trials of a wide range of vessels.

Documenting the status and achievements for hard-chine, semi-planing, or planing monohulls is better served when you can compare them with other hull concepts. To make credible comparisons among the various configurations, you must adopt a consistent approach to account for hull drag, interactive factors, and propulsive characteristics.

In 1950, Gabrielli and von Karman presented a landmark paper for comparing maximum velocity performance for single vehicles in level motion utilizing specific tractive force (ε) versus velocity for which they developed a limiting tangent line approximated by $\varepsilon = 0.000155(V_{MPH})^{1.02}$.

The utility of this format is enhanced when velocity is replaced by a dimensionless speed coefficient such as volume Froude number (F_{nV}) to relate vessel size with absolute speed.

Full-scale trial performance defined by speed, total of propulsive and dynamic lift power and vessel weight may be combined into a transport efficiency ($E_T = 1/\varepsilon$) and a dimensionless speed (F_{nV}). For comparison, preference for E_T vice its reciprocal specific tractive force (ε) allows the largest numerical value to represent the most efficient mode of transport.

This format is useful for addressing relative performance of different advanced craft concepts as well as the separate trends of bare-hull resistance to weight ratio and overall propulsive efficiency.

From Equations 8-1 and 8-2, it is clear that E_T increases by improving overall propulsive efficiency (η) and reducing the bare-hull resistance-to-weight ratio (R/W).

$$E_T = W(V)/P_{DL}(326) = \eta/(R/W) \quad \text{(FPS System)}$$
Equation 8-1

$$E_T = \Delta_{mt}(V_{m/s})/P_{DL\,KW}(0.102) = \eta/(R/W) \quad \text{(Metric)}$$
Equation 8-2

Figure 8-2 presents calm-water trial data available to me as E_T versus F_{nV} for selected hard-chine craft to demonstrate the value of this format and achievements made by designers and shipbuilders. The Pareto curve represents the upper boundary of the most efficient performance in evidence circa 1993.

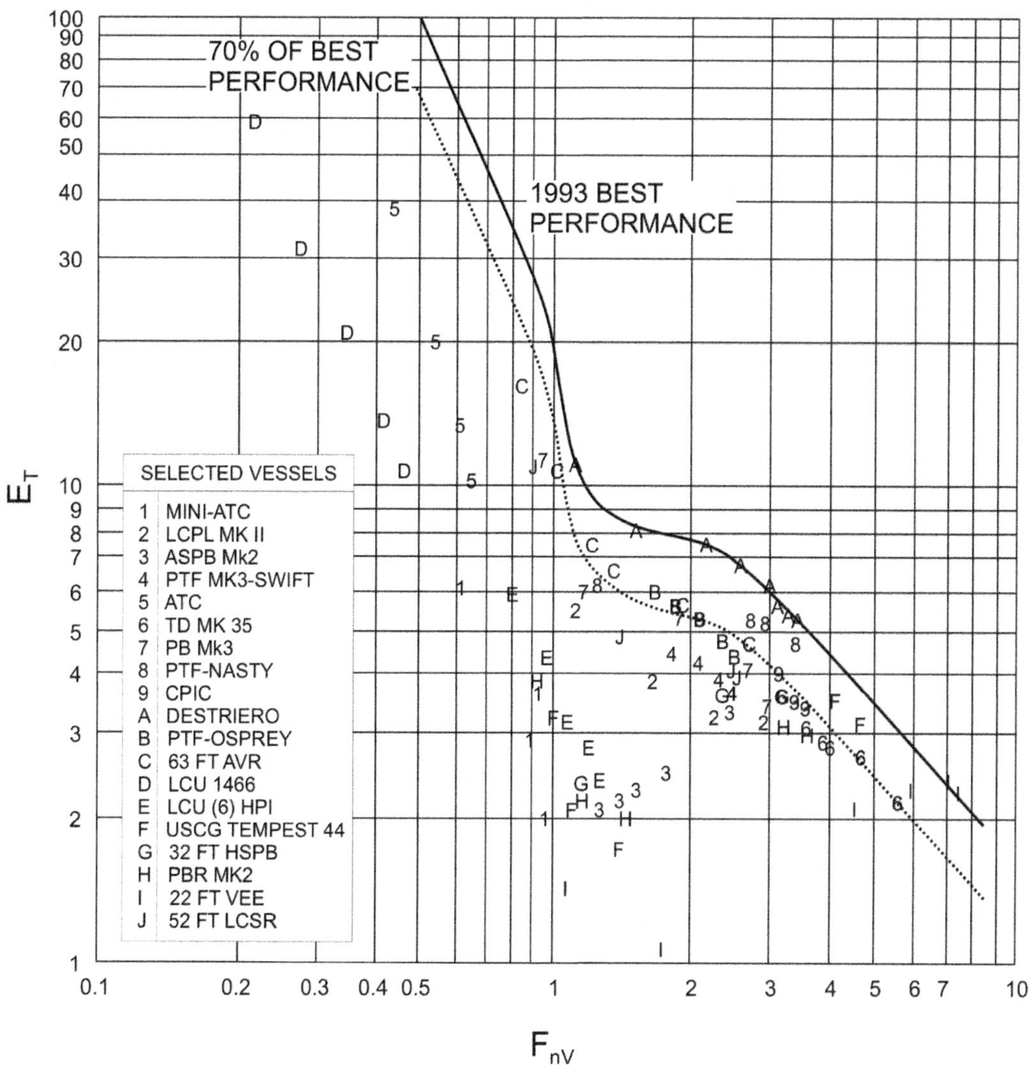

Figure 8-2: Transport efficiency for selected hard-chine craft circa 1993

As of the publication date of this book, typical performance of hard-chine craft in all types of service attained approximately 84 percent of the best calm-water transport efficiency. Some of these less successful vessels may be the result of inflexible design requirements or constraints which prohibit optimization of state-of-the-art technology into various components of a vessel.

Thus, early in a project, you must identify the sensitivity of design features that may improve overall vehicle efficiency to help your client understand that—with alteration of those requirements—performance may see significant improvements and construction time and operational costs may be reduced.

I have included in Figure 8-3 the relative calm-water performances of various craft concepts for comparison where reliable full-scale trial data have been made available. These data relating to craft from 5.5 to 68 m (18 to 222 ft) in length have been reduced to a dimensionless format, but with no Reynolds' number corrections applied. The Pareto curves of E_T for each hull concept in Figure 8-3 were obtained from full-scale, calm-water trial data of numerous military, commercial, and recreational craft.

These curves represent the upper boundaries (circa 2008) of efficient calm-water performance available for each concept in format similar to the example for hard-chine craft in Figure 8-2.

Note that these upper boundaries of efficient performance change over time, hopefully trending toward improved efficiency as new technologies become available. Data for each hull form defining E_T versus F_{nV} in Figure 8-3 are provided in Table 8-B.

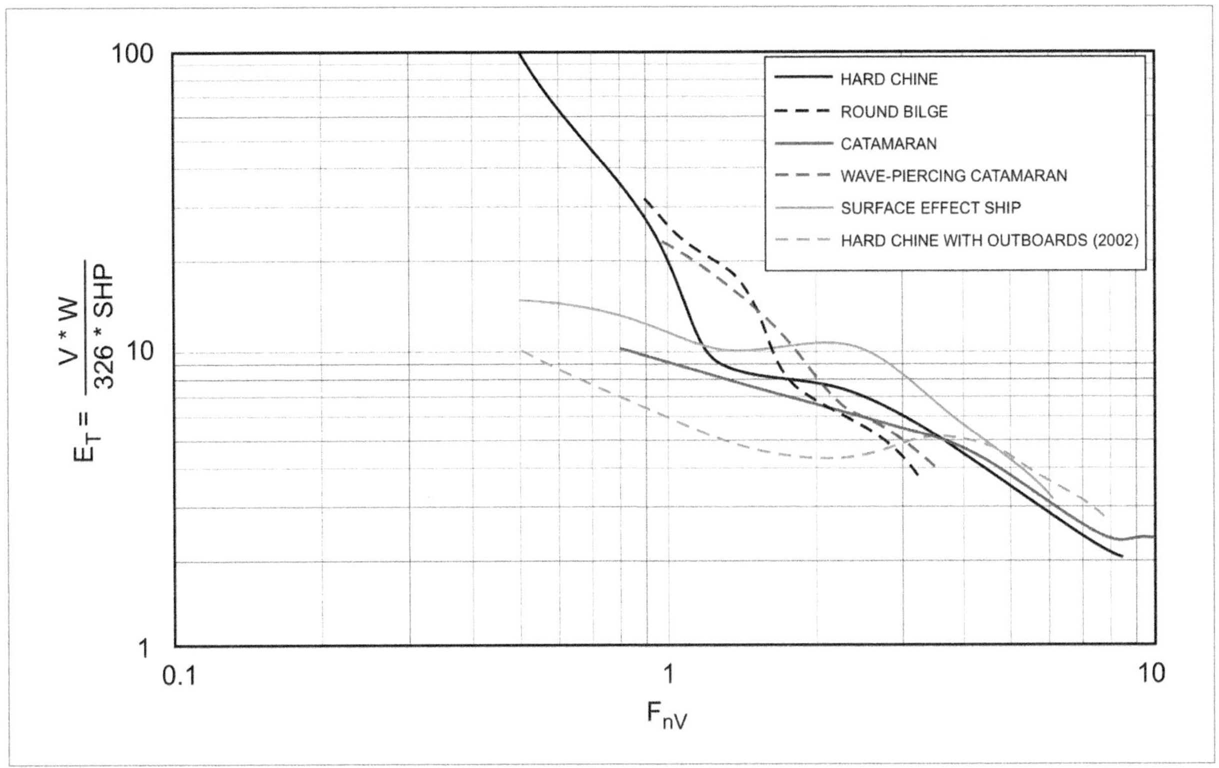

Figure 8-3: 2008 State-of-the-art calm-water performance except for outboards

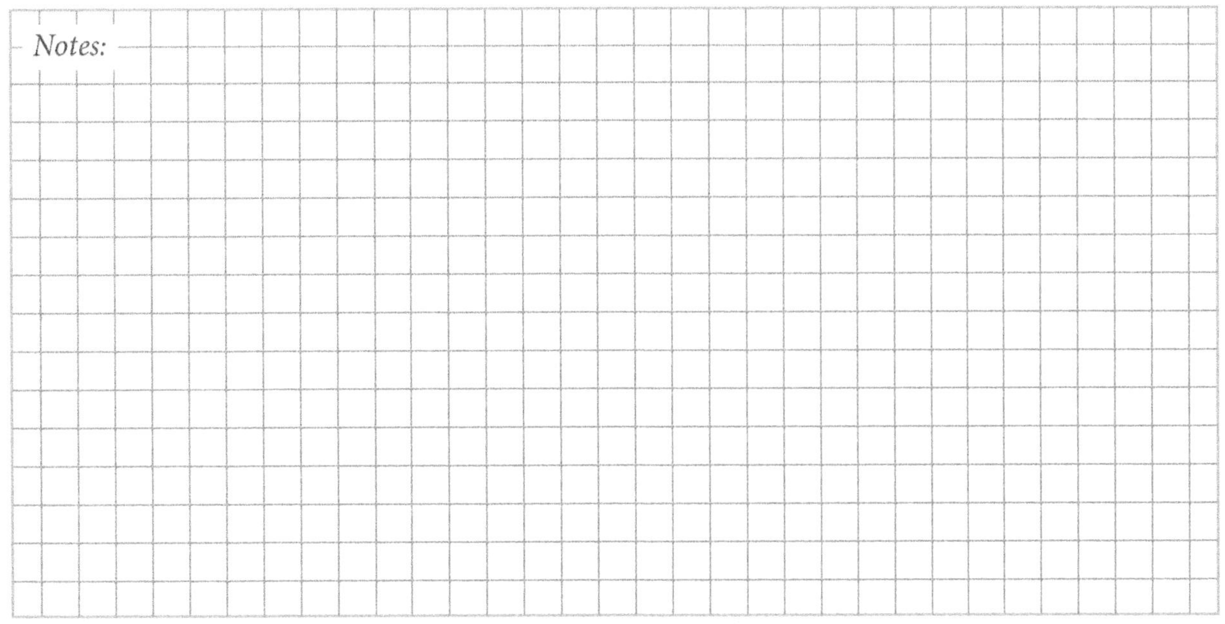

Table 8-B: 2008 State-of-the-Art Calm Water Performance						
F_{nV}	E_T Hard Chine	Round Bilge	Catamaran	SES	Stepped Hull	Hard Chine with Outboard Engines*
0.5	100.0	---	---	14.9	---	10.0
0.6	64.5	---	---	14.4	---	8.5
0.7	47.5	---	---	13.9	---	7.5
0.8	36.0	---	10.0	13.0	10.0	6.8
0.9	28.0	31.00	9.45	12.3	7.90	6.4
1	20.0	26.00	9.00	11.4	6.40	5.8
1.2	9.75	20.10	8.25	10.2	4.75	5.2
1.4	8.51	16.90	7.69	9.99	4.00	4.7
1.6	8.11	10.40	7.23	10.2	3.75	4.5
1.8	7.90	7.49	6.80	10.4	3.62	4.4
2	7.70	6.80	6.48	10.5	3.60	4.3
2.2	7.45	6.15	6.23	10.5	3.62	4.3
2.4	7.11	5.75	5.97	10.2	3.65	4.4
2.6	6.70	5.29	5.75	9.70	3.70	4.5
2.8	6.30	4.85	5.55	9.00	3.75	4.7
3	5.90	4.40	5.36	8.11	3.80	4.8
3.2	5.61	3.80	5.20	7.45	3.84	5.0
3.4	5.29	---	5.08	6.85	3.86	5.1
3.6	4.99	---	4.95	6.35	3.88	5.1
3.8	4.70	---	4.81	5.90	3.88	5.1
4	4.45	---	4.70	5.60	3.85	5.0
4.5	3.92	---	4.15	4.80	3.72	4.7
5	3.50	---	3.71	4.18	3.55	4.4
5.5	3.15	---	3.35	3.68	3.31	3.9
6	2.82	---	3.06	3.20	3.05	3.6
7	2.40	---	2.61	---	2.65	3.1
8	2.09	---	2.28	---	2.32	2.6
8.4	2.06	---	2.28	---	2.20	---
9	---	---	2.28	---	2.08	---
10	---	---	2.28	---	1.80	---
18	---	---	2.28	---	---	---

* 2002 S-O-A for hard-chine boats with outboard engines

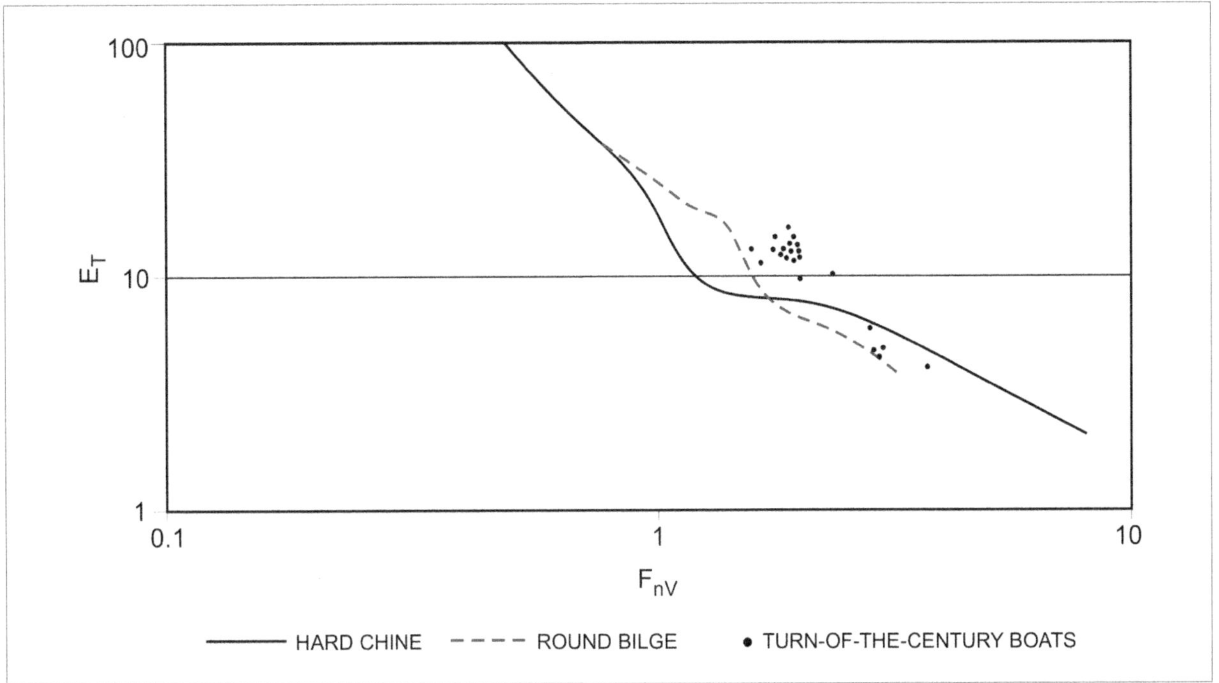

Figure 8-4: Comparison of 1993 E_T data for round-bilge hulls with that attained circa 1898

Transport efficiency, however, can trend toward reduced efficiency, such as when availability of internal combustion engines circa 1900 resulted in hulls with reduced length-to-beam ratios and lower slenderness ratios.

Data collected by William Durand in 1898 (see Figure 8-4) compared E_T for round-bilge hulls with the upper boundaries derived around 1993. Steam-engine-powered vessels of the late 1890s indicate superior transport efficiency relative to modern round-bilge hulls. Has the application of technology regressed since then?

In an absolute sense: Yes. Practical considerations and applications of vessels since that time, however, have evolved into hulls of different proportions. For example, the weight and volume required for propulsion machinery became a much-reduced proportion of a powered vessel's weight and volume.

In the late 1800s, extremes in length-to-beam ratios and slenderness ratios were essential to achieving high speeds with the available power to weight ratings of steam engines, boilers, and fuel.

And so it is that today's hull proportions are substantially different. Contemporary hulls are best suited to applications such as commercial cargo, military missions, yachting, and general arrangements, but are not proportioned for minimum hull resistance. Since high power-to-weight ratio internal-combustion engines, and gas turbines are now available to propel these modern craft, some hull-resistance increase may be readily traded for more utilitarian hull proportions.

Steam-engine-powered vessels of the late 1890s indicate superior transport efficiency relative to modern round-bilge hulls. Has the application of technology regressed since then?

Figure 8-3 represents the upper boundaries of available transport efficiency for various hull concepts comprised of the best combination of hulls and propulsors. Since hull hydrodynamics tend to be closely related to Froude number and propulsors are closely related to cavitation number, designers must consider actual vessel size and speed when selecting and integrating a hull form with a propulsor concept.

When you are developing a craft to meet an owner's requirements, remember that the size and speed of the various configurations play a significant part in the complex marriage of hull form and propulsion technologies. This is demonstrated in Figure 8-3. At $F_{nV} = 2.0$, those hull hydrodynamic factors sensitive to Froude scaling are the same at displacements for a 1 mt SES at 12 knots and a 1,000 mt SES at 38 knots. Several propulsors are available for the smaller 1 mt vessel operating at $\sigma_O = 5.0$; however, a limited number of practical choices exist for the 1,000 mt vessel operating at $\sigma_O = 0.5$.

From 1975 to early 1993, significant improvements occurred in transport efficiency for hard-chine monohulls in all speed ranges with some continuing advances being recorded through 2010. The progress between 1975 and 1993 is indicated in Figure 8-5. You can see the results most clearly by evaluating the percentage improvements in E_T. An analysis of the individual craft indicates that reduction in R/W generally contributed to increasing E_T for $F_{nV} \leq 1.5$. For $F_{nV} > 1.5$, improved E_T was achieved almost entirely by increasing η.

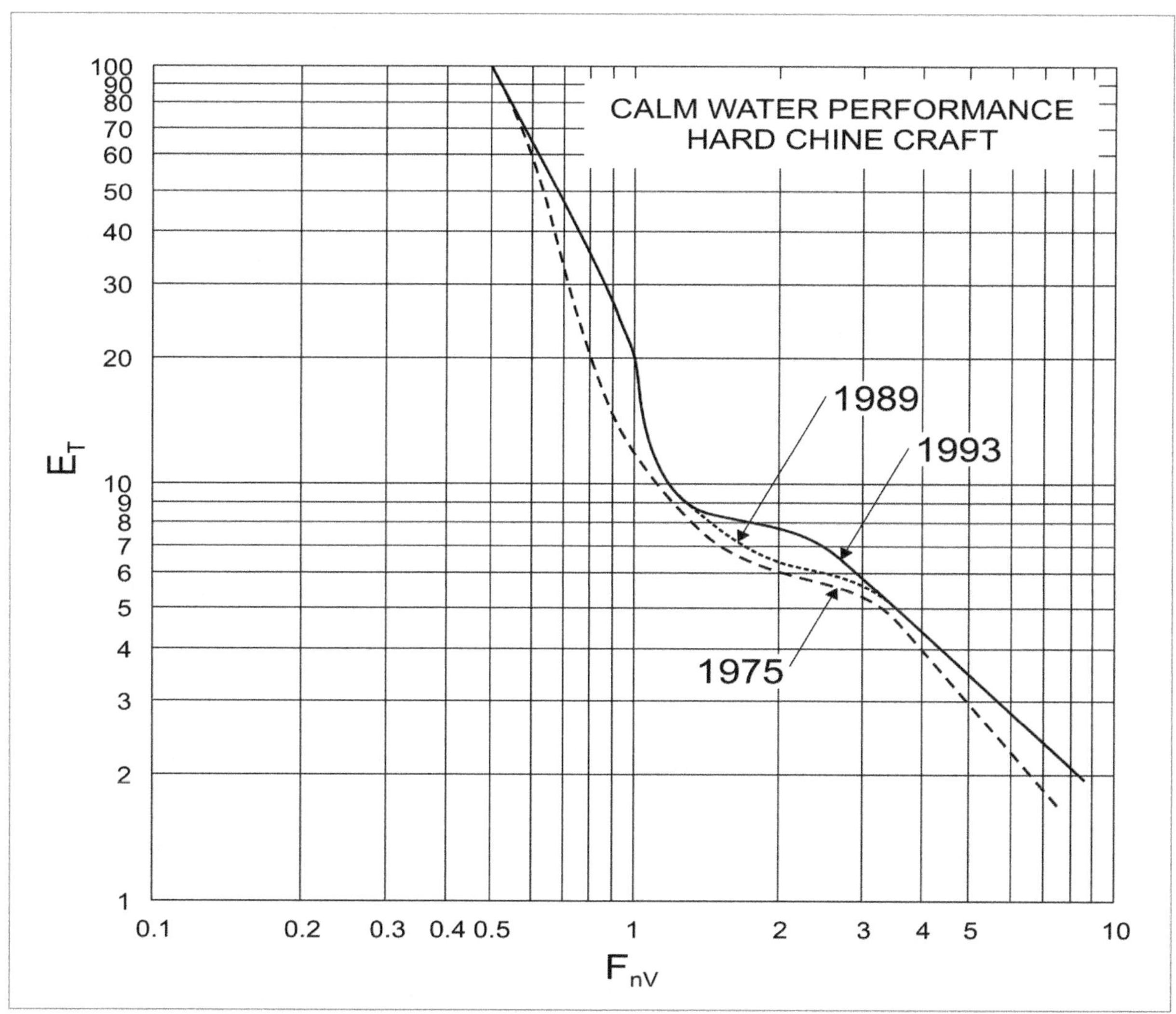

Figure 8-5: Increasing trends in transport efficiency for hard-chine craft over time

Hydrodynamic Viewpoint

Designers of planing and dynamically supported craft need to understand significant factors influencing the dimensions and shape of the hull bottom.

The overriding design principle for these craft is that of defining the hull boundaries whereby flow separation must reliably occur for all operating conditions of load and sea conditions.

The consequences of requirements and early design decisions which result in the quantification of the longitudinal distribution of projected chine beam and transom for full load—and its LCG—may well control the ultimate success of the as-built boat.

To repeat, the overriding design principle of planing craft is to define the hull boundaries whereby flow separation must reliably take place. The outer-hull boundaries for flow separation, when planing, occur at the chine and transom. Among the other features which may be incorporated onto hulls, longitudinal spray strips and transverse steps are notable and become increasingly important as operating boat speeds become much higher than the threshold of planing.

I believe it's much more realistic to associate the design process of planing craft with that for aircraft than to reach back to the approach followed for displacement vessels.

Confirm Client Requirements are Feasible

The approach of aesthetics first, just described, is highly inappropriate in the current environment of very mature science for design of planing craft. My intent here is to address a hydrodynamic technology-based design process. However, before pulling out a clean sheet of paper to design a vessel for your prospective client, the first step is to evaluate the feasibility of the task, both in verbal discussions and written requirements.

Then you can begin with a rough sketch of plan-form arrangements below as well as on and above the main deck to understand an approximate overall length of a vessel which satisfies volume and deck area needed for your client's requirements. Should a client specify a maximum overall length, your sketches will provide guidance as to whether or not all requirements are conflict-free.

You can allocate internal and deck spaces, areas and volumes with your rough sketches. And then, since performance craft technology is sufficiently mature, you can approximate—with engineering accuracy—with size boundaries of LOA and BOA to confirm client requirements are satisfied.

The approach that I am going to describe was followed to define major the feasible dimensions for *Destriero*, before engineering design had been initiated (Blount and Codega, 1991). The technical points we assessed included hard chine and volume, calm- and rough-water hull resistance, pitch, heave and vertical acceleration in waves; human factors; propulsive efficiency for variations in waterjet size; and dynamic stability. We could not consider maneuvering and control characteristics for high-speed waterjet-propelled craft operating in a seaway in the feasibility study as technology was inadequate at that time.

The format of the solution for feasible dimensions for *Destriero*'s record run is shown in Figure 8-6. The feasibility study bounded the combinations of waterline length and beam for monohulls which could meet the requirements when powered by the gas turbines selected.

This figure shows regions of primary and secondary solutions which offered hull dimensions with acceptable ride quality in design operating conditions. Combinations in the secondary region resulted in a slight reduction in potential speed; i.e., bare-hull drag was more than five percent greater than minimum.

It is worthy of note that *Destriero* was delivered having dimensions within the *region of primary design solutions* defined by the initial feasibility study. I find that a graphic presentation like the summary shown for *Destriero* is meaningful for clients, and it also stimulates discussions about the relative priorities of their requirements.

Using assumed requirements for a military craft, which are given in Table 8-C, I will work through this process for you with an example with dimensions in feet, pounds, and seconds.

Without much effort, designers working with current technology can effectively determine if client requirements are feasible. Making a few assumptions or approximations having negligible impact on accuracy of calculations can lead to the decision to proceed with the design task or identify where some research may be needed to extend technology boundaries before requirements may be met.

Table 8-C provides the assumed requirements for a military craft that I have selected to serve as the basis of a feasibility exercise. Take note that the craft to be designed must be transported by aircraft to be deployed very quickly in response to rapidly emerging military circumstances.

You will find that client requirements often include some type of external-restriction components such as shallow draft, dock length, bridge clearance, transit via canals (which may limit LOA and BOA), classification rules, flag rules, transport over the road by trailer, and hoisting by davits.

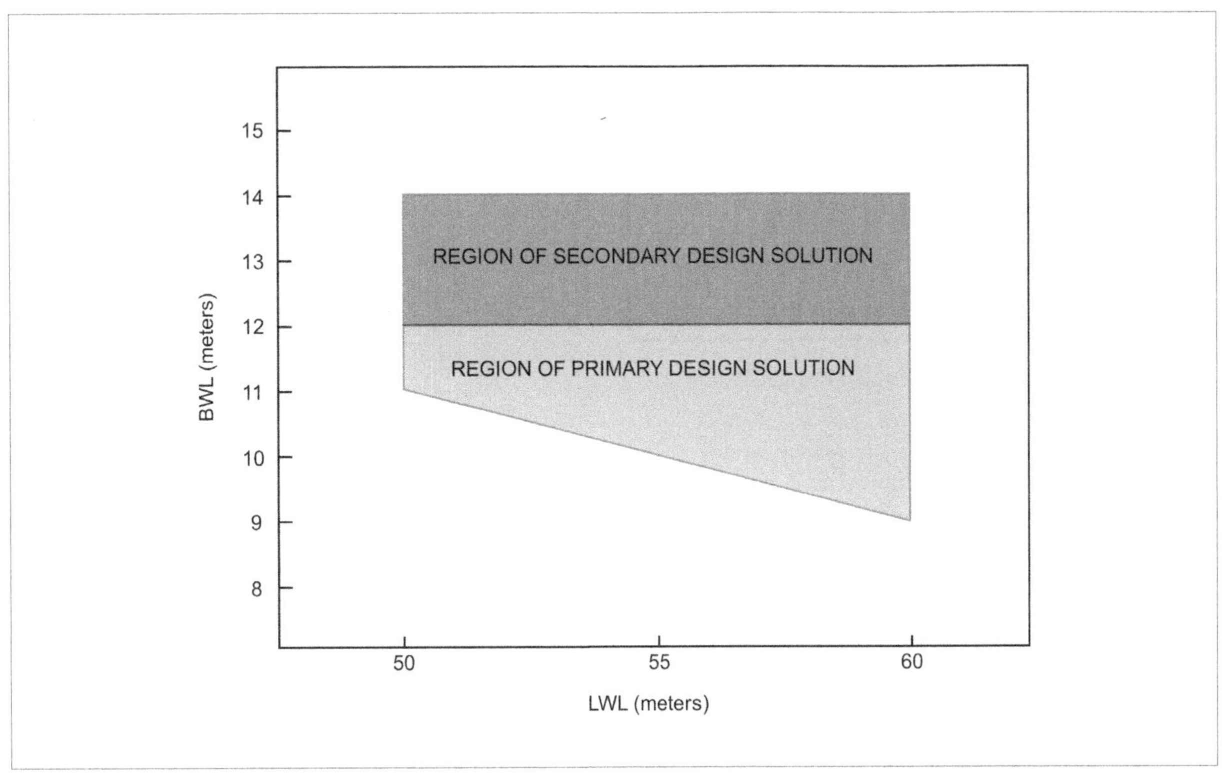

Figure 8-6: Major dimensions for feasible monohull solutions for Destriero's *record run (Blount and Codega, 1991)*

TABLE 8-C: EXAMPLE REQUIREMENTS FOR A MILITARY CRAFT	
Hull Form	Hard chine monohull
LOA	40 ft ≤ LOA ≤ 60 ft
Speed	Minimum of 40 kt at full load in deep water in Sea State 2, $H_{1/3}$ = 2.9 ft with $T_W \approx$ 4.8 seconds
Ride Quality	$\eta_{1/10th} \leq$ 1.5 g in irregular waves
Operating Environment	Deep and shallow salt water
Propulsion	Twin diesel engines with flush inlet waterjets
Below deck area for mission equipment and personnel	Minimum 470 ft^2 – maximum 550 ft^2
Transportability	By military cargo aircraft, maximum weight of craft at light weight plus wheeled transporter – 100,000 lb
aircraft-loadable dimensions:	
width – 18 ft	
height – 14 ft	
cargo length – 70 ft	

TABLE 8-D: HULL GEOMETRIC RATIOS FOR THIS EXAMPLE	
Relationship used	Typical multiplier range
$A_P = 0.83 (L_P B_{PX})$	0.81 to 0.85
$L_P = 0.93$ (LOA)	0.90 to 1.00
$B_{PX} = 0.90$ (BOA)	0.82 to 0.95

(Note: Each designer should use values typical of his or her own experience.)

Figures 8-7a to 8-7f show how client requirements individually contribute to major dimensions within which the craft will ultimately be delivered. Figure 8-7a begins defining the restrictions based on loading the boat into the aircraft and describing its overall length limitations of the requirements.

In Figure 8-7b, maximum practical hydrodynamic hull loading for planing craft is assumed to be $A_P/\nabla^{2/3} = 5.5$. To relate planing area (A_P) to overall dimensions, the following assumed designer-specific ratios are used:

For salt water and $A_P/\nabla^{2/3} = 5.5$

$$LOA = 0.4898\, W^{2/3}/(BOA) \qquad \text{Equation 8-3}$$

For this table, LOA and BOA were calculated with Equation 8-3 for displacements of 50,000; 60,000; 70,000; and 80,000 lb. and added in Figure 8-7b.

From the initial study sketches (not shown), the below-deck areas for mission equipment and personnel were estimated to be:

$$\text{Minimum } LOA = 723/(BOA) \qquad \text{Equation 8-4A}$$

$$\text{Maximum } LOA = 846/(BOA) \qquad \text{Equation 8-4B}$$

These dimensions for minimum and maximum below-deck areas were then added to Figure 8-7b as dashed lines. These lines terminated at BOA = 18 ft and LOA = 60 ft which have completely boxed-in possible geometric solutions.

The below-deck areas for mission equipment and personnel in Figure 8-7b indicated that full-load displacements (W) must be approximately between 55,000 lb and 70,000 lb. With these two displacements at 40 knots, average $F_{nV} = 3.8$ and for this speed ET = 4.70 from Table 8-B.

Beginning with the definition of E_T, accounting for transmission efficiency, and multiplying by 0.84 to reduce E_T SOA to typical value currently being attained by designer and builders, results in Equation 8-5.

$$BHP = WV/E_T\,(264.2) \qquad \text{Equation 8-5}$$

Total calm-water BHP for displacements 55,000 lb and 70,000 lb is found to be 1,772 and 2,255, respectively. Added drag due to waves was found to be four percent using Hoggard (1979).

Curves for nominal BHPs of 1,500, 1,800 and 2,100 were added as black lines to Figure 8-7c. With engine power defined, width of stock engines including space access established for service and maintenance is accounted for with fixed minimum-craft width at 12 ft in Figure 8-7c.

Next the ride quality requirement of $\eta 1/10\text{th} \leq 1.5$ g was calculated assuming trim control and using Hoggard and Jones (1980) for $\tau = 2.5$ degrees. This was added in Figure 8-7d as well as the boundary for $\eta 1/10\text{th} = 1.0$ g to see if potential change to hull dimensions were possible to achieve improved ride quality.

The many curves in Figure 8-7d can make this confusing. Therefore in the region of dimensional feasibility, lines were removed in Figure 8-7e and within the trapezoidal-shaped "pink-colored space" design solutions are possible which meet requirements.

Finally, Figure 8-7f is offered with a small light-blue addition space to the previous figure which might be brought to the attention of the client that with some LOA increase improved ride quality is possible.

This example is based on a single-chine, monohull relating BOA (0.90) = B_{PX} which would be a relatively slender L/B hull. Increased BOA craft would have more utility and static stability, if needed. This could be attained with a double-chine hull with B_{PX} of this example assumed to be the lower chine beam.

The requirement specified flush inlet waterjets. Figure 8-8 is provided to show for comparison that overall propulsive coefficients with submerged and surface-piercing propellers, that waterjets had the highest potential efficiency at 40 knots which confirmed that the specification was technically well founded.

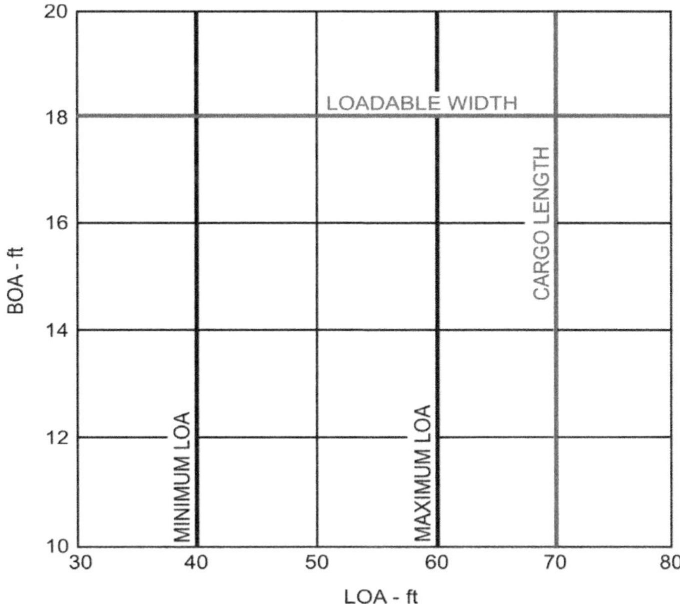

Figure 8-7a: Maximum aircraft loading dimensions along with limitations for military craft length

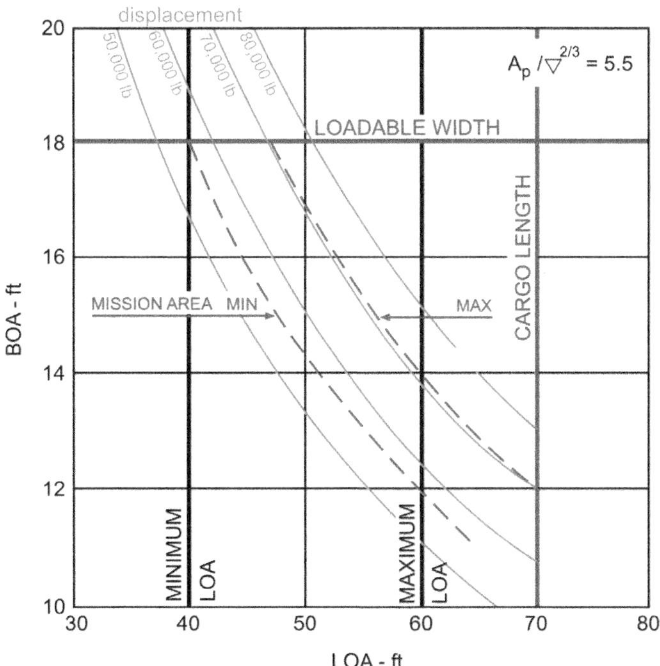

Figure 8-7b: Combinations of LOA and BOA for several displacements for hull loading $A_p/\nabla^{2/3} = 5.5$ with limitations for below deck areas

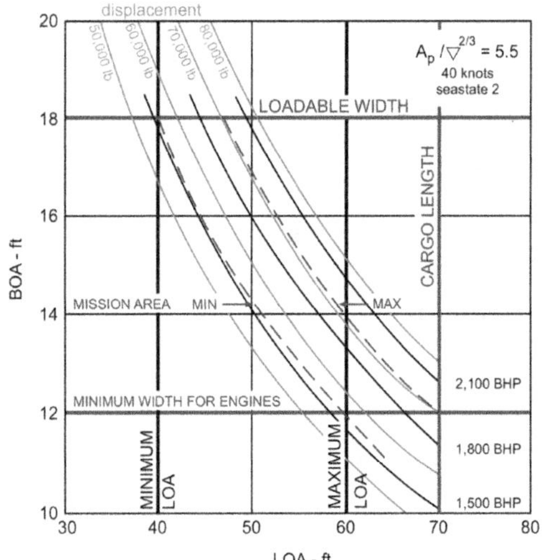

Figure 8-7c: Contours of constant BHP are added for a speed of 40 knots in Sea State 2 along with minimum beam for engines

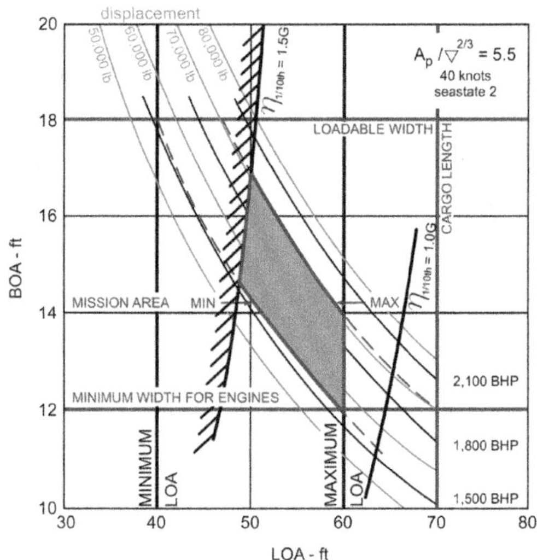

Figure 8-7e: Pink space indicates feasible dimensions where requirements can be met

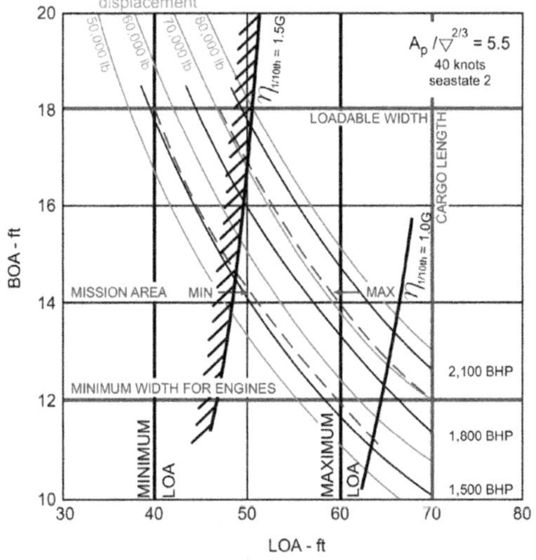

Figure 8-7d: Prediction for vertical acceleration at the LCG ($\eta_{1/10th}$ = 1.5 g) and locus for $\eta_{1/10th}$ = 1.0 g are indicated

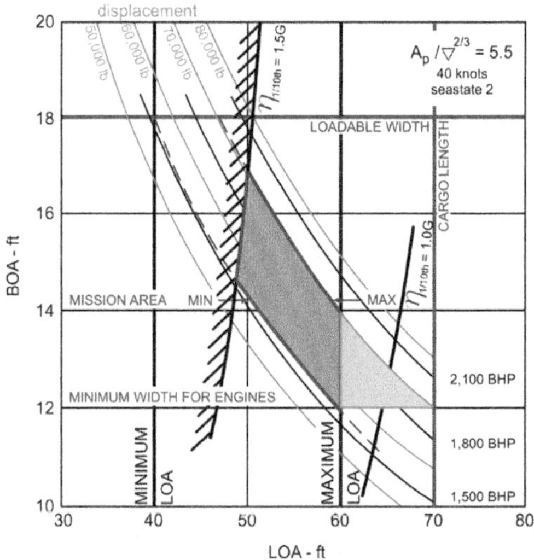

Figure 8-7f: Extended blue area where ride-quality requirement can be exceeded if maximum LOA may be increased

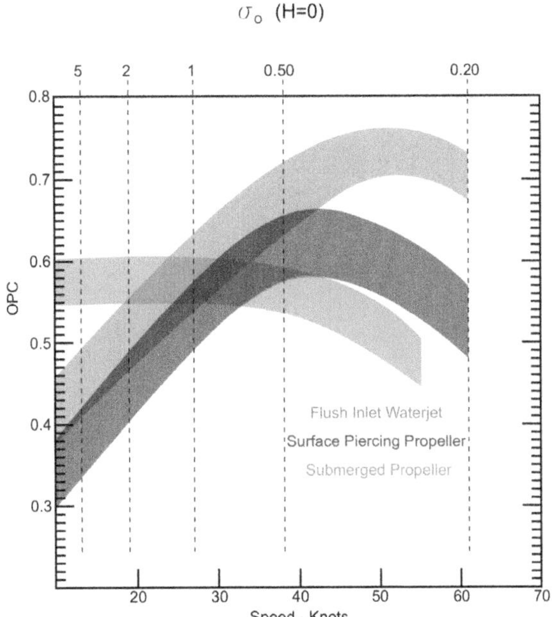

Figure 8-8: Comparison over a range of speeds of overall propulsive coefficients for flush inlet waterjets, submerged and surface propellers

Useful Load Fraction (ULF)

In Figure 8-3, the data masks information regarding relative useful load-carrying capability. Useful load is defined as the total weight required for fuel and payload. The useful load fraction (ULF) is the ratio of useful load divided by full-load displacement which provides input to the designer to make unbiased comparisons between craft concepts.

This approach is helpful in evaluating the suitability of a craft when you are considering its operational requirements; duration, range, payload, personnel and equipment.

You should consider ULF during selection of hull concepts for commercial, paramilitary, or military vessels which tend to have variable mission-loading requirements.

If the operation requires the transport of a large, low-density payload, then ULF should be further expanded to include consideration for hull volumetric characteristics or the available deck area. The ULF based on weight is defined as follows:

ULF = (Fuel + Payload)/(Full Load Displacement)

Equation 8-6

The ULF in Figure 8-9 represents the best load-carrying capability achieved (circa 1994) for hard-chine, monohull vessels operating in calm, deep water. These numbers were derived from building yards' weight reports and calm, deep-water sea trials (Blount, 1994).

In this study, weight estimates for selected vessels, which had been validated by load cells or draft readings at the time of inclining experiments or sea trials were significant in developing Figure 8-9.

In many cases, when instrumented sea trials were conducted, speed and power measurements were taken for light, design, and overload displacements. Thus, two curves are depicted for ULF versus F_{nV}: one for design load and one for an overload condition.

You can assume that ULF will be lower than that depicted in Figure 8-6 for craft that are not optimized for load-carrying.

The same is true for complex craft such as hydrofoils and air-cushion vehicles requiring additional systems for dynamic support as well as additional structure related to each concept.

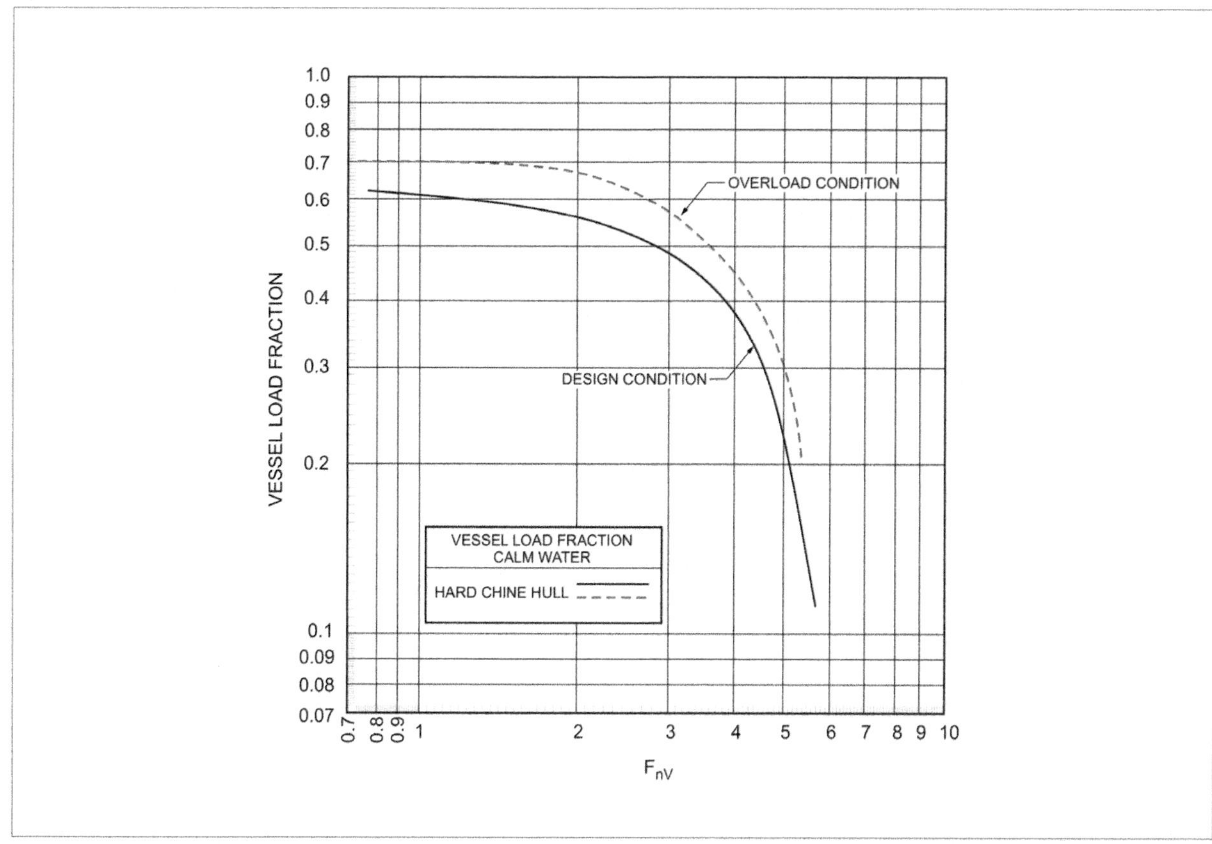

Figure 8-9: Achievable useful load fractions for hard-chine monohulls

These latter factors increase the light ship weight fraction relative to that achievable by monohulls. With regard to carrying capacity, the achievements of displacement, semi-displacement, and planing vessels are quite remarkable.

Useful load fractions near 50 percent have been demonstrated for $F_{nV} = 3.0$ and even ULF = 22 percent at $F_{nV} = 5.0$.

The rapid drop of ULF with increasing F_{nV} is not an exclusive characteristic of hard-chine craft.

The same trend is observed for other dynamically supported craft such as SES, which has ULF ≈ 0.45 up to $F_{nV} = 4.0$, but drops rapidly for $F_{nV} > 5.0$.

For new construction of advanced craft, ULF tends to improve at higher speeds when hull structures and propulsion machinery are carefully crafted to be weight efficient. Small gains in ULF also result when engines with low fuel rates are selected.

Safety Has Priority Over Aesthetics

Design is an amalgamation of science and art, and—as implied in this book's title—performance and therefore science for safety during the design process of high-performance craft must take priority.

Thus, the development of the hull form becomes the initial focus of the task at hand.

In essence, while the sheer line in profile can be a pure art form to be admired, it must be designed for function, and its final shape must be defined during the development of the hull lines.

It is not simply a visual component of a creative concept.

Estimating Full-Load Displacement

To make a first estimate of the displacement (weight) of the vessel, you will need an initial indication of overall hull length by the client or information from designer sketches.

When specified maximum speed is required at some displacement (full-load, half-load or light-ship) and range for a cruise speed, you can define the technology for that hull form. In a macro sense, performance of a vessel is fundamentally controlled by displacement, speed, and power.

The intent of this design approach is to initially develop the hull form and dimensional geometry to be as close as possible to the contract build size while also having the best possible hydrodynamic characteristics with the minimum amount of rework or revision other than what might result from client change orders.

With regard to obtaining minimum calm-water resistance and minimizing risk of dynamic instability for full-load conditions, round bilge hulls are recommended for maximum speeds up to $F_{nL} = 0.75$.

Above this speed, single- and double-chine hulls should be used for attaining dynamic stability. Round-bilge hulls might be considered for operation up to $F_{nL} = 1.0$ in calm water if designs are verified to be dynamically stable above speeds of 22 knots.

Concept Weight Estimates

Development of below-water hull lines should only begin when there is a full-load displacement objective as well as LCG required for speed and seakeeping. When sufficiently detailed information is unavailable for the magnitude of displacement, consider the following resources:

- If you are an experienced designer, you should have *files of as-built weight data* for various length vessels which may be adjusted by ratios to approximate a displacement for the new design.
- If you are a novice designer, you may consult *at-test articles* from boat and yacht magazines for published data for vessels of similar length and speed. Confirm that displacement data are identified as either light ship, mean, or full-load. A useful reference by Dawson (1997) provides new designers with experinced guidance.
- In either case, collecting a large *database of displacement versus LOA* allows you to make a regression analysis providing representative trends with vessel length. My experience indicates that displacement is proportional to $(LOA)^{2.3}$—not $(LOA)^3$—as hull beam and depth increase at a lower ratio than increase of length. Thus, weight would be proportional to a constant (X_N) multiplied by $(LOA)^{2.3}$.

Every design office should calculate, tabulate, and track typical values of X_N using as-built, full-load weights for vessels of their designs delivered by various builders. My experience for X_N trends is in Table 8-E.

For consistent definition of LOA, features overhanging the bow and stern of the hull, such as swim platforms aft and bow pulpits forward, should be excluded from this measurement of overall length. Tracking X_N over time also provides a measurement of design and technology progress with regard to design weight reduction.

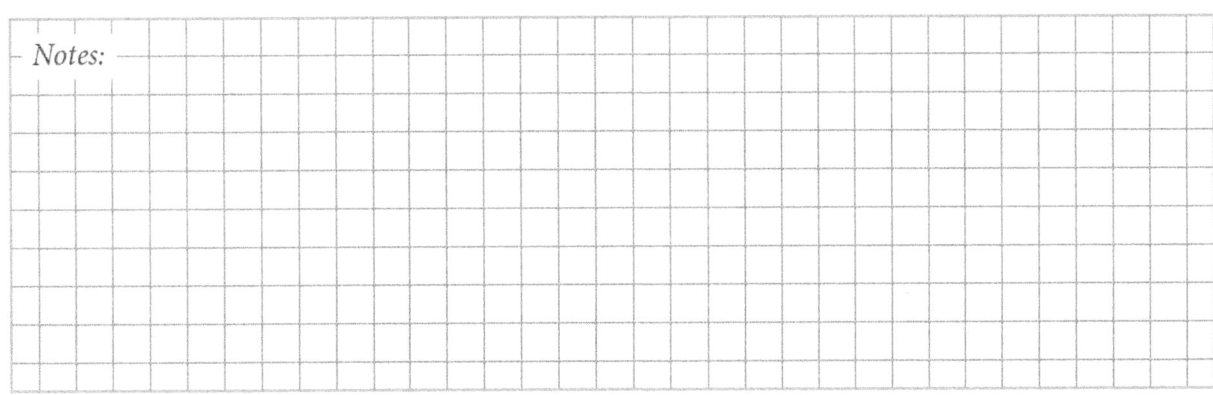

Weight Estimate (Option 1)

With no more information than LOA a first full-load weight estimate may be made with Equation 8-7 (a and b).

$W = X_N (LOA)^{2.3}$ (lb) Equation 8-7a

$\Delta = X_N (LOA)^{2.3}$ (mt) Equation 8-7b

TABLE 8-E: AUTHOR'S VALUES OF X_N FOR FULL-LOAD DISPLACEMENT

Performance Speeds	X_N for Full-Load	
	for LOA dimension in ft	for LOA dimension in m
Displacement	15.0	0.1046
Semi-displacement	10.0	0.0697
Semi-planing	8.5	0.0593
Planing	5.0	0.0349

Weight Estimate (Option 1A)

A variation of Option 1 for estimating the full-load displacement only for hulls designed for semi-displacement speeds is based on minimizing calm-water hull resistance for speeds just above $F_{nL} = 0.40$, before considering bulbous bows discussed in Chapter 4.

Taking $X_N = 10.0$ or 0.0697 for appropriate dimensions results in typical full-load displacements of as-built vessels which are often too heavy for minimum resistance for their length.

The best length for minimum calm water resistance for displacement and semi-displacement vessels occurs when slenderness ratio is $L/\nabla^{1/3} \geq 7.0$ where L is static waterline length.

For this option, I recommend using the typical as-built weight (for example, $X_N = 10.0$) for semi-displacement vessels delivered but increasing station spacing of the hull lines longitudinally so that L results in $L/\nabla^{1/3} = 7.0$ in order to approach minimum hull resistance.

Weight Estimate (Option 2)

Should a client specify propulsion engines by make, model, and power with a maximum speed at full-load displacement, then you must obtain the initial recommended design weight for the vessel recommended by using transport efficiency (E_T) from Figure 8-3 and Table 8-B.

I recommend that an initial full-load displacement be for 0.84 E_T at the expected F_{nV} for maximum speed as client requirements often preclude optimizing all aspects of a design.

Since displacement (W) is used for calculating both E_T and F_{nV}, Equation 8-10, for salt water, has been developed to solve for the value of W with data from Table 8-B.

Previously defined definitions for E_T and F_{nV} are repeated here for convenience.

$E_T = WV/SHP(325.9)$ Equation 8-8

$F_{nV} = V1.6878/(g\nabla^{1/3})^{1/2}$ Equation 8-9

Assumption: $\eta_T = 0.965$ and W in pounds

Calculations: $W = E_T\,(SHP)\,325.9/V$

$W = 0.84\,E_T\,(BHP)\,0.965(325.9/V)$

$W = E_T\,(BHP)\,264.2/V$

$V = F_{nV}\,[g^{1/2}(W/64)^{1/6}]/1.6878$

$V = F_{nV}\,(W^{1/6})\,1.680$

Combining equations

$W = E_T\,(BHP)\,264.2/F_{nV}\,(W^{1/6})1.680$

$W^{7/6} = 157.2\,(BHP)\,(E_T/F_{nV})$ Equation 8-10

Table 8-B provides data for calculating the ratio E_T/F_{nV} for the hull form being considered and BHP is the total power of propulsion engines specified.

Taking the 6/7 root of $W^{7/6}$ calculated with Equation 8-10 for each of several values of F_{nV} provides full-load displacement in pounds (W) which is used to calculate V from F_{nV}. When you have found agreement with owner's specified speed with interpolation, then you can use the W corresponding with that speed to develop hull lines. An example of this method for estimating weight is given in Table 8-F.

TABLE 8-F: EXAMPLE OF ESTIMATING FULL-LOAD DISPLACEMENT FOR REQUIRED SPEED WHEN ENGINES ARE SPECIFIED

Example: For a planing boat
Required: V = 40 knots using 3 engines rated at 1,500 BHP
Reference: Table 8-B

	F_{nV} (1)	E_T (1)	E_T/F_{nV}	$W^{7/6}$ (2)	W lb	V kt (3)
Notes: 1. From Table 8-B for hard-chine hulls 2. Calculated with Equation 8-10 3. Calculated with W and F_{nV}	3.0 3.2 3.4	5.90 5.61 5.29	1.97 1.75 1.56	1,394,000 1,238,000 1,104,000	184,700 166,800* 151,200	38.0 39.9* 41.7

* Solution: Design full-load, W = 166,800 lb

Hull-Sizing Method Based on ULF (Option 3)

Another option for predicting full-load displacement for hard-chine planing craft is well-founded when client specifications define fuel volume and payload weight. Knowing the weight of fuel plus payload permits you to calculate, based on useful load fraction (ULF), minimum chine length (L_P) for values of hull loading ($A_P/\nabla^{2/3}$ and L_P/B_{PX}) for required full-load design speed (F_{nV}).

For hard-chine monohulls, consider that for full-load displacement, the bottom-area coefficient $A_P/\nabla^{2/3} = 5.5$ offers an opportunity for achievable performance of a minimum-size craft. A design value of 6.0 provides lower risk considering the possibility that a slightly larger as-built craft might be delivered heavier than design full-load.

There are boats in service loaded to $4.5 \leq A_P/\nabla^{2/3} \leq 5.5$. This is a challenging range of area coefficients in which a designer might venture, a design range only for very experienced professionals. Boats designed to operate in this very heavy range of bottom loading have, on the plus side, improved ride quality in rough seas. On the downside, the likelihood of heavy spray and dynamic instabilities increases.

With this guidance regarding design values for $A_P/\nabla^{2/3}$, you may investigate an engineering approach early in the design process and begin estimating dimensions of projected chine length (L_P) and chine beam (B_{PX}) with a first approximation of A_P for hard-chine monohulls.

$$A_P = 0.83\,(L_P)\,B_{PX} \qquad \text{Equation 8-11a}$$

Refer to Table 8-D for the range of multipliers for Equations 8-11 (a and b) for hard-chine hulls designed for speeds of $F_{nV} \geq 3.0$. Designers who have an existing portfolio of well-performing planing boats might calculate multipliers from their hull lines.

$$\text{multiplier} = A_P/(L_P)B_{PX.} \qquad \text{Equation 8-11b}$$

In 2002, Dean Schleicher developed the following formulation which relates L_P,B_{PX} and Useful Load developed from operational requirements:

$$\text{Payload + Fuel} \leq X_P\,(\text{ULF})(L_P)^3/(L_P/B_{PX})^{1.5} \qquad \text{Equation 8-12}$$

Where dimensions for Payload + Fuel are pounds and L_P and B_{PX} are feet, the constant X_P has dimensions of lb/ft³ and is a function of water density and hull loading ($A_P/\nabla^{2/3}$) as seen in Table 8-G.

Payload + Fuel versus minimum L_P may be calculated directly for constant F_{nV} for values of $A_P/\nabla^{2/3}$, L_P/B_{PX}, water density, and ULF using Equation 8-13.

$$\text{Payload + Fuel} \leq [(\text{ULF})(\rho g)0.7562\; L_P^3]/[(L_P/B_{PX})(A_P/\nabla^{2/3})]^{1.5} \qquad \text{Equation 8-13}$$

With input from Tables 8-G and 8-H and Equation 8-13, you can calculate the weight of useful load for the minimum-size craft for chine lengths (L_P) and ratios of L_P/B_{PX} for a range of speeds (F_{nV}). When values of ULF less than given in Table 8-H for each F_{nV} are used, larger craft dimensions are defined.

An example calculation using ULF values for design load from Table 8-H for $A_P/\nabla^{2/3} = 6.0$ and $L_P/B_{PX} = 4.0$ is shown graphically in Figure 8-10.

To determine the non-fuel component of useful load, you will need to take into consideration client requirements; the number of persons, provisions, mission equipments; the number of un-replenished days of operation, etc.

Factors that influence the design weight of fuel are the type of propulsion power and its specific fuel consumption (SFC) including the operational profile at various speeds, range at speed, sea conditions, full-load displacement, and design margins.

TABLE 8-G: VALUES OF X_P FOR $A_P/\nabla^{2/3}$ AND WATER DENSITY

$A_P/\nabla^{2/3}$	X_P	
	Salt Water	Fresh Water
6.5	2.92	2.85
6.0	3.29	3.21
5.5	3.75	3.66
5.0	4.33	4.22
4.5	5.07	4.94

TABLE 8-H: STATE-OF-THE-ART USEFUL LOAD FRACTIONS (CIRCA 1994) VERSUS F_{NV} FROM FIGURE 8-9

F_{nV}	ULF	
	Design Load	Overload
1	0.60	0.70
2	0.57	0.67
3	0.49	0.57
4	0.38	0.44
5	0.22	0.30
6*	0.12*	0.14*

* extrapolated

Notes:

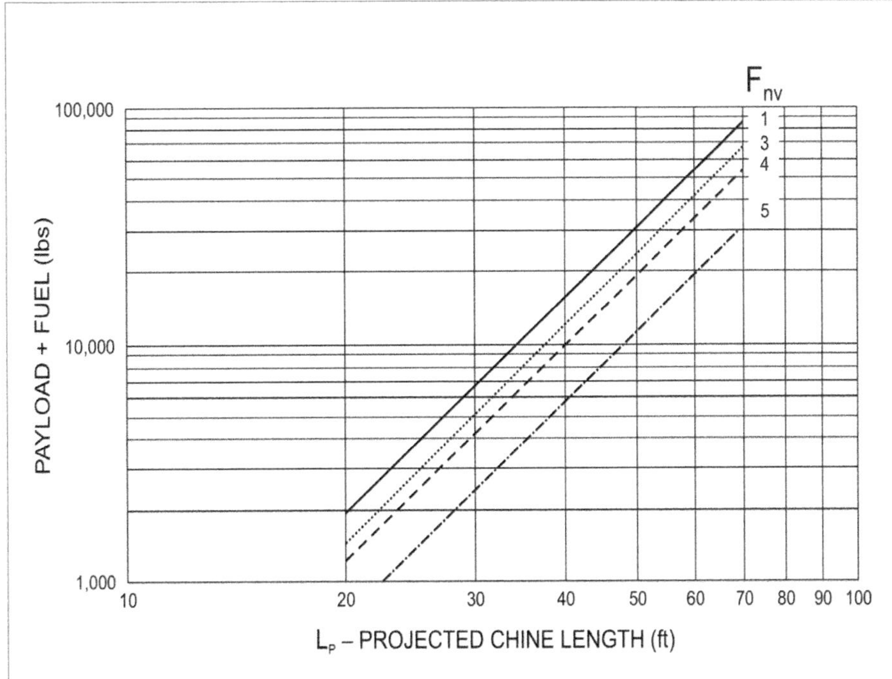

Figure 8-10: Example for design payload + fuel vs. project chine length for $A_P/\nabla^{2/3} = 6.0$ and $L_P/B_{PX} = 4.0$

In addition to being a useful design-guidance tool, Equation 8-13 has application when a naval architect is tasked with reviewing proposals offered in response to a design-build contract to purchase boats and craft which meet performance specifications.

In combination, Equation 8-13 and ULF in Figure 8-9 define the minimum-size proposed craft which will carry the weight of fuel plus payload of the performance specifications. Since this tool is a very good discriminator of the dimensions, boats, and craft offered in response to requests for proposals to meet performance specifications; the resulting proposed craft are either potentially feasible or they are not. For those that are feasible, resources can be directed toward detailed analysis of those designs to identify and rank the suitable contract offerings.

Useful Transport Efficiency

Considering the real measure of economy indicated by Gabrielli and von Karman (1950), Useful Transport Efficiency becomes UE_T with data trends for E_T given in Figure 8-3 and Table 8-B.

$$UE_T = E_T\,(ULF) \qquad \text{Equation 8-14}$$

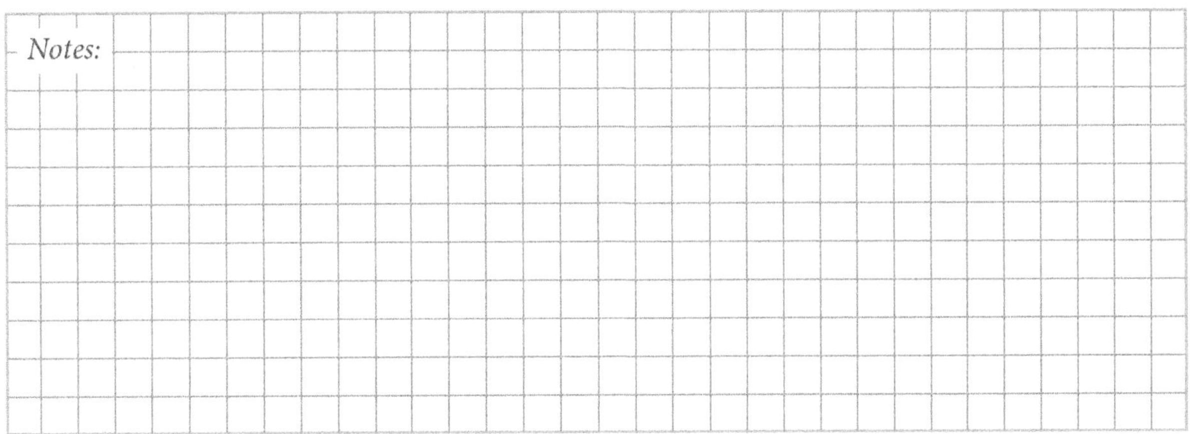

Trim and LCG Control by Operators

In marine design, strong views have expressed both praise and condemnation for installing the capability to control running trim. Some argue that including trim control on a delivered boat indicates a lack of designer or builder skill.

While this could be true for some boats, the need for trim control may result from improper loading by the owner or operator; a factor over which the designer or builder has no control. Personally, I believe trim control is as important as throttle for speed and helm for steering. Trim control should be considered for craft as a means for the operator to extend the boat's use with regard to variations in weight, loading, and sea conditions.

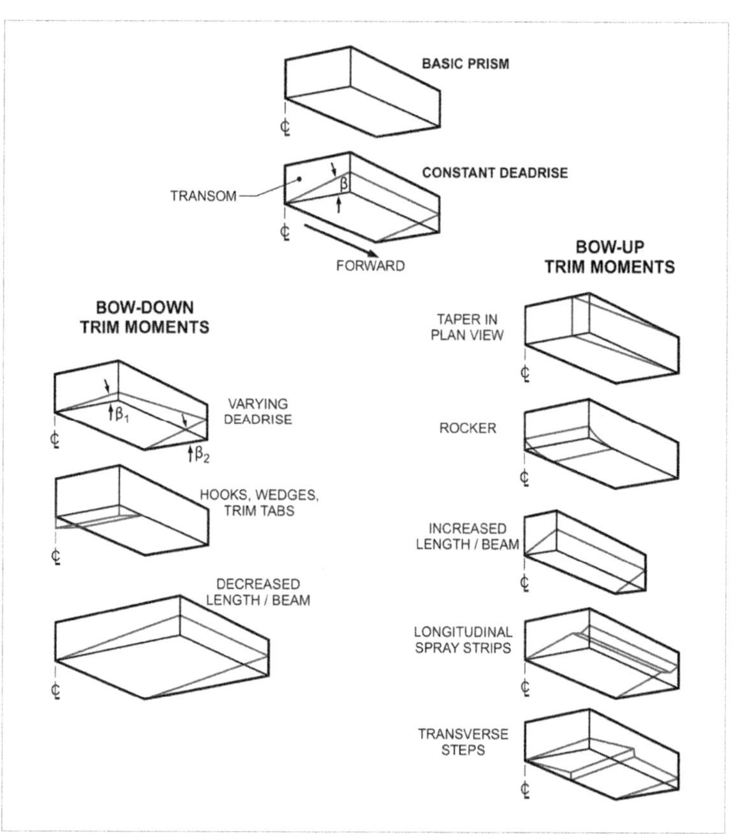

Figure 8-11: Trim moments created by design with relation to hydrodynamic planing hull geometry

Why consider controlling running trim? And how is it accomplished? Experienced boaters are well aware of possible improvements in visibility, speed, and ride quality achieved by operating at the best trim. To assign some direction to the change of trim, reducing trim angle is caused by developing a bow-down moment, and an increasing trim angle is achieved by effecting a stern-down moment. For example, a bow-down moment is created by a lifting force at the stern as when trim tabs or interceptors are deployed downward increasing pressure on the aft hull bottom.

You have the greatest opportunity to develop a balanced boat design when you have unrestricted freedom which, however, often conflicts with the reality of client requirements. At the beginning of a project, you have a variety of trim control techniques to consider, but the number of design tools diminish as construction approaches completion and running trim option are limited by "designed-in" fixed features and operator-controlled systems. Still, you have the option to make subtle, designed-in geometric refinements which in turn can make significant contributions to dynamic trim and thus to resistance.

Figure 8-11 depicts simple hull diagrams for bottom geometry variations and shows how they result in dynamic bow-up or bow-down trim moments. Hull geometric features which develop bow-up trimming moments are typically used for design speeds of $F_{nV} \geq 4.0$. Features to develop bow-down trimming moments generally are used for speeds of $F_{nV} \leq 2.5$, or as sea conditions worsen. These suggested values for F_{nV} are approximate because hull loading and LCG contributions also need to be considered as well as the operational environment (such as mostly in calm water or in the open ocean).

Devices which give the operator control of running trim include trim tabs, interceptors, outboard engines or inboard-outboard lower unit trim, and movement of liquid ballast. These all may be added to boats already in service, with the exception of freely locating ballast tanks and transfer pumps.

Trim Tabs and Interceptors

Trim tabs and interceptors are two different devices that serve the same function. They generate a lift force at the stern of a boat when underway, and this produces a bow-down trimming moment. When used differentially, port and starboard, these devices also produce a heeling moment, which causes the boat to level an off-center load or beam wind.

In designing or establishing the size of trim tabs or interceptors, your decision criteria is very important. The issue is to permit adequate operator trim management while minimizing the danger of driving a boat bow-down to a point of possible loss of craft control. For high-speed boats, bow-down trim control devices can improve visibility and acceleration through the hump speed. There are also benefits to reducing trim in a seaway for ride quality and for transverse leveling.

For this reason, I size trim-control systems to deliver 1.0 to 1.5 degrees bow-down trim at the speed of the boat having the greatest running trim angle. This criteria results in trim devices which are operator friendly with minimal opportunity to cause too much trouble at very high speeds. For boats with a maximum design speed near hump, it is convenient to incorporate some aft hook into the buttocks in combination with trim devices of moderate size. This will provide refined control as consumable loads vary.

Equation 8-15 provides design data for moment to trim one degree at hump speed. This equation has been developed from three planing-hull series having midship β between 12.5 and 25 degrees with some assumptions to simplify the design process. (Estimating lift and drag force of trim devices is described in Chapter 4.) Dawson and Blount (2002).

$$\text{Moment to trim } 1.0° \text{ in ft lb} = W/100{,}000)^{4/3}[12{,}000(L_p/\nabla^{1/3})^2 - 33{,}000(L_p/B_{PX})]$$

Equation 8-15

Note: Equation 8-15 applies to the speed of a boat at its greatest running trim angle, which is generally between F_{nV} of 2.0 and 3.0. For planing boats of normal proportions and loading, the speed of greatest trim will be near $F_{nV} = 2.8$.

Evaluating Performance with Trim versus Speed

Measuring running trim for a full range of boat speeds during sea trials provides a valuable database for evaluating and diagnosing operating characteristics of existing craft. Change of trim with speed can be easily measured with electronic angle-measuring devices. Figure 2-9 shows a speed-trim graph obtained from free-to-trim model tests and scaled to a 100,000-lb, 75-ft (45.4 mt, 22.8 m) boat.

By showing the range of trim angle variations with speed, this figure will help you understand the significance of having operator trim control throughout all speeds. At hump speed, a boat's agility (acceleration to planing) is improved by a bow-down trimming moment; at planing speeds, operating conditions may be improved by reducing the bow-down moment.

There are, of course, safety limits to which bow-down trim control may be applied. Too high of a running trim angle may cause porpoising, and too low of an angle will result in loss of dynamic stability in yaw, roll or pitching down (stuffing).

Design Management of LCG for Operational Flexibility

The longitudinal center of gravity (LCG) of a marine vessel is a significant factor influencing the development of hull lines. Specific operational requirements often include at least overall length, speed, range, and sea conditions while larger vessels may specify maximum gross tonnage.

When the vessel is delivered, maintaining LCG with related trim may well have equal or greater influence than meeting design displacement for best design performance. During design development and construction, the naval architect and builder are responsible for controlling the best displacement and LCG. Then throughout the vessel's service life, the owner and crew are in control of these operating conditions.

When it is possible for radical variations of operational LCG to occur, design features should be given to provide operators with trim-control capability to mitigate off-design loading conditions. Trim tabs or

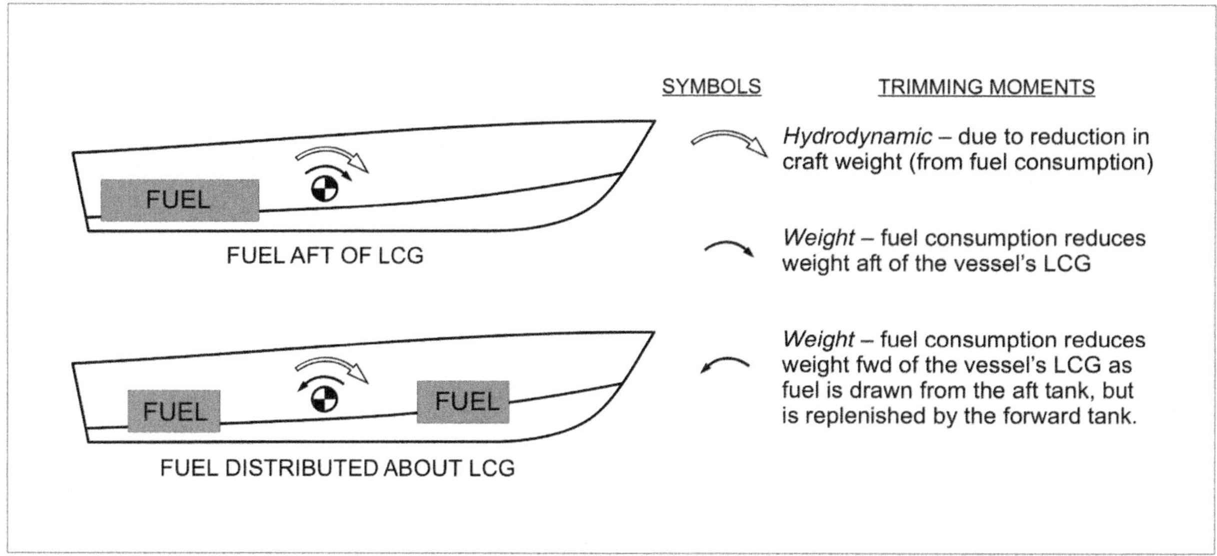

Figure 8-12: Significance of location of consumable loads for planing craft

interceptors for small craft provide a simple capability for crew control when underway, as do multiple, longitudinally distributed fuel tanks with a fuel transfer system in combination with salt water ballast tank(s) for large yachts, military and commercial vessels.

Destriero (67m) is an example of a vessel with a small salt-water ballast tank forward that could take on water to maintain optimum LCG when the distributed fuel load was low. Ballast tank capacity could hold water weight of about 3.5 percent of design full load.

You should also avoid relying on consumable load location to attain proper trim or LCG to meet requirements for static and dynamic stability necessary for project success. There are two conditions to consider. First is control of the LCG and resulting trim changes when craft weight drops due to consumption of fuel and water. Second is when operational conditions necessitate trim changes as a result of the operational environment such as head or following seas.

A recommended way to manage the effects of reduced weight due to fuel consumption for planing craft is to have an appropriate longitudinal location of interconnecting tanks, except for small boats with one fuel tank. In this latter case the single tank should be located with its centroid at or slightly forward of the LCG.

Consider the diagrams in Figure 8-12 simply relating the fuel weight and moment shifts as fuel is consumed relative to changes of hydrodynamic planing lift. For planing craft, an undesirable example of design management of weight shift as fuel is consumed by engines and generators is when the centroid of fuel weight is aft of the LCG. Using fuel from aft tanks makes the stern lift as weight reduces, which results in a bow-down moment while at the same time the total weight is reducing.

With less total boat weight, a planing boat needs less hydrodynamic lift for support at high speeds. Due to hydrodynamic principles, planing boats then naturally trim down by the bow to maintain equilibrium.

When returning to home port with a low fuel load, especially in a following sea condition, the boat condition with only aft fuel tanks as I just described can be dangerous. There are possibilities for yaw instability (bow steering), transverse (roll) instability or pitch instability (stuffing). Safe operational control may only be maintained by reducing speed.

The second example in Figure 8-12 is preferred for tank arrangements for planing craft which allow the centroid of total fuel at full load to be slightly forward (up to 0.04 L_P) of the lightship LCG while the day tank(s) supplying the engines are aft of the lightship LCG. Fuel tank arrangements consisting of two or more tanks with either gravitational or powered pumps to transfer fuel on high-speed craft, is the desired approach.

As engines consume fuel from aft day tank(s), fuel is transferred from forward storage tank(s) and, thus, weight is reduced forward of the LCG to produce a bow-up moment. Again, as the total weight of the boat is reduced by consumed fuel, less planing lift is necessary, and the running trim lowers.

The combination of bow-up moment due to the centroid fuel weight moving aft and bow-down hydrodynamic lift moment resulting from a lower total craft weight can offset each other. With appropriate location of the centroid of the fuel tanks relative to the lightship LCG, it is possible to have a planing boat design which will operate at high speed and have virtually the same running trim angle versus speed, regardless of fuel load.

Take note that the appropriate location of the centroid of the fuel tanks is also dependent on both the weight of fuel load relative to the lightship displacement and engine fuel-rate demand.

To achieve the ride-quality and structural-load requirements at high speeds in irregular seas, you must design the hull lines to minimize pitch oscillations in response to waves, and to give the crew operational control of setting LCG and trim angle for the best performance for all displacements and sea conditions.

Fuel and water tank arrangements provide capacity for the range and operational duration. Tank arrangements with liquid transfer capability allows the desired LCG to be achieved even when the fuel load is low. On large vessels, designing for some limited extra fuel-storage tank volume to permit the option to transfer fuel forward or aft—and to make it possible to take on salt-water ballast separately—offers opportunities to achieve the LCG for best performance for a wide range of displacements and sea conditions.

Should a client later add fuel to the extra volume fuel-storage tanks, losing some LCG control in an overloaded condition, a quick study can evaluate operational time, range, and extended range at lower speeds. When the extra tank volume becomes available due to burn-off, fuel transfer may be used to return LCG for best performance. The full range of LCG control will then be restored.

Design Criteria

Your own design criteria will evolve over time as knowledge increases with personal experiences and as you develop a growing database of works published by others. Keep track of both successes and failures of designs that are ultimately validated during both builder's and acceptance trials. Well-conducted trials are great learning experiences.

More often than not, analyzing and correcting a problem contributes tremendously to a positive, personal learning experience. Granted, some faults are the result of designers extrapolating information beyond known limits or using inappropriate criteria to support a design decision. Keeping notes in the spaces provided in this book about these problems, solutions and fixes is the beginning of building your personal designer's notebook of procedures to follow—as well as decision criteria and "things not to do that way again."

As for molding an approach for developing and conducting experimental research programs, I have found it to be very beneficial to accept design projects of vessels with requirements that are outside of the boundaries of existing technologies. Rather than just model-testing a design to validate or confirm compliance with requirements, this is an opportunity for experimental research extending conditions well beyond the design needs of a current project to find the practical limits of the application of the technology.

This approach provides an extension of additional knowledge, develops criteria for current and future programs, and builds a unique database. If your design office has access to the most advanced technical database available, then—when a prospective client requests a design proposal with ever-increasing and demanding requirements—you will be able to respond with a confident, low-risk offer.

Chapter 9 summarizes and documents design criteria for guidance that I have evolved over many years. *Note: Some of this guidance may be at variance with reference material published in the past, but the guidance herein is my current thinking.* The order in which each criteria is briefly presented may not be representative of the process you use for developing a hull design, or making hydrodynamic performance predictions. However it may assist rethinking approaches to revise your design process.

Weight Estimating and Management

Developing hull lines beginning at the earliest conceptual design cycle must be based on a target full-load displacement and LCG, both of which must be realistic estimates for satisfying all client requirements. As detail grows through preliminary and contract design phases it is imperative that you document weight and track the location of their centers to increase accuracy of results.

The Navy's Ship Work Breakdown Structure (SWBS) is a widely employed system to account for the weight of individual components and their 3-D locations relative to defined references for all items necessary for construction, outfitting, and operation of craft.

In an excellent 2007 reference published by the Society of Allied Weight Engineers, Cimino and Tellet document the principles for estimating weights and line items of the SWBS system. The SWBS system accounts for weights in general groups as defined in Table 8-I. These major groups are further subdivided down in SWBS documentation to the smallest weight components making up the light ship weight of a vessel.

TABLE 8-I: WEIGHT GROUPS BY FUNCTIONAL CLASSIFICATION

Group	Function
100	Hull structure
200	Propulsion plant
300	Electrical systems
400	Command and control systems
500	Auxiliary systems
600	Outfit and furnishings
700	Armament (para-military and military vessels)
Loads	Variable and mission-specific loads

Tables 8-J and 8-K provide examples of percent full load weight grouping of contract weight estimates blended from a small number of craft designed since the year 2000 for two types of service. Table 8-J is representative of percent weight grouping for planing sportfishing boats and yachts between 65 and 115 feet (20 and 35m).

Table 8-K is representative of weight grouping for semi-planing motor yachts between 115 and 180 feet (35 and 55m). As these examples denote a point in time both owner requirements and technology will continue to change, resulting in future reallocation of the percentage distribution of weight grouping and loads.

TABLE 8-J: PERCENT OF FULL-LOAD WEIGHT GROUPING EXAMPLE FOR PERFORMANCE (PLANING) SPORTFISHING YACHTS BETWEEN 65 AND 115 FEET

Weight Group	% of Full Load
100	22.4
200	23.3
300	5.3
400	1
500	5.2
600	13.7
Subtotal	70.7
8% contract design margin	5.6
Lightship total	76.3
Loads	
Personnel & effects	1.8
Provisions	0.4
Potable water	2.8
Fuel	16.5
Fishing gear	1.6
Subtotal loads	23.7
Full load	100

TABLE 8-K: PERCENT OF FULL-LOAD WEIGHT GROUPING EXAMPLE FOR PERFORMANCE (SEMI-PLANING) MOTOR YACHTS BETWEEN 115 AND 180 FEET

Weight Group	% of Full Load
100	22
200	15.7
300	5
400	2
500	9
600	19
Subtotal	72.7
8% contract design margin	5.8
Lightship total	78.5
Loads	
Personnel & effects	2
Provisions	0.8
Potable water	2.2*
Fuel	13
Miscellaneous liquids	2
Water toys	1.5
Subtotal loads	21.5
Full load	100
*Vessels with water makers	

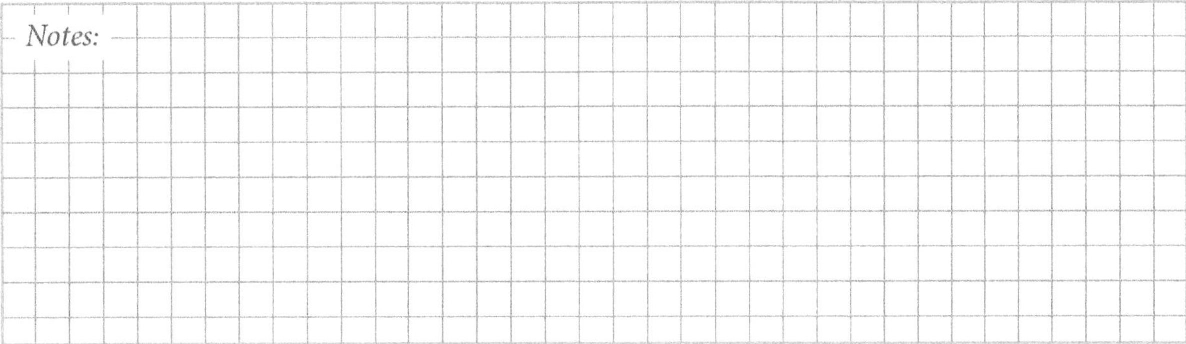

Margins in the Design Process

Due to the lack of definition in the early stages, the application of proper margins will mitigate inaccuracy of the estimates, according to Hatchell and Wilson (1998). Ideally, the weight estimate becomes more refined as the number of iterations around the design spiral increases.

Should a craft be found to differ from the target full-load weight at some uncompleted point in the design process, the craft may have to be redesigned for the revised displacement or the design process paused to reassess requirements with the client. It may become necessary to consider alternatives such as dimensions or payload changes which must be presented to the client for consideration and decision.

To accurately reflect the finished craft, you must include weight margins in the weight estimate throughout the design process to avoid inadequacies in the design. The weight margin is greatest in the earlier design stages, and it decreases in the later design stages as the design becomes more refined and complete. The weight margins reflect the degree of confidence you have that your weight estimate will accurately reflect the finished product.

LCG Margin

LCG location has a significant effect on performance craft. In addition, the 3-D dimensions of CG affects intact and damages stability as well as hoisting conditions.

For this reason, in the feasibility and preliminary designs, you must analyze variants of CG and their impact carefully in order to correct possible adverse effects before contract design commences.

In preliminary design, you should evaluate any change of LCG positions greater than ±1% of L or L_P for its effect on performance and stability. This highlights aceptable limitations of CG on the craft as it progresses from feasibility to preliminary and then to contract design.

Be well aware of CG problem areas before making changes in contract design and necessary equipment, arrangements, or the hull form.

High-performance craft in operation rely on having a clean and smooth underwater hull, as well as having design weight and center of gravity maintained for consistent performance.

If the original requirements need to be changed in the future, any resulting alterations to the vessel should not reduce performance or stability unless you have advised the client in advance of any construction.

No service life allowance for performance vessels should be incorporated in the design, engineering, or construction of a vessel having specific requirements unless required by the client.

Calm-Water Design LCG

Development of hull lines begins with estimated values of full-load displacement and LCG. I discussed concept design displacement in previous sections of this book; here I am presenting guidance on LCG design. LCG affects hull resistance, motions, and accelerations in irregular waves as well as dynamic stability at high speeds.

When developing the initial hull lines for safe and best performance, LCG is equally as important as satisfying the range of operational displacements. The preferred LCG also has dependence on design speed and hull form.

For performance vessels in calm water, design LCG is primarily related to dimensionless speed. Hull loading is a secondary factor. With speed being the primary factor for establishing LCG, you must have some information regarding the operational profile of the craft being designed.

Operational profile may be stated in the requirements, obtained in your discussions with the client, or assumed by the designer based on patterns consistent with existing craft or similar type currently in service.

The purpose of some craft is just to go fast. For example, a race boat or a bass boat would be designed only for one speed—the maximum. Another craft might operate in a search-and-rescue profile: 80 percent at cruising/displacement speeds and 20 percent at top speed.

You might also pursue an operational profile when there is an equal probability of operating at any speed between zero and maximum speed. I frequently use this latter approach for operational profile when requirements are unstated or the client has been silent on a preference.

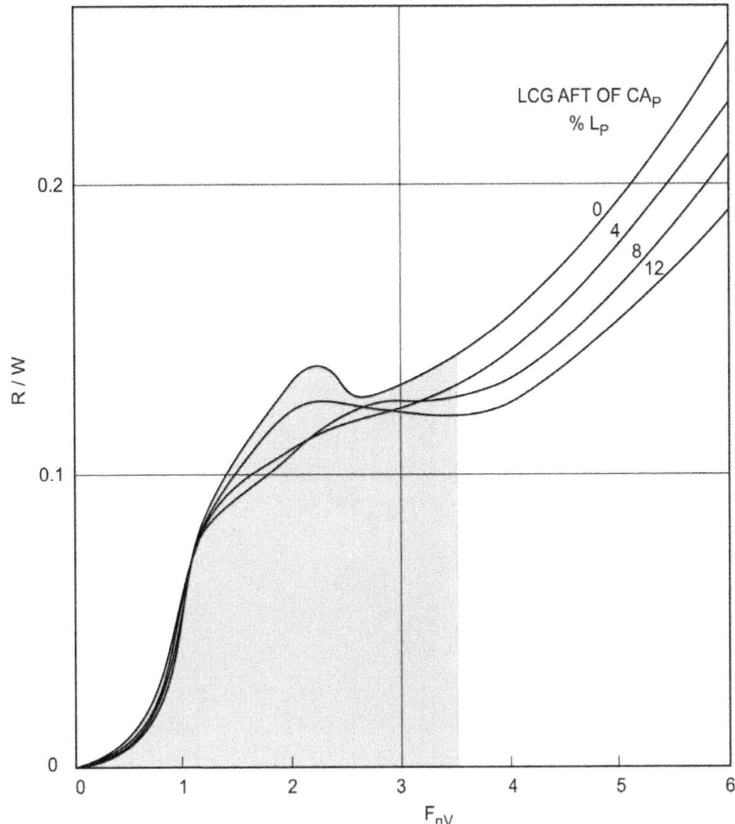

Figure 8-13: Resistance data used for establishing LCG for minimum power for a craft having equal probability of operating at any speed

When assuming equal probability of operating at any speed, the LCG for minimum power can be readily established when resistance data for constant displacement are available for three or more LCGs. As an example, Figure 8-13 from Series 62 R/W data for one model is shown versus F_{nV} for four different LCGs. For $F_{nV} \leq 1.0$ differences of R/W due to LCG are not significant and above $F_{nV} = 3.5$ R/W follows orderly theoretical patterns.

All four LCGs at these high speeds, $F_{nV} \geq 3.5$ permit the free-to-trim planing hull to be in equilibrium below the optimum trim angle. Thus, the aft most LCG has the highest trim angle, which is nearest to the optimum angle for the tested hull loading and therefore the lowest R/W.

For $1.0 \leq F_{nV} \leq 3.5$ the R/W curves cross each other. As an example, take maximum speed of $F_{nV} = 3.5$ (the shaded area in Figure 8-13): To find LCG for minimum power when there is equal probability of operating at any speed, you must compare the total areas under each resistance curve. The area under R/W versus F_{nV} curve for each LCG is proportional to power. Therefore, LCG for minimum area is the LCG for minimum power.

$$R/W\,(F_{nV}) \propto SHP \qquad \text{Equation 8-16}$$

This calculation procedure has been followed for Series 62, $\beta = 12.5°$ and Delft $\beta = 25°$ model data for speed ranges from zero up to $F_{nV} = 4.0$. LCG for minimum R/W is found as well as the LCG band width for 1.02 (R/W). The results of this calculation procedure are presented in Figure 8-14 indicating recommended bandwidth for design LCG for minimum to plus two percent R/W.

For semi-displacement speeds Blount and McGrath (2009) reported results of a design study of a large model test database of hulls for the speed range of $0.3 \leq F_{nL} \leq 1.0$. Figure 2-13 provides LCB/L from that study for minimum resistance for different speeds.

Recall that LCB ≡ LCG and measurement is forward of the hydrodynamic transom. The data in Figure 2-13 is the recommended initial LCG for developing hull lines for round-bilge hulls when optimizing a design for a single speed.

A semi-displacement vessel having an operational profile of 80 percent cruise at $F_{nL} = 0.40$ and with 20 percent dash at $F_{nL} = 0.60$ will, for minimum calm water resistance have two different design recommendations for LCB/L of 0.45 and 0.40 respectively.

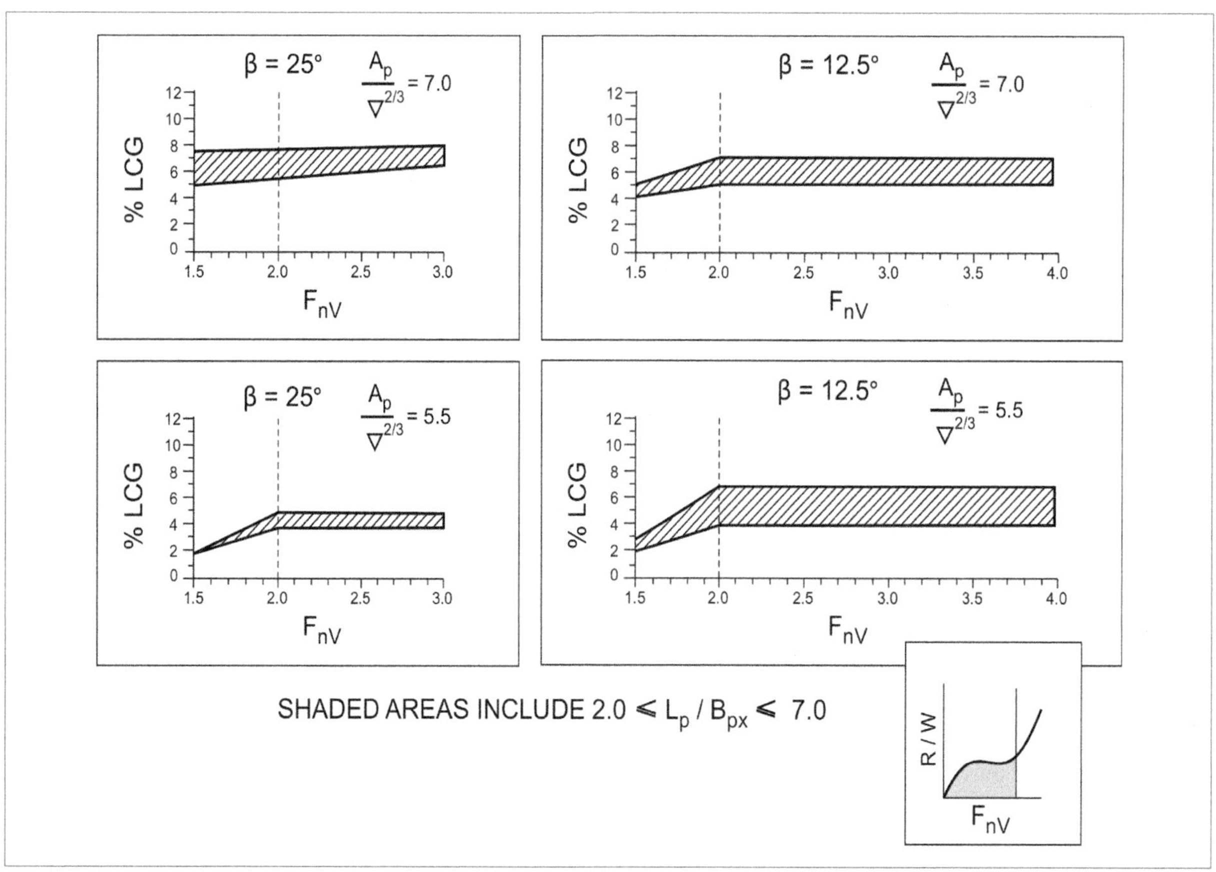

Figure 8-14: Calm water optimum LCG for integrated minimum to +2% R/W for speeds from zero up to indicated F_{nV}

This offers a difficult design optimization decision because it is in a speed range of potentially wide variations in resistance. For this two-speed design condition for semi-displacement vessels, I generally optimize the hull for the higher speed as R/W is not very sensitive to LCG at low-speed operation.

Otherwise, if the goal is to minimize life cycle operating fuel costs, designing the hull for the 80 percent cruise condition is preferred so long as the 20 percent dash speed can be attained.

A bulbous bow is a worthy consideration for this 80 to 20 percent operational profile for maximum speed ($F_{nL} \leq 0.6$) and hull loading is $6 \leq L/\nabla^{1/3} \leq 7$. Should $L/\nabla^{1/3} < 6.0$, a longer waterline length is urged.

Hard-chine planing craft selection of a calm-water design LCG for planing speeds, $F_{nV} \geq 3.0$ may be made using Figure 8-14 as can be seen by the relatively constant bandwidths which includes wide variations at hump speed and yet keep within two percent of minimum resistance.

Fortunately, these near-optimum LCGs result in trim angles which can be reduced, relatively easily, by trim tabs or interceptors to improve ride quality when in a seaway.

CHAPTER NINE

Design Criteria

From a holistic perspective, to design is to originate or to extend one's vision. The focus of this book, however, is about a component of development of vessels in order to get designers near to the best technical underwater hull form, proportions, and mass distribution in minimal time. After that, the designer can appeal to the client's desires by completing the comprehensive design with clever arrangements and crowning the exterior with styling much like the sportfishing yacht shown in Figure 9-1. Its purpose as a sportfishing yacht is visually obvious while at the same time its above-water appearance has been harmonized at the hands of a stylist.

Figure 9-1: 105-ft sportfishing yacht built by Jim Smith Boats, Inc.

The marine industry is in a phase where minimum motion is a highly desirable quality for vessels when at anchor or operating at slow speeds. There are now a number of motion-control systems available which mitigate roll motions. Some devices, such as gyro-stabilizers, are mounted internally to the hull, and others have some external components that create restoring roll moments in reaction to water flow. Regardless of the vessel's hull form, the designer must satisfy any regulatory small-craft stability requirements which may apply and also provide a comfortable roll period for the passengers and crew. With regard to roll period, Figure 9-2 offers a design goal highlighting the comfort zone of many recreational boaters.

In this chapter, I summarize design criteria for the minimum hull resistance. This summary includes criteria for both calm-water resistance and rough-water ride quality. Most criteria discussed here are for deep water; however, shallow-water design guidance is only for calm water as wave characteristics vary considerably from deep water waves. Finally, I want you to keep in mind that as your experience grows and industry knowledge evolves, your decision criteria may change.

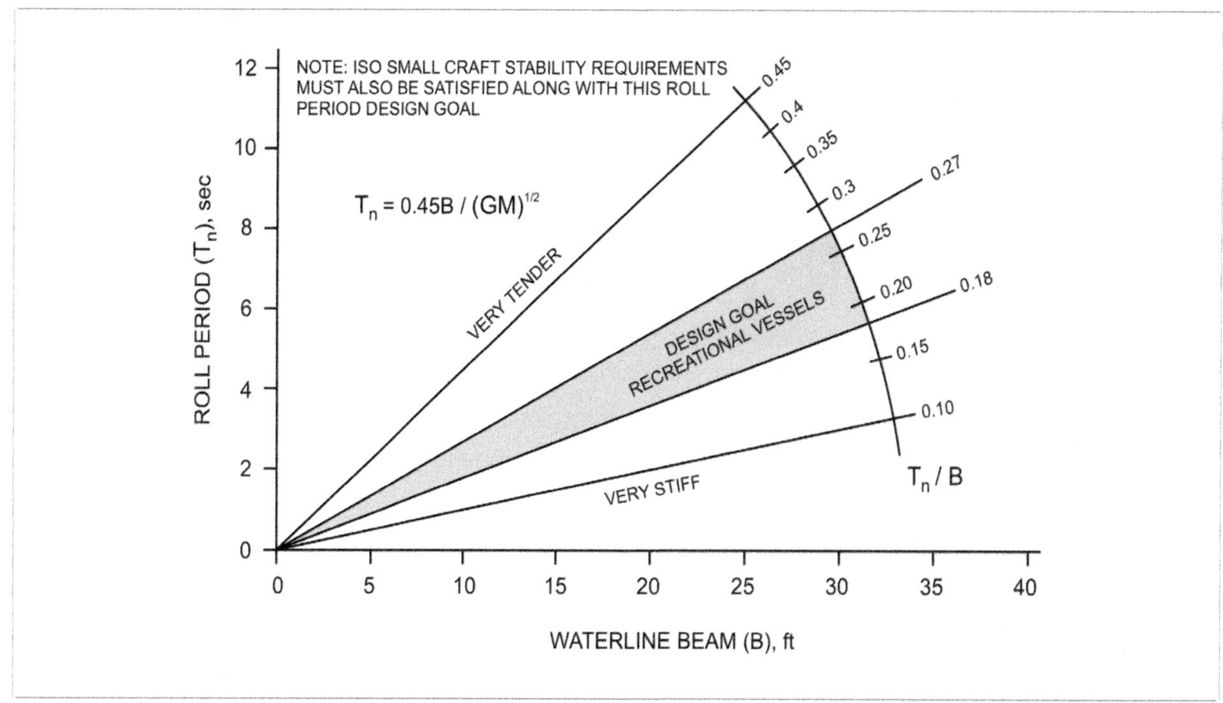

Figure 9-2: Roll period design goals for zero vessel speed (Note: ISO Standards are intended for craft of LOA less than 24 m)

Calm, Deep Water

I have included four data sheets (Tables 9-A through 9-D) to provide design guidance for hull-form conditions for minimum to +2% hull resistance. For planing craft, data is provided for two hull loadings for $A_p/\nabla^{2/3}$ of 5.5 and 7.0 respectively in Tables 9-C and 9-D.

For each dimensionless design speed, significant hull form parameters are listed. Variations from these tables for reducing $L/\nabla^{1/3}$ or selecting a different LCG for semi-displacement or semi-planing speeds generally results in increased hull resistance.

For semi-planing and planing speeds $A_p/\nabla^{2/3}$ and LCG variations can result in changes to resistance with possible complications when trim control is taken by the craft operator. For hard-chine hulls, there will be a speed above which the craft will slow down when a bow-down trim moment is effected by the operator.

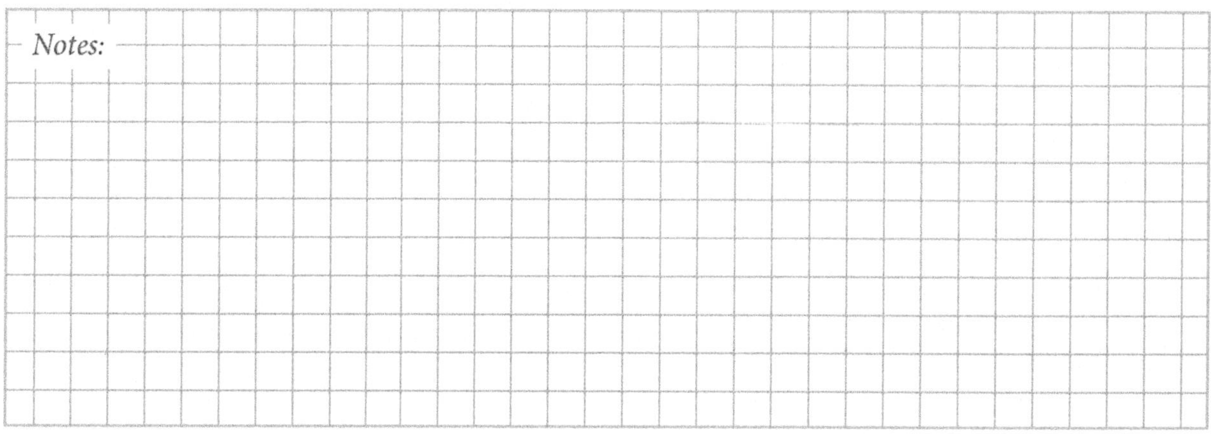

TABLE 9-A: DESIGN DATA FOR ROUND-BILGE HULLS OPERATING IN CALM, DEEP WATER					
Dimensionless speeds	**Design speed[1]**		**Hull conditions**		
	F_{nL} [6][7][8]	F_{nV}	$L/\nabla^{1/3}$ [2]	C_P [4]	L_{CG}/L [3]
Semi-displacement $0.40 \leq F_{nL} \leq 0.65$	0.40	1.0	8.0	0.65	0.450
	0.50	1.3	8.0	0.69	0.420
	0.60	1.6	8.0	0.74	0.405
	0.65	1.7	8.0	0.76	0.395
Semi-planing $F_{nL} > 0.65$ to $F_{nV} < 3.0$	0.65	1.7	0.85	0.76	0.395
	0.76	2.0	0.85	---	0.385
	0.94	2.5	0.80	---	0.375
	1.13	3.0	---	---	---

Notes:

1. $F_{nV} = F_{nL} (L/\nabla^{1/3})^{1/2}$ assuming $L/\nabla^{1/3} = 7.0$
2. Refer to Figure 2-16
3. Refer to Figure 2-13
4. Refer to Figure 2-6
5. Consider bulbous bow for $0.40 \leq F_{nL} \leq 0.60$ having bulb geometry coefficients $L_B/L = 0.038$; $A_B/A_X = 0.11$; $Z_B/T_{FP} = 0.57$ and a ∇-type cross section. Refer to Figures 4-17 and 4-18, Kracht (1978) and Hoyle et al. (1988).

   Vessels with bulbous bows have almost the same seakeeping qualities compared to non-bulbous vessels. Thus, bulbs may be optimized with calm water studies with the provision that extremely large bulbs be avoided.
6. Round-bilge hulls need a knuckle or longitudinal flow separator in this speed range. Flow separators begin at the bow above the static full-load waterline and continue aft underwater, crossing the bilge radius to about mid-ships to inhibit outboard heel in turns.
7. Hook in the afterbody buttocks will mitigate stern squat for $0.50 \leq F_{nL} \leq 0.65$.
8. Transom sterns are preferred for $F_{nL} \geq 0.45$

TABLE 9-B: DESIGN DATA FOR DOUBLE-CHINE HULLS [7] OPERATING IN CALM, DEEP WATER							
Dimensionless speeds	Design speed[1]		Hull conditions				
	F_{nL}	F_{nV}	$L/\nabla^{1/3}$ [2]	C_P [4]	L_{CG}/L [3]	$A_P/\nabla^{2/3}$ [5]	% LCG [6]
Semi-displacement $0.40 \leq F_{nL} \leq 0.65$	0.40	1.0	8.0	0.65	0.450		
	0.50	1.3	8.0	0.69	0.420		
	0.60	1.6	8.5	0.73	0.405		
	0.65	1.7	8.5	0.75	0.395		
Semi-planing $F_{nL} > 0.65$ to $F_{nV} < 3.0$	0.65	1.7	8.5	0.75	0.395		
	0.76	2.0	8.0	---	0.350		
	0.94	2.5	7.5	---	0.380		
	1.13	3.0	6.5	---	0.380		
Planing $3.0 \leq F_{nV} \leq 5.0$		3.0			0.380	5.5	7.0
		3.5			0.350	5.5	8.0
		4.0			0.330	5.5	10.0

Notes:

1. $F_{nV} = F_{nL} (L/\nabla^{1/3})^{1/2}$ assuming $L/\nabla^{1/3} = 7.0$
2. Refer to Figure 2-16
3. Refer to Figure 2-13
4. Refer to Figure 2-6
5. A_P is projected planing area bounded by the lower chine beam
6. % LCG = $(CA_P - LCG)\ 100/L_P$
7. For lower chine $C_\Delta = 0.85$

TABLE 9-C: DESIGN DATA FOR SINGLE CHINE, PLANING HULLS OPERATING IN CALM, DEEP WATER $A_p/\nabla^{2/3} = 5.5$							
Dimensionless speeds	Design speed[1]		Hull conditions				
	F_{nL}	F_{nV}	$L/\nabla^{1/3}$ (2)	C_p (4)	L_{CG}/L (3)	$A_p/\nabla^{2/3}$ (5)	% LCG (5)
Semi-displacement $0.40 \leq F_{nL} \leq 0.65$	0.40	1.0				5.5	4.0
	0.50	1.3				5.5	3.0
	0.60	1.6				5.5	2.0-3.0
	0.65	1.7				5.5	2.7-4.5
Semi-planing $F_{nL} > 0.65$ to $F_{nV} < 3.0$	0.65	1.7				5.5	2.7-4.5
	0.76	2.0				5.5	4.0-6.0
	0.94	2.5				5.5	4.0-6.0
	1.13	3.0				5.5	4.0-6.0
Planing $3.0 \leq F_{nV} \leq 6.0$		3.0				5.5	4.0-6.0
		3.5				5.5	4.0-6.0
		4.0				5.5	4.0-6.0
		4.5				5.5	8.0-10.0
		5.0				5.5	8.0-10.0
		5.5				5.5	8.0-10.0
		6.0				5.5	8.0-10.0

Notes:

1. $F_{nV} = F_{nL} (L/\nabla^{1/3})^{1/2}$ assuming $L/\nabla^{1/3} = 7.0$
2. Refer to Figure 2-16
3. Refer to Figure 2-13
4. Refer to Figure 2-6
5. Refer to Figure 8-14 and Appendix 6 where % LCG = $(CA_P - LCG)\ 100/L_P$

TABLE 9-D: DESIGN DATA FOR SINGLE-CHINE PLANING HULLS OPERATING IN CALM, DEEP WATER $A_P/\nabla^{2/3} = 7.0$							
Dimensionless speeds	Design speed[1]		Hull conditions				
	F_{nL}	F_{nV}	$L/\nabla^{1/3}$ [2]	C_P [4]	L_{CG}/L [3]	$A_P/\nabla^{2/3}$ [5]	% LCG[5]
Semi-displacement $0.40 \le F_{nL} \le 0.65$	0.40	1.0				7.0	4.0
	0.50	1.3				7.0	4.0
	0.60	1.6				7.0	4.0-6.0
	0.65	1.7				7.0	4.0-6.5
Semi-planing $F_{nL} > 0.65$ to $F_{nV} < 3.0$	0.65	1.7				7.0	4.0-6.5
	0.76	2.0				7.0	5.0-7.5
	0.94	2.5				7.0	5.0-8.0
	1.13	3.0				7.0	6.0-8.0
Planing $3.0 \le F_{nV} \le 6.0$		3.0				7.0	6.0-8.0
		3.5				7.0	5.0-7.0
		4.0				7.0	5.0-7.0
		4.5				7.0	5.0-7.0
		5.0				7.0	6.0-10.0
		5.5				7.0	6.0-10.0
		6.0				7.0	6.0-10.0

Notes:

1. $F_{nV} = F_{nL} (L/\nabla^{1/3})^{1/2}$ assuming $L/\nabla^{1/3} = 7.0$
2. Refer to Figure 2-16
3. Refer to Figure 2-13
4. Refer to Figure 2-6
5. Refer to Figure 8-14 and Appendix 6 where % LCG = $(CA_P - LCG)\, 100/L_P$

Rough Water

The most realistic predictor I have used for vertical acceleration at the LCG to account for ride quality and calculate added drag due to waves has been Hoggard and Jones (1980) and Hoggard (1979) respectively. These are represented by equations 9-1 and 9-2 where $\eta_{1/10CG}$ has dimension in gs.

$$\eta_{1/10CG} = 7\,(H_{1/3}/B_{PX})\,(1 + \tau/2)^{0.25}\, F_{nV}/(L_P/B_{PX})^{1.25} \quad \text{Equation 9-1}$$

$$R_A/W = 1.3\,(H_{1/3}/B_{PX})^{0.5}(L_P/\nabla^{1/3})^{-2.5}\, F_{nV} \quad \text{Equation 9-2}$$

In recent years, significant research has begun to refine predictive capability for planing craft at high speeds in irregular waves. The distribution of planing craft vertical accelerations in irregular waves has been shown (Grimsley, 2010) to follow the Gumbel distribution rather than the widely accepted exponential distribution. The existing database in RMS format mentioned in Chapter 7 can now be analyzed statically with the relationships in Table 9-E from Grimsley (2010). As additional experimental data becomes available, these linear relations will be refined.

TABLE 9-E: EMPIRICALLY DERIVED RELATIONSHIPS BETWEEN PARENT AND PEAK STATISTICS (FROM GRIMSLEY, 2010)			
Peak data statistic	**Relationship to parent data**	**Correlation coefficient, R**	**Coefficient of determination, R^2**
Average	2.39 (RMS) -0.19	0.96	0.92
Average of 1/3rd highest	4.58 (RMS) -0.40	0.97	0.93
Average of 1/10th highest	7.11 (RMS) -0.67	0.96	0.94
Average of 1/100th highest	9.58 (RMS) -0.77	0.93	0.87
Note: *For use with RMS vertical acceleration data obtained experimentally with planing craft in irregular waves.*			

Figure 9-3 is a combination of data from three references: Hoggard and Jones (1980), Blount and Funkhouser (2009) and Grimsley (2010) to give a speed-wave height diagram for single-chine planing hulls, which is achievable for vertical acceleration limited to $\eta_{1/10} = 1.5$ g at the LCG without shock-mitigating features.

Considering crew tolerance, I believe this level of vertical acceleration should be limited to one or two hours of exposure. To mitigate vertical acceleration exposure, designers should consider hydrodynamic hull features, shock seats, and dynamic motion-control systems.

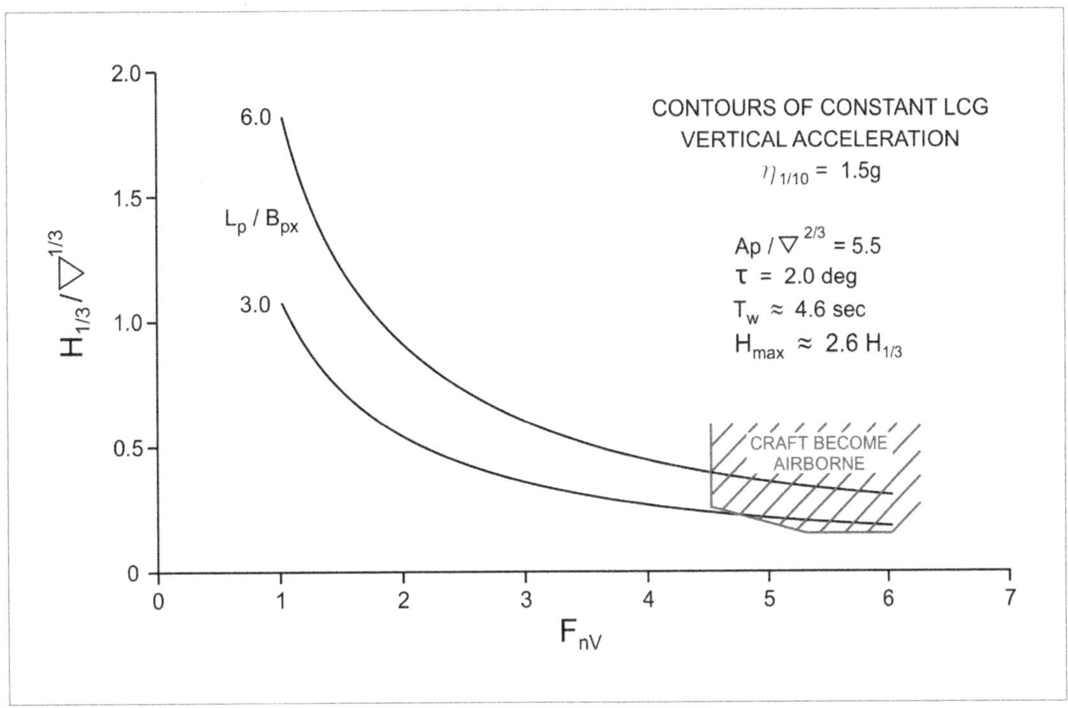

Figure 9-3: Achievable vertical acceleration limits at the LCG for hard-chine and round-bilge hulls

Dynamic Stability

The stability of dynamically supported planing craft in roll and pitch needs to be validated. Analytical solutions are yet to be widely accepted, however, although sea-trial and model-test data provide some guidance as shown in Figure 9-4. Experience suggests that LCG is a primary factor with less dependence based on VCG.

Figure 9-4 provides a quick check with regard to LCG. You may download a free calculator for dynamic stability representing this figure from the following website: http://www.dlba-inc.com/dlba_interactive.htm. In the next chapter, "Model Testing," I will discuss an experimental procedure for evaluating the possibility of dynamic instability, based on the criteria described in Figure 10-2.

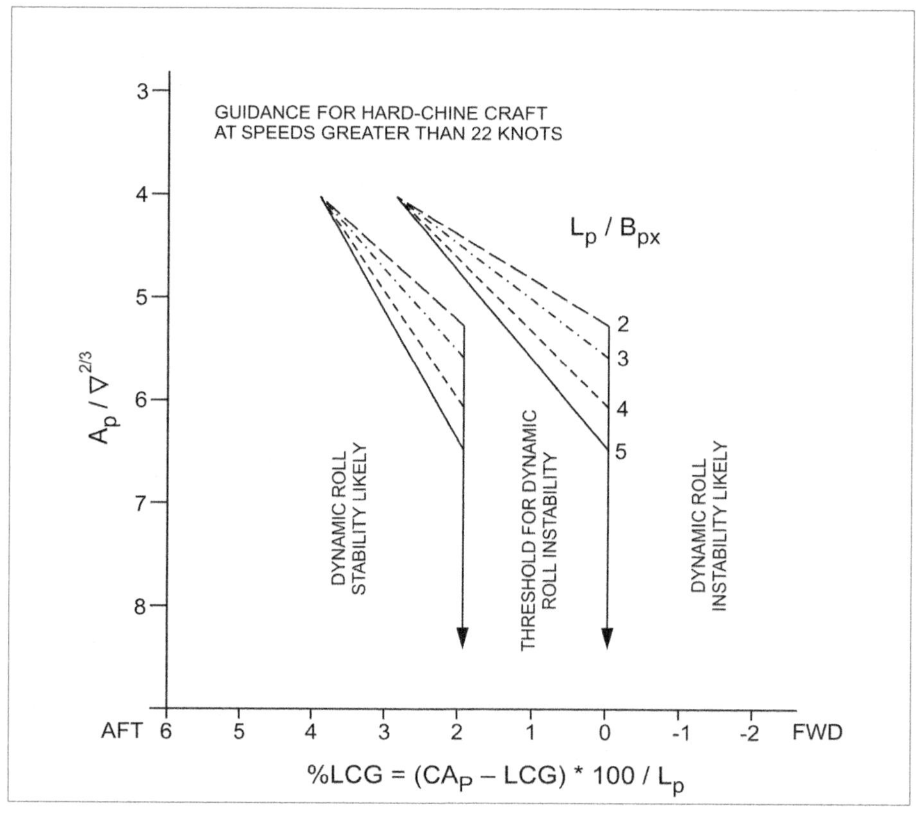

Figure 9-4: Indicator of design conditions when dynamic roll instability is unlikely

Shallow Water

The effects on performance of operating in shallow water are dramatic in the critical region ($0.70 \le F_{nh} \le 1.20$) as discussed in detail in Chapter 3. For craft having significant periods of operation in the critical region, designers should carefully understand the reaction of hull resistance while transiting shallow water ($h/LOA < 0.8$).

CHAPTER TEN
MODEL TESTING

As early as 1775, model experiments were carried out in France. In the following years between 1793 and 1798 Colonel Mark Beaufoy conducted some 1,700 experiments with various shapes in open water in the Greenland Dock, in London with the towing facility depicted in Figure 10-1. Data for the many shapes tested were published and tabulated in Beaufoy (1834).

Beaufoy's tests were conducted with a gravity apparatus which consisted of a three-legged strand from which a box for holding weights was suspended by rope. When the box was released to drop, the rope connected to the body/model on trial was drawn through the water. For each falling weight, he measured the velocity for the model. After each test, a mule pulled the next weight up to the strand in preparation for the next test run. By obtaining towed-model velocities for different falling weights, Beaufoy computed resistance, assuming it to be proportional to some power of velocity. As a mean value, he found resistance to be proportional to the 2.106 power of velocity.

In the early 1870s, William Froude developed the present-day approach for predicting resistance, based on towing models of full-scale marine vessels. He defined the complex resistance connection between models and full-scale ships that have the same dimensionless hull geometry and speed when operating at the water surface.

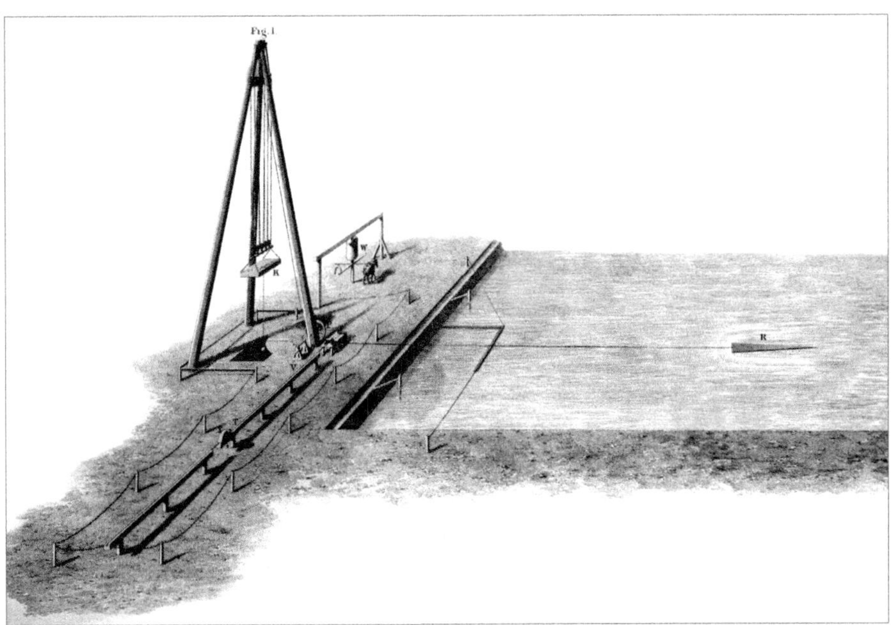

Figure 10-1: Col. Mark Beaufoy's towing facility at the Greenland Dock in London

Simply stated, Froude's Law indicates that the frictional resistance determined by calculation is to be subtracted from the total model experimental resistance; the remaining model resistance in a dimensionless form is assumed to be of the same quantity for full-size ships which have the same dimensionless hull geometry, hull loading and speed. Then calculated full-size ship frictional resistance is added to the remaining resistance to obtain the total ship resistance. Since some marine craft operate at high speeds, they may experience cavitation or ventilation of propulsors, hull and appendages. These latter effects then need to be accounted for by techniques other than following Froude Law.

Different types of testing are likely to be conducted for a variety of purposes during a program to design, construct and operate marine craft. During the design phase, hydrodynamic and aerodynamic model testing may be conducted to develop or extend technology to confirm performance prediction or for risk reduction, such as exploring the possibility of dynamic instabilities.

During construction, builders often test critical equipments prior to installation to confirm they comply with specifications. Should new materials or construction methods be involved, builders are likely to experimentally evaluate structural characteristics of samples using new production techniques.

Upon completion of marine craft, builder's and acceptance trials are conducted. A variety of circumstances support the need for these trials. Operating the vessel in both daylight and darkness may expose equipment installation problems, faulty design arrangements affecting personnel access for movement about the boat or impeding maintenance, electro-magnetic interference between navigation and communication equipment, light sources which may interfere with the night vision of operating crew, unacceptable vibration and noise, engine back pressure resulting from a wrong-sized exhaust system, etc.

It is important from the designer's perspective, that speed-power sea trials be conducted in both calm and rough, deep water to evaluate if the craft meets the owner's requirements. Also, sea trial data may be used to confirm or refine designer's performance prediction techniques.

These builder's and acceptance trial data establish the baseline for the operational life of the craft. Should future operational performance degrade, these trial records will support problem identification and resolution.

Hydrodynamic Model Testing

The designer must evaluate and select the appropriate test program to enhance the success of the craft to be constructed. In some instances no test program is necessary. In other cases such as a heavily loaded hull with a forward LCG cries out for experiments to investigate dynamic roll and yaw stability.

Self-propulsion tests should be approached with caution as they require knowledgeable personnel and proper equipment to obtain meaningful results. Whatever the motivation or objective, whenever a model test program is appropriate, the designer must take maximum advantage of the opportunity to reduce performance risk of the design to be constructed, enhance the designer's technology base and evaluate the quality of performance of the craft relative to other boats. Thus, growth in design quality will lead to enhanced performance and reliability for future craft.

The focus of this book relates to the design of high-performance surface marine craft with hydrodynamic and aerodynamic test planning included as it is a tool for risk reduction as well as technology extension. In addition to, and more importantly, an approach is offered so that experimental results not only provide important information for a specific design, but also to build an ever-expanding data base for future design studies. My personal approach to planning hydrodynamic model tests was influenced in my early years when working at the U.S. Navy's David Taylor Model Basin at Carderock, MD. I prefer to organize model test programs so that experimental results can have dual use; both to confirm performance predictions of a specific design, and to extend or advance development of new technology.

The experience from the Navy's Series 62 planing hull research program, Clement and Blount (1963), laid the groundwork for developing this dual use approach. The latter component may add some small cost to a program, but by following a similar test organizational approach over an extended period, hydrodynamic technology trends become clearly defined with the possibility of new discoveries. Clearly-defined objectives are usually necessary to support the logic of a well-prepared memorandum of cost benefits and risk reduction to obtain the client's authorization for a model test program.

The objectives to be achieved through model testing will influence the scope of work ordered from the research facility. Frequently the purpose of model testing is to increase confidence of the probable success of the performance of a newly designed vessel prior to construction. Other reasons include exploring the fringes where available technology might be ill-defined or non-existent, extending one's own design data base and possibly supplementing a marketing campaign with photographs and videos of models being tested.

A designer may not elect to conduct a model test program when the craft to be constructed is similar to one of the same size and power already in service, the design is hydrodynamically similar to an existing available experimental data base or can be reliably replicated by analytical techniques. It is, however, important for designers to know limitations with regard to hydrodynamic subtleties.

Otherwise they might unwisely avoid recommending a model test program when there could be risk of under or unsafe performance of the craft when delivered, such as might occur when propulsion power is significantly increased in an existing hull designed for a slower speed.

There are details of test models worthy of note that may minimize or avoid the influence of scale effects which can distort realistic trends or magnitude of data when expanded to full-size vessels.

Some significant references on scale effect include Moore and Hawkins (1969), a David Taylor Model Basin (DTMB) study of planing boat scale effects on trim and drag; Sottorf (1933), investigation of seaplane float scale effects; Tanaka et al. (1991), a multi-faceted reference of many aspects of the influence of hard-chine model size on experimental results; and Hadler et al. (2007) report on the capabilities of the model basin at Webb Institute. The first three references reported on hard-chine craft, and the latter focused on round-bilge, semi-displacement vessels.

While there are some variations in references, my recommendations to minimize or avoid the influence of scale effects are summarized below:

MINIMUM MODEL LENGTH:

- L and $L_P \geq 5.0$ ft (1.5m) for model test speeds resulting in $R_n \geq 5 \times 10^6$.

HARD-CHINE MODELS:

- All design flow separating edges including chines, transom, longitudinal spray strips, transverse steps, etc. of models must be sharp, radius ≤ 0.02 in (0.0008 mm), to achieve appropriate flow separation.
- Turbulence stimulation not required.

ROUND-BILGE MODELS:

- Turbulence stimulation with Hama triangles, Hama et al. (1957) are recommended.

EXPECTATIONS AT HIGH SPEED:

- Resistance versus speed may be predicted which will be representative of full-scale vessels.
- Repeatability of model test trim angles may be ± 0.25° for $F_{nV} \geq 1.5$.
- Lack of sharpness of chines and other flow separating edges at planing speeds reduces dynamic lift, increases hull resistance, reduces lift to drag ratio, results in virtually no change to CG rise.
- Little change to the longitudinal center of pressure.

For hard-chine models which are smaller than the recommended minimum length, the following scale effects have been reported for planing speeds when test data are scaled to full-size:

- The maximum height of spray separating from the chine will be too low.
- The transverse position of the maximum height of spray separating from the chine will be too close to the hull.
- Trim angles will be too high.
- The hull resistance will be too high.

It has been observed that best correlation of model and full-scale resistance and trim seems to occur for speeds higher than when flow has separated at the transom. Dry transom speed was discussed previously in Chapter 2 suggesting that this occurs as low as $F_T = 1.95$ with a more conservative estimate of $F_T = 2.55$.

Calm-Water Testing

Whenever it becomes appropriate to conduct a model test program, maximum benefit should be derived with thoughtful definition of test conditions and data analysis. Calm water tow tank tests of a model are frequently ordered requesting EHP and trim as a function of speed for expected LCGs and displacements for light and full-load conditions.

With this minimal data and an assumed overall propulsive coefficients (OPC) the designer can predict speed for the installed engine power. However, calm water resistance tests can have extended design benefits whenever a well-planned program is executed. But first, a most important issue regarding testing of performance hull forms needs to be addressed, whether round bilge or hard chine. Models must be free to trim, heave and **ROLL** when being towed.

Unrestricted motion in all planes except yaw and sway are necessary so that models are allowed freedom of movement in order that they attain dynamic equilibrium before steady-state data may be recorded. Also, freedom of motion allows a model to exhibit speed-related unstable oscillating or non-oscillating conditions which forewarn of problems needing correction. Common dynamic instabilities include porpoising, a pitch-heave oscillation, and chine walking, a roll oscillation. Non-oscillating instabilities mostly are seen above $F_{nL} > 0.75$ and $F_{nV} > 1.5$ when speed-related hydrodynamic forces may cause a craft to assume an unexpected, unnatural attitude or position relative to the surface of the water.

This unnatural position of the hull is non-oscillatory and often is in a speed-initiated equilibrium state which is a combination of hydrodynamic and hydrostatic forces. With increased speeds, these types of instabilities include heeling to one side (also known as chine riding or loll) and remaining heeled until the vessel slows down. Also the sudden pitch down of the bow of a boat from one steady trim angle to another, lower, steady trim angle is a non-oscillatory instability. And occasionally the bow pitch-down continues with enormous increase of spray until the boat slows.

Directional instability (bow steering) is hydrodynamically coupled with heel (loll) which is also a non-oscillatory instability. However, this cannot be readily predicted with a model in tow in the manner used for resistance testing which requires that models be restrained in yaw to properly follow the direction of the towing carriage. This can be predicted with tests of a restrained model with measurements made of forces and moments.

Model tests for a particular boat are generally conducted for the expected operational displacements with the longitudinal centers of gravity appropriate for each displacement. Test results are often scaled to the specific size of the full-scale craft. In general, the test conditions are unrelated from boat to boat; data of this kind are not well suited for making merit comparisons of different hull forms.

TABLE 10-A: POTENTIAL VARIATION OF R DUE TO MODEL SIZE AND FRICTION FORMULATION		
Lowest Model Test Speed Reynolds Number	**Potential Difference in Predicted Full-Scale Total Bare-Hull Resistance**	$R_{ATTC} \approx R_{ITTC}$ ( )
1×10^6	3%	1.03
1×10^7	1%	1.01
1×10^8 and higher	0%	1.00

Should a designer wish to start collecting model test data from tests at standard conditions to establish a personal database, it is recommended that the ITTC (International Towing Tank Conference) 1957 friction line be used. The difference of correcting total bare-hull resistance to a 100,000-lb (45,340-kg) full-scale displacement when employing the ATTC 1947 versus the ITTC 1957 friction line is dependent on model size; the Table 10-A quantifies the variation for this displacement.

Fortunately, differences arising from tank tests only at specific operational loading conditions can be avoided by always testing a model at a "standard condition" of hull loading and LCG location, with the resistance data expanded to the same full-scale displacement. The data from standard-condition tests can then yield merit or comparative differences in performance between various hull forms.

The standard model-test condition established for evaluating various planing-hull designs, and in developing the parent hull form for Series 62, can be expressed as follows:

Hull Loading (area coefficient): $A_P/\nabla^{2/3} = 7.0$

LCG Location: $(CA_P - LCG)100/L_P = 6.0\%$

Correlation Allowance: $C_A = 0$

Resistance for Series 62 was corrected to 100,000 lb (45,359 kg) displacement for salt water at 59°F (15°C), using the 1947 ATTC (American Towing Tank Conference) friction line and reported as R/W versus F_{nV}. Note that various planing model data sheets can be purchased from SNAME (The Society of Naval Architects and Marine Engineers), and are based on the ATTC 1947 friction line.

Fifty-plus years after the standard-condition concept was established for Series 62, I now think that two changes in the definition should be updated to contemporary practice which is representative of most recreational, military, and commercial craft built in the last 15 years.

The proposed calm-water resistance test standard condition for 2014 follows:

Hull Loading (area coefficient): $A_P/\nabla^{2/3} = 5.5$

LCG Location: $(CA_P - LCG)100/L_P = 4.0\%$

Correlation Allowance: $C_A = 0$

Thus, just as there are two friction lines, 1947 and 1957, there could be 1963 and 2014 standard conditions for planing-craft resistance data. For direct comparison with the 1963 standard condition (the year our Series 62 paper was published), my proposed "2014 standard condition" is one of the test conditions for which calm-water resistance data were reported for the original Series 62 hulls, and later by Keuning et al. for the Delft hulls having additional deadrise angles of 25° and 30°.

Defining a standard condition for displacement and semi-displacement hulls would also benefit designers specializing in vessels operating at speeds up to $F_{nL} = 1.0$ as test data could provide comparative differences in performance between various hull forms. Blount and McGrath (2009) found, upon review of a number of references reporting experimental data, that a representative standard full-scale displacement for this speed range could be 500 mt. Resistance data for many references were scaled to 500 mt vessels, and with analysis, minimum values were established for round-bilge, double-chine and hard-chine hull forms. (See Appendix Table A1-G and Figure 2-16.) Data analysis of Blount and McGrath (2009) suggests that for performance displacement/semi-displacement vessels a standard condition for comparison purposes would be hull loading of $L/\nabla^{1/3} = 6.0$ and LCG = 0.40L forward of the hydrodynamic transom with resistance scaled to a displacement of 500 mt for $C_A = 0$.

Table 10-B provides a suggestion for test program conditions for displacement and semi-displacement vessels with data to be recorded as indicated in Table 10-C.Table 10-D provides suggested test program conditions for planing craft.

For extended benefits from a calm-water resistance program each hull design should be tested for at least three displacements, three LCGs and for a range of dimensionless speeds, rather than for specific craft speeds. The maximum speed range should extend beyond that appropriate for hydrodynamic shape and loading of the hull form and the operational requirements for the craft. Maximum extended benefit of model experimental data can be achieved when test results are reported in the three formats of Table 10-E and related comments.

TABLE 10-B: SUGGESTED TEST CONDITIONS FOR PERFORMANCE DISPLACEMENT AND SEMI-DISPLACEMENT VESSELS		
Displacements	**LCGs**	
A – Lightship B – Full Load C – 1.25 X Full Load	I – Design Full Load II – I + 4% L III – I – 4% L	
Standard Condition		
$L/\nabla^{1/3} = 6.0$	LCG = 0.40L fot	
Suggested Calm Water Resistance Test Conditions		
A – I A – II A – III	B – I B – II B – III	C – I C – II C - III
Standard Condition		
Suggested Test Speeds		
F_{nL}*	Hydrodynamic Speeds	
0.30 0.40	Displacement	
0.45 0.50 0.55 0.60	Semi-displacement	
0.70 0.80 0.90 1.00	Semi-planing	
*If safe for the test model run at least one speed, F_{nL} higher than the maximum expected lightship speed each hull design.		

TABLE 10-C: RECORDED DATA	
Record Static Condition Data	**Record Model Test Data for Every Speed**
Displacement	Speed
LCG	Total resistance
Trim	Trim or trim change from zero speed
Waterline length	CG rise or CG rise change from zero speed
*Wetted length of keel	*Wetted length of keel
*Wetted length of chine	*Wetted length of chine
Tow tank water properties	Above-water HD videos or DVDs
Temperature	Above-water digital photos
Mass density	Underwater digital photos
Kinematic viscosity	
* - Record these data only for planing craft in calm water	
Record Model Test Observations	
Unusual spray, wake or other flow patterns	
Dynamic behavior, i.e. inception speeds for porpoising or chine walking, hull heeling or bow suddenly dropping with increasing speed, etc.	

TABLE 10-D: SUGGESTED TEST CONDITIONS FOR HARD-CHINE PLANING CRAFT

Suggested Test Conditions for Hard-Chine Craft	
Displacements	LCGs
A – Lightship B – Full Load C – 1.25 x Full Load	I – Design Full Load II – I + 4% L_P III - I – 4% L_P
Standard Condition	
$A_P/\nabla^{2/3} = 5.5$	$C_{AP} - LCG)100/L_P = 4.0\%$

Suggested Calm Water Resistance Test Conditions		
A – I A – II A – III	B – I B – II B – III	C – I C – II C – III

Standard Condition	
Suggested Test Speeds	
F_{nV}	Hydrodynamic Speeds
0.5 - 1.0	Displacement
1.2 - 1.5	Semi-displacement
1.8 - 2.2 - 2.8	Semi-planing
3.5 - 4.5 - 6.0	Planing

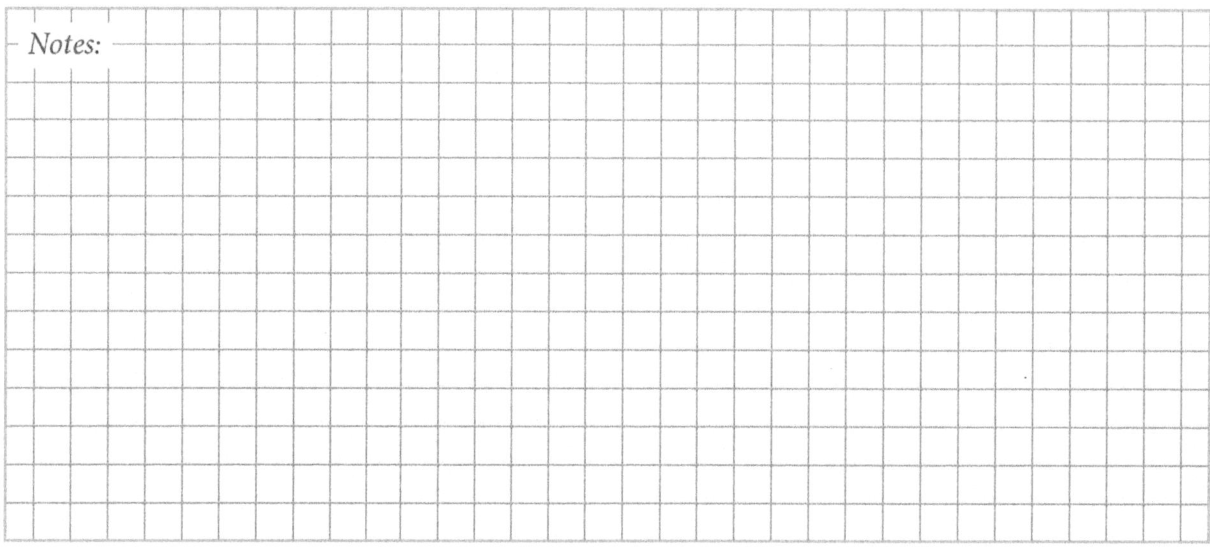

TABLE 10-E: REPORTED DATA AND ANALYSIS FOR HARD-CHINE PLANING CRAFT		
ITEM	DATA ANALYSIS	FORMAT
1.	Data scaled to full size, craft operating water (F/W or S/W) and correlation allowance of choice	Tabular and graphic hard copy and electronic data files as well as HD videos or DVDs and digital photographs (designer and client data)
2.	Data scaled to a displacement of 100,000 pounds and operating in salt water with a zero correlation allowance.	Tabular - Hard copy and electronic data files (designer data)
3.	Report model scale test data without analysis	Tabular - Hard copy and electronic data files (designer data)

Item 1

With the combination of nine conditions (three displacements times three LCGs), the analysis and format suggested a number of direct applications and advantages are possible for the vessel program funding the model tests. First, the optimum LCG as well as the speed impact for non-optimum loading can be established for each displacement and required operational condition. Thus, during the design stage arrangement of equipments and consumables may be configured to best suit design requirements or correct problems indicated by observations. Second, data exists for speed and power predictions which can be made for any condition of loading for the as-constructed vessel. In addition, data exists for making correlation analysis for various sea trial conditions when the vessel becomes operational.

The model test photographs would indicate any potential bottom ventilation problem emanating at the stagnation line, suggest functional locations for underwater exhausts as well as longitudinal spray strips and need for improving chine detail should that be necessary. The speed-trim curve in the hump speed range can give an indication of possible dynamic instability (roll and/or yaw) at high speeds, trim along with CG rise is invaluable in addressing operator visibility requirements. Videos or DVDs may indicate spray and wake conditions representative of full-scale craft when models are at least 1.5m long (Tanaka et al. 1991). HD videos, DVDs and digital photographs also serve as an excellent visual source for validating unforeseen test results as well as for valuable marketing purposes.

Item 2

Data analyzed for a 100,000 pound displacement provides an excellent basis for a standardized comparison of hard-chine planing craft to evaluate the quality of the hull as well as to serve as a measure of the designer's progress. Data scaled to this displacement may be compared to Series 62 in Figure 2-20 or with other hull forms as a baseline to evaluate hulls of equivalent loading and length-beam ratio. This type of comparison can also be used to show when owner requirements may impose operational conditions inhibiting attainment of an optimal design solution.

For displacement and semi-displacement, designers would also benefit by using the suggested standard condition defined in Tale 10-B and scaling data to 500 mt for $C_A = 0$.

Item 3

Obtaining model scale test data has the long term benefit to the designer for establishing an ever-expanding experimental resource which can be analyzed by a regression technique or in any format as future needs arise. For example, developing a larger design with similar hull proportions would have a reliable basis for speed prediction when the model test data are scaled appropriately. Another example of future worth could be evaluating subtle differences on vessels of similar proportions when there are variations to the afterbody buttock shape, i.e. hook, straight, rocker or transverse section shape, i.e. concave, convex, bell or ox bow.

Dynamic Stability Tests

With towing gear appropriate for resistance testing, it is possible to organize a model loading condition which foretells of possible dynamic instabilities at high speeds. Non-oscillating instabilities tend to occur for craft free to trim, heave, and roll when a hull's geometric size and shape are too heavily loaded and/or LCG is too far forward to achieve equilibrium at all steady operating speeds. Just referring to the trim versus speed (F_{nV}) curves for $0.5 \leq F_{nV} \leq 2.5$ from monohull resistance tests provides a clue of potential dynamic instabilities. In Figure 10-2 the curves marked B and C, zero and negative slopes respectively, indicate that there is potential for dynamic instability at higher speeds.

A minimum of one test condition is usually sufficient to document that a non-oscillatory instability is not inherent in a particular craft design. The initial calm water test condition should be at the heaviest displacement with the LCG at the forward-most position likely to be operated by an owner of the vessel. Model ballast should be offset transversely to achieve a static heel angle of about 5° to 7°.

In addition to data recorded for resistance tests, Table 10-C, also records change of heel angle and bow rise with speed. Even though the model is heavy, test speed should exceed that for the maximum lightship displacement, if safe for the towing gear. The increase of heel with speed is one decision factor while bow sinkage continuing with increasing speed is the other factor indicating possible dynamic instability at high speeds. Should tests indicate dynamic instability at operating speeds, additional tests for other LCG, displacements and speeds are suggested to define safe conditions or to study design changes to correct the situation.

The dynamic instability is especially significant for high-speed, round-bilge vessels which tend to be more susceptible to this form of instability than single or double-chine hulls.

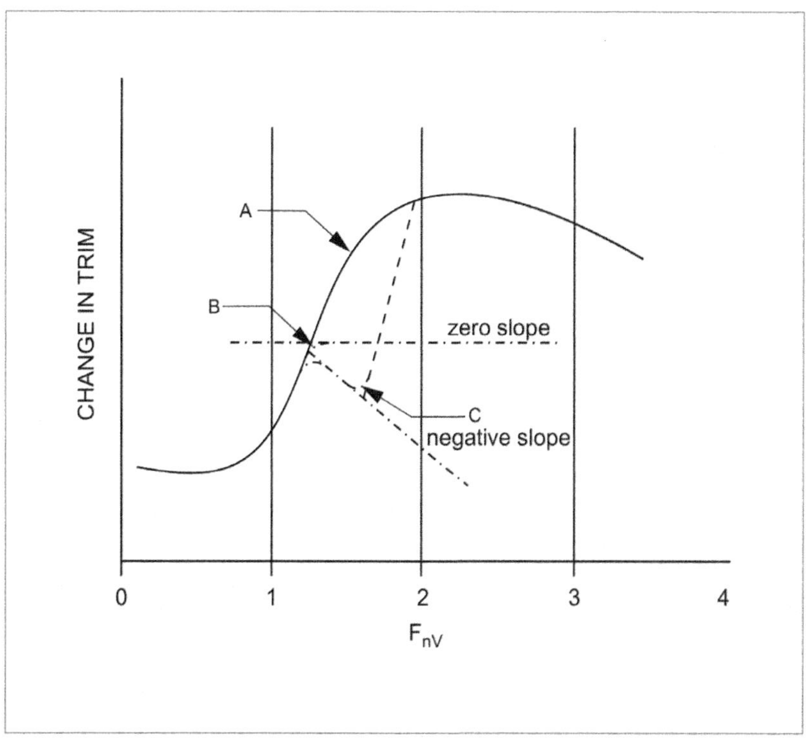

Figure 10-2: At semi-planing speeds, dynamic instability may occur at higher speeds

Propulsion Testing

Testing models which can be self-propelled with propellers and waterjets provide predictions for vessel power versus speed as well as characteristic data regarding the interaction of propulsors in proximity with the hull. With regard to the latter issue, both hulls and propulsors have unique pressure fields which, due to their nearness, interact causing changes to their respective characteristics compared to when they are operated individually in open water.

In addition, major components of model resistance, friction and residual resistance, are obtained by testing at Froude scaled speeds to obtain C_R. It is, therefore, necessary to conduct model propulsion tests at full-scale vessel propulsion equilibrium. One way this can be accomplished is for the model resistance to be reduced to account for the difference between model and full-scale friction coefficients, C_F.

For each speed, model resistance reduction is calculated as follows to augment the force resulting from the difference between model and full-scale Reynolds numbers.

Data from resistance test – v, L, S, ν, ρ

λ – model linear ratio

$$R_{n,m} = VL/\nu \qquad \text{Equation 10-1}$$

$$R_{n,f} - R_{n,m} (\lambda^{3/2})(v_m/v_f) \qquad \text{Equation 10-2}$$

$$\text{ITTC 1957 } C_F = 0.075/[(\text{Log}_{10} R_n)\text{-}2]^2 \qquad \text{Equation 10-3}$$

$$\delta C_F = C_{F,m} - C_{F,f} \qquad \text{Equation 10-4}$$

$$\delta R_m = \delta C_F (1/2) \rho_m S_m v_m^{\ 2} \qquad \text{Equation 10-5}$$

By applying the appropriate δR_m resistance reduction for each speed, the model propulsors (propellers or waterjets) will be operating at the correct thrust loading of the full-scale vessel. Thus, from this test condition, full-scale propulsive efficiency can be predicted along with appropriate propulsive factors.

As might be inferred from this discussion, it is strongly recommended that towing tank professionals be involved with propulsion test planning for high-performance vessels.

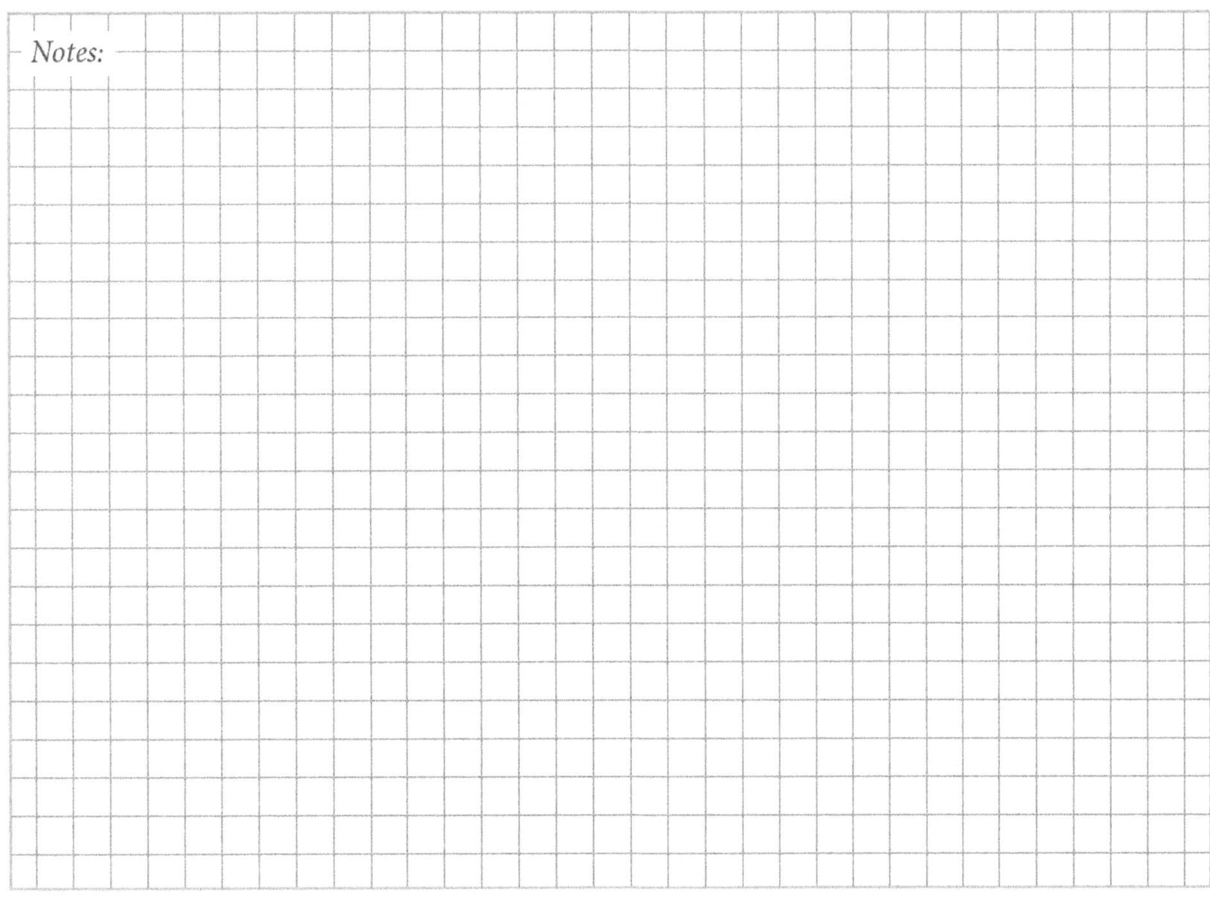

Rough Water Testing

Seakeeping tests provide extended opportunities for dual use of test data as did calm water resistance tests. Typically, vertical acceleration measurements are analyzed for statistical inferences, i.e., maximum, one-hundredth, one-tenth or one-third highest accelerations with primary application for structural design and ride quality for personnel onboard. Analyzing these same measurements in a one-third octave RMS format provides frequency-sensitive components relative to human-factor criteria regarding fatigue decreased proficiency and motion sickness. With three, or better yet four, longitudinally placed accelerometers, the longitudinal position of minimum vertical motion can be defined should this information be vital for placement of personnel or sensitive equipment.

Of vessel seakeeping characteristics, added drag due to waves (R_{AW}) is difficult to predict with confidence. Collecting these data experimentally with analysis in the form of R_A/W, $H_{1/3}/\nabla^{1/3}$, trim and volume Froude number ($F_{n\nabla}$) provide designer benchmarks to evaluate hull forms to determine speed loss in waves.

The response motions of dynamically-supported craft operating at high speeds in irregular seas are non-linear. At this time, model testing in irregular seas is, in my opinion, the preferred method for obtaining predictions of craft vertical motions and accelerations.

The linkage between the carriage and the model should be automatically adjusting so that the tow force remains in alignment with the thrust line of propulsors regardless of range of motions of the model responding to irregular waves. The mass distribution of the model should be scaled to replicate the full-scale values of CG and pitch radius of gyration (gyradius). The client must agree on the test wave spectrum, significant wave height ($H_{1/3}$) and wave period.

Vertical accelerations and motions of surface craft are primary functions of vessel geometry, mass and its distribution, the spectrum of waves, $H_{1/3}$ and modal wave period. The significance of the latter, modal wave period is often not stated as an owner's requirement or might be selected as the annual average for the expected operational area of the vessel. Often, rough water model testing is only conducted for one modal wave period appropriate for the required $H_{1/3}$ of the spectrum. My experience strongly suggests that two modal periods for each $H_{1/3}$ be used for seakeeping model tests since both motor yachts and military craft can operate globally in international waters. In addition to data recorded during calm water resistance testing, rough water tests require additional sensors indicated in Table 10-F.

In order to evaluate various vessel designs based on response to irregular seas, I recommend that a standard condition be established for rough water testing similar to that indicated for calm water resistance tests. The following is offered for consideration as a standard to be used for comparison/evaluation of vertical acceleration for different designs. Two or more designs could be compared by graphs of normalized vertical acceleration data in RMS, peak or standard G formats (Riley et al., April 2010).

A proposed standard test condition for evaluating relative seakeeping characteristics for designs of planing boats operating in irregular head seas is as follows:

$H_{1/3}/\nabla^{1/3} = 0.25$

T_w - 4.6 sec

Suggested test speeds: $F_{nv} = 1.0, 1.8, 2.2, 3.5, 5.0$

The dimensionless ratio $H_{1/3}/\nabla^{1/3}$ relates significant wave height to the size of the vessel as ∇ is the volume of displacement. The ratio is important, as vessel size has a meaningful influence on the vertical acceleration response to surface waves of dynamically supported craft.

Some typical wave period must also be established as it has a measurable effect on the hull's response of vertical acceleration to the surface of the sea. T_w = 4.6 sec is an approximate representation for Sea State 3. A longer wave period might have been considered, however, vertical acceleration would be reduced as maximum wave slope is less.

TABLE 10-F: DATA IN ADDITION TO CALM WATER RESISTANCE TESTS TO RECORD DURING ROUGH WATER TESTS	
Record Model Test Data for Every Speed	
Digitize data at a high sample rate	**Location**
Vertical acceleration (forward of transom)	
Minimum of three longitudinal positions	$0.1L_p$, LCG, $0.9L_p$
four longitudinal positions	$0.1L_p$, LCG, helm, $0.9L_p$
Pitch	LCG
Heave	LCG
Test wave spectrum	Model test facility
Observations	
Examples: Wetness?	
Does entire hull become airborne?	
Does bow dive?	
Analysis	
VERTICAL ACCELERATION: RMS, peak, averages of 1/100th, 1/10th, 1/3 highest, 1/3 octave analysis, longitudinal position of minimum RMS acceleration	
Pitch	RMS, mean, peaks, 1/3 highest
Heave at LCG	RMS, mean, peaks, 1/3 highest
Added drag in waves	mean
Waves	$H_{1/3}$, modal wave period, maximum peak and trough

Aerodynamic Model Testing

With regard to performance craft, wind tunnel testing is divided into two categories – external and internal. Modeling the external geometry of a vessel is often thought to focus only on minimizing aerodynamic drag to benefit maximum possible speed.

The purpose of this testing is, however, much more than just documenting air drag. The value of having the hull with deckhouse of very high-speed motor yachts in the wind tunnel serves to outline the technical advantages to be gained which have broad application.

The aerodynamic technologies studied especially for gas-turbine-powered motoryachts having speeds greater than 50 knots include safety on open deck areas, aerodynamic characteristics affecting maneuvering and control, interaction of helicopter take-off and landing, control of gas turbine exhaust (both fumes and heat), and streamline flow of air from water surface over sheer line, deckhouse and antennae.

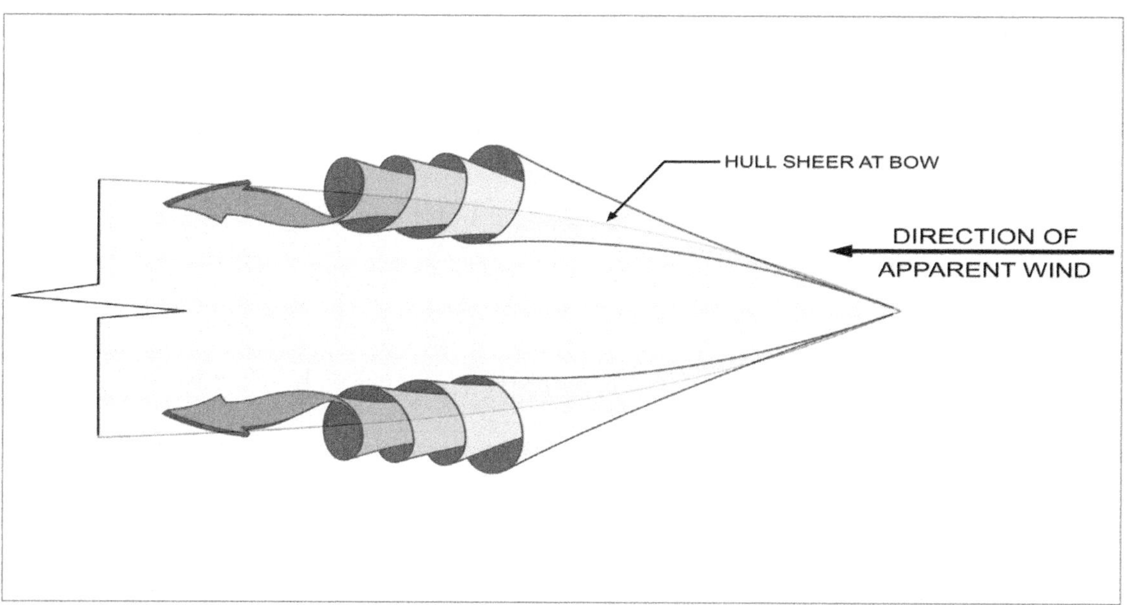

Figure 10-3: Combination of bow flare of hullsides and deck plan form can generate dual vortices, much like a delta wing aircraft

External Aerodynamic Testing

Safety and Open Deck Areas

Gale winds range from 34 to 47 knots and steady winds of 64 knots are the threshold for a hurricane. Whether on a bass boat or a fast motoryacht, personnel at locations exposed to gale and hurricane force winds can be at risk for injury. Wind tunnel studies provide a means to evaluate changes to details to improve safety of the design upstream of open deck locations where personnel may have operational responsibilities or owner and guests are likely to gather when the vessel is underway at high speeds. The desire of the client and designer is to have these open deck locations quiescent; free of exposure to high-speed winds.

Aerodynamic Characteristics

Air drag of vessels increases in proportion to speed squared. There is ever-increasing client demand to go faster. So the need to minimize the air drag coefficient becomes very important for high-speed craft as it affects installed horsepower. With increase in speed, maneuvering and control become more difficult as relative velocity of the vessel speed and ambient wind become an input to control of coursekeeping and steering. Based on this, both hydrodynamic and aerodynamic characteristics of drag, side force, and yaw moment coefficients versus relative wind direction are required input to the specification of steering-control systems.

Interaction with Helicopters

Maritime vessels large enough to allow helicopters to safely land and take off have interaction between the rotor down-wash with deckhouse shapes. Also, vessels having gas turbine engines must have the heated exhaust outside of the landing approach to avoid loss of helicopter rotor lift. Both the shape of the deckhouse and hot exhaust source near a helicopter landing pad can make control of the aircraft difficult.

Evaluating Longitudinal Streamlines

The combination of the plan form shape of the sheer and the flare of the hull sections at the bow of high-speed craft, produce spiral vortex sheets suggestive of air flow of a delta wing aircraft. (*See Figure 10-3.*)

It has been observed during wind tunnel tests that these port and starboard spiral vortices generated at the bow can be coincident with the passageways along the sides of the deckhouse and make it difficult for personnel to traverse forward and aft at high speed.

Spiral vortices from the bow have also been observed to pass over the flybridge, interfering with the comfort of personnel in the social/sunning area of the upper open deck.

Internal Aerodynamic Testing

Engine Room

Large performance vessels often have a dense machinery space dominated by high-powered engines, whether diesels or gas turbines. The flow of combustion and ventilation air, along with exhaust, must be controlled with regard to volume flow rate, heat extraction, and acoustics.

Relatively large-scale models are necessary to study the air flow entering from outside of the vessel to exiting via engine exhausts as well as ventilation systems which remove heat both when underway and during cool-down when deckside. Ducting systems move air flow through filters to remove spray, salt and particulant matter while intake silencers attenuate sounds from gas turbines. Therefore, air flow for cooling and combustion must be divided. Gas turbines often have lightweight educting exhaust systems sensitive to the pressure relationship between very hot gas and extracting ventilation air being pulled through the machinery space for cooling.

Modeling the air flow from outside and its distribution in the machinery space is important to avoid creating vorticies to gas turbine compressors, from encountering hot spots having inadequate cooling ventilation, under-performing engines due to lack of combustion, and poorly performing educting exhaust systems which can occur when pressures are found to be out of balance.

APPENDICES

Table of Contents

APPENDIX 1: Data Tables

A1-A: Mass Density of Water

A1-B: Kinematic Viscosity of Water

A1-C: Vapor Pressure of Water

A1-D: Mass Density of Air at 1 Atmosphere

A1-E: Kinematic Viscosity of Air at 1 Atmosphere

A1-F : Adjusting C_R to be Consistent with Friction Formulation Being Used for Full-Scale Resistance Predictions

A1-G: Deep, Calm, Salt Water Minimum Bare Hull R/W for Notional 500 mt Hull Forms

A1-H : Sea State Table

TABLE A1-A: ρ, MASS DENSITY OF WATER					
Temperature °C	ρ kg sec^2/m^4		Temperature °F	ρ lb sec^2/ft^4	
	Fresh	Salt*		Fresh	Salt*
0	101.95	104.83	32	1.9399	1.9947
5	101.96	104.79	41	1.9401	1.9939
10	101.93	104.71	50	1.9396	1.9924
15	101.87	104.61	59	1.9384	1.9905
20	101.78	104.49	68	1.9367	1.9882
25	101.66	104.34	77	1.9344	1.9854
30	101.52	104.18	86	1.9317	1.9823
35	101.34	104.02	95	1.9283	1.9793
40	101.17	103.84	104	1.9250	1.9758

*salinity 3.5%

References: ITTC 2006; Sharaqawy et al. (2010)

TABLE A1-B: ν, KINEMATIC VISCOSITY OF WATER					
Temperature °C	ν $(m^2/sec)10^6$		Temperature °F	ν $(ft^2/sec)10^5$	
	Fresh	Salt*		Fresh	Salt*
0	1.7867	1.8284	32	1.9231	1.9681
5	1.5170	1.5614	41	1.6329	1.6807
10	1.3064	1.3538	50	1.4062	1.4572
15	1.1390	1.1883	59	1.2260	1.2791
20	1.0037	1.0537	68	1.0804	1.1342
25	0.8929	0.9425	77	0.9611	1.0145
30	0.8009	0.8493	86	0.8621	0.9142
35	0.7295	0.7719	95	0.7851	0.8309
40	0.6580	0.6945	104	0.7083	0.7476

*salinity 3.5%

References: ITTC 2006; Sharaqawy et al. (2010)

TABLE A1-C: P_V, VAPOR PRESSURE OF WATER					
Temperature °C	P_v kPa		Temperature °F	P_v lb/ft²	
	Fresh	Salt*		Fresh	Salt*
0	0.611	0.599	32	12.8	12.5
5	0.920	0.901	41	19.2	18.8
10	1.228	1.203	50	25.6	25.1
15	1.784	1.747	59	37.2	36.5
20	2.339	2.291	68	48.8	47.8
25	3.293	3.226	77	68.8	67.4
30	4.247	4.160	86	88.7	86.9
35	5.816	5.697	95	121.4	119.0
40	7.384	7.233	104	154.2	151.1

*salinity 3.5%

References: Sharaqawy et al. (2010)

TABLE A1-D: ρ_{AIR}, MASS DENSITY OF AIR AT 1 ATMOSPHERE			
Temperature		ρ_{AIR}	
°C	°F	kg sec²/m⁴	lb sec²/ft⁴
0	32	0.1317	0.002506
5	41	0.1294	0.002462
10	50	0.1271	0.002418
15	59	0.1249	0.002377
20	68	0.1227	0.002337
25	77	0.1207	0.002296
30	86	0.1187	0.002259
35	95	0.1168	0.002222
40	104	0.1149	0.002186

TABLE A1-E: ν_{AIR}, KINEMATIC VISCOSITY OF AIR AT 1 ATMOSPHERE			
Temperature		ν_{AIR}	
°C	°F	$m^2/sec \times 10^5$	$ft^2/sec \times 10^4$
0	32	1.332	1.434
5	41	1.376	1.481
10	50	1.421	1.530
15	59	1.466	1.578
20	68	1.511	1.626
25	77	1.557	1.676
30	86	1.604	1.726
35	95	1.651	1.777
40	104	1.698	1.828

A1-F: ADJUSTING C_R TO BE CONSISTENT WITH FRICTION FORMULATIONS BEING USED FOR FULL-SCALE RESISTANCE PREDICTIONS

When available, C_R versus R_n data have been calculated using either the ATTC 1947 or the ITTC 1957 friction formulations, the C_R data may be converted to conform with values consistent with the other friction formulation. The table below provides for R_n a sample of the differences between these two friction formulations (δC_F) which you can readily calculate utilizing Equations 3-4 and 3-5.

The equation for adjusting the C_R to be consistent with the use of the friction formulation appropriate for full-scale resistance predictions is based on the differences at the same model test scale R_n of C_F calculated with ATTC 1947 and ITTC 1957. At Reynold's numbers below 1 x 10^9, the ITTC 1957 friction formulation yields lower values of C_R than are realized when ATTC 1947 is used.

The equations for correcting C_R derived using ATTC 1947 to C_R derived from ITTC 1957 and vice versa are provided in equations A1-1 and A1-2.

$C_{R/ITTC} = C_{R/ATTC} + \delta C_F$ Equation A1-1

$C_{R/ATTC} = C_{R/ITTC} - \delta C_F$ Equation A1-2

Example to correct C_R calculated using ATTC 1947 to C_R calculated using ITTC 1957:

Example for Rn = 1 x 10^6

$C_{R/CF1957} = 0.002700 + [-0.000278] = 0.002422$

Example to correct C_R calculated using ITTC 1957 to C_R calculated using ATTC 1947:

Example for Rn = 4 x 10^6

$C_{R/CF1947} = 0.006644 - [-0.000118] = 0.006762$

δC_F BETWEEN ATTC 1947 AND ITTC 1957

R_n	$C_{F\ ATTC}$	$C_{F\ ITTC}$	$[C_{F\ ATTC} - C_{F\ ITTC}] = \delta C_F$
1 x 10^5	0.007179	0.008333	-0.001154
4 x 10^5	0.005294	0.005780	-0.000486
1 x 10^6	0.004410	0.004688	-0.000278
4 x 10^6	0.003423	0.003541	-0.000118
1 x 10^7	0.002934	0.003000	-0.000066
4 x 10^7	0.002365	0.002390	-0.000025
1 x 10^8	0.002072	0.002083	-0.000011
4 x 10^8	0.001718	0.001721	-0.000003
1 x 10^9	0.001531	0.001531	0
4 x 10^9	0.001298	0.001298	0
1 x 10^{10}	0.001172	0.001172	0

For your convenience, Equations 3-4 and 3-5 for friction resistance coefficients are repeated here.

ATTC 1947 Schoenherr Friction Resistance Coefficient

$0.242/(C_F)^{1/2} = \text{Log}_{10}(R_n \times C_F)$ Equation 3-4

ITTC 1957 Model-Ship Correlation Line

$C_F = 0.075/[(\text{Log}_{10} R_n)-2]^2$ Equation 3-5

TABLE A1-G: MINIMUM BARE HULL R/W FOR NOTIONAL 500 MT HULL FORMS IN DEEP, CALM SALT WATER								
Hull Form	$L/\nabla^{1/3}$	F_{nL}						
		0.4	0.5	0.6	0.7	0.8	0.9	1.0
Hard Chine	4.0	0.0380	0.0895	0.1400	0.1475	0.1400	0.1335	0.1260
	4.5	0.0380	0.0850	0.1185	0.1205	0.1175	0.1130	0.1055
	5.0	0.0375	0.0785	0.1050	0.1090	0.1085	0.1075	0.1025
	5.5	0.0375	0.0690	0.0895	0.0970	0.1005	0.1030	0.1020
	6.0	0.0380	0.0640	0.0800	0.0875	0.0930	0.0990	0.1010
	6.5	0.0370	0.0605	0.0745	0.0820	0.0875	0.0960	0.1005
	7.0	0.0370	0.0585	0.0710	0.0780	0.0850	0.0945	0.1015
	7.5	0.0375	0.0555	0.0665	0.0755	0.0850	0.0965	0.1040
	8.0	0.0370	0.0525	0.0605	0.0720	0.0865	0.0985	0.1085
	8.5	0.0350	0.0485	0.0570	0.0690	0.0850	0.1025	0.1135
Double Chine	6.5	0.0400	0.0625	0.0700	0.0765	0.0835	0.0935	0.1025
	7.0	0.0400	0.0565	0.0650	0.0740	0.0815	0.0910	0.1020
	7.5	0.0360	0.0510	0.0615	0.0695	0.0775	0.0875	0.1030
	8.0	0.0350	0.0475	0.0585	0.0665	0.0765	0.0880	0.1055
	8.5	0.0350	0.0475	0.0570	0.0650	0.0765	0.0915	0.1085
Round Bilge	4.5	0.0210	0.0690	0.1080	0.1160	---	---	---
	5.0	0.0250	0.0685	0.0930	0.1000	0.1070	0.1135	0.1320
	5.5	0.0250	0.0625	0.0810	0.0885	0.0940	0.1000	0.1090
	6.0	0.0240	0.0565	0.0700	0.0800	0.0865	0.0935	0.1035
	6.5	0.0230	0.0510	0.0625	0.0725	0.0810	0.0890	0.1015
	7.0	0.0230	0.0455	0.0565	0.0665	0.0765	0.0850	0.1005
	7.5	0.0210	0.0415	0.0523	0.0620	0.0725	0.0820	0.1020
	8.0	0.0200	0.0375	0.0480	0.0590	0.0700	0.0795	0.1035
	8.5	0.0195	0.0355	0.0460	0.0565	0.0683	0.0785	---
	9.0	0.0185	0.0335	0.0445	0.0550	0.0665	0.0790	---
	9.5	0.0180	0.0325	0.0435	0.0535	0.0650	0.0805	---
	10.0	0.0175	0.0320	0.0430	---	---	---	---
Notes: For 500 mt $R_n > 1 \times 10^8$; Thus, C_F is essentially equivalent for both ATTC 1947 and ITTC 1957 formulations.				C_R was calculated using ATTC 1947 formulations. $C_A = 0$ Some conditions may be dynamically unstable.				

TABLE A1-H: SEA STATE TABLE USED FOR MILITARY CRAFT

SEA STATE[1]	FULLY RISEN SEAS[2] – WAVE HEIGHT (feet): AVERAGE	SIGNIFICANT (avg. of 1/3 highest)	AVG. OF 1/10 HIGHEST	SIGNIFICANT RANGE OF PERIODS (seconds)	T_{max} (period of max energy of spectrum, sec)	$\bar{T}$ (average period, sec.)	$\bar{L}$ (avg. wave length, ft.)	WIND[3]: WIND SPEED (knots)	MINIMUM FETCH (nmi)	MINIMUM DURATION (hrs)	BEAUFORT NUMBER[4]	WIND DESCRIPTION	WIND SPEED (knots)	APPEARANCE OF THE SEA[5]
0	0	0	0	–	–	–	–	–	–	–	0	CALM	<1	Sea like a mirror.
	0.05	0.08	0.10	up to 1.2 sec.	0.7	0.5	0.83	2	5	0.3	1	LIGHT AIRS	1-3	Ripples with the appearance of scales are formed, but without foam crests.
1	0.18	0.29	0.37	0.4-2.8	2.0	1.4	6.7	5	8	0.7	2	LIGHT BREEZE	4-6	Small wavelets, still short but more pronounced; crests have glassy appearance and do not break.
	0.6	1.0	1.2	0.8-5.0	3.4	2.4	20	8.5	9.8	1.7	3	GENTLE BREEZE	7-10	Large wavelets. Crests begin to break. Foam of glassy appearance. Perhaps scattered white horses.
2	0.88	1.4	1.8	1.0-6.0	4.0	2.9	27	10	10	2.4				
	1.4	2.2	2.8	1.0-7.0	4.8	3.4	40	12	18	3.8	4	MODERATE BREEZE	11-16	Small waves, becoming longer; fairly frequent white horses.
	1.8	2.9	3.7	1.4-7.6	5.4	3.9	52	13.5	24	4.8				
3	2.0	3.3	4.2	1.5-7.8	5.6	4.0	59	14	28	5.2				
	2.9	4.6	5.8	2.0-8.8	6.5	4.6	71	16	40	6.6				
4	3.8	6.1	7.8	2.5-10.0	7.2	5.1	90	18	55	8.3	5	FRESH BREEZE	17-21	Moderate waves, taking a more pronounced long form; many white horses are formed. (Chance of some spray.)
	4.3	6.9	8.7	2.8-10.6	7.7	5.4	99	19	65	9.2				
5	5.0	8.0	10	3.0-11.1	8.1	5.7	111	20	75	10				
	6.4	10	13	3.4-12.2	8.9	6.3	134	22	100	12	6	STRONG BREEZE	22-27	Large waves begin to form; the white foam crests are more extensive everywhere. (Probably some spray.)
	7.9	12	16	3.7-13.5	9.7	6.8	160	24	130	14				
6	8.2	13	17	3.8-13.6	9.9	7.0	164	24.5	140	15				
	9.6	15	20	4.0-14.5	10.5	7.4	188	26	180	17				
	11	18	23	4.5-15.5	11.3	7.9	212	28	230	20	7	MODERATE GALE	28-33	Sea heaps up and white foam from breaking waves begins to be blown in streaks along the direction of the wind. Spindrift begins.
7	14	22	28	4.7-16.7	12.1	8.6	250	30	280	23				
	14	23	29	4.8-17.0	12.4	8.7	258	30.5	290	24				
	16	26	33	5.0-17.5	12.9	9.1	285	32	340	27				
	19	30	38	5.5-18.5	13.6	9.7	322	34	420	30	8	FRESH GALE	34-40	Moderately high waves of greater length; edges of crests break into spindrift. The foam is blown in well-marked streaks along the direction of the wind.
	21	35	44	5.8-19.7	14.5	10.3	363	36	500	34				
	23	37	46.7	6.0-20.5	14.9	10.5	376	37	530	37				
	25	40	50	6.2-20.8	15.4	10.7	392	38	600	38				
8	28	45	58	6.5-21.7	16.1	11.4	444	40	710	42				
	31	50	64	7.0-23.0	17.0	12.0	492	42	830	47	9	STRONG GALE	41-47	High waves. Dense streaks of foam along the direction of the wind. Sea begins to roll. Spray may affect visibility.
	36	58	73	7.0-24.2	17.7	12.5	534	44	960	52				
9	40	64	81	7.0-25.0	18.6	13.1	590	46	1110	57				
	44	71	90	7.5-26.0	19.4	13.8	650	48	1250	63	10	WHOLE[6] GALE	48-55	Very high waves with long overhanging crests. The resulting foam in great patches is blown in dense white streaks along the direction of the wind. On the whole the surface of the sea takes a white appearance. The rolling of the sea becomes heavy and shock like. Visibility is affected.
	49	78	99	7.5-27.0	20.2	14.3	700	50	1420	69				
	52	83	106	8.0-28.2	20.8	14.7	736	51.5	1560	73				
	54	87	110	8.0-28.5	21.0	14.8	750	52	1610	75				
	59	95	121	8.0-29.5	21.8	15.4	810	54	1800	81				
	64	103	130	8.5-31.0	22.6	16.3	910	56	2100	88	11	STORM[6]	56-63	Exceptionally high waves. (Small and medium-sized ships might for a long time be lost to view behind the waves.) The sea is completely covered with long white patches of foam lying along the direction of the wind. Everywhere the edges of the wave crests are blown into froth. Visibility affected.
	73	116	148	10.0-32.0	24	17.0	985	59.5	2500	101				
	>80[7]	>128[7]	>164[7]	10-(35)[7]	(26)[7]	(18)[7]	–	>64	–	–	12	HURRICANE[6]	64-71	The air filled with foam and spray. Sea completely white with driving spray; visibility very seriously affected.

From a table compiled by Wilbur Marks, David Taylor Model Basin

[1]Sea states refer only to wind waves. Swells from distant or old storms are often superimposed on the wind wave pattern.

[2]Practical Methods of Observing and Forecasting Ocean Waves, Pierson, Neuman, James, H.O.Pub. 603, 1955.

[3]Wind required to create a fully risen sea. To attain a fully risen sea for a certain wind speed, the wind must blow at that speed over a minimum distance (fetch) for a minimum time (duration).

[4]The Beaufort Number is a wind force scale. While wind and seas are causally related, Beaufort Number and sea state are not the same. For example, it is common to have force 7 winds, but because of limited fetch or duration, a sea state of only 2.

[5]Manual of Seamanship, Vol. II, Admiralty, H.M. Stationary Office, 1952.

[6]For whole gale, storm, and hurricane winds (50 knots or more) the required durations and fetches are rarely attained. Seas are therefore not fully arisen.

[7]For such high winds the seas are confused. The wave crests are blown off, and the water and air mix.

APPENDIX 2

Designing Rudders, Stock, and Steering Gear

Designing appropriately sized rudders is a hydrodynamic exercise that takes into consideration the vessel's full-load displacement and maximum calm, deep-water speed (Blount and Dawson 2002). This appendix combines the hydrodynamic design of rudders with sizing the diameter of the rudder stock for strength, which is an interaction of the hydrodynamic loading, center of application of the loads, and fabrication material properties.

The most common, but not necessarily the most efficient, propulsion system for larger boats—including yachts, commercial vessels, and military craft—is two inboard engines driving submerged propellers on inclined shafts. Because such systems do not have inherent steering capabilities, the naval architect and/or the builder must take on the responsibility to design, optimize, and integrate a steering system with the propulsion system. Rudders are generally the steering system of choice.

The design and sizing of rudders and their associated steering gear is the focus of this appendix. This design process applies to low-speed boats as well as to the highest-speed performance boats and craft. "High speed" generally means greater than 20 knots, the speed at which the potential for rudder ventilation becomes a factor in the design process. In designing rudders for high-performance craft, the distinguishing factor is that the designer must consider high-speed hydrodynamic characteristics of different rudder sections in order to make informed choices with regard to cavitation and ventilation which affect both directional and dynamic stability.

In earlier times, simple guidelines were sufficient for designing rudders, which were often cast with airfoil sections, or fabricated from a piece of flat plate. For many craft, a rudder's total blade area was usually equal to three to four percent of underwater projected lateral plane of the hull. Faster boats required less area, heavier boats more. Builders occasionally sized rudder stocks by using pieces of leftover propeller shafting, with diameters based on maximum propeller torque, not hydrodynamic loads and moments of the rudder.

Sophisticated boats necessarily call for better design solutions. Because high-performance boats often must meet speed and range targets, they also have weight constraints and can be sensitive to ventilation and dynamic instability problems. These steering systems must be the best possible. Rules of thumb are inadequate.

Steering-system design follows a logical process that includes not only the rudders, but their placement relative to the boat's hull and propellers, along with their connection to the steering gear inside the boat. All these factors can have significant effects on turning forces and controllability.

Sizing Rudders

Note: Specific to Appendix 2, Table A2-A provides notation and graphic definitions to clarify computations.

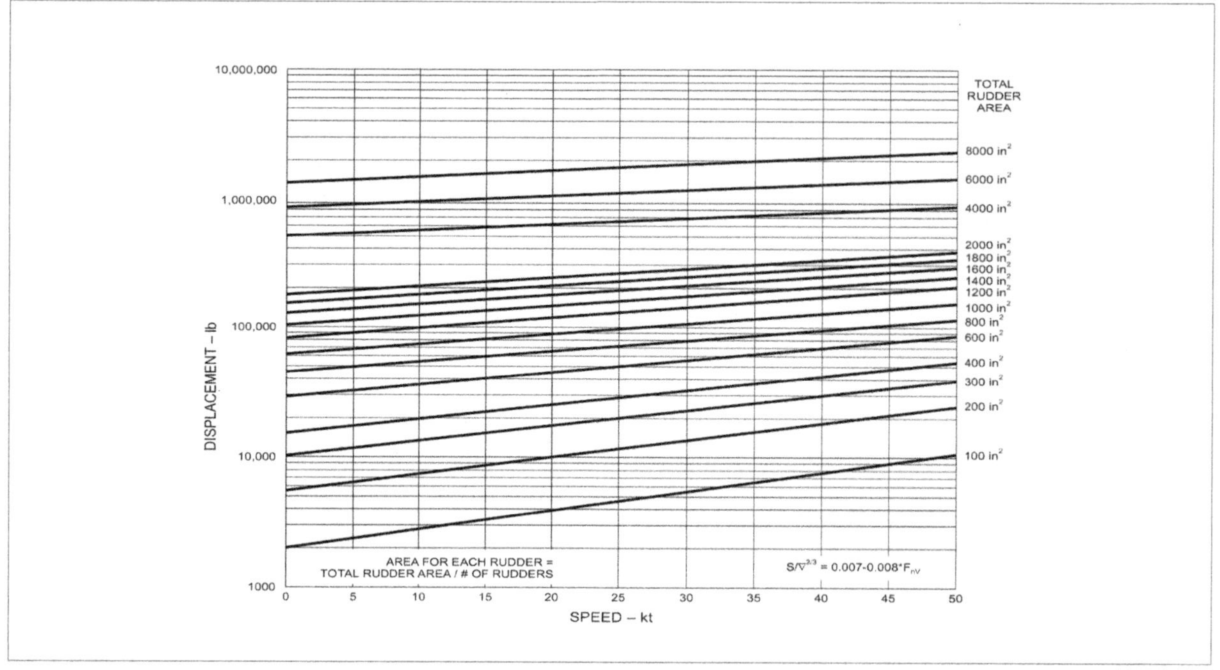

Figure A2-1: Rudder-size Chart

Before you can determine rudder size, you must make the necessary propulsion system decisions to establish the best combination of propeller dimensions and reduction-gear ratio. While there are several methods for sizing rudders, an appropriate approach should take into consideration the boat's full-load displacement and maximum calm-water speed. The criterion that I have developed over many years is provided by Figure A2-1, which is based on Equation A2-1. The area S, derived is the total projected area of all rudders. When multiple rudders are installed, divide area S by the number of rudders, N, to determine the area of individual rudders to be used in calculating the diameter of the stock.

$$S/\nabla^{2/3} = 0.07 - 0.008\,(F_{nV}) \qquad \text{Equation A2-1}$$

There are a variety of approaches to manufacturing rudders. They may be cast, machined, fabricated weldments, or molded from composites. The characteristics of rudder cross-section shapes at different design speeds will influence choices. Fortunately, information has been made available by Gregory and Dobay (1973) which includes data from hydrodynamic tests over a range of speeds ($0.5 \leq \sigma \leq 4.0$) conducted for some of the most frequently used rudder-section shapes: Airfoil, flat plate, wedge, cubic, and parabolic section rudders with characteristcs shown in Figures A2-2 to A2-6. These characteristics provide guidance for making choices between steering/directional-control performance and steering-system acquisition costs.

The characteristics of rudders with different sections can be compared when described with dimensionless coefficients. The side-force coefficient C_{SF} describes the rudder's turning force. The rudder-moment coefficient $C_{M\ C/4}$ relates to the torque needed to turn and/or hold the rudder's position. The drag coefficient C_D is associated with the resistance force retarding boat speed. These coefficients have been obtained from tests in a cavitation tunnel with rudders having a geometric aspect ratio of 1.5.

Characteristics for airfoil, flat-plate, wedge, cubic, and parabolic rudders operating up to approximately 40 knots are given. It's important to understand that the data are based on an ideal flow situation, in which a rudder is mounted on a large flat plate, with no transom downstream nor propeller upstream.

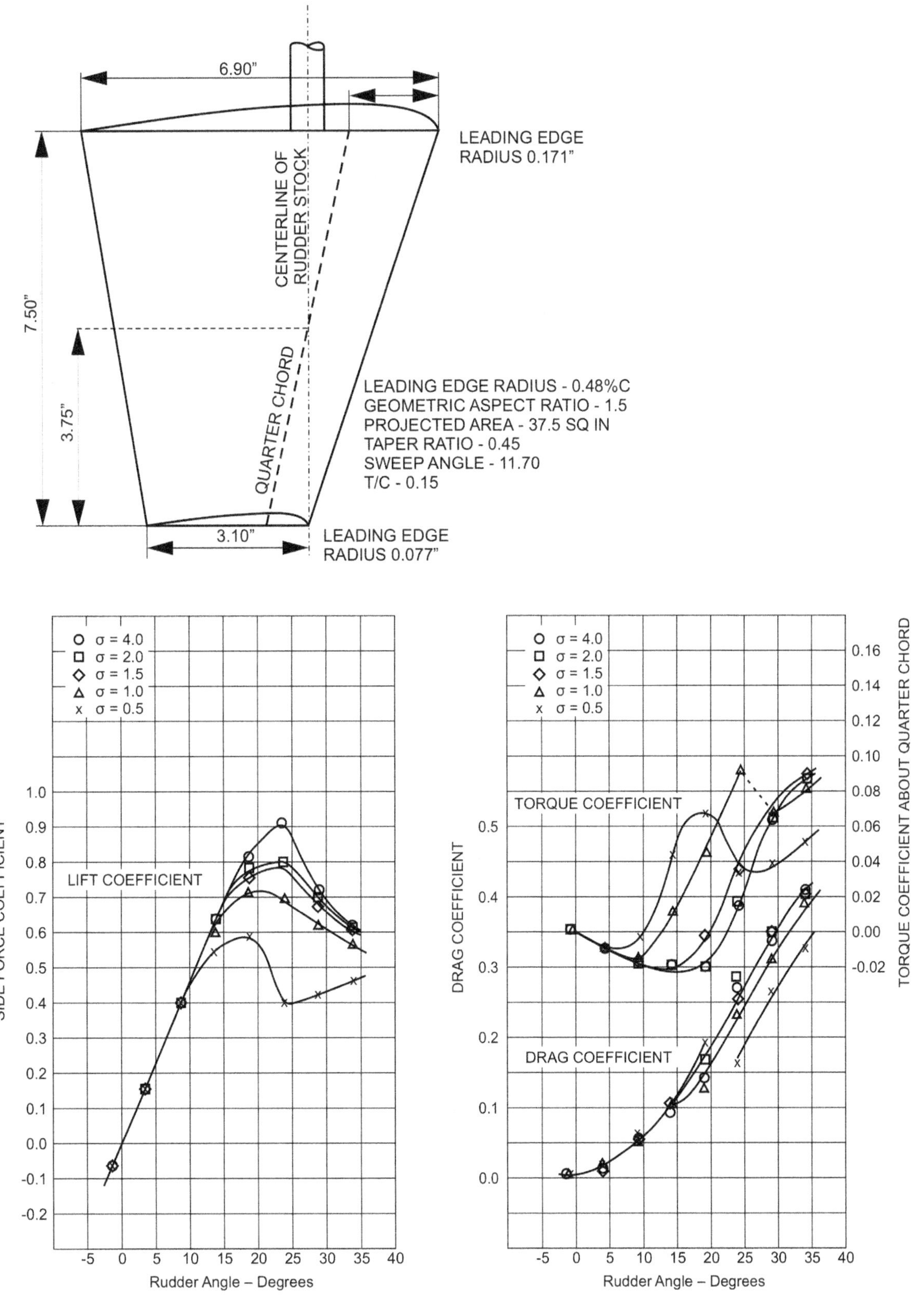

Figure A2-2: Airfoil rudder and characteristics

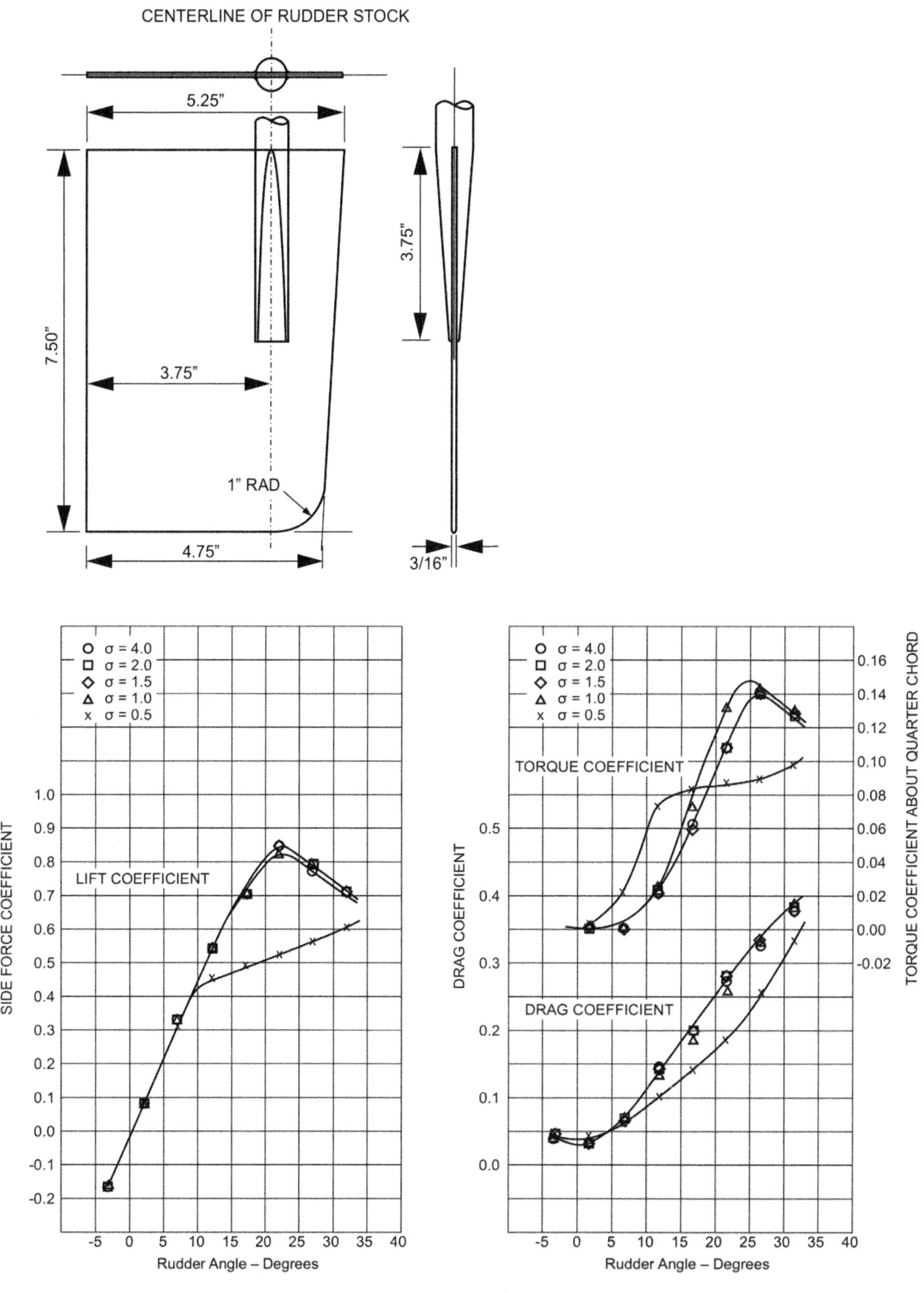

Figure A2-3: Flat-plate rudder and characteristics

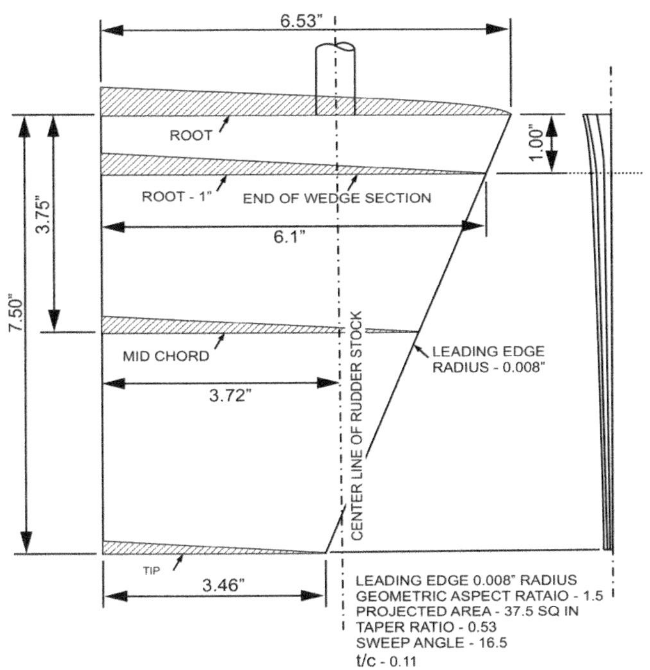

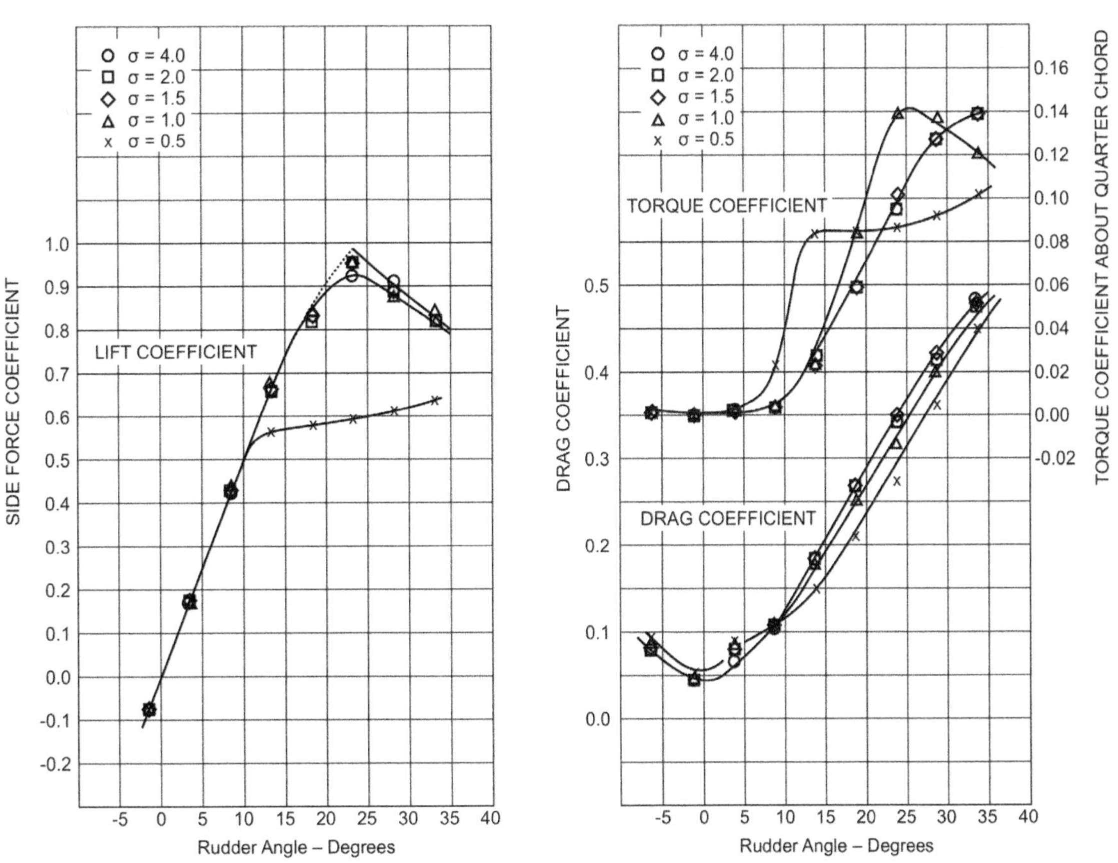

Figure A2-4: Six-degree wedge section rudder and characteristics

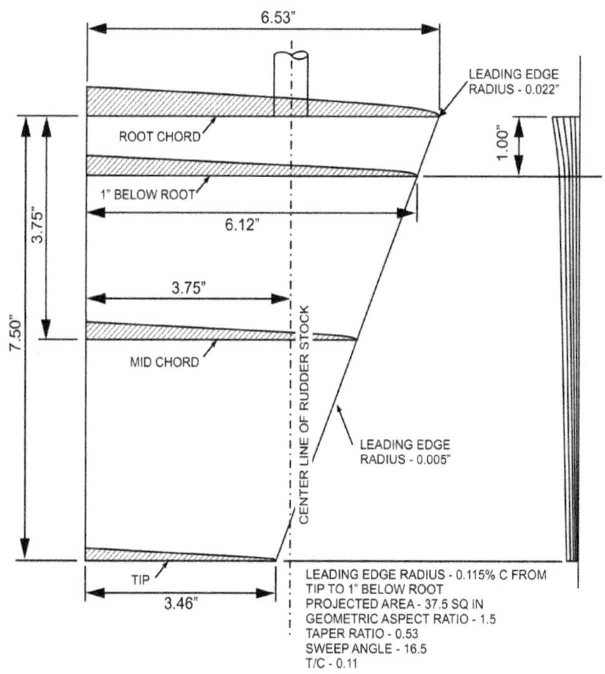

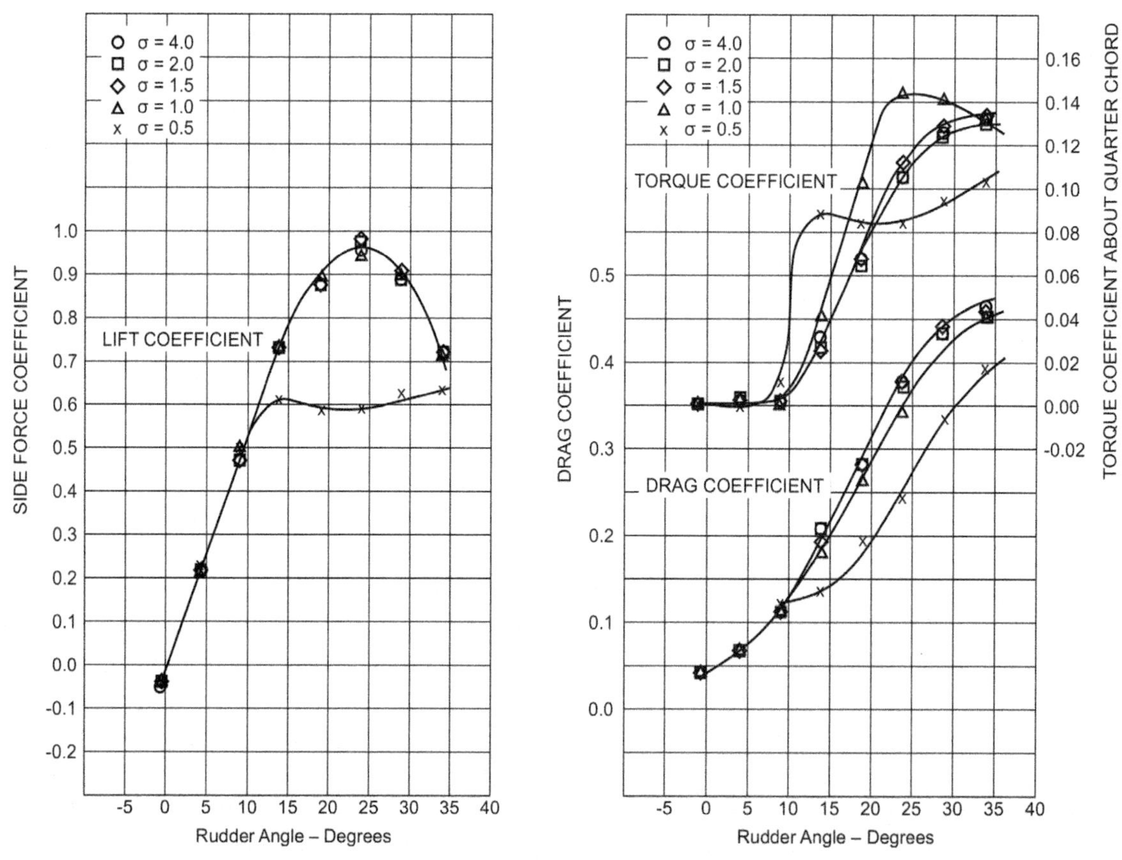

Figure A2-5: Cubic section rudder and characteristics

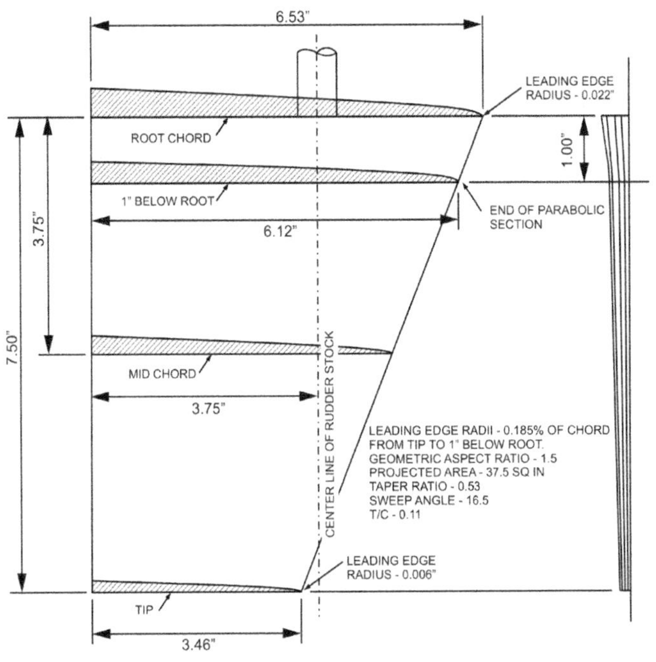

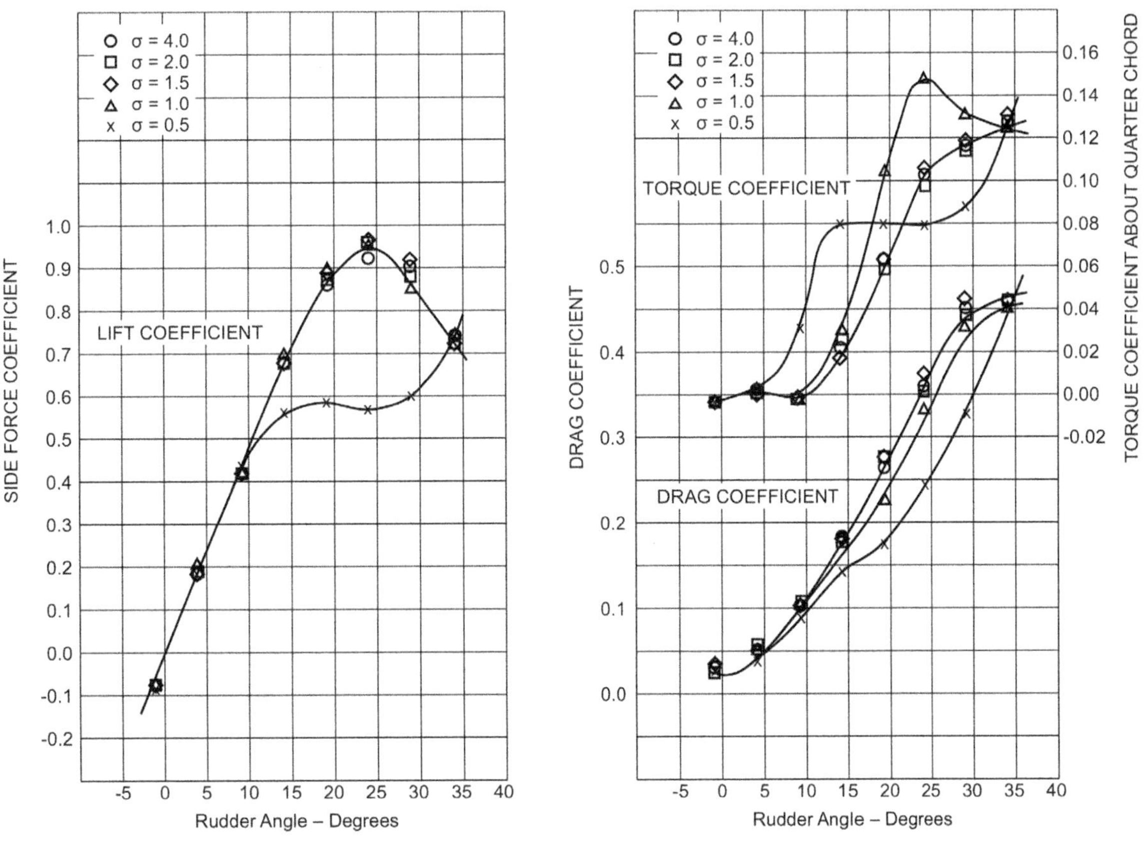

Figure A2-6: Parabolic rudder and characteristics

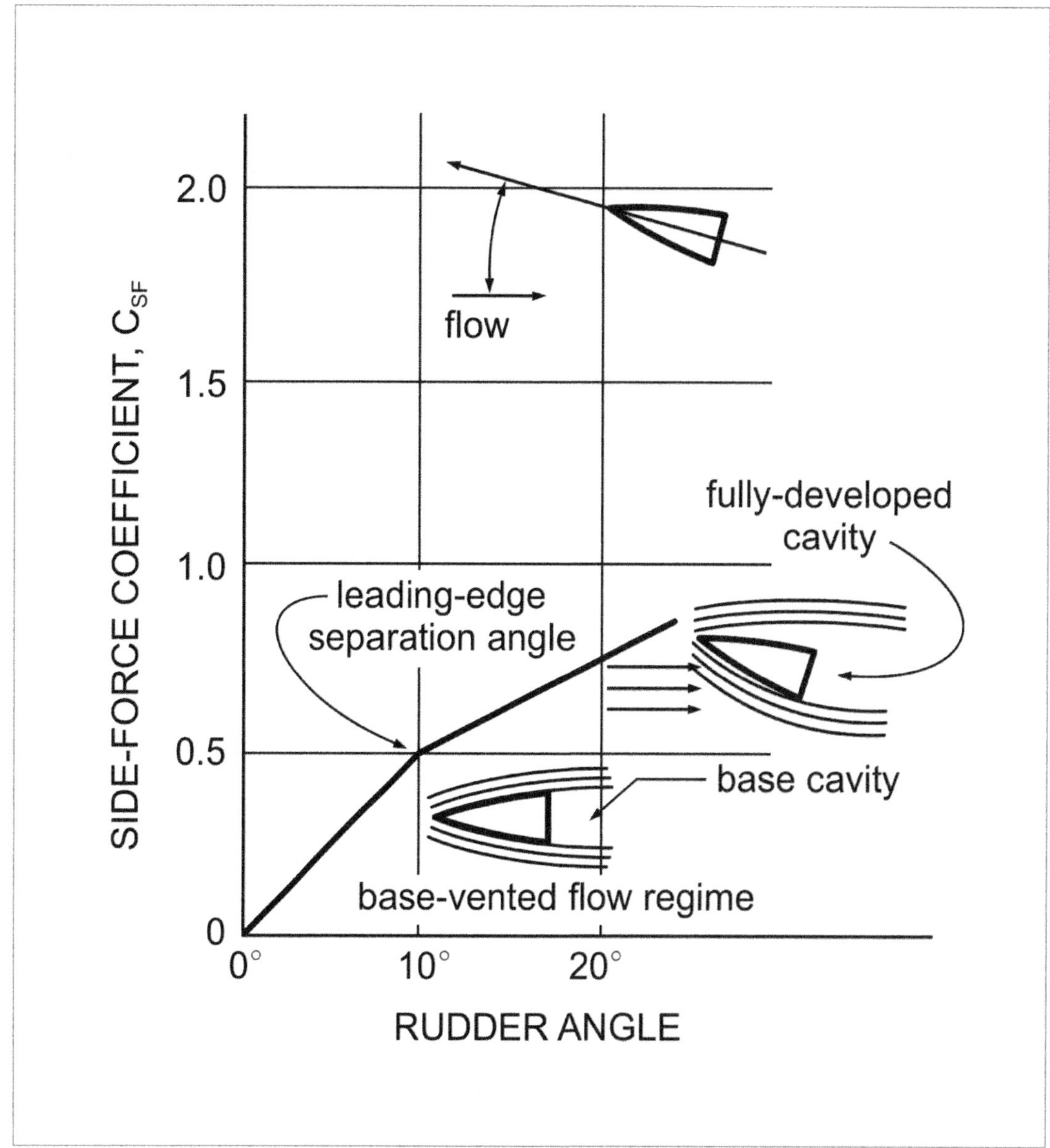

Figure A2-7: Typical effect of cavity development on the side-force coefficient

When boats increase speed up to the inception of cavitation, rudders develop different cavity patterns depending on airfoil sections vice blunt base sections with sharp leading edges. When these blunt base rudders are turned, the cavity changes, shifting from trailing edge separation, affecting the rudder's C_{SF} and $C_{M\ C/4}$.

The general effect on the slope of the side-force coefficient when the cavity separates from the leading edge is shown in Figure A2-7 as a reduction in slope. The vapor pressure in the cavity is very low when the cavity is not ventilated. The pressure in the cavity acting on the bottom of the hull, if it has an open path to atmospheric pressure, can also cause an unexpected reaction to the trim and direction of a boat at high speed. (For example, approximately 0.3 psi in a vapor cavity as compared to atmospheric pressure of 14.7 psi when ventilated.)

Relative to blunt-based, sharp, leading-edge section rudders, airfoil rudders have distinctly different cavitation characteristics than when fully wetted. And the differentiation between cavitation and ventilation phenomenon has to be accounted for at intermediate to high speeds. It has been observed that sail boats have experienced loss of steering control resulting from rudder ventilation at speeds below 15 knots. As rudders are usually mounted near the transom, air paths can follow a cavity behind the rudder stock, rudder-tip vortex, or even in the wake of the rudder. Airfoil rudders, due to their section shapes having convex curved surfaces from leading edge to tail, are much more prone to ventilation than is experienced by flat plate or wedge rudders which have flat surfaces on their pressure sides.

When an airfoil-section rudder cavitates, pressure within may be 0.3 psi; even so, some small pressure differential can remain with limited rudder force for steering control. Should the cavity side of an airfoil rudder ventilate, however, the pressure jumps from 0.3 psi to atmospheric pressure, 14.7 psi. Thus, the remaining cambered wetted rudder surface is inadequate to oppose the side force generated due to steering angle. At small angles, a ventilated airfoil section rudder can cause a craft to heel or even cause side-force reversal. In this latter situation, a boat will turn to the opposite direction to that expected. This sensitivity to side-force reversal due to ventilation is specific to airfoil section rudders. This is not so for rudders with section shapes when one remaining wetted side always generates positive pressures at positive angles of attack.

Sizing Rudder Stock

In general, the point of maximum material stress is the location where the rudder stock and the lowest rudder bearting mounted in the hull meet. For hydrodynamic considerations, the smallest stock diameter consistent with adequate strength is desired to blend the geometry of the rudder with the stock so as to minimize the disturbance of water flow. Thus, the stock diameter is required before the designer can make a complete construction drawing for a rudder and develop the details of propulsion arrangements locating rudder and bearings in relation to propellers and hull lines. For these reasons, I am including strength of materials calculation procedures for sizing rudder stock diameters here even though this book is primarily about hydrodynamics of performance vessels.

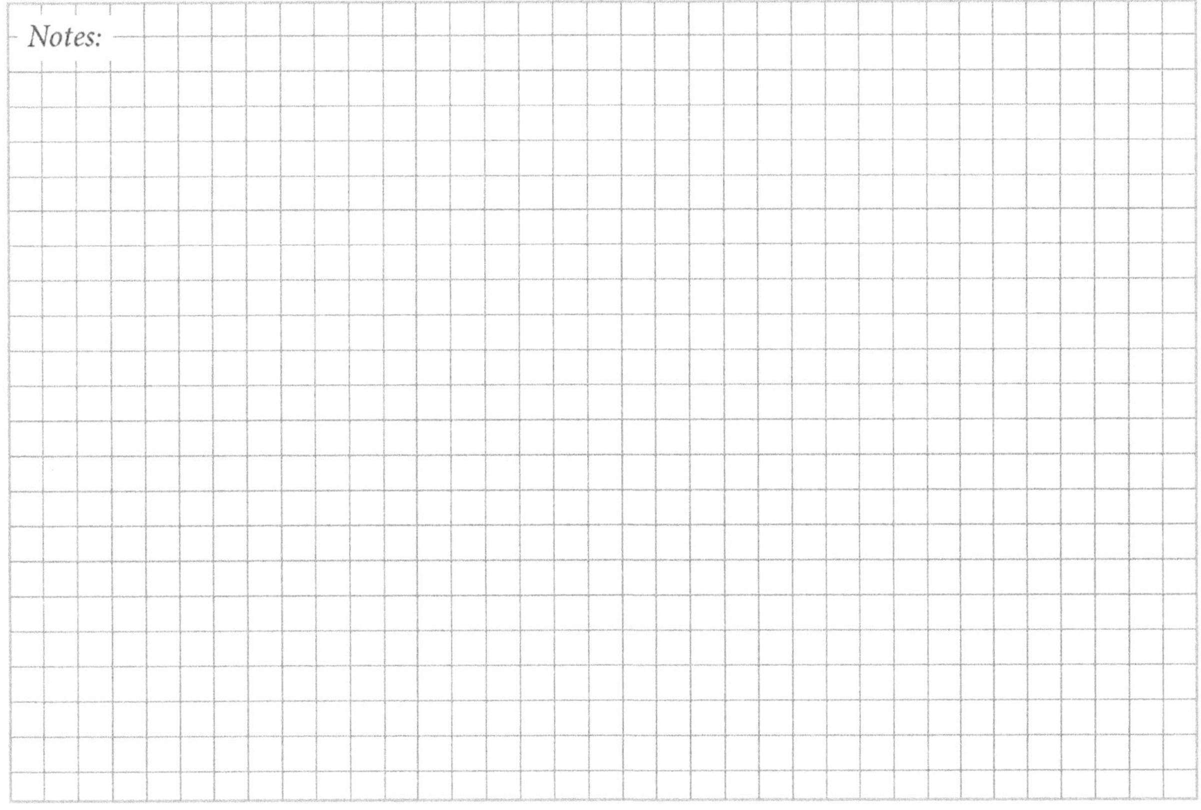

TABLE A2-A: NOTATION SPECIFIC TO APPENDIX 2			
A	span, ft, $[S_N (AR)]^{1/2}$	N	number of rudders
AF	projected area forward of rudder stock, ft^2	R	resultant hydrodynamic force on rudder, lb
AR	geometric aspect ratio, A^2/S	RB	rudder balance, (AF/S)
AS	allowable stress in rudder stock, lbs/sq ft Designer's choice, 0.6 (US) or 0.5 (YS)	S	total projected area of all rudders, ft^2
B	minimum chord of rudder, ft	S_N	individual rudder projected area, ft^2, $S_N = S/N$
c	mean chord, S_N/A, ft	SF	rudder side force, lb, $C_{SFC}(1.424\ \rho S_N V^2)$
C_D	rudder drag coefficient, lb, $D_R/1.424\rho SV^2$	T	rudder torque, ft-lb
C_{DC}	rudder drag coefficient corrected for AR	t/c	thickness/chord ratio
$C_{M\ C/4}$	rudder moment coefficients reference quarter chord, $M/1.424\rho cSV^2$	US	ultimate strength of rudder-stock material, lbs/sq ft
CP	center of pressure	V	speed, knots
C_S	combined stress in rudder stock, ft-lb	YS	yield strength of rudder-stock material, lbs/sq ft
C_{SF}	rudder side-force coefficient, lb, $SF/1.424\rho SV^2$	X	distance between CP and stock axis (+ means CP is aft of axis), ft
C_{SFC}	rudder side force coefficient corrected for AR	US	ultimate strength of rudder-stock material, lbs/sq ft
d	rudder-stock diameter, ft	V	speed, knots
D_R	rudder drag, lb $C_{DC}\ (1.424\ \rho S_N V^2)$	YS	yield strength of rudder-stock material, lbs/sq ft
D	diameter of propeller, ft	X	distance between CP and stock axis (+ means CP is aft of axis), ft
E	maximum chord of rudder, ft	Y	Tiller-arm length, ft
F	actuator stroke force for one rudder, lb	Z	distance between lower rudder bearing and top of rudder, ft
FS	factor of safety	Z_B	section modulus
H	assume water depth for cavitation number is equal to span, A	Z_P	polar section modulus
L	actuator stroke length, ft	ρ	mass density of water, $lb\text{-}sec^2/ft^4$ (salt water 1.99, fresh water 1.94)
M	rudder resultant moment (rudder torque), reference: quarter chord, ft-lb	σ	cavitation number
M_B	vertical bending moment in the stock, about lower rudder bearing, ft-lb	∇	value of displacement, ft^3

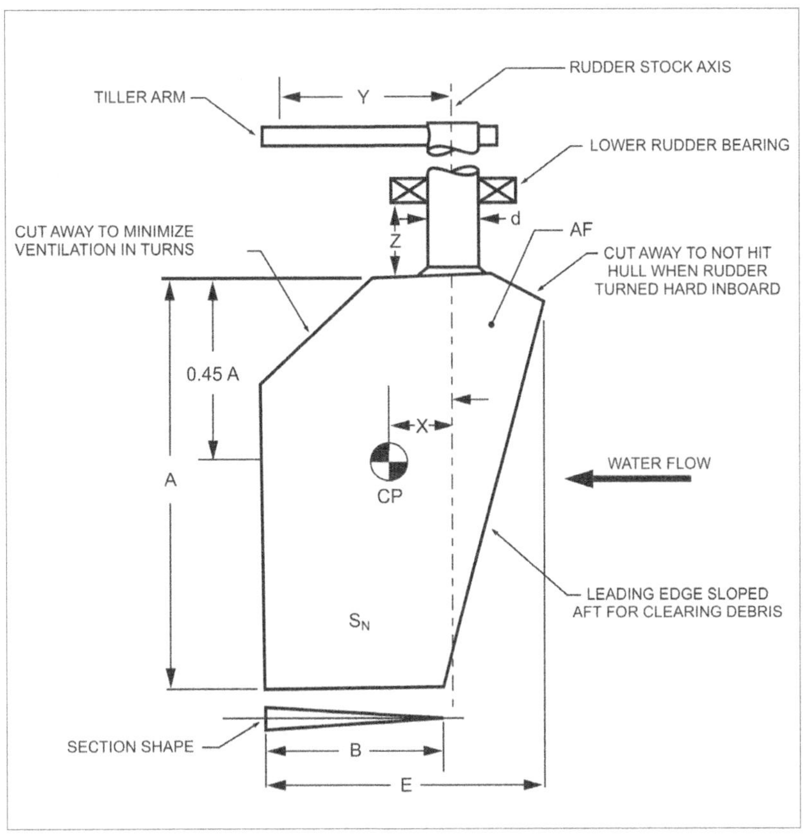

Figure A2-8: Graphic depiction of notation

Calculating Rudder Loads

The notation in Table A2-A and the graphic depiction in Figure A2-8 combined define the dimensions used in the equations of Appendix 2. You may also use the data in Figures A2-2 to A2-6 to calculate the hydrodynamic loads on rudders with geometric aspect ratio of 1.5. Side force produces a bending moment which tends to dominate the stock design.

Begin by select the rudder angle for maximum value for C_{SF} for the design speed. Use this same rudder angle for C_D and $C_{M\ C/4}$.

Correction of coefficients for small variations (AR from 1.0 to 2.0) from the geometric aspect ratio of 1.5 may be approximated. The side-force coefficients from Figures A2-2 to A2-6 may be multiplied by an approximate correction factor of 0.4(1+AR). Similarly, drag coefficients from Figures A2-2 to A2-6 can be multiplied by an approximate correction factor of 1.75-(0.5AR).

It is sometimes desirable to drop the rudder away from the hull a short distance, for example, to avoid or correct an air drawing or ventilation condition. Theoretically, this should reduce the effective aspect ratio, but experiments with a wedge rudder mounted near a transom do not confirm this for speeds up to 50 knots. The only measurable characteristic change was to the drag coefficient near zero degrees, the drag coefficient increased as the exposed length of the rudder stock becomes longer. When dropping a rudder away from the hull, a non-turning fairing fixed to the hull can minimize the drag increase that results from an exposed rudder stock.

To locate the longitudinal (chord-wise) center of pressure (CP) at any rudder angle for the rudders in Figures A2-2 to A2-6, use the following relationship:

$$X = C_{M\ C/4}\ (c)/(C_{SF}^2+C_D^2)^{1/2} \qquad \text{Equation A2-1}$$

X is the lever arm, i.e., the distance between the CP and the stock axis. X is positive when the CP is aft of the stock axis and the rudder tends to return to zero angle (trail) if the tiller arm breaks or the rudder actuator fails.

Note that Figures A2-2 to A2-6 provide characteristics for rudders with the stock located for a 25 percent balance. I recommend something less, generally between 15 and 20 percent. The X dimension should be adjusted accordingly, by the actual dimension difference, if the as-designed stock is not located at 25% balance. Rudder balance is the ratio in percent of the area forward of the rudder stock to total rudder area.

The size of the rudder stock depends on the combined stress resulting from the bending moment of the side force (SF) and drag (D_R) as well as the torsional stress necessary to turn the rudder to the desired angle. The resultant force (R) developed by the rudder from SF and D_R forces, is:

$$R = [(SF)^2+D_R^2]^{1/2} \qquad \text{Equation A2-2}$$

The vertical (spanwise) center of pressure (CP) for these rudders is approximated by 0.45A, measured down from the top of the rudder. Thus, the bending moment in the stock when the lower bearing is very near the top of the rudder becomes:

$$M_B = R(0.45)A \qquad \text{Equation A2-3}$$

If the rudder is lowered away from the hull and bearing such that the unsupported distance is Z, then Equation A2-3 is revised to be:

$$M_B = R(0.45A+Z) \qquad \text{Equation A2-4}$$

Torque required to rotate the rudder depends on the distance between the center of pressure and the stock axis as well as the resultant hydrodynamic force (R). Rudder torque is also affected by the friction in the bearings and stuffing box. One way to account for this friction is to increase the calculated hydrodynamic torque by a margin of 25 percent. The experience of other designers may indicate that a rudder torque margin other than 25 percent might be appropriate.

The rudder torque (T) becomes:

$$T = [X+(0.25-RB)c](R)(1.25) \qquad \text{Equation A2-5}$$

T is the minimum torque value the steering equipment must provide to rotate each rudder; double that value when two rudders are connected with a tie bar. The tiller arms, in conjunction with hydraulic or mechanical equipment, must develop the force and have the stroke to achieve the total angular rotation of the rudders.

The combined stock stress resulting from the bending moment (M_B) and rudder torque (T) becomes

$$C_S = (M_B/2Z_B) + [(M_B/2Z_B)^2 + (T/Z_P)^2]^{1/2} \qquad \text{Equation A2-6A}$$

Where for the stock

$Z_B = \pi d^3/32$ and $Z_P = \pi d^3/16$.

A convenient form of this equation for calculations is obtained by multiplying both sides by Z_B and rearranging to become Equation A2-6B.

$$(Z_B)\, C_S = M_B/2 + [(M_B/2)^2 + (T/2)^2]^{1/2} \qquad \text{Equation A2-6B}$$

The quantity $(Z_B)C_S$ is input into Equation A2-8 for calculating rudder stock diameter (d).

Selecting rudder stock material and the factor of safety (FS) are the remaining choices in the design process. The allowable stress (AS) may range from a conservative value of 50 percent of yield strength (FS = 2.00) to the value of 60 percent of the ultimate strength (FS = 1.67). I suggest the latter value.

$$AS = 0.5\,(YS) \qquad \text{Equation A2-7A}$$

or

$$AS = 0.6\,(US) \qquad \text{Equation A2-7B}$$

The stock diameter (d) becomes:

$$d = [(Z_B)C_S(103.8)/(AS)]^{1/6} \qquad \text{Equation A2-8}$$

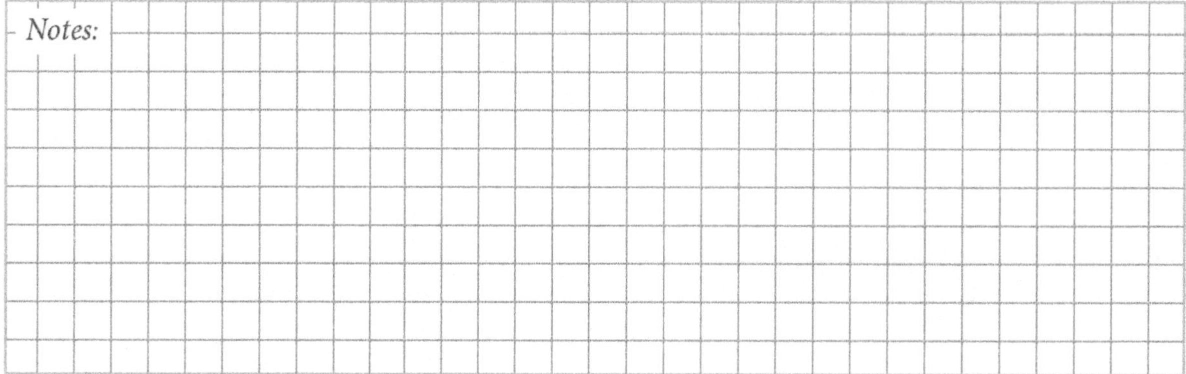

Sizing and Setting Up the Steering Gear

After you have completed the rudder design, sizing the steering gear is straightforward, as seen in Figure A2-9. The factors for selecting components include: maximum rudder angle, desired rudder-turn rate, the force to turn one or more rudders (if connected by a tie bar), the force to rotate the helm, and the number of helm turns to swing the rudder lock to lock. Most large boats have power-assisted steering gear, while small boats may have mechanical or manual-hydraulic steering systems.

I suggest the following criteria for steering gear:

- Maximum rudder angle: ±35°
- Rudder turn-rate: 3° per second
- Number of helm turns, lock to lock: 3 to 5 for power-assisted systems, but may depend on client preferences.
- Torque allowance including a margin to account or friction in system: Equation A2-5.

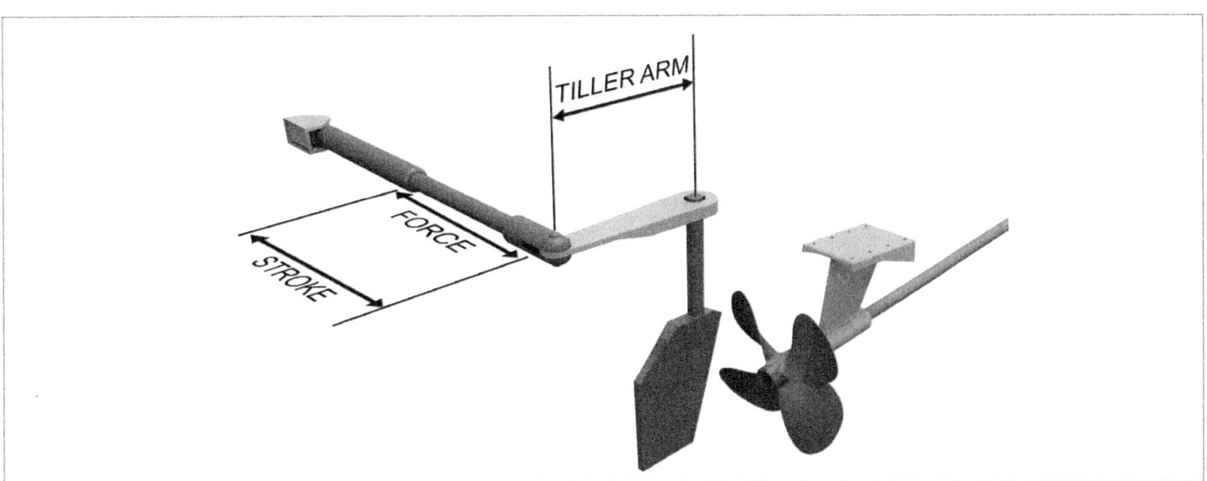

Figure A2-9: Design components of the steering gear

With a tiller-arm length of Y, the minimum actuator stroke (L) required to achieve a full 70° arc of rudder swing is:

$$L = 2Y \sin 35°$$ Equation A2-9

The required actuator force, F, to turn each rudder is:

$$F = T/Y$$ Equation A2-10

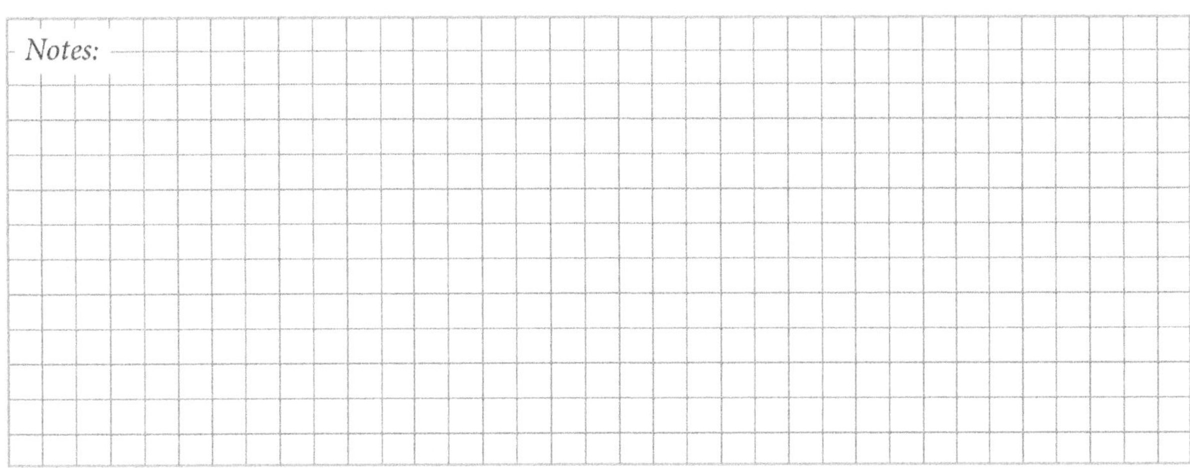

Alignment of a Rudder Steering System

The installation of a rudder on a conventional inclined-shaft propeller-driven powerboat appears to be a simple process. Many builders think all they need to do is mount the rudder aft of the propeller, place a tiller arm on the rudder stock inside the boat, and then set up the appropriate hydraulic ram so that the rudder moves to the desired angle when the boat is operating at speed.

It's not that simple! Subtle changes in the installation can make significant differences in the maneuvering and control characteristics of the boat. Thus, a significant detail still remains: Setting up the rudders and steering equipment on the boat.

Inattention to the setup process can result in a good boat fitted with good rudders behaving badly. To mitigate this possibility, pay close attention to these three factors:

1. aligning rudders with the flow using tow-in or tow-out (usually the trailing edges of the rudders are farther apart than the leading edges when propellers have outboard rotation);
2. having the trailing edges of the rudders under the hull to avoid ventilation; and
3. using differential steering to improve high-speed turning characteristics on high-performance craft.

Sea-trial experience indicates that the best criterion for establishing the position of the rudders is specifying the distance of the trailing edge, vertically and longitudinally, relative to the transom. Since both rudder profile and balance may vary from one design to another, the current recommendation on rudder placement relative to the transom and propeller is given in Figure A2-10. Note that the propeller shaft may be removed while leaving the rudder in place when the rudder stock is offset 0.10 D.

Water-flow angle at the leading edge of rudders on twin-screw boats is generally at a small angle relative to the centerline of the hull. When running straight, best boat speed is achieved along with least steering torque when rudders are aligned with water flow. By having an adjustable length tie rod connecting rudder tiller arms, alignment with flow may be readily achieved during builder's trials. Typical flow alignment on twin-screw boats with top blades of propellers turning outward occurs when trailing edge of rudders is outboard of leading edge. This generally results in zero rudder angle being 1° to 3° relative to the centerline of the hull. This offset angle also prevents rudder chattering.

For high-performance boats, the best turning characteristics are achieved with differential angles between inboard and outboard rudders. Differential angles are achieved by having the respective centerlines of the rudder and tiller arms having angular offset. For example, as I discussed in a 1997 *Professional BoatBuilder* article, an approximate 11° offset between till arm and rudder centerline results in about a 4° angle difference for inboard and outboard rudders when a boat is turning.

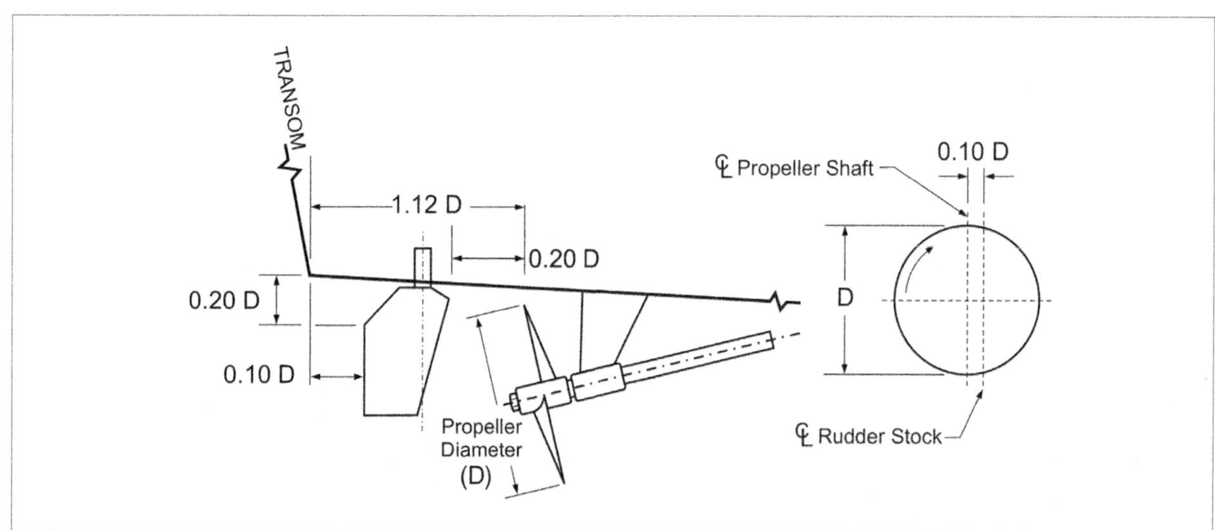

Figure A2-10: Shaft line buttock recommened minimum clearance dimensions for high-speed boats

APPENDIX 3

Parent Body Plans of Systematic Series

DATA STUDIED BY BLOUNT, MCGRATH (2009)

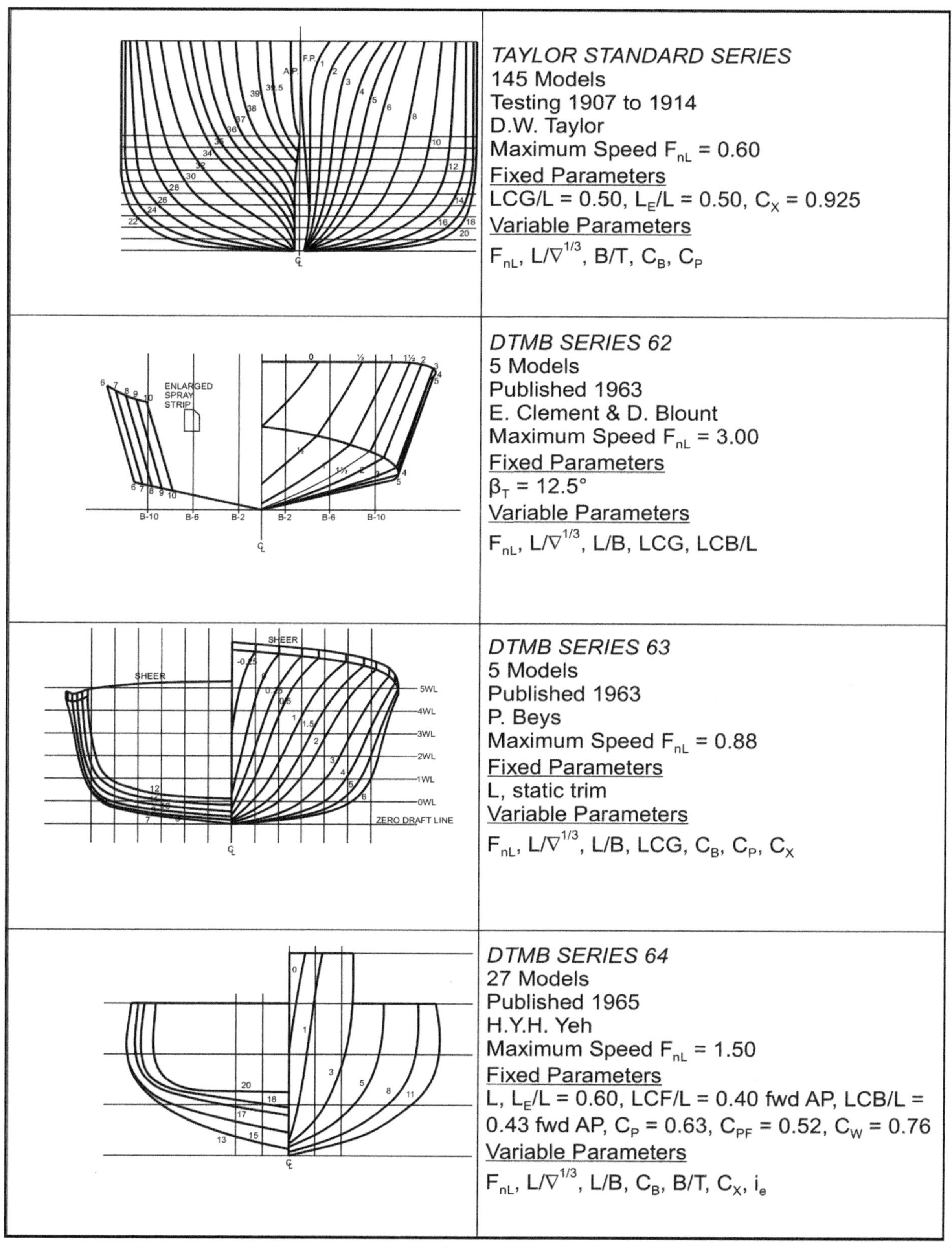

TAYLOR STANDARD SERIES
145 Models
Testing 1907 to 1914
D.W. Taylor
Maximum Speed F_{nL} = 0.60
<u>Fixed Parameters</u>
$LCG/L = 0.50$, $L_E/L = 0.50$, $C_X = 0.925$
<u>Variable Parameters</u>
F_{nL}, $L/\nabla^{1/3}$, B/T, C_B, C_P

DTMB SERIES 62
5 Models
Published 1963
E. Clement & D. Blount
Maximum Speed F_{nL} = 3.00
<u>Fixed Parameters</u>
$\beta_T = 12.5°$
<u>Variable Parameters</u>
F_{nL}, $L/\nabla^{1/3}$, L/B, LCG, LCB/L

DTMB SERIES 63
5 Models
Published 1963
P. Beys
Maximum Speed F_{nL} = 0.88
<u>Fixed Parameters</u>
L, static trim
<u>Variable Parameters</u>
F_{nL}, $L/\nabla^{1/3}$, L/B, LCG, C_B, C_P, C_X

DTMB SERIES 64
27 Models
Published 1965
H.Y.H. Yeh
Maximum Speed F_{nL} = 1.50
<u>Fixed Parameters</u>
L, $L_E/L = 0.60$, LCF/L = 0.40 fwd AP, LCB/L = 0.43 fwd AP, $C_P = 0.63$, $C_{PF} = 0.52$, $C_W = 0.76$
<u>Variable Parameters</u>
F_{nL}, $L/\nabla^{1/3}$, L/B, C_B, B/T, C_X, i_e

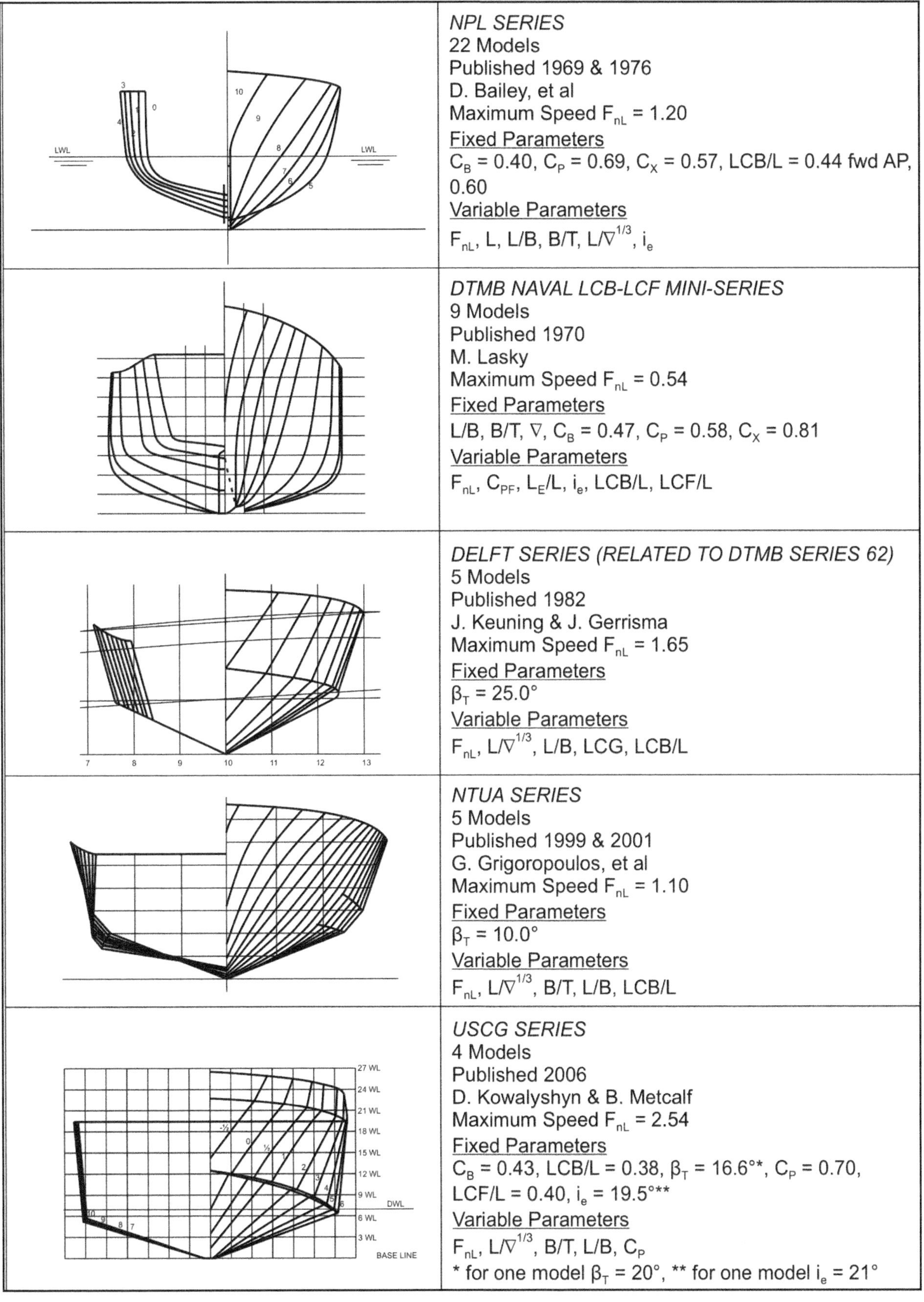

NPL SERIES
22 Models
Published 1969 & 1976
D. Bailey, et al
Maximum Speed F_{nL} = 1.20
<u>Fixed Parameters</u>
$C_B = 0.40$, $C_P = 0.69$, $C_X = 0.57$, LCB/L = 0.44 fwd AP, 0.60
<u>Variable Parameters</u>
F_{nL}, L, L/B, B/T, $L/\nabla^{1/3}$, i_e

DTMB NAVAL LCB-LCF MINI-SERIES
9 Models
Published 1970
M. Lasky
Maximum Speed F_{nL} = 0.54
<u>Fixed Parameters</u>
L/B, B/T, ∇, $C_B = 0.47$, $C_P = 0.58$, $C_X = 0.81$
<u>Variable Parameters</u>
F_{nL}, C_{PF}, L_E/L, i_e, LCB/L, LCF/L

DELFT SERIES (RELATED TO DTMB SERIES 62)
5 Models
Published 1982
J. Keuning & J. Gerrisma
Maximum Speed F_{nL} = 1.65
<u>Fixed Parameters</u>
$\beta_T = 25.0°$
<u>Variable Parameters</u>
F_{nL}, $L/\nabla^{1/3}$, L/B, LCG, LCB/L

NTUA SERIES
5 Models
Published 1999 & 2001
G. Grigoropoulos, et al
Maximum Speed F_{nL} = 1.10
<u>Fixed Parameters</u>
$\beta_T = 10.0°$
<u>Variable Parameters</u>
F_{nL}, $L/\nabla^{1/3}$, B/T, L/B, LCB/L

USCG SERIES
4 Models
Published 2006
D. Kowalyshyn & B. Metcalf
Maximum Speed F_{nL} = 2.54
<u>Fixed Parameters</u>
$C_B = 0.43$, LCB/L = 0.38, $\beta_T = 16.6°$*, $C_P = 0.70$, LCF/L = 0.40, $i_e = 19.5°$**
<u>Variable Parameters</u>
F_{nL}, $L/\nabla^{1/3}$, B/T, L/B, C_P
* for one model $\beta_T = 20°$, ** for one model $i_e = 21°$

APPENDIX 4

DTMB Series 62 and 65 Prediction Method

FOR RESISTANCE OF STEPLESS, HARD-CHINE HULLS

RANGE OF APPLICATION $0.5 \leq F_{nV} \leq 4.0$ and $4.0 \leq L_P/\nabla^{1/3} \leq 10.0$

References: Hubble (1974), Hubble (1982)

In the range of semi-displacement and semi-planing speeds, slenderness ratio ($L_P/\nabla^{1/3}$) is the dominant factor affecting R/W. With this reasoning, data for 22 hard-chine models has been analyzed by Hubble (1974) to define the relationship of (F_{nV}, $L_P/\nabla^{1/3}$) and R/W for craft having a displacement of 100,000 lb (45.4 mt). The range of parameters represented by these models is given in Table A4-A, and the geometric characteristics, lines, and body plans of these models are described in Table A4-B and in Figures A4-1 to A4-3.

TABLE A4-A: HULL GEOMETRY AND LOADING OF DATABASE	
Speed	$0.5 \leq F_{nV} \leq 4.0$
Slenderness ratio	$4.0 \leq L_P/\nabla^{1/3} \leq 10.0$
Chine length to beam ratio	$2.0 \leq L_P/B_{PX} \leq 9.38$
LCG to chine length ratio: fot	$0.30 \leq LCG/L_P \leq 0.52$
Transom to maximum chine beam ratio	$0.35 \leq B_{PT}/B_{PX} \leq 0.99$
Deadrise angle at 0.5 L_P	$13° \leq \beta_{mid} \leq 37°$
Deadrise angle at transom	$12.5° \leq \beta_T \leq 30.4°$
Deadrise angular warp rate* from 0.5 L_P to transom $(\beta_{mid} - \beta_T)/(0.5L_P/B_{PX})$: deg per beam	$0.26 \leq Warp \leq 7.57$

*Warp rate is defined as the change of deadrise angle from 0.5 L_P aftward to the transom, divided by the ratio $0.5L_P/B_{PX}$ which represents deadrise angle change for the after half of the hull bottom measured in maximum beams.

The results of this analysis are presented graphically in Figure A4-4 giving the mean values of R/W versus F_{nV} for constant values of $L_P/\nabla^{1/3}$. In addition, Figure A4-5 provides wetted area coefficient, $S/\nabla^{2/3}$ corresponding to the results seen in Figure A4-4. Data for these two figures is in Tables A4-C and D to facilitate making bare hull resistance predictions for vessels having both larger and smaller displacements than provided here for 100,000 lbs.

This particular analysis of resistance data for hard-chine hulls produced a most useful and realistic predictive approach for use when minimal information exists regarding hull geometry such as for concept designs and feasibility studies for speeds up to F_{nV} = 4.0. It can also be used to evaluate the sensitivity of resistance to hull loading for different vessel lengths and speeds.

The only input data needed to use this resistance prediction method is vessel displacement, projected chine (or its approximation 1.022 times waterline length) and a speed range of $F_{nV} \leq 4.0$. This prediction method can output mean, minimum, or other correlated resistance values of this model data set dependent on the standard deviation factor (SDF) selected.

This method, however, cannot be used analytically to minimize resistance by varying LCG, nor to predict running trim angle. Guidance for LCG and trim may be determined manually with data and graphics in Hubble's 1974 346-page report.

The calm-water, bare-hull resistance prediction procedure was developed in Hubble 1982. It is slightly modified here to calculate R_n using mean wetted length (L_M) to be consistent with data analyzed in Hubble 1974 which is provided in Table A4-D and Figure A4-4.

Table A4-B – Geometric Characteristics of Series 65 and Series 62 Models TABLE A4-D: GEOMETRIC

	Model	L_P ft	A_P ft^2	B_{PA} ft	B_{PT} ft	B_{PX} ft	$\frac{L_P}{B_{PA}}$	$\frac{L_P}{B_{PX}}$	$\frac{B_{PX}}{B_{PA}}$	$\frac{B_{PT}}{B_{PX}}$	$\bar{x}_P$	β_{II} deg	β_T deg
Series 65-A	5251	6.105	8.19	1.341	0.648	1.852	4.55	3.30	1.38	0.35	0.496	16.0	14.8
	5249	8.635	11.58	1.341	0.648	1.852	6.44	4.66	1.38	0.35	0.496	16.0	14.8
	5198	7.302	8.28	1.134	0.548	1.566	6.44	4.66	1.38	0.35	0.496	22.1	20.5
	4966-1	8.635	8.19	0.949	0.458	1.309	9.10	6.60	1.38	0.35	0.496	22.1	20.5
	5204	8.635	5.79	0.671	0.324	0.925	12.87	9.34	1.38	0.35	0.496	22.1	20.5
	5250	8.635	8.19	0.949	0.458	1.309	9.10	6.60	1.38	0.35	0.496	29.9	27.9
	5248	8.635	5.79	0.671	0.324	0.925	12.87	9.34	1.38	0.35	0.496	29.9	27.9
Series 65-B	5237	6.142	11.85	1.929	2.600	2.618	3.18	2.35	1.36	0.99	0.397	21.2	16.3
	5240	6.142	8.38	1.364	1.838	1.852	4.50	3.32	1.36	0.99	0.397	21.2	16.3
	5239	8.687	11.85	1.364	1.838	1.852	6.36	4.69	1.36	0.99	0.397	21.2	16.3
	5186	6.142	8.38	1.364	1.838	1.852	4.50	3.32	1.36	0.99	0.397	28.7	22.5
	5184	6.142	5.92	0.965	1.300	1.309	6.36	4.69	1.36	0.99	0.397	28.7	22.5
	5167	8.687	8.38	0.965	1.300	1.309	9.00	6.64	1.36	0.99	0.397	28.7	22.5
	5236	6.142	5.92	0.965	1.300	1.309	6.36	4.69	1.36	0.99	0.397	37.4	30.4
	5208	8.687	8.38	0.965	1.300	1.309	9.00	6.64	1.36	0.99	0.397	37.4	30.4
	5238	8.687	5.92	0.682	0.919	0.925	12.73	9.38	1.36	0.99	0.397	37.4	30.4
Series 62	4665	3.912	6.47	1.654	1.565	1.956	2.36	2.00	1.18	0.80	0.475	13.0	12.5
	4666	5.987	9.72	1.623	1.386	1.956	3.69	3.06	1.21	0.71	0.482	13.0	12.5
	4667-1	8.000	12.80	1.600	1.250	1.956	5.00	4.09	1.22	0.64	0.488	13.0	12.5
	4668	8.000	9.52	1.190	0.934	1.455	6.72	5.50	1.22	0.64	0.488	13.0	12.5
	4669	8.000	7.48	0.935	0.734	1.143	8.56	7.00	1.22	0.64	0.488	13.0	12.5
	DL 62-A	2.994	2.38	0.795	0.958	0.978	3.77	3.06	1.23	0.98	0.424	20.0	13.0

Notes:

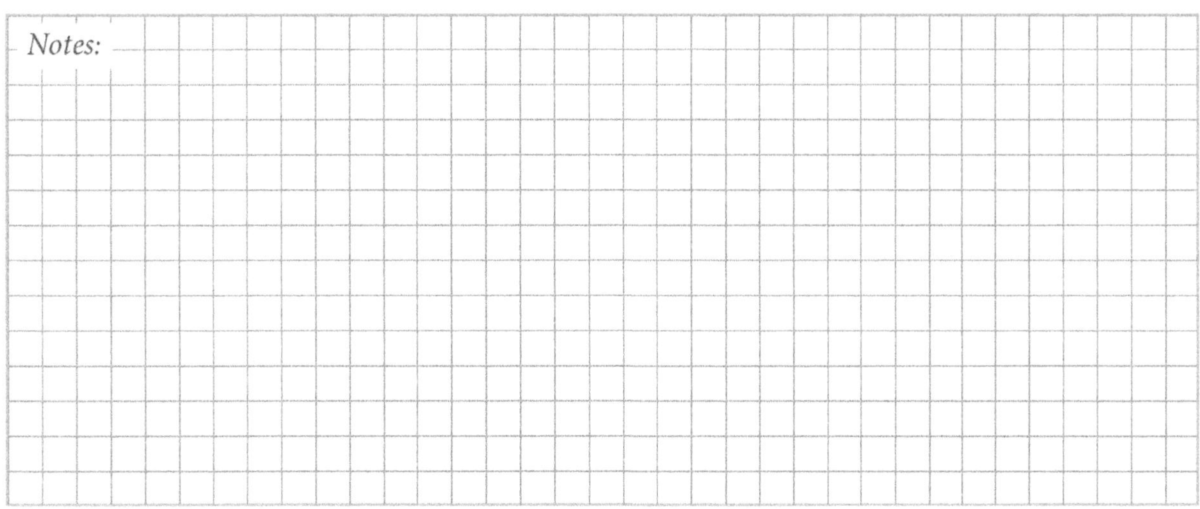

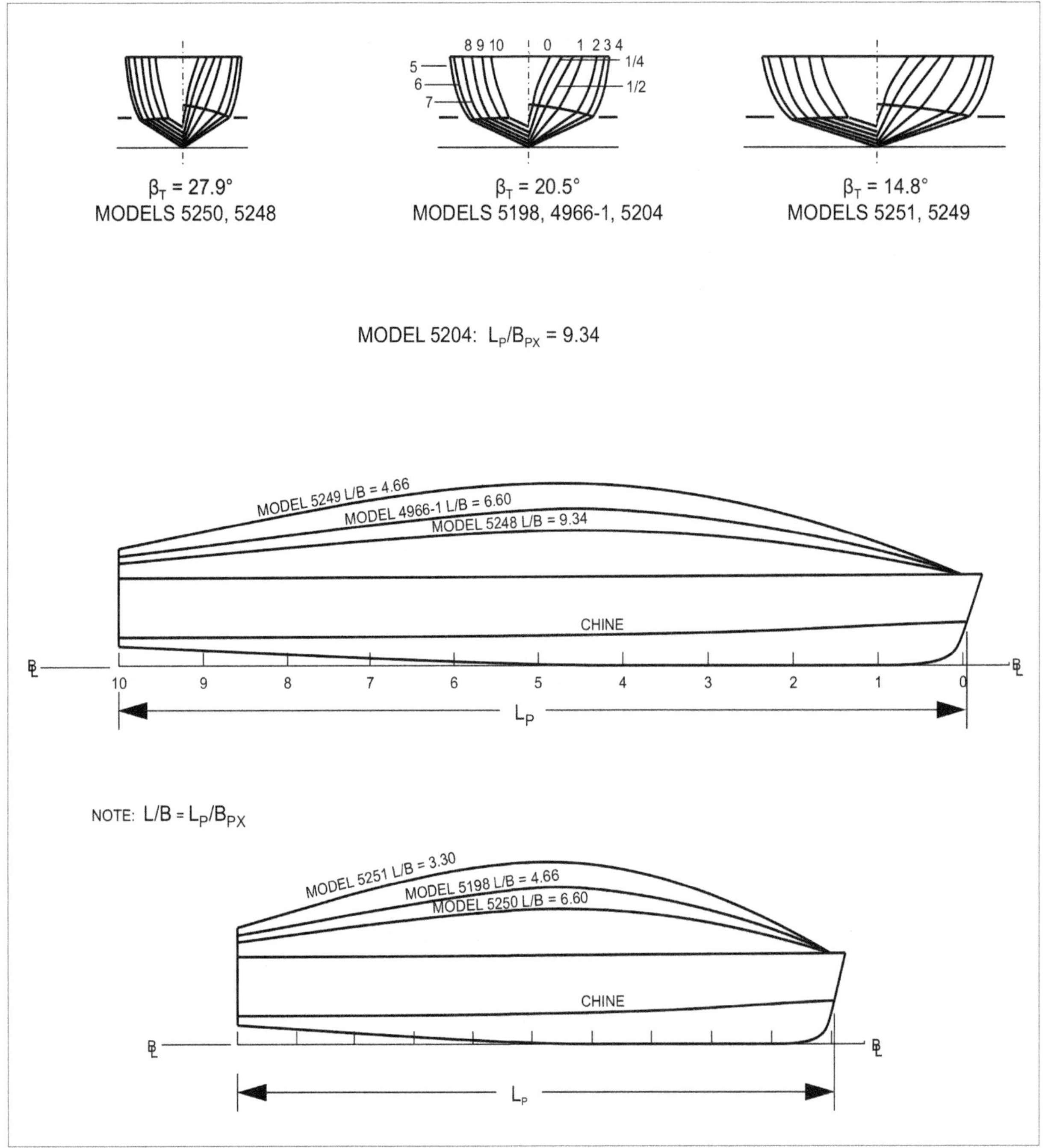

Figure A4-1:Lines and body plans of Series 65-A hulls

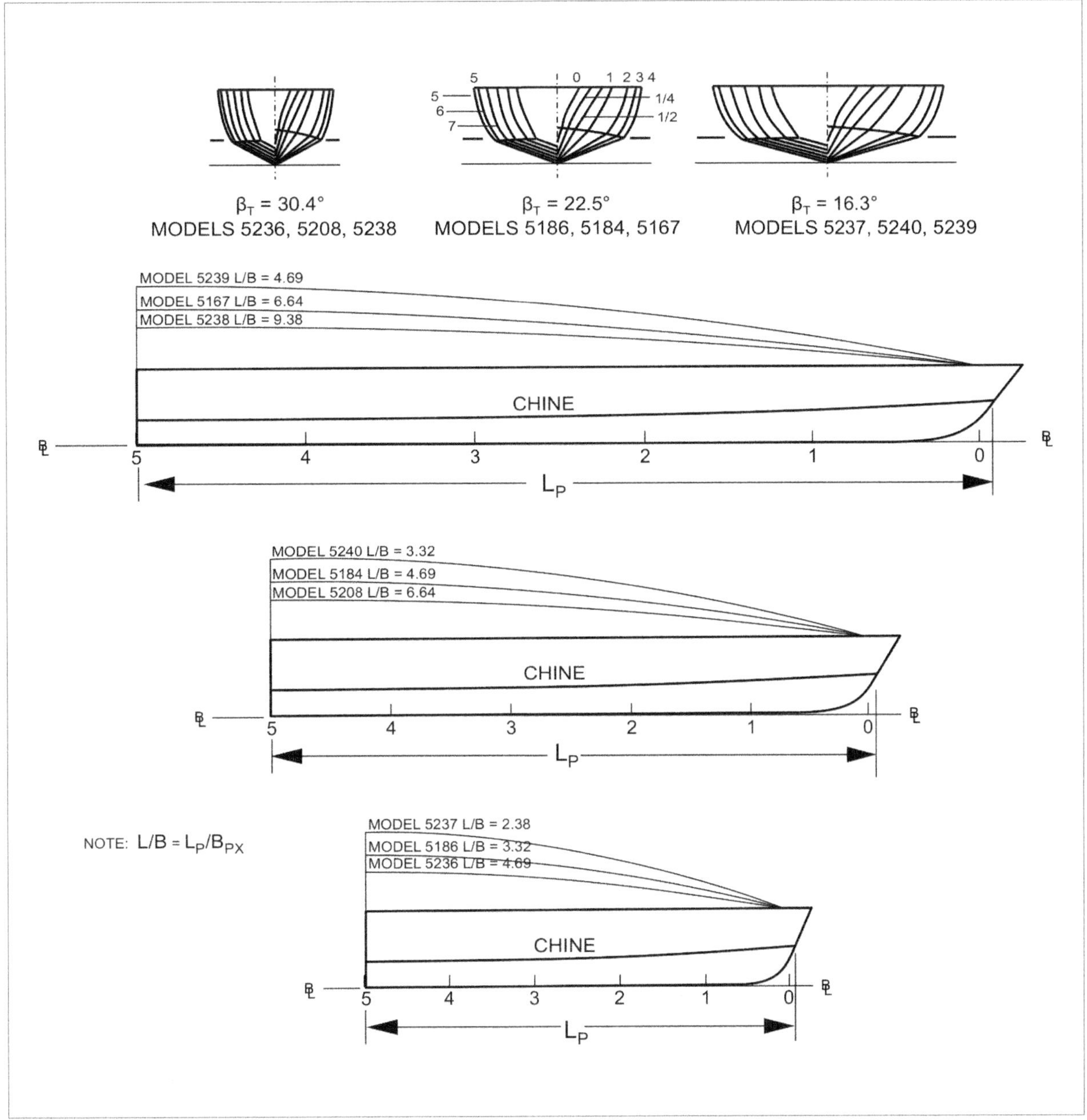

Figure A4-2: Lines and body plans of Series 65-B hulls

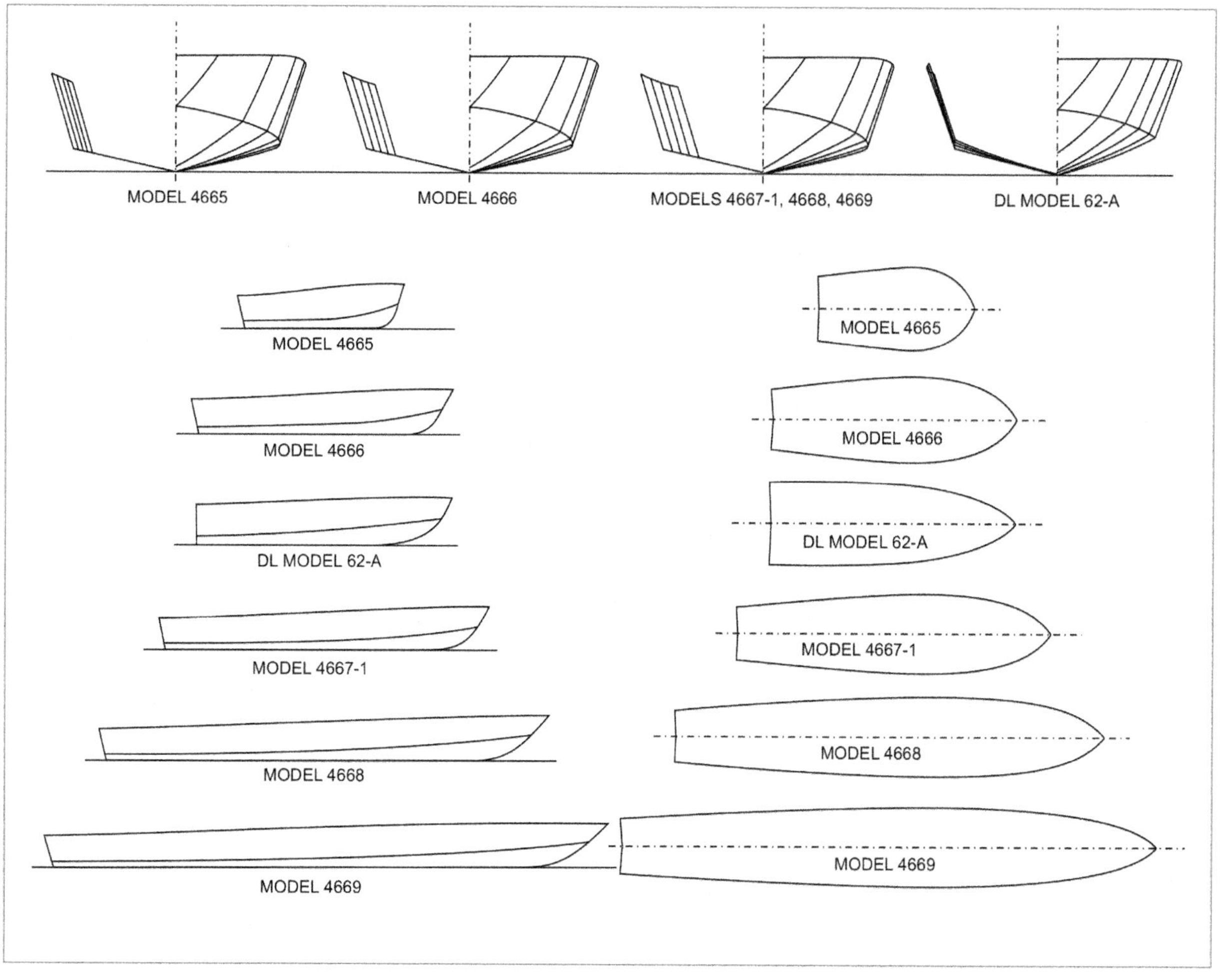

MODEL 4665
MODEL 4666
MODELS 4667-1, 4668, 4669
DL MODEL 62-A

MODEL 4665
MODEL 4666
DL MODEL 62-A
MODEL 4667-1
MODEL 4668
MODEL 4669

Figure A4-3: – Lines and body plans of Series 62 hulls and DL Model 62-A

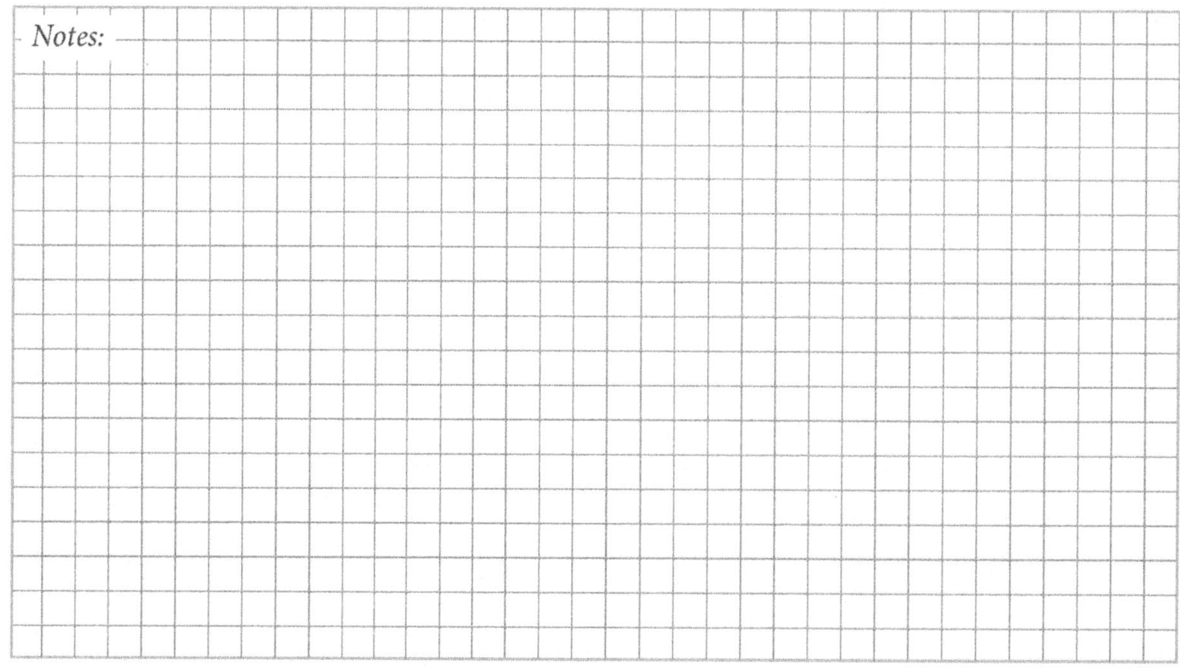

TABLE A4-C: MEAN VALUE OF R/W FOR 100,000 LB (45.4 MT) HARD-CHINE CRAFT

SPEED (KNOTS)	$F_{n\nabla}$	L_P (FT) 46.4 $L_P/\nabla^{1/3}$ 4.0	52.2 4.5	58.0 5.0	63.8 5.5	69.6 6.0	75.4 6.5	81.2 7.0
0.00	0.00	0.0000	0.0000	0.0000	0.0000	0.0000	0.0000	0.0000
5.72	0.50	0.0120	0.0100	0.0085	0.0075	0.0070	0.0065	0.0060
8.59	0.75	0.0420	0.0345	0.0280	0.0235	0.0200	0.0170	0.0150
11.45	1.00	0.1050	0.0875	0.0715	0.0580	0.0480	0.0405	0.0350
14.31	1.25	0.1800	0.1420	0.1140	0.0940	0.0795	0.0675	0.0585
17.17	1.50	0.1980	0.1550	0.1255	0.1065	0.0930	0.0815	0.0730
20.03	1.75	0.1995	0.1602	0.1350	0.1165	0.1025	0.0910	0.0820
22.89	2.00	0.1900	0.1630	0.1430	0.1275	0.1135	0.1020	0.0930
25.76	2.25	0.1775	0.1642	0.1505	0.1375	0.1260	0.1150	0.1060
28.62	2.50	0.1690	0.1645	0.1575	0.1475	0.1375	0.1280	0.1200
31.48	2.75		0.1620	0.1610	0.1550	0.1480	0.1405	0.1330
34.34	3.00			0.1610	0.1590	0.1565	0.1520	0.1465
37.20	3.25				0.1590	0.1595	0.1600	0.1585
40.06	3.50					0.1610	0.1665	0.1695
42.93	3.75						0.1735	0.1795
45.79	4.00							0.1890

SPEED (KNOTS)	$F_{n\nabla}$	87.0 7.5	92.8 8.0	104.4 9.0	116.0 10.0	Standard Deviation σ
0.00	0.00	0.0000	0.0000	0.0000	0.0000	0.0000
5.72	0.50	0.0057	0.0055	0.0050	0.0045	0.0065
8.59	0.75	0.0135	0.0125	0.0110	0.0100	0.0080
11.45	1.00	0.0305	0.0270	0.0220	0.0190	0.0089
14.31	1.25	0.0510	0.0450	0.0360	0.0305	0.0095
17.17	1.50	0.0660	0.0600	0.0500	0.0425	0.0100
20.03	1.75	0.0755	0.0700	0.0610	0.0530	0.0106
22.89	2.00	0.0855	0.0795	0.0705	0.0630	0.0112
25.76	2.25	0.0985	0.0915	0.0815	0.0745	0.0121
28.62	2.50	0.1125	0.1060	0.0950	0.0880	0.0132
31.48	2.75	0.1270	0.1210	0.1110	0.1040	0.0148
34.34	3.00	0.1415	0.1365	0.1280	0.1205	0.0170
37.20	3.25	0.1560	0.1530	0.1465	0.1400	0.0199
40.06	3.50	0.1700	0.1700	0.1670	0.1620	0.0231
42.93	3.75	0.1825	0.1840	0.1850	0.1830	0.0266
45.79	4.00	0.1930	0.1960	0.2005	0.2030	0.0300

Notes:

1. From data published in Hubble 1974 for Series 62 and 65 model tests
2. LCG ranging from 1/3 to 1/2 (L_P) forward of transom

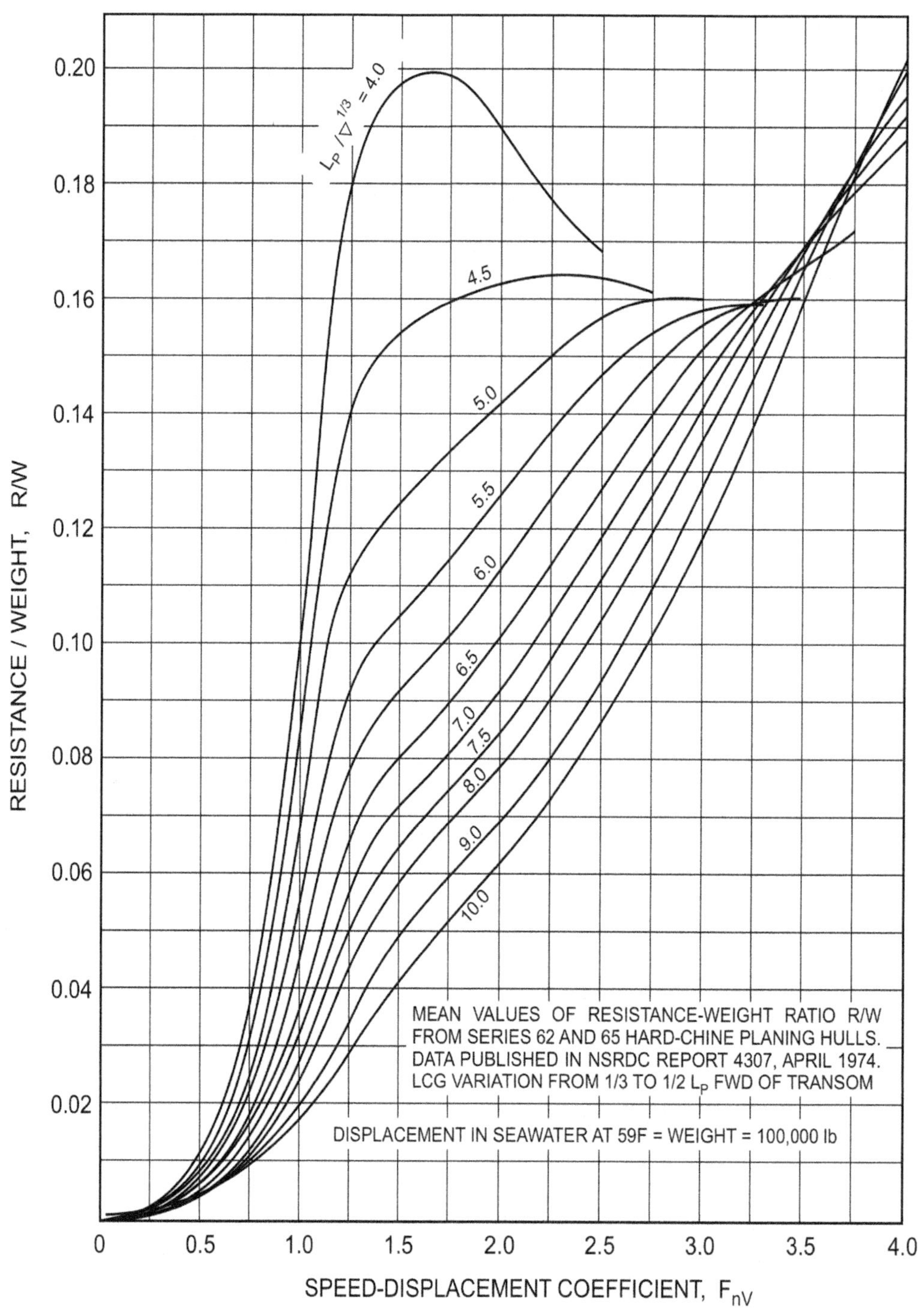

Note: Data scaled using ATTC 1947 friction formulation with zero correction allowance.

Figure A4-4: Mean values of R/W from Series 62 and 65 data

TABLE A4-D – MEAN VALUES OF $S/\nabla^{2/3}$ FOR HARD-CHINE CRAFT

$F_{n\nabla}$ \ $L_p/\nabla^{1/3}$	4.0	4.5	5.0	5.5	6.0	6.5	7.0	7.5	8.0	9.0	10.0
0.00	5.80	6.15	6.50	6.85	7.20	7.55	7.90	8.25	8.60	9.30	10.00
0.50	5.95	6.33	6.70	7.07	7.43	7.77	8.09	8.42	8.75	9.42	10.10
0.75	5.99	6.38	6.77	7.15	7.50	7.85	8.18	8.50	8.82	9.48	10.15
1.00	5.99	6.40	6.80	7.20	7.57	7.90	8.23	8.56	8.88	9.54	10.21
1.25	5.92	6.37	6.80	7.22	7.60	7.93	8.27	8.61	8.93	9.60	10.28
1.50	5.76	6.29	6.78	7.21	7.60	7.95	8.30	8.65	8.97	9.65	10.34
1.75	5.51	6.16	6.72	7.17	7.59	7.94	8.29	8.67	9.00	9.70	10.41
2.00	5.20	5.97	6.59	7.08	7.54	7.92	8.27	8.65	9.01	9.75	10.48
2.25	4.76	5.70	6.41	6.97	7.46	7.85	8.23	8.62	9.01	9.78	10.55
2.50	4.20	5.37	6.18	6.81	7.35	7.75	8.15	8.56	8.99	9.80	10.62
2.75		4.95	5.89	6.60	7.17	7.61	8.04	8.48	8.94	9.80	10.68
3.00			5.55	6.35	6.94	7.42	7.89	8.37	8.85	9.79	10.75
3.25				6.06	6.65	7.17	7.68	8.21	8.73	9.76	10.80
3.50					6.30	6.87	7.43	8.01	8.58	9.71	10.85
3.75						6.53	7.10	7.75	8.37	9.62	10.88
4.00							6.70	7.40	8.10	9.50	10.90

Notes:

1. From data published in Hubble 1974 for Series 62 and 65 model tests
2. LCG ranging from 1/3 to 1/2 (L_p) forward of transom

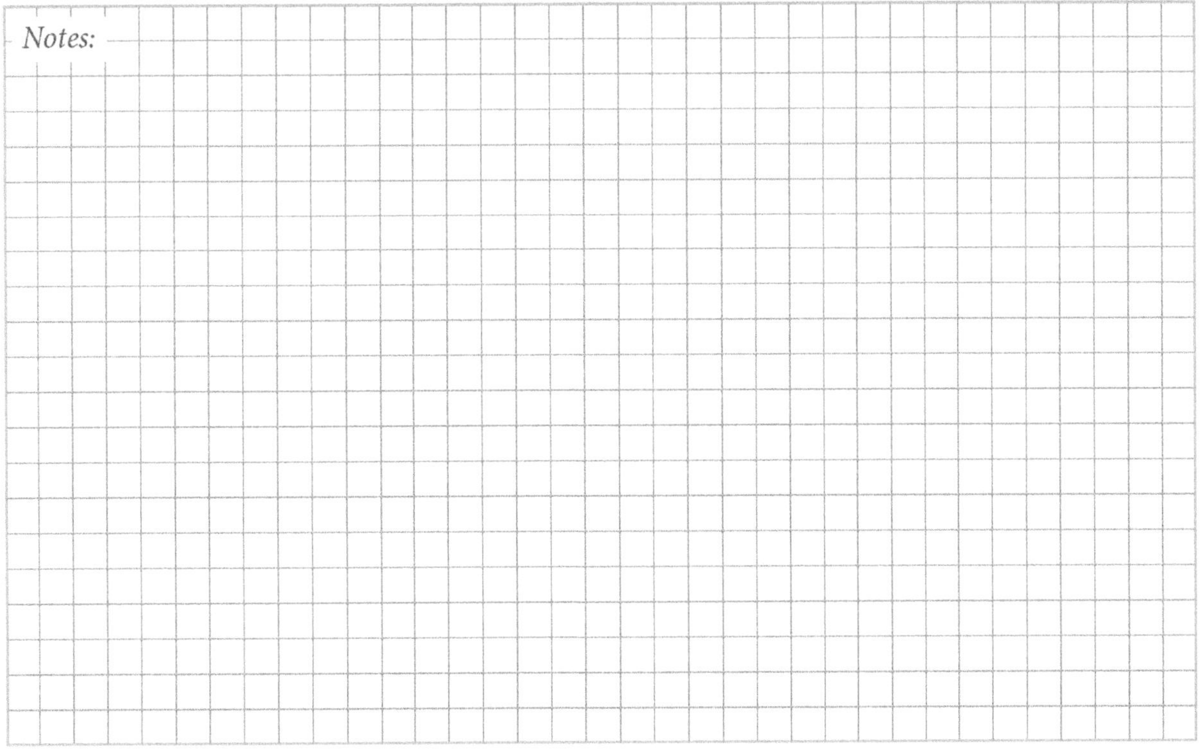

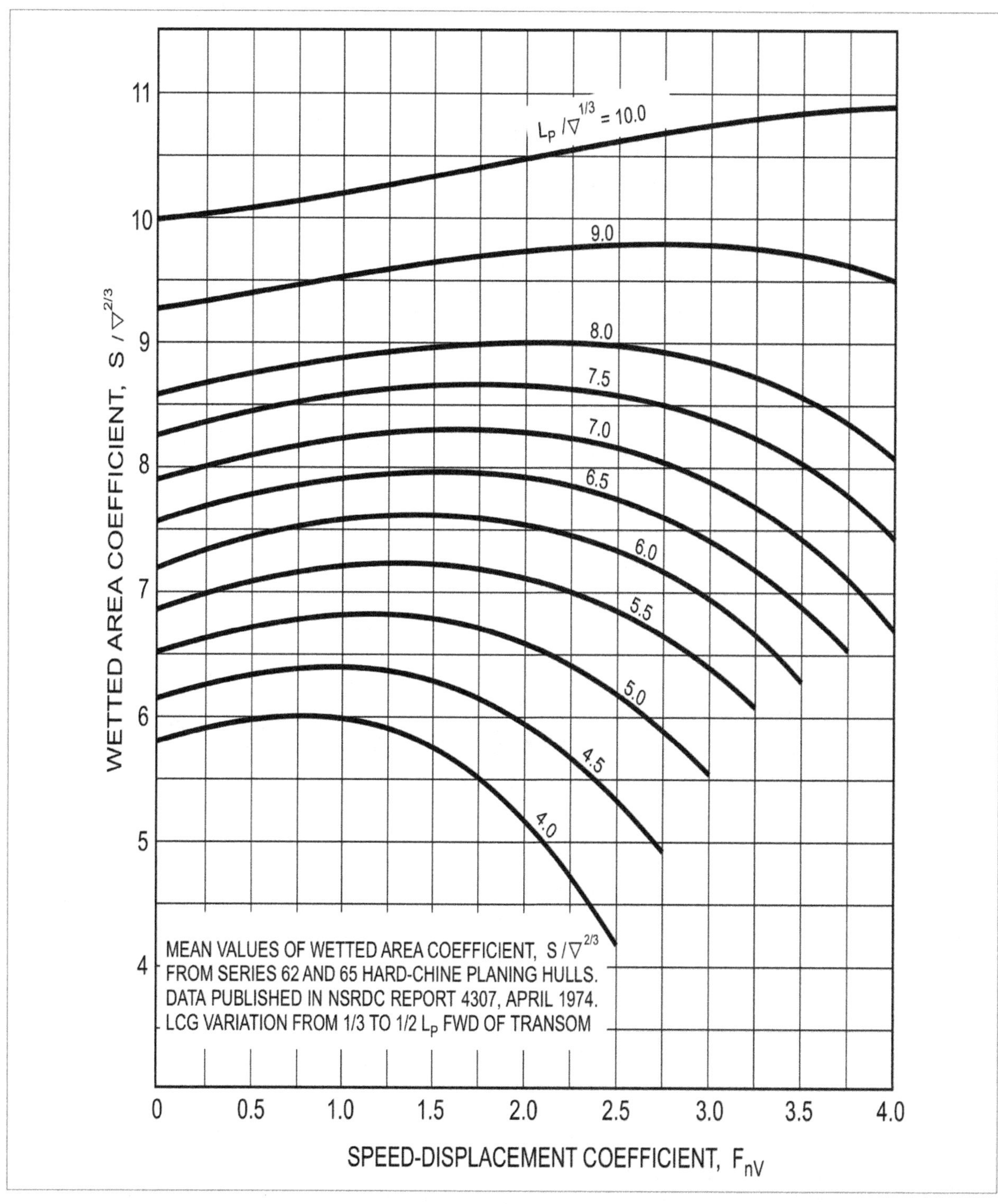

Figure A4-5: Mean value of wetted area coefficient, $S/\nabla^{2/3}$ from Series 62 and 65 data

Calculation Procedure with Example

The procedure defined in steps 1 to 24 provides the calculation steps which yield full-scale resistance versus speed predictions for the actual vessel using the ATTC 1947 friction formulation for scaling data. Continuing calculations from Steps 25 to 34 provides full-scale approximate resistance versus speed predictions for the actual vessel using the ITTC 1957 friction formulation for scaling data.

Table 4A-E: INPUT FOR CALCULATION EXAMPLE FOR TWO VESSELS			
Input for example (59°F salt water)		**Large Vessel**	**Small Vessel**
Vessel displacement, W in lb.	W	1.1025×10^6 lb (500 mt)	50,000 lb
Projected chine length, L_P in ft	L_P	155 ft	50.6 ft
Maximum speed, V in kt	V	26 kt	40 kt
Select standard deviation factor (SDF): 0 for mean R/W, 1.645 for minimum R/W, or an interim value to improve resistance prediction accuracy as found by correlation study with experimental model test data.	SDF	0	1.645
Calculate input values: $L_P/\nabla^{1/3}$ & F_{nV}	$L_P/\nabla^{1/3}$ F_{nV}	6.0 1.52	5.5 3.92
		Desired Output for Example	
Desired output: $V = F_{nV} (W)^{1/6}/0.5933$ in knots $R = W$ (mean R/W – SDF X σ) for each F_{nV}	F_{nV} $L_P/\nabla^{1/3}$	1.25 6.0	3.00 5.5

TABLE A4-F: CALCULATION PROCEDURE WITH EXAMPLES FOR TWO VESSELS				
PROCEDURE				
1	From Table A4-C prepare a matrix of mean R/W as a function of F_{nV} and $L_P/\nabla^{1/3}$			
2	From Table A4-C prepare an array of σ as a function of F_{nV}			
3	From Table A4-D prepare a matrix of mean wetted area coefficients $S/\nabla^{2/3}$ as a function of F_{nV} and $L_P/\nabla^{1/3}$			
4	For each F_{nV} of the mean R/W matrix fit a curve through the data or interpolate between three $L_P/\nabla^{1/3}$ values bounding the input $L_P/\nabla^{1/3}$ to obtain the mean R/W.	Mean R/W	0.0795	0.1590
5	For each F_{nV} and corresponding $L_P/\nabla^{1/3}$ obtain σ by curve fit of the array or interpolation.	σ	0.095	0.0170
6	For each F_{nV} using σ to statistically adjust R/W to another desired value other than mean. Adjusted R/W = mean R/W – SDF x σ.	Adjusted R/W	0.0795	0.1310
7	For each F_{nV} of the mean wetted area coefficients $S/\nabla^{2/3}$ matrix fit a curve through the data or interpolate between three $L_P/\nabla^{1/3}$ values bounding the input $L_P/\nabla^{1/3}$ to obtain the mean value of $S/\nabla^{2/3}$.	$S/\nabla^{2/3}$	7.60	6.35
8	Calculate the linear ratio, for the actual vessel. $\lambda = (W/100{,}000)^{1/3}$	λ	2.226	0.7937
9	Calculate speed of 100,000 lb vessel in ft/sec $V_m = 19.32(F_{nV})$	V_m	24.15	57.96
10	Calculate speed of actual vessel in ft/sec $V_f = V_M(\lambda^{1/2})$	V_f	36.03	51.64
11	Calculate length of 100,000 lb vessel in ft $L_{P,M} = 11.604$ (input $L_P/\nabla^{1/3}$)	$L_{P,M}$	69.6	63.8
12	Calculate mean wetted length of 100,000 lb vessel in ft $L_{M,M} = L_{P,M}\,(e)^{-0.23\,FnV}$	$L_{M,M}$	52.2	32.0
13	Calculate length of actual vessel in ft $L_{P,f} = L_{P,M}(\lambda)$	$L_{P,f}$	154.9	50.6
14	Calculate Reynold's number of 100,000 lb vessel $R_{n,M} = V_M L_{M,M}/\nu_M$ Note: $\nu_M = \nu_f = 1.279 \times 10^{-5}$ ft²/sec for 59°F salt water	$R_{n,M}$	$0.9857\mathrm{x}10^8$	$1.450\mathrm{x}10^8$

15	Calculate Reynold's number of actual vessel $R_{n,f} = R_{n,M} (\lambda^{3/2})(v_M/v_f)$	$R_{n,f}$	3.274x10[8]	1.025x10[8]
16	Calculate frictional resistance coefficients using ATTC 1947 formulation for 100,000 lb craft. $C_{F,M}$ and actual vessel, $C_{F,f}$ An explicit equation approximating ATTC 1947 is given in Chapter 3, Equation 3-7.	$C_{F,M}$ $C_{F,f}$	2.076×10^{-3} 1.764×10^{-3}	1.968×10^{-3} 2.065×10^{-3}
17	Calculate wetted area of 100,000 lb vessel in ft^2 $S_M = 134.65 (S/\nabla^{2/3})$	S_M	1,023	855.0
18	Calculate wetted area of actual vessel in ft^2 $S_f = S_M (\lambda^2)$	S_f	5,069	538.6
19	Calculate total bare hull resistance of 100,000 lb vessel in lb $R_M = (R/W)\ 100{,}000$ Note: R/W is from Procedure 6	R_M	7,950	13,100
20	Calculate total bare hull resistance coefficient of 100,000 lb craft $C_{T,M} = R_M/(v_M^2\ S_M\ \rho_M/2)$ Note: $\rho_M = \rho_f = 1.9905$ lb sec^2/ft^4 for 59° salt water	$C_{T,M}$	13.388×10^{-3}	4.583×10^{-3}
21	Calculate residual resistance coefficient of 100,000 lb craft $C_R = C_{T,M} - C_{F,M}$	$C_{R/ATTC}$	11.312×10^{-3}	2.615×10^{-3}
22	Calculate total resistance coefficient of actual vessel $C_{T,f} = C_{F.f} + C_R + C_A$ Note: See the discussion in Chapter 3 about considerations for use of correlation allowance, C_A, or as an alternative use of SDF values between 0 and 1.645 may be obtained by correlating this prediction procedure with model test resistance data for specific designs scaled to a vessel displacement of 100,000 lb.	$C_{T,f}$ For $C_A = 0$	13.076×10^{-3}	4.680×10^{-3}
23	Calculate total bare hull resistance of actual vessel in lb $R = C_{T,f}\ v_f^2\ S_f\ \rho_f/2$	R	85,636	6,690
24	Calculate actual vessel speed in knots $V = V_f/1.6878$	V	21.3	30.6

25	Predict total resistance of actual vessel assuming model test data had been analyzed using ITTC 1957 frictional resistance coefficients.			
26	Calculate approximate R_n of tow tank model $R_n = [0.85 + 0.8\ F_{nV}]10^6$	R_n	1.85×10^6	3.25×10^6
27	Calculate $C_{F/ITTC}$ for R_n in Procedure 26	$C_{F/ITTC}$	4.119×10^{-3}	3.684×10^{-3}
28	Calculate $C_{F/ATTC}$ for R_n in Procedure 26	$C_{F/ATTC}$	3.927×10^{-3}	3.549×10^{-3}
29	$\delta C_F = [C_{F/ATTC} - C_{F/ITTC}]$ Reference Table A1-6	δC_F	-0.192×10^{-3}	-0.135×10^{-3}
30	$C_{R/ITTC} = C_{R/ATTC} + \delta C_F$ $C_{R/ATTC}$ from Procedure 2	$C_{R/ITTC}$	11.120×10^{-3}	2.480×10^{-3}
31	Calculate $C_{F,f/ITTC}$ for actual vessel using $R_{n,f}$ from Procedure 15	$C_{F,f/ITTC}$	1.767×10^{-3}	2.076×10^{-3}
32	Calculate total resistance coefficient for actual vessel using ITTC friction scaling $C_{T,f/ITTC} = C_{F,f/ITTC} + C_{R/ITTC} + C_A$	$C_{T,f/ITTC}$ for $C_A=0$	12.887×10^{-3}	4.556×10^{-3}
33	Calculate total resistance in lb for actual vessel using ITTC friction scaling $R_{T/ITTC} = (C_{T/f/ITTC})v_f^2 S_f \rho_f/2$	$R_{T/ITTC}$	84,398	6,513
34	Repeat actual vessel speed in knots from Procedure 24	V	21.3	30.6

APPENDIX 5

Gawn-Burrill Propeller Characteristics

Formatted η_0 and J_T Versus K_T/J_T^2

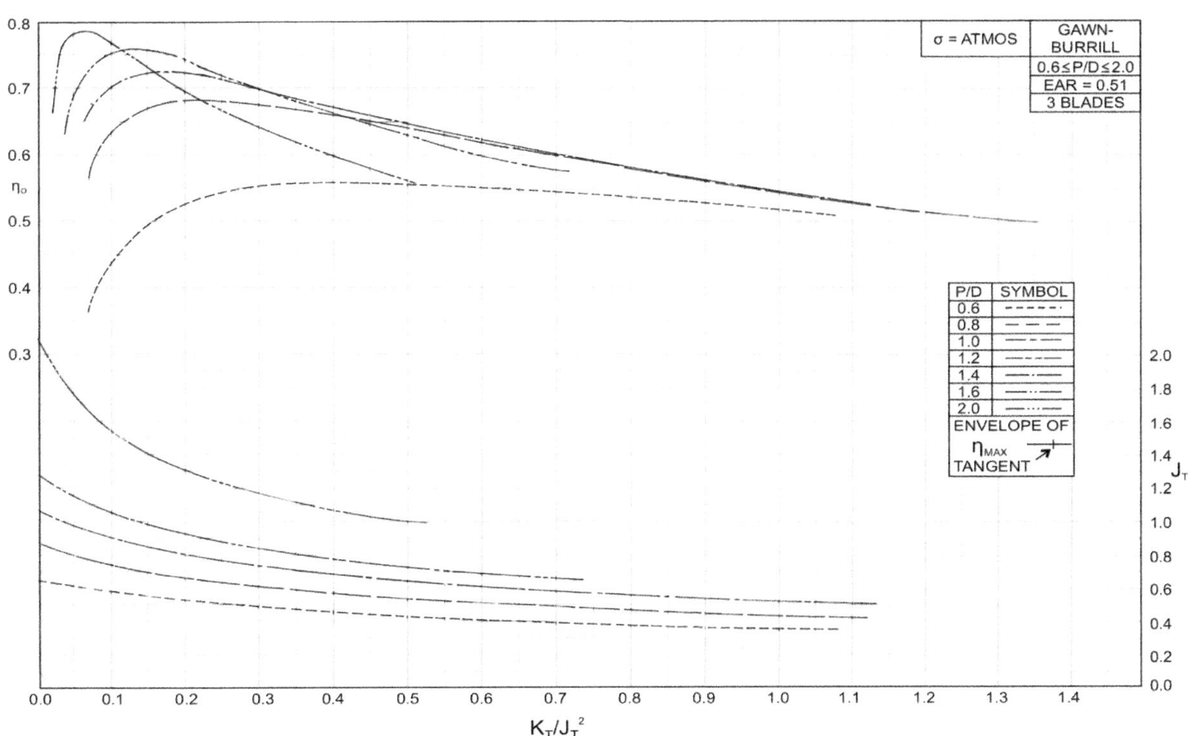

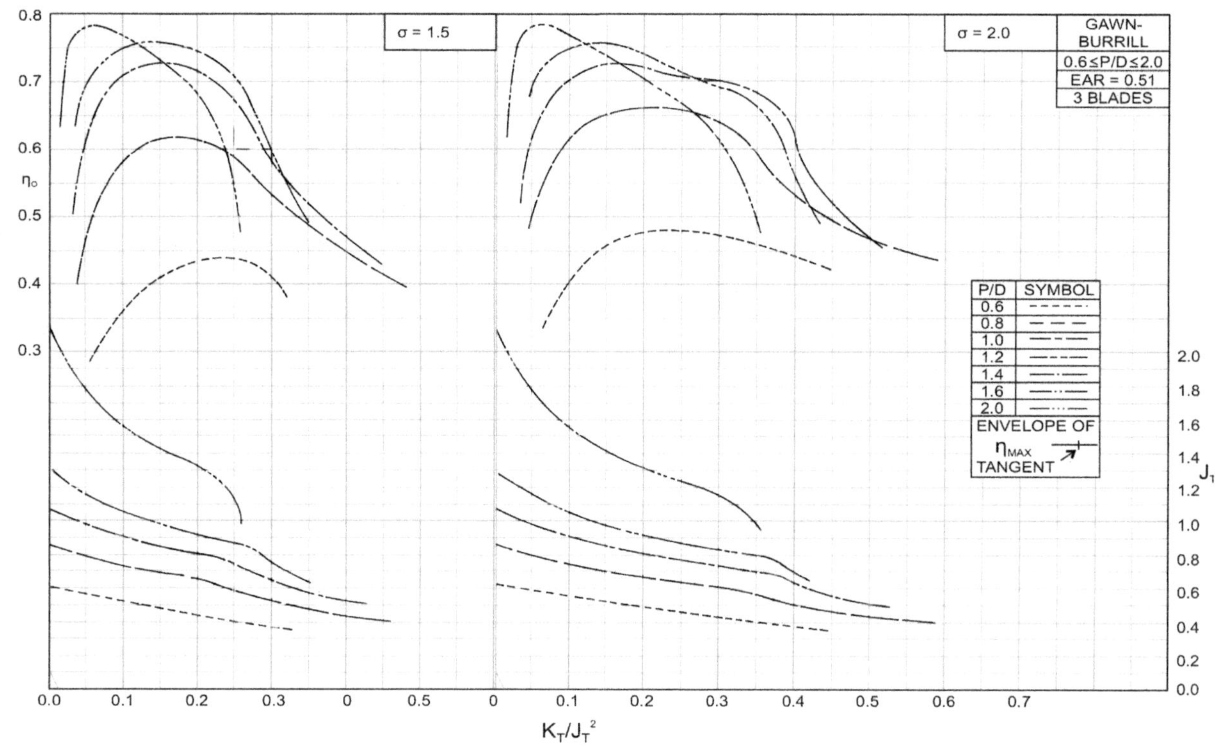
σ = 1.5
σ = 2.0
GAWN-BURRILL
0.6≤P/D≤2.0
EAR = 0.51
3 BLADES
P/D SYMBOL
0.6
0.8
1.0
1.2
1.4
1.6
2.0
ENVELOPE OF
η_{MAX}
TANGENT
η_o
J_T
K_T/J_T^2

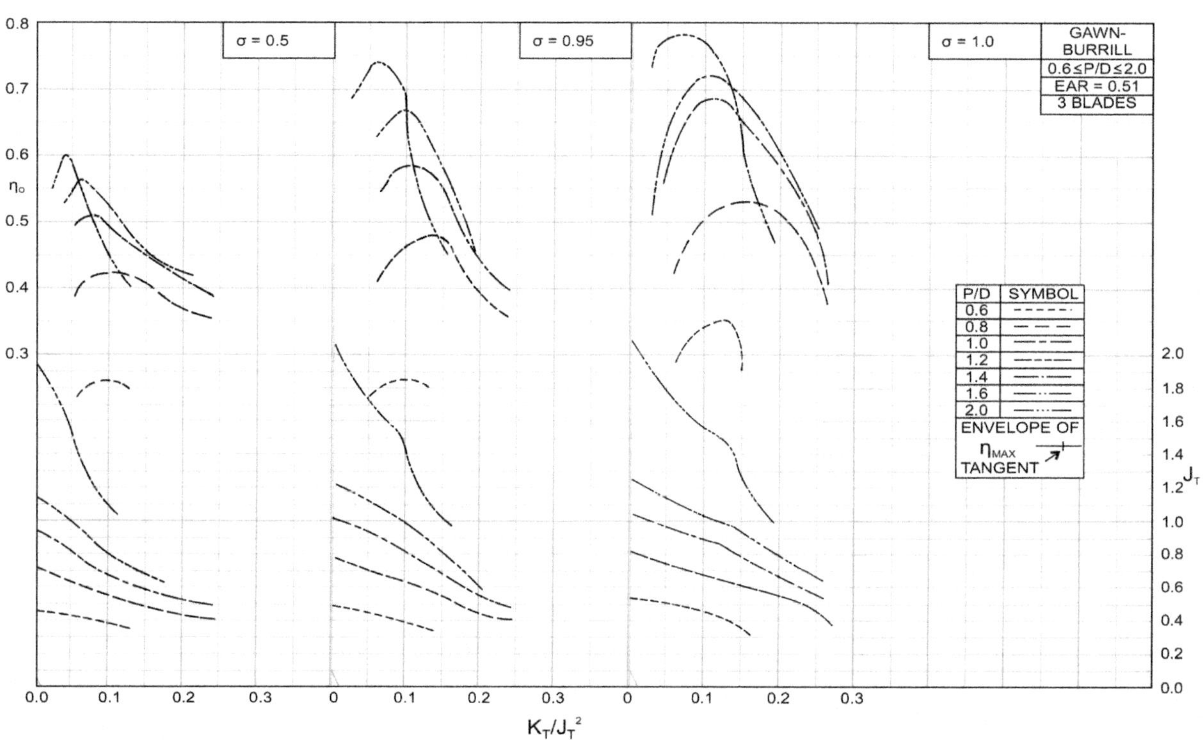
σ = 0.5
σ = 0.95
σ = 1.0
GAWN-BURRILL
0.6≤P/D≤2.0
EAR = 0.51
3 BLADES
P/D SYMBOL
0.6
0.8
1.0
1.2
1.4
1.6
2.0
ENVELOPE OF
η_{MAX}
TANGENT
η_o
J_T
K_T/J_T^2

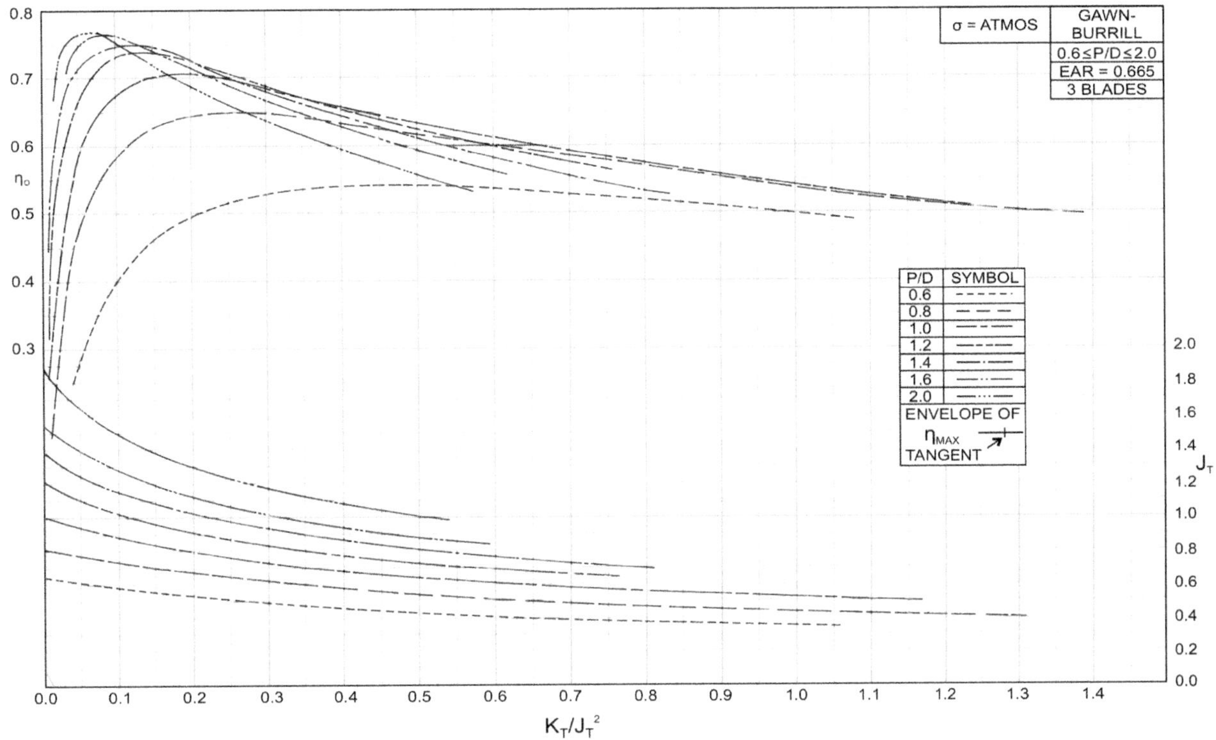
σ = ATMOS
GAWN-BURRILL
0.6≤P/D≤2.0
EAR = 0.665
3 BLADES
P/D SYMBOL
0.6
0.8
1.0
1.2
1.4
1.6
2.0
ENVELOPE OF ηMAX TANGENT
ηo
JT
KT/JT²

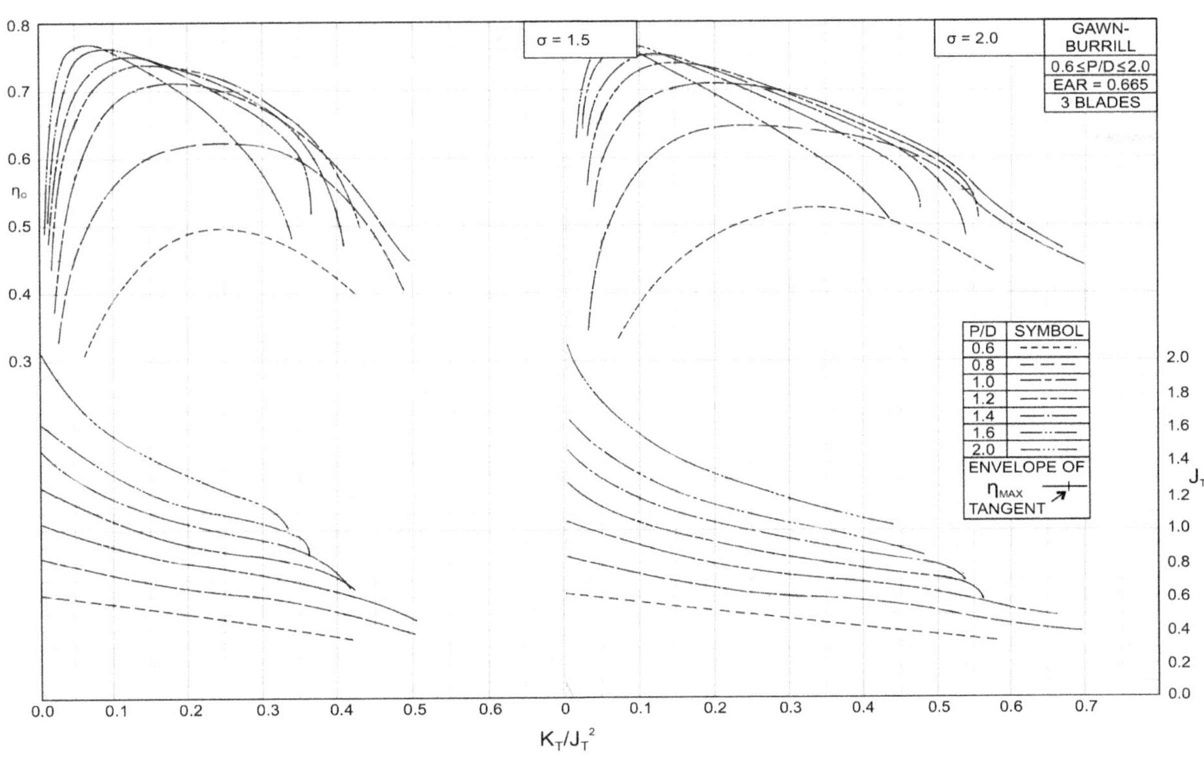
σ = 1.5
σ = 2.0
GAWN-BURRILL
0.6≤P/D≤2.0
EAR = 0.665
3 BLADES
P/D SYMBOL
0.6
0.8
1.0
1.2
1.4
1.6
2.0
ENVELOPE OF ηMAX TANGENT
ηo
JT
KT/JT²

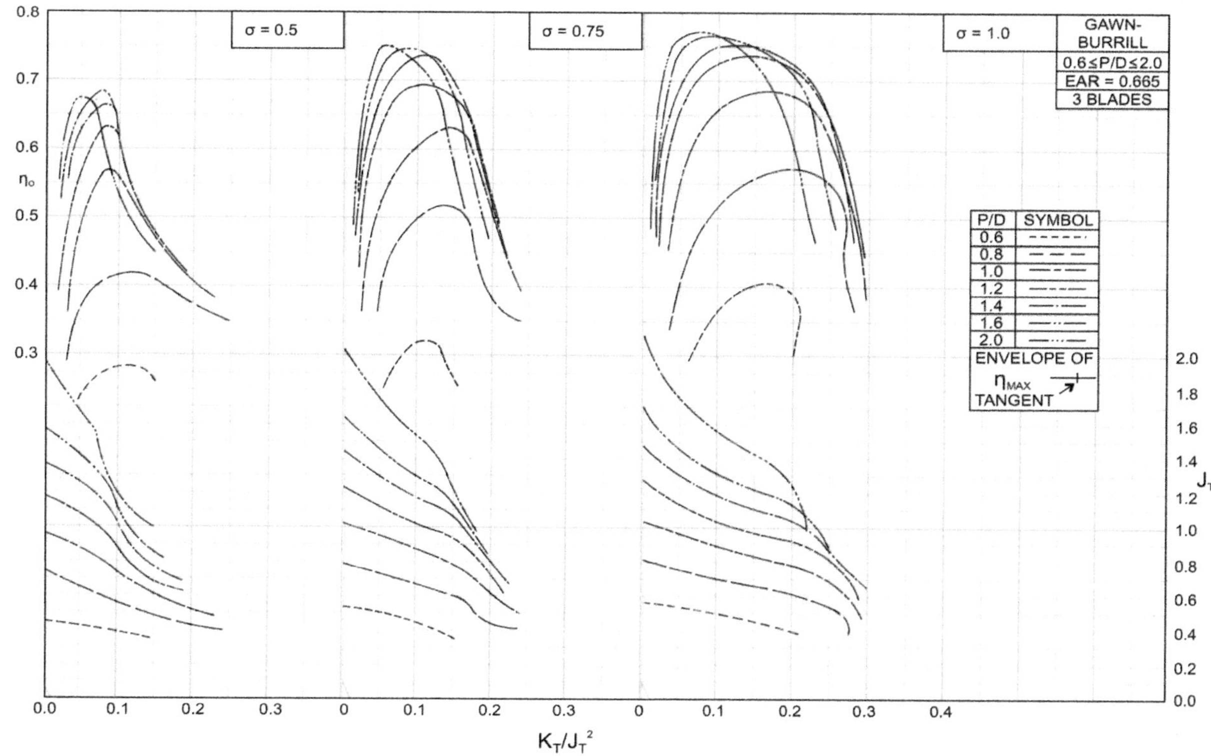
σ = 0.5
σ = 0.75
σ = 1.0
GAWN-BURRILL
0.6≤P/D≤2.0
EAR = 0.665
3 BLADES
P/D SYMBOL
0.6
0.8
1.0
1.2
1.4
1.6
2.0
ENVELOPE OF
η_{MAX} TANGENT
η_o
J_T
K_T/J_T²

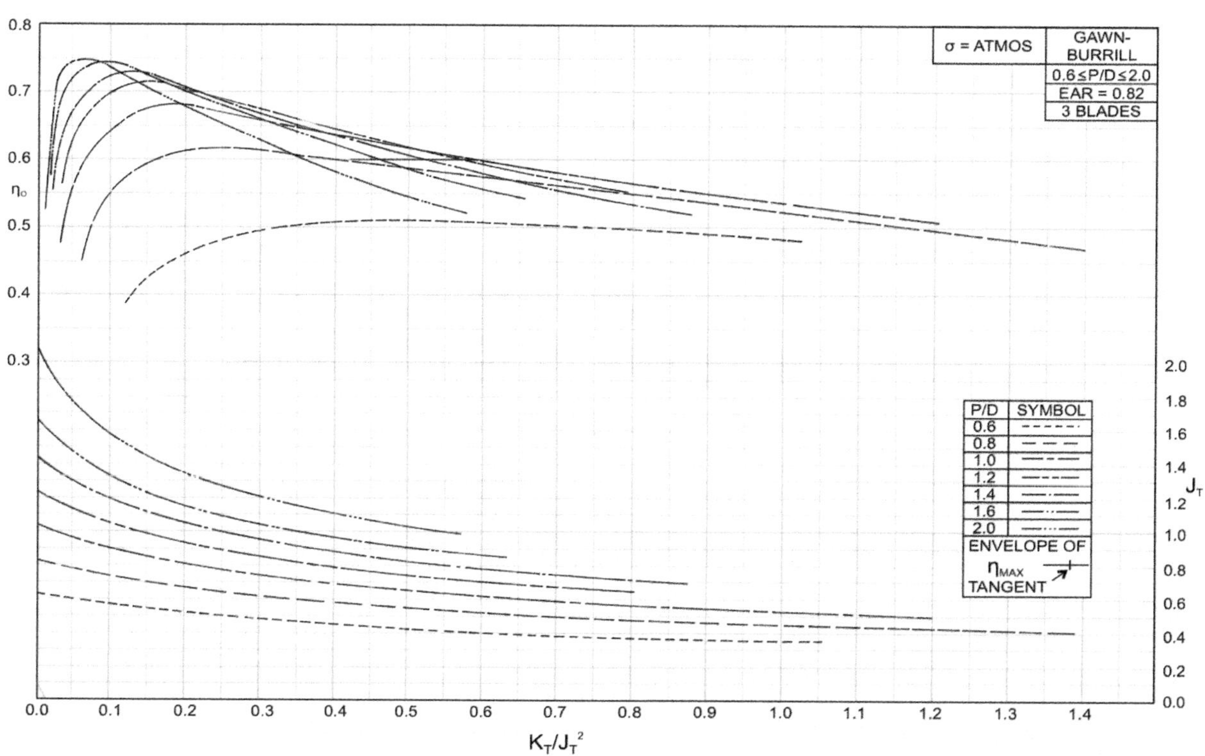
σ = ATMOS
GAWN-BURRILL
0.6≤P/D≤2.0
EAR = 0.82
3 BLADES
P/D SYMBOL
0.6
0.8
1.0
1.2
1.4
1.6
2.0
ENVELOPE OF
η_{MAX} TANGENT
η_o
J_T
K_T/J_T²

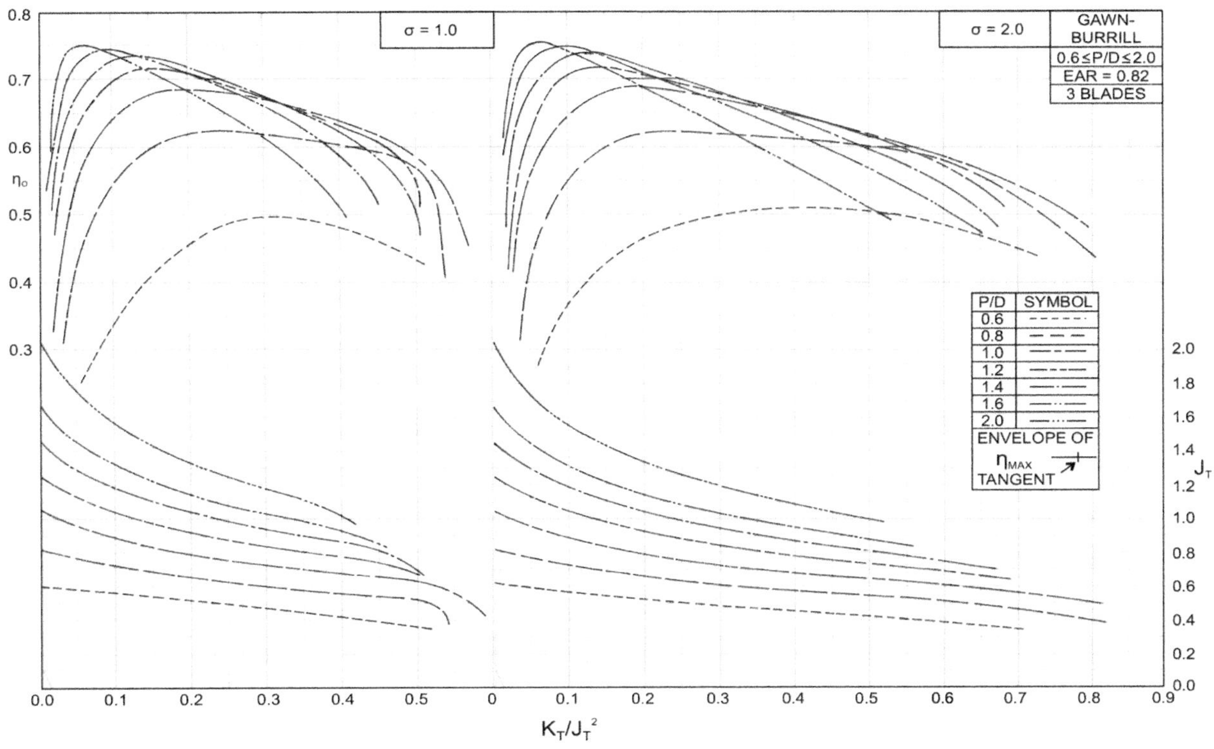
σ = 1.0
σ = 2.0
GAWN-BURRILL
0.6≤P/D≤2.0
EAR = 0.82
3 BLADES
P/D SYMBOL
0.6
0.8
1.0
1.2
1.4
1.6
2.0
ENVELOPE OF η_{MAX} TANGENT
η_o
J_T
K_T/J_T^2
0.0 0.1 0.2 0.3 0.4 0.5 0.6 0.7 0.8 0.9

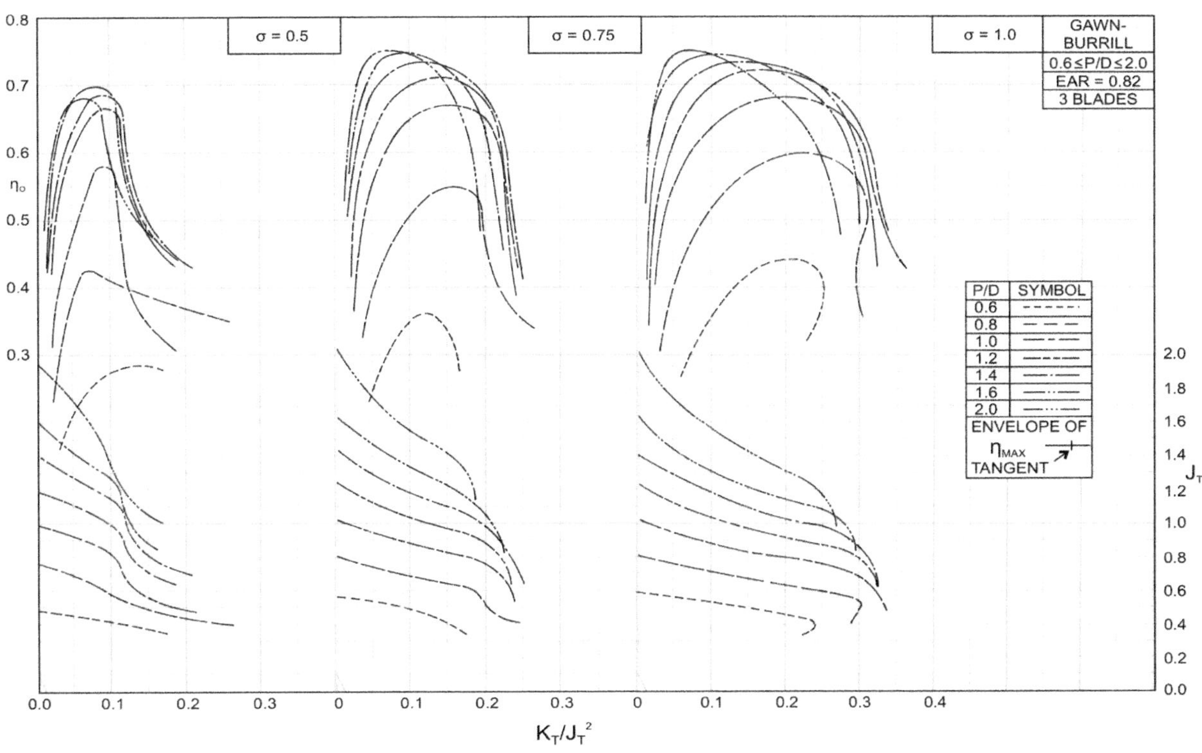
σ = 0.5
σ = 0.75
σ = 1.0
GAWN-BURRILL
0.6≤P/D≤2.0
EAR = 0.82
3 BLADES
P/D SYMBOL
0.6
0.8
1.0
1.2
1.4
1.6
2.0
ENVELOPE OF η_{MAX} TANGENT
η_o
J_T
K_T/J_T^2

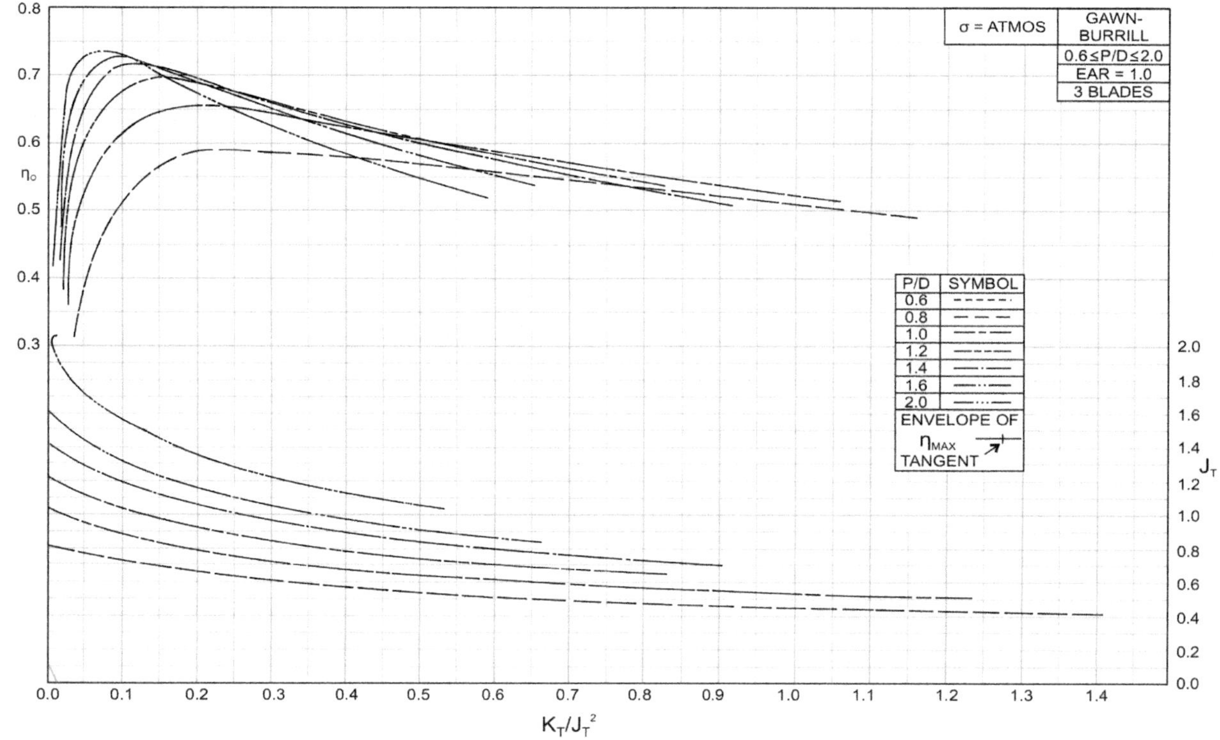
σ = ATMOS
GAWN-BURRILL
0.6≤P/D≤2.0
EAR = 1.0
3 BLADES
η₀
P/D SYMBOL
0.6
0.8
1.0
1.2
1.4
1.6
2.0
ENVELOPE OF
ηMAX
TANGENT
J_T
K_T/J_T^2
0.0 0.1 0.2 0.3 0.4 0.5 0.6 0.7 0.8 0.9 1.0 1.1 1.2 1.3 1.4
0.3 0.4 0.5 0.6 0.7 0.8
0.0 0.2 0.4 0.6 0.8 1.0 1.2 1.4 1.6 1.8 2.0

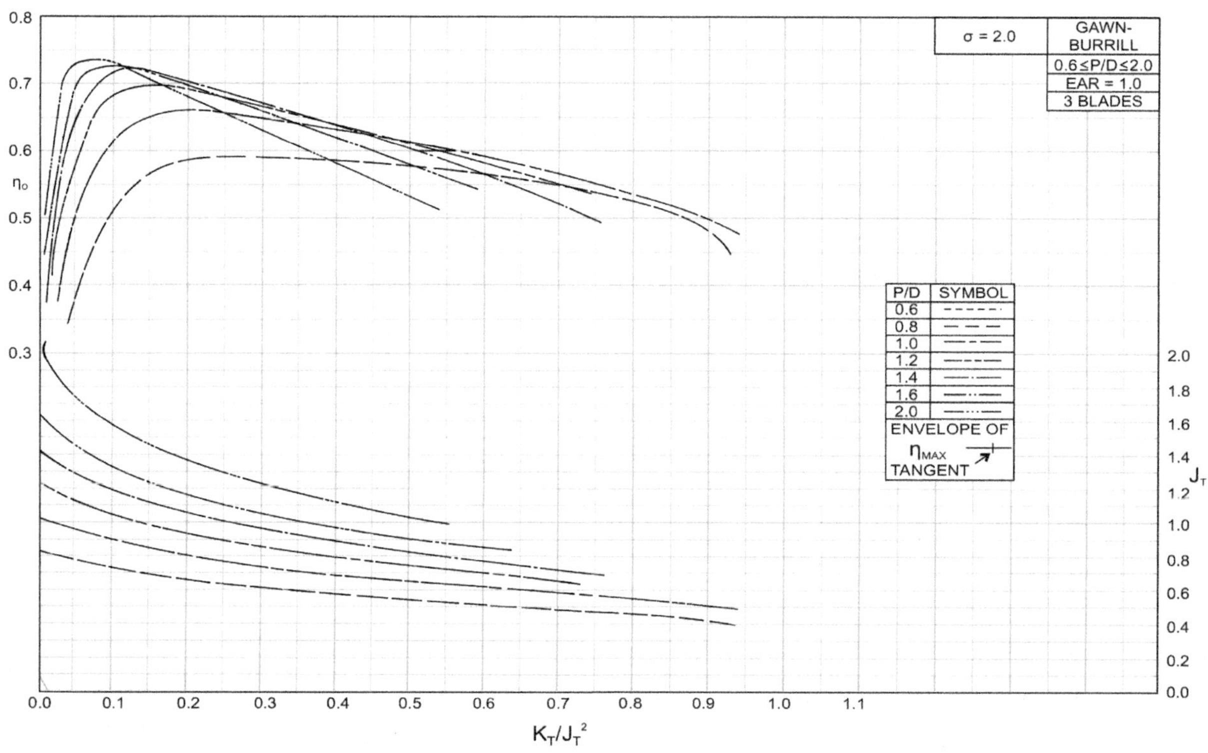
σ = 2.0
GAWN-BURRILL
0.6≤P/D≤2.0
EAR = 1.0
3 BLADES
η₀
P/D SYMBOL
0.6
0.8
1.0
1.2
1.4
1.6
2.0
ENVELOPE OF
ηMAX
TANGENT
J_T
K_T/J_T^2
0.0 0.1 0.2 0.3 0.4 0.5 0.6 0.7 0.8 0.9 1.0 1.1
0.3 0.4 0.5 0.6 0.7 0.8
0.0 0.2 0.4 0.6 0.8 1.0 1.2 1.4 1.6 1.8 2.0

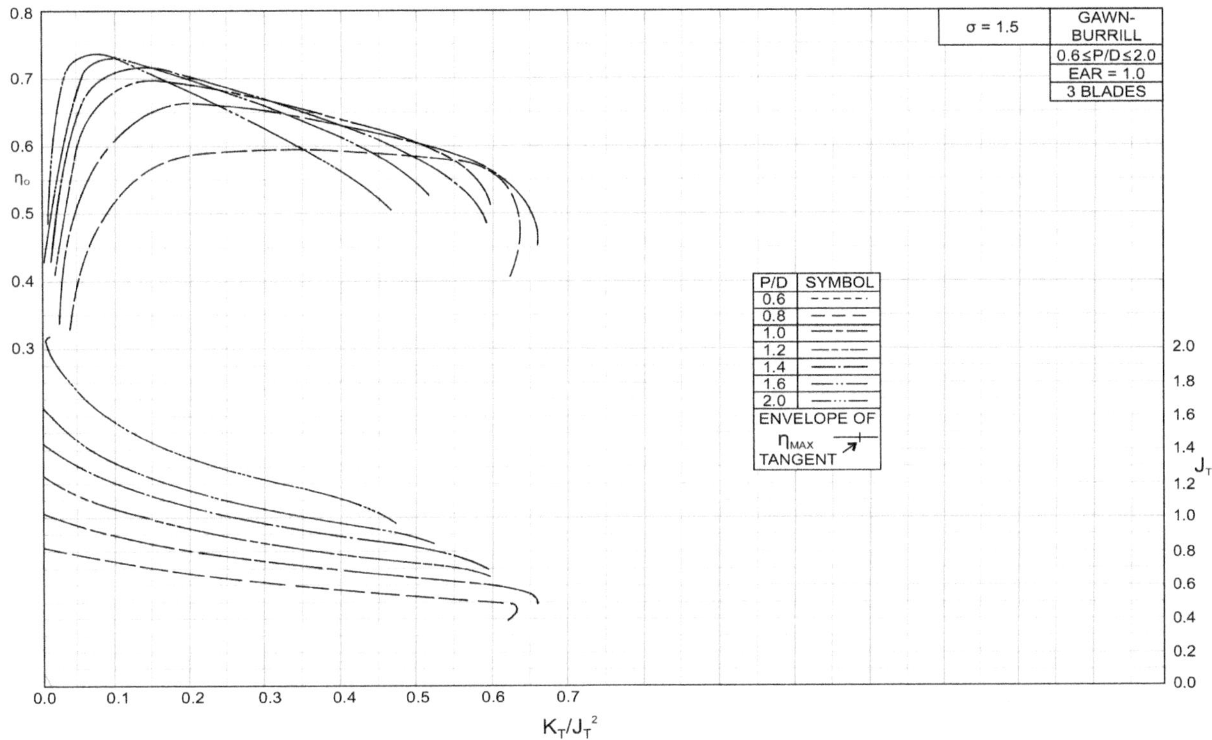
σ = 1.5
GAWN-BURRILL
0.6≤P/D≤2.0
EAR = 1.0
3 BLADES
P/D SYMBOL
0.6
0.8
1.0
1.2
1.4
1.6
2.0
ENVELOPE OF
ηMAX
TANGENT
ηo
JT
KT/JT2
0.0
0.1
0.2
0.3
0.4
0.5
0.6
0.7
0.8
1.0
1.2
1.4
1.6
1.8
2.0

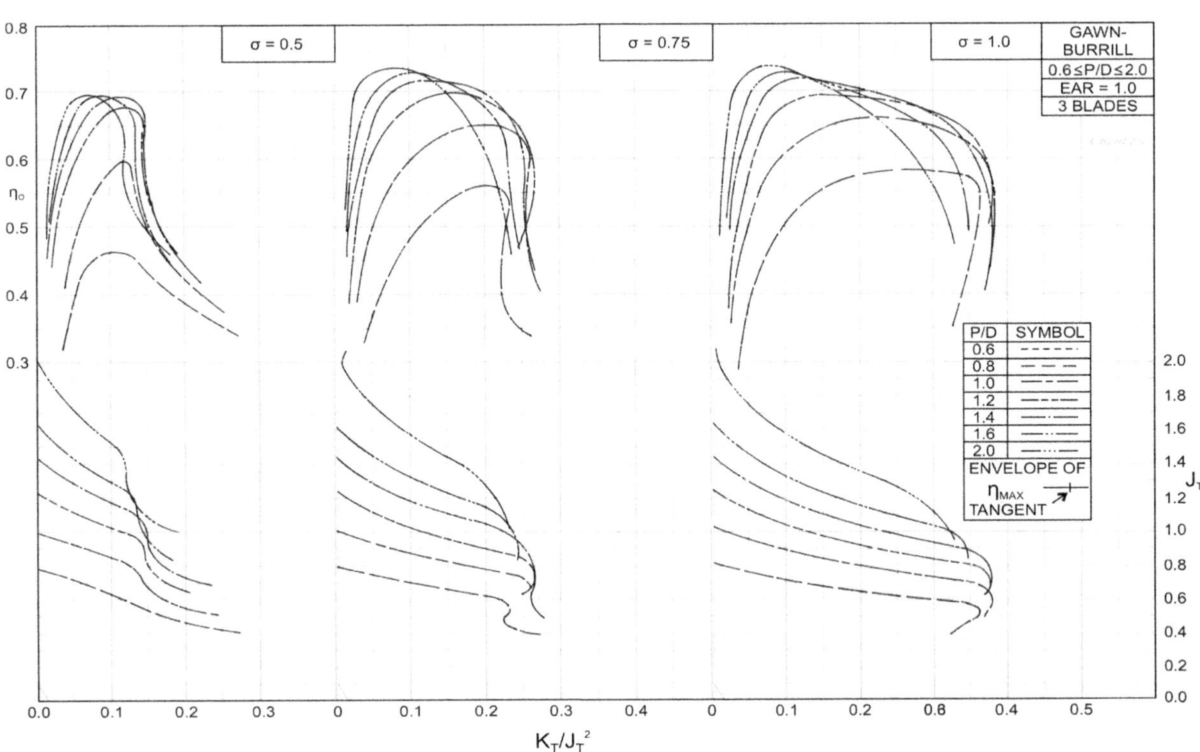
σ = 0.5
σ = 0.75
σ = 1.0
GAWN-BURRILL
0.6≤P/D≤2.0
EAR = 1.0
3 BLADES
P/D SYMBOL
0.6
0.8
1.0
1.2
1.4
1.6
2.0
ENVELOPE OF
ηMAX
TANGENT
ηo
JT
KT/JT2

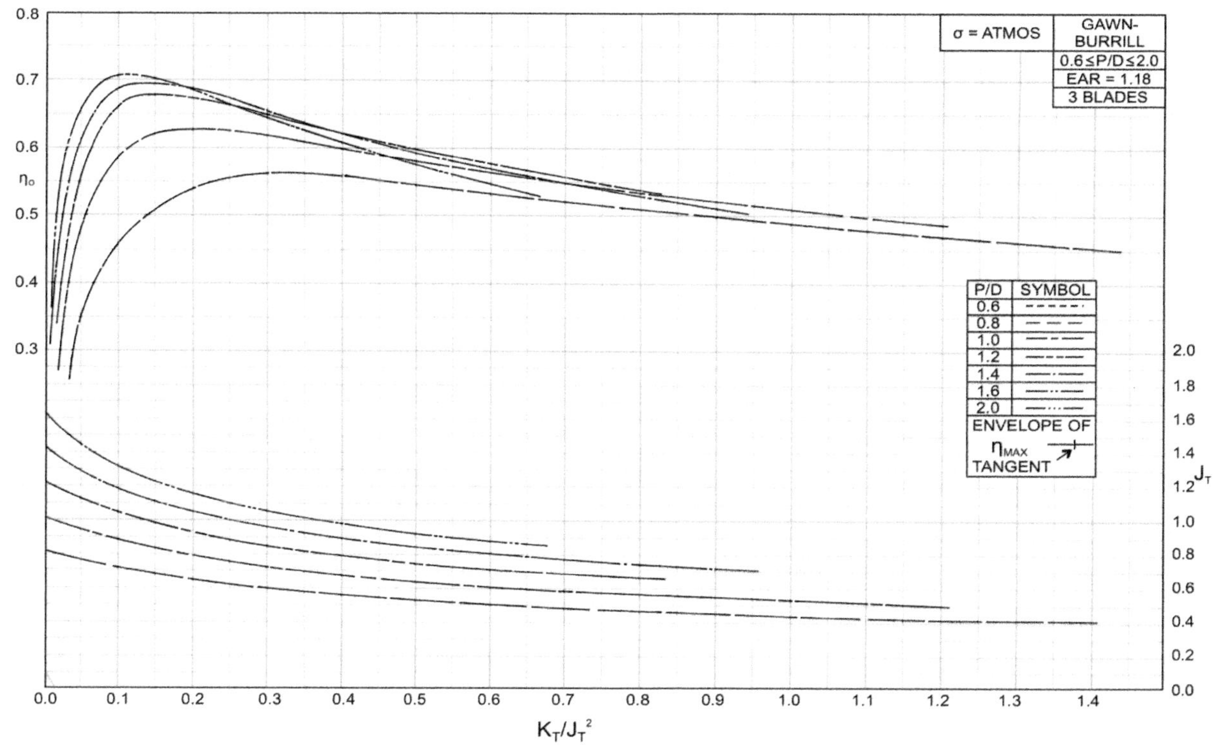
σ = ATMOS
GAWN-BURRILL
0.6≤P/D≤2.0
EAR = 1.18
3 BLADES
P/D SYMBOL
0.6
0.8
1.0
1.2
1.4
1.6
2.0
ENVELOPE OF η_{MAX} TANGENT
η_o
J_T
K_T/J_T^2

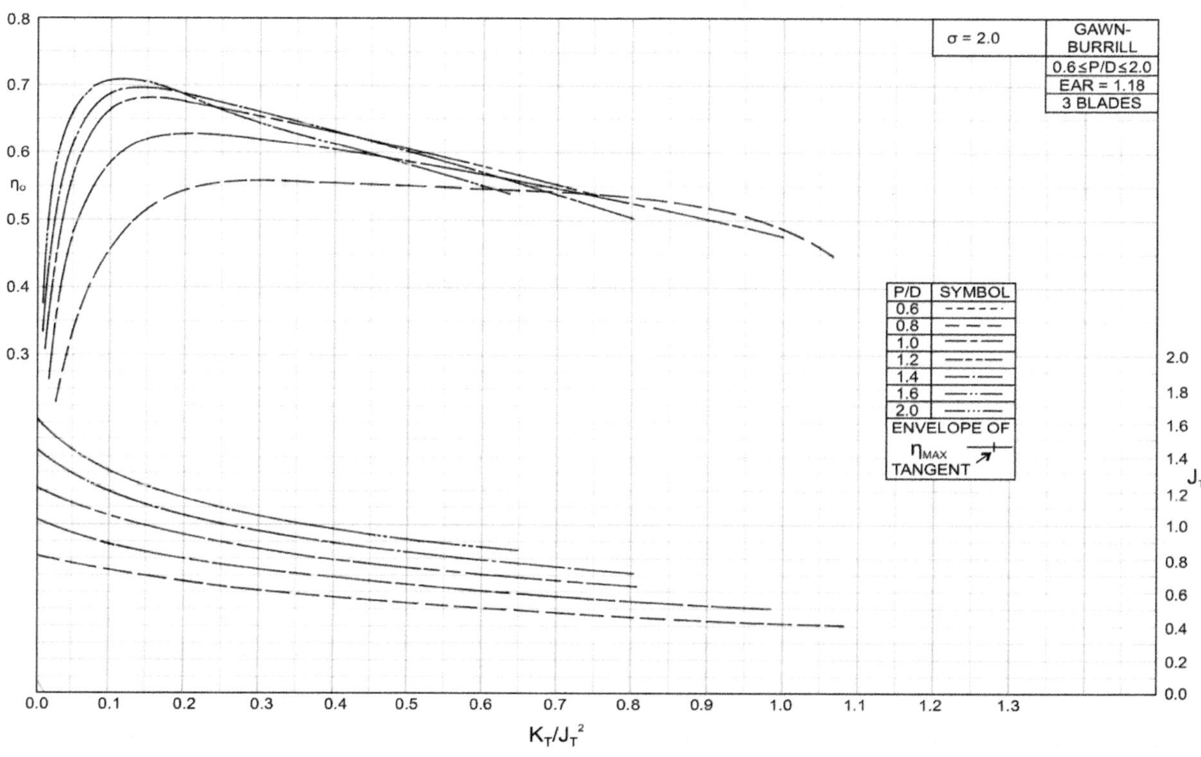
σ = 2.0
GAWN-BURRILL
0.6≤P/D≤2.0
EAR = 1.18
3 BLADES
P/D SYMBOL
0.6
0.8
1.0
1.2
1.4
1.6
2.0
ENVELOPE OF η_{MAX} TANGENT
η_o
J_T
K_T/J_T^2

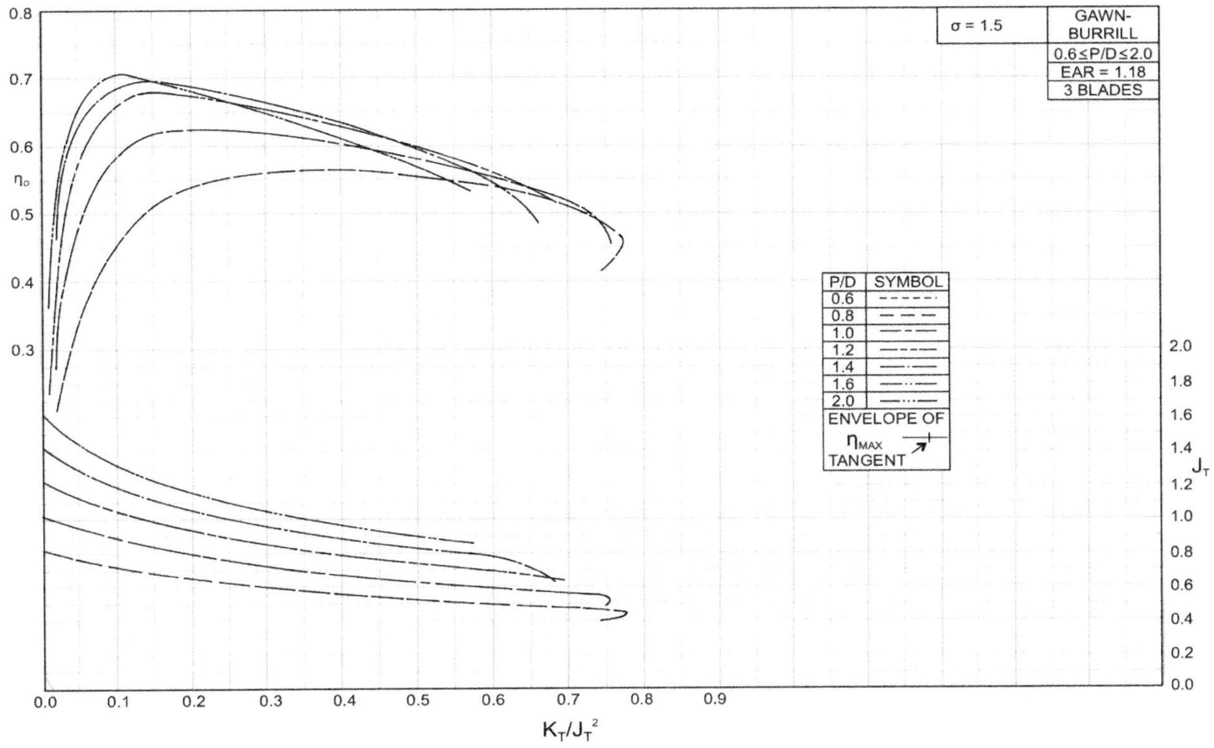
σ = 1.5
GAWN-BURRILL
0.6≤P/D≤2.0
EAR = 1.18
3 BLADES
P/D SYMBOL
0.6
0.8
1.0
1.2
1.4
1.6
2.0
ENVELOPE OF
ηMAX
TANGENT
ηo
JT
KT/JT²

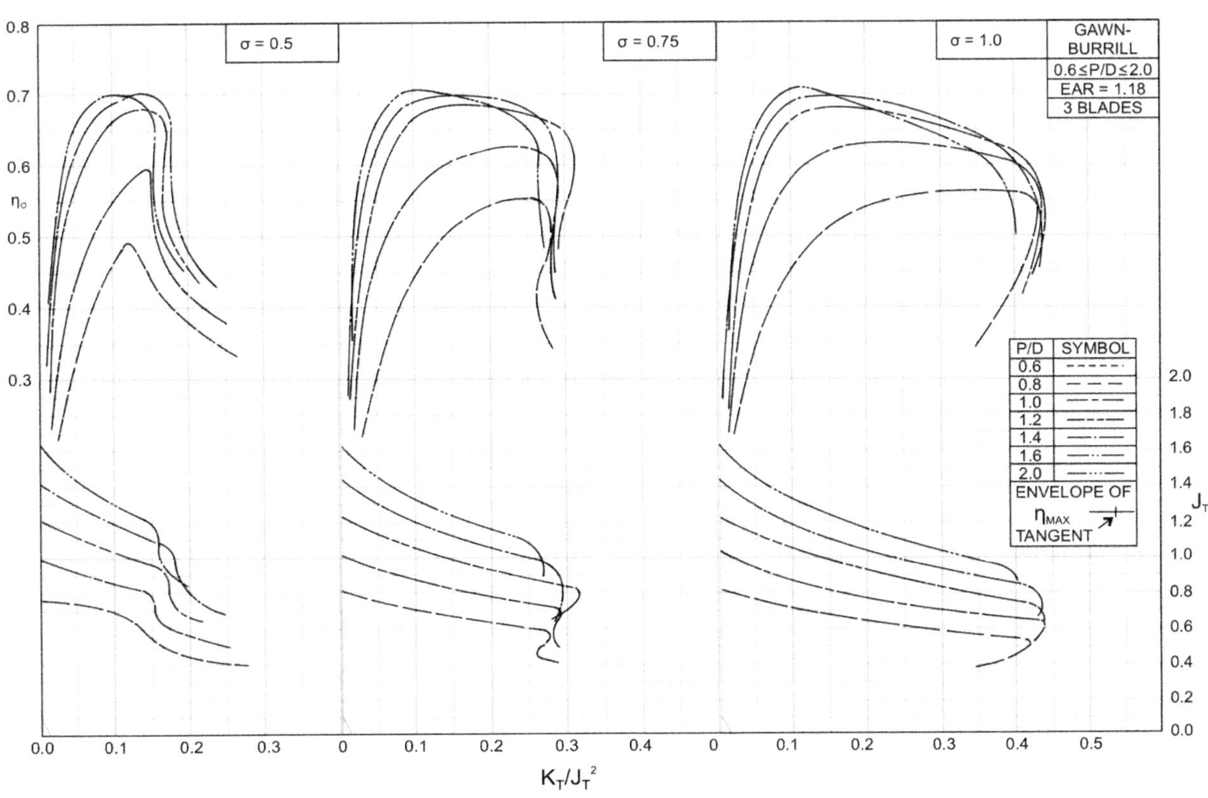
σ = 0.5
σ = 0.75
σ = 1.0
GAWN-BURRILL
0.6≤P/D≤2.0
EAR = 1.18
3 BLADES
P/D SYMBOL
0.6
0.8
1.0
1.2
1.4
1.6
2.0
ENVELOPE OF
ηMAX
TANGENT
ηo
JT
KT/JT²

APPENDIX 6

Variation of R/W Relative to L_P/B_{PX}, $A_P/\nabla^{2/3}$, %LCG, and F_{nV}

from Series 62 Model Tests. W = 100,000 lb (45.4 mt).

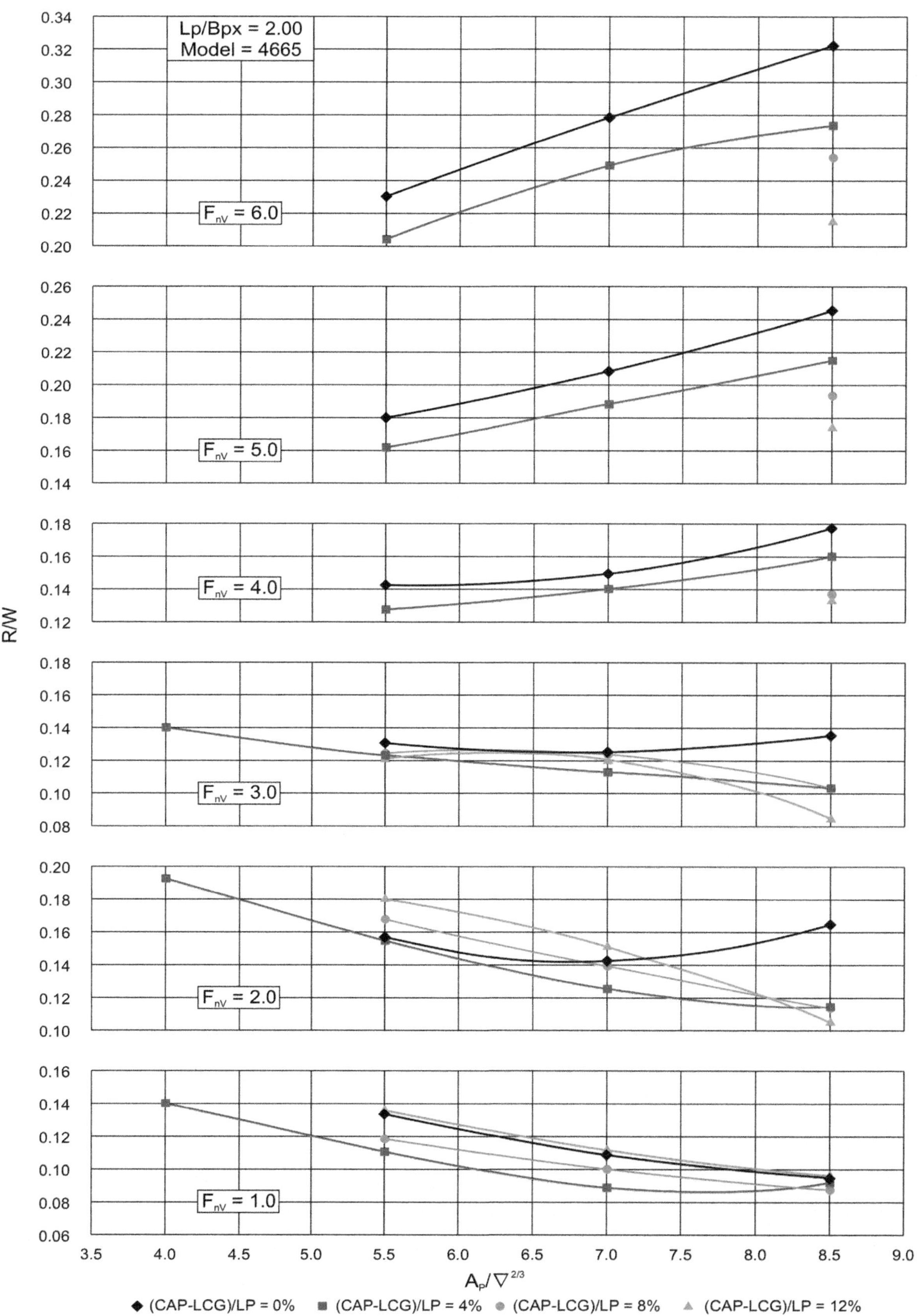

Lp/Bpx = 2.00
Model = 4665
F$_{nV}$ = 6.0
F$_{nV}$ = 5.0
F$_{nV}$ = 4.0
F$_{nV}$ = 3.0
F$_{nV}$ = 2.0
F$_{nV}$ = 1.0
R/W
A$_P$/∇$^{2/3}$
(CAP-LCG)/LP = 0%
(CAP-LCG)/LP = 4%
(CAP-LCG)/LP = 8%
(CAP-LCG)/LP = 12%

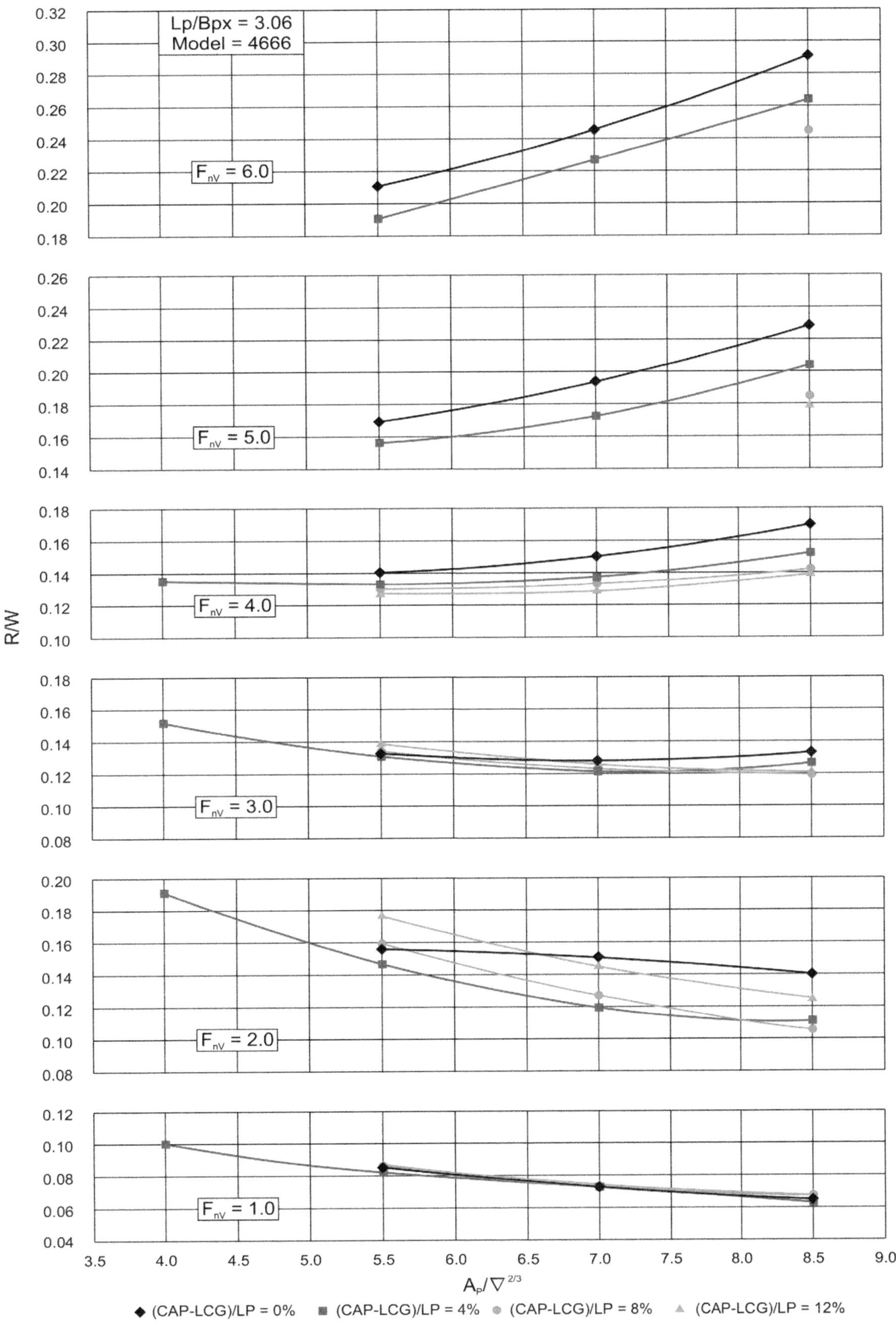
Lp/Bpx = 3.06
Model = 4666
F_nV = 6.0
F_nV = 5.0
F_nV = 4.0
F_nV = 3.0
F_nV = 2.0
F_nV = 1.0
R/W
A_P/∇^2/3
(CAP-LCG)/LP = 0%
(CAP-LCG)/LP = 4%
(CAP-LCG)/LP = 8%
(CAP-LCG)/LP = 12%

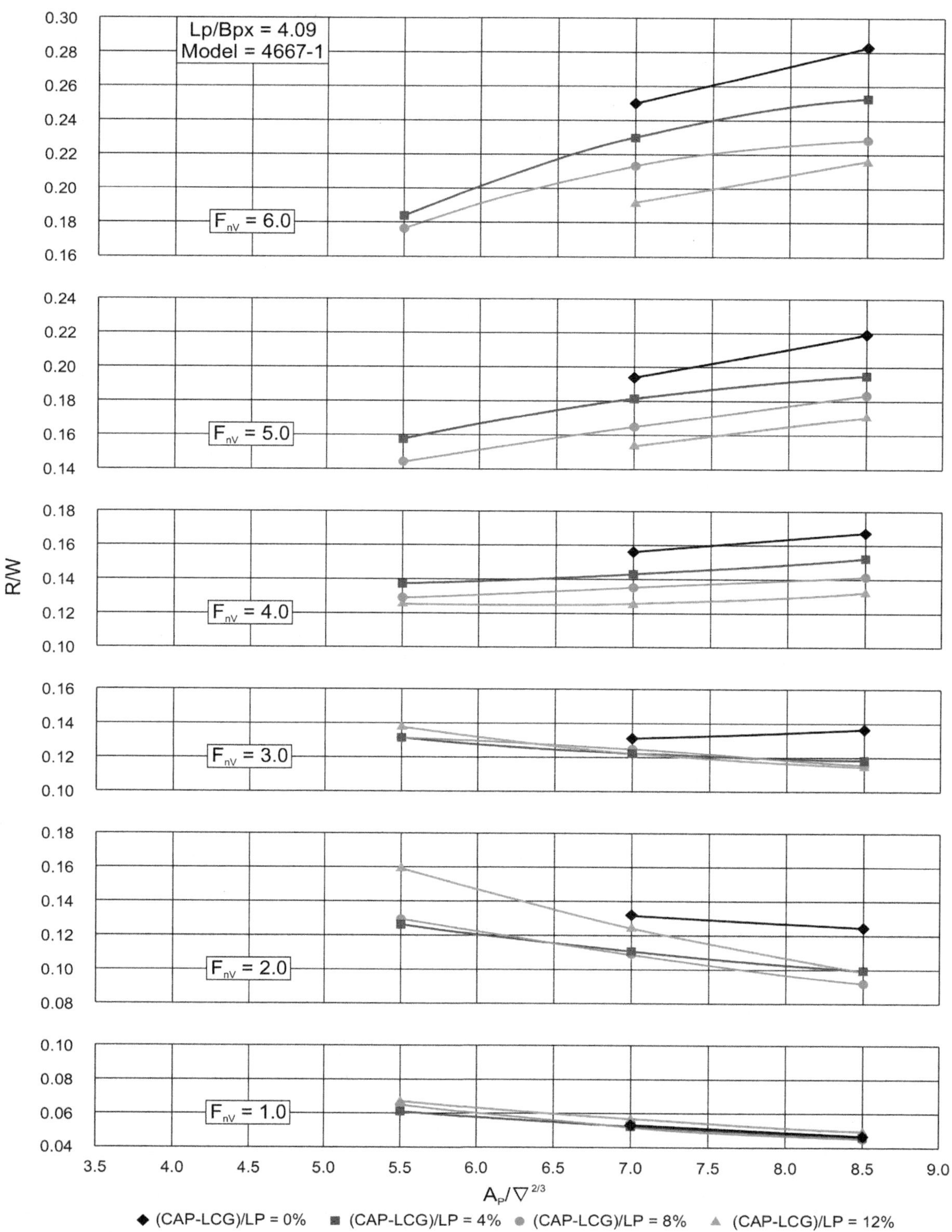

Lp/Bpx = 4.09
Model = 4667-1
F_nV = 6.0
F_nV = 5.0
F_nV = 4.0
F_nV = 3.0
F_nV = 2.0
F_nV = 1.0
R/W
A_P/∇^2/3
(CAP-LCG)/LP = 0%
(CAP-LCG)/LP = 4%
(CAP-LCG)/LP = 8%
(CAP-LCG)/LP = 12%

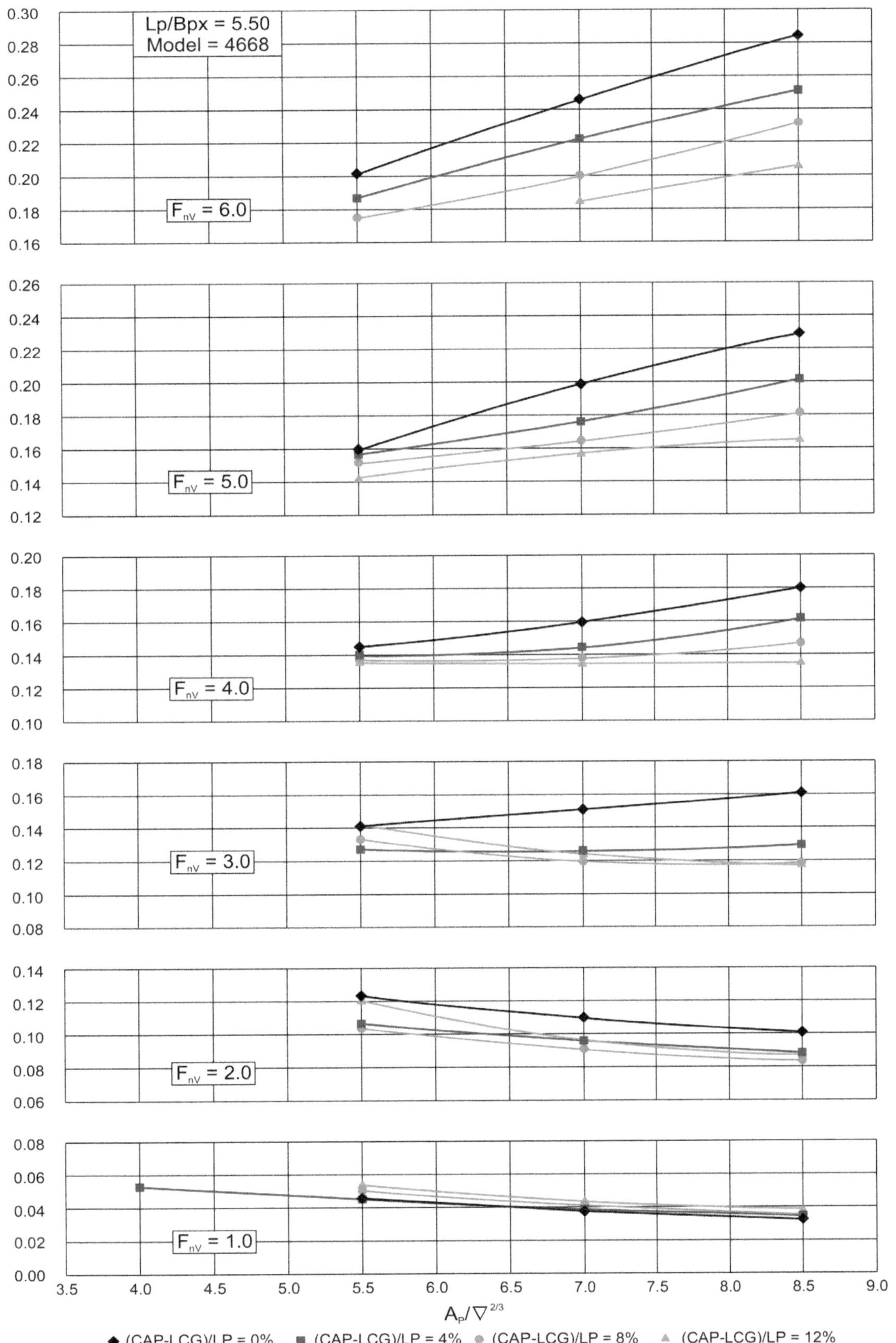

Lp/Bpx = 5.50
Model = 4668
FnV = 6.0
FnV = 5.0
FnV = 4.0
FnV = 3.0
FnV = 2.0
FnV = 1.0
0.30
0.28
0.26
0.24
0.22
0.20
0.18
0.16
0.14
0.12
0.10
0.08
0.06
0.04
0.02
0.00
3.5
4.0
4.5
5.0
5.5
6.0
6.5
7.0
7.5
8.0
8.5
9.0
Ap/∇2/3
(CAP-LCG)/LP = 0%
(CAP-LCG)/LP = 4%
(CAP-LCG)/LP = 8%
(CAP-LCG)/LP = 12%

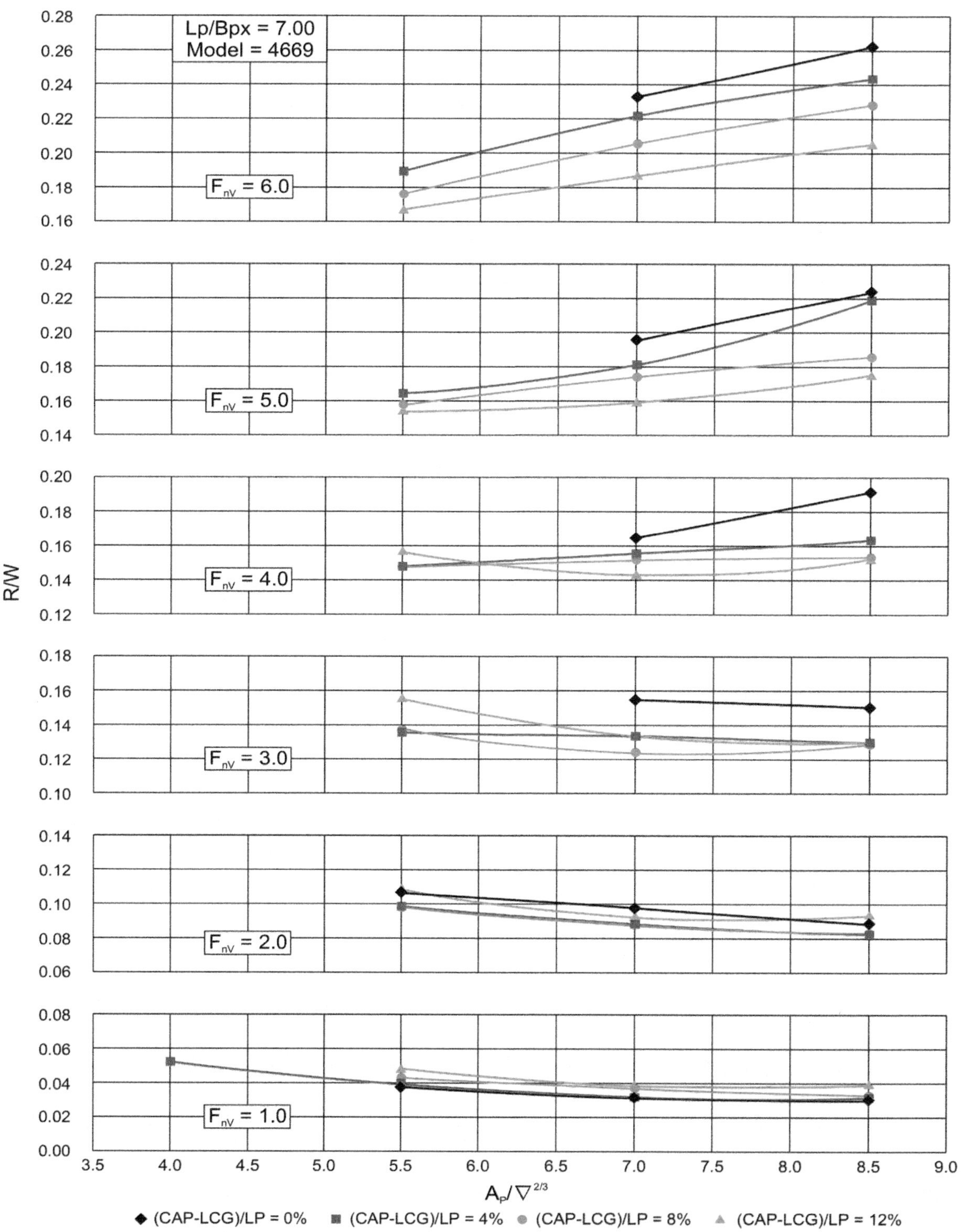
Lp/Bpx = 7.00
Model = 4669
F_{nV} = 6.0
F_{nV} = 5.0
F_{nV} = 4.0
F_{nV} = 3.0
F_{nV} = 2.0
F_{nV} = 1.0
R/W
$A_P/\nabla^{2/3}$
(CAP-LCG)/LP = 0%
(CAP-LCG)/LP = 4%
(CAP-LCG)/LP = 8%
(CAP-LCG)/LP = 12%

APPENDIX 7

Nomenclature, Definitions, and Abbreviations

Nomenclature

Nomenclature for technology related to high-performance vessels differs considerably from references published throughout the world. I have included notation for this book which has been meaningful for me in recent years. You may, however, find that there are terms in this listing which are different than those found by many readers to be "common usage."

a	longitudinal acceleration
A	frontal projected area of seawater inlet
A_E	EAR = expanded area of propeller blades
A_B	projected area of propeller blades
A_{BB}	transverse section area of bulbous bow at the FP
A_L	lateral projected area of the superstructure and hull above water surface (considering pitch and heave at design speed)
A_O	$\pi D^2/4$ = disk area of propeller
A_P	projected area of hull bottom bounded by chines and transom
A_T	projected frontal area of superstructure and hull above water surface (considering pitch and heave at design speed)
A_X	maximum section area of hull
B	static waterline beam
B_B	maximum width of bulbous bow
BHP	brake horsepower
BOA	overall beam
B_P	projected chine beam
B_{PX}	maximum projected chine beam
B_{PXi}	maximum projected inner chine beam of double-chine hull
$B_{PX}/4$	quarter beam buttock
B_{PXO}	maximum projected outer chine beam of double-chine hull
B_{PT}	projected chine beam at transom
B_T	static waterline beam at transom
CA_P	centroid of A_P forward of transom
C_A	correlation allowance
C_B	block coefficient
C_D	drag coefficient for course-keeping fins
C_{DL}	$\nabla/(0.1L)^3$ = volume of displacement coefficient
C_{DO}	drag coefficient for hull-mounted appendages
C_{DP}	drag coefficient for strut palm
C_F	frictional resistance coefficient
CG	center of gravity
C_{Lb}	lift coefficient based on B_{PX}

C_{LS}	hydrodynamic lift coefficient based on projected wetted area
C_N	aerodynamic yawing moment coefficient, normalized with (A_L x LOA)
CP	center of pressure for hydrodynamic and hydrostatic lift measured forward of the transom
C_P	prismatic coefficient
C_R	residuary resistance coefficient
C_{SF}	side force coefficient
C_T	total resistance coefficient
C_V	same as F_{nB}
C_X	aerodynamic drag coefficient normalized with A_T
C_Y	aerodynamic lateral force coefficient normalized with A_L
C_Δ	$\nabla/B_{PX}{}^3$ = beam load coefficient
d	diameter of shaft or strut barrel
D	propeller diameter
D_I	waterjet inlet diameter
D_k	skeg drag
D_O	drag of seawater strainers
D_P	drag of strut palm
D_{SH}	drag of inclined shaft or strut barrel
D_T	diameter of bow thruster
DWL	design waterline
EAR	A_E/A_O = expanded area ratio
EHP	$R_Tv/550$ = total effective horsepower (FPS System)
EHP_{BH}	$Rv/550$ = effective horsepower, bare hull (FPS System)
E_T	$\eta_D/(R/W)$ = transport efficiency
F	accelerating force
F_{nB}	$v/(gB_{PX})^{1/2}$ = beam Froude number
F_{nh}	$v/(gh)^{1/2}$ = depth Froude number
F_{nL}	$v/(gL)^{1/2}$ = length Froude number
F_{nV}	$v/(g\nabla^{1/3})^{1/2}$ = volume Froude number
fot	forward of transom measured from its intersection with the hull bottom on centerline
FP	forward perpendicular
F_T	$v/(gT_T)^{1/2}$ = transom draft Froude number
F_V	vertical propeller force
F_X	force normal to propulsor shaft line
F_Y	lateral (side) force
g	acceleration due to gravity
GM	metacentric height
GR	gear ratio
h	water depth
H	depth of propulsor below water surface
HP	manufacturer's rated horsepower
h_P	strut palm thickness
$H_{1/3}$	significant wave height
J_A	v/nD = propeller advance coefficient based on boat speed
J_T	$v(1\text{-}W_T)/nD$ = propeller advance coefficient based on thrust measurements
J_Q	$v(1\text{-}W_Q)/nD$ = propeller advance coefficient based on torque measurements
K	hump speed resistance correlation factor
K_T	$T/\rho n^2D^4$ = thrust coefficient
K_Q	$Q/\rho n^2D^5$ = torque coefficient
l	wet length of shaft or strut barrel
L	static waterline length
L_B	length of bulbous bow forward of FP
L_C	wetted length of chine of planing craft
L_{CP}	distance of hydrodynamic force measured forward of transom
L_K	wetted length of keel of planing craft
L_M	mean wetted length of planing craft
L_{MC}	mean wetted length of convex longitudinal curved planing surface
L_P	projected chine length
L_W	length of wake hollow aft of the hull
LCB	longitudinal center of buoyancy (fot)
LCB_V	longitudinal center of buoyancy (fot) of volume of hull below plane of water level

LCG longitudinal center of gravity forward of transom (fot)

LOA overall length

$L/\nabla^{1/3}$ slenderness ratio

M Blount/Fox "M" factor

M' revised Blount/Fox "M" factor

m mass of vessel plus entrained water

N propeller or propulsor shaft rpm

N_e engine rps

N_F force normal to planing surface due to hydrodynamic lift

N_{PR} number of propulsors

n propulsor rotational speed (rps)

OPC overall propulsive coefficient = η_D

P propeller pitch

P_A atmospheric pressure

P_{DL} total power for propulsion and dynamic lift (horsepower)

$P_{DL\,KW}$ total power for propulsion and dynamic lift (kilowatts)

P_H ρgh = static water pressure

P_v vapor pressure of water

Q propulsor torque

Q_C torque load coefficient

r radius of curvature

R bare hull resistance

R_A added resistance in waves

R_{APP} appendage resistance

R_{AIR} aerodynamic resistance

R_B resistance of hull with bulbous bow

R_F frictional resistance

R_n Reynolds number = $v(L)/\nu$

R_o resistance due to opening of bow thruster

R_P pressure resistance ($R_P \approx R_R$ at planing speeds)

R_R residual resistance

R_T total resistance

rpm engine rotational speed in revolutions per minute

s transverse projected area of rudder or strut

S wetted area

s_A frontal projected area of seawater inlets

SFC specific fuel consumption (lb/hp hr) or metric (gm/kwhr)

SHP total shaft horsepower

s_K projected area of skeg

t thrust deduction fraction

t_i time

T total craft thrust

T_{FP} draft of the hull at the FP

T_H hull draft

T_T static transom draft

TCG transverse center of gravity relative to centerline of craft

T_n roll period

T_P thrust of each propulsor

T_W modal wave period

t/c thickness to chord ratio for rudders or struts

ULF useful load fraction

UE_T useful transport efficiency

V velocity of craft (knots)

V_{av} average speed taken over a short interval (knots)

VCG vertical center of gravity relative to low point of centerline of keel of craft

V_{JB} jet velocity at bollard conditions

$V_{m/s}$ velocity of craft (meters/sec)

$V/(L)^{1/2}$ speed-length ratio (FPS System)

v velocity of craft (ft/sec)

v_{AIR} resultant velocity of wind and boat speed

v_M mean velocity over wetted planing surface

v_R resultant velocity of flow at tip of propeller

v_{TW} true wind velocity

$v_{0.7R}$ resultant velocity of flow at 0.7 radius of propeller

W weight of craft at rest in pounds

Warp	deadrise angular warp rate from 0.5 L_P to transom; $(\beta_{mid} - \beta_T)/(0.5L_P/B_{PX})$
W_Q	torque wake fraction
W_T	thrust wake fraction
X	distance vessel transits during acceleration
X_P	distance from stagnation line to appendage
X_N	computational defined constants in several chapters
YM	yawing moment
y	width of strut palm
Z	number of propeller blades
$\dot{Z}$	vertical velocity
Z_B	elevation above the baseline of the forward-most profile point of the bulbous bow
α	transom shape parameter
α_X	maximum wave slope
β_{AW}	relative wind direction
β_e	effective deadrise angle
β_{mid}	deadrise angle at $L_P/2$
β_T	deadrise angle at the transom
β_{TW}	true wind direction relative to vessel centerline
Δ	displacement of craft at rest in long, or metric, tons
δ	boundary layer thickness
δC_F	difference in frictional resistance coefficients between model and full-scale Reynolds numbers
δR_M	model scale resistance due to δC_F
∇	volume of displaced water at rest
∇_v	volume of hull below plane of water level for speeds greater than zero
ϵ	angle of shaft relative to buttock
ε	specific tractive force; $\varepsilon = 1/E_T$
η	propulsive efficiency
η_A	appendage drag factor
η_{CG}	RMS vertical acceleration at the center of gravity
η_D	overall propulsive coefficient
η_H	hull efficiency
$\eta_{1/N}$	average of the $1/N_{tn}$ highest vertical accelerations
η_o	efficiency of propulsor in absence of hull influence
η_{PEAK}	maximum vertical acceleration
η_R	relative rotative efficiency
η_T	transmission efficiency
η_{RMS}	RMS (root mean square) vertical acceleration
λ	linear scale ratio
ν	kinematic viscosity of water
ν_{AIR}	kinematic viscosity of air
$\Theta_{B/4}$	longitudinal slope of $B_{PX}/4$ relative to the keel of warped planing surfaces
ρ	mass density of water
ρ_{AIR}	mass density of air
σ	standard deviation
σ_H	cavitation number based on water velocity for the plane of the propeller at a depth of H below static water surface
σ_O	cavitation number at water depth for H = 0
σ_R	cavitation number based on resultant velocity at propeller tip
$\sigma_{0.7R}$	cavitation number based on resultant velocity at 0.7 radius of propeller
τ	trim angle relative to the straight afterbody buttock at $B_{PX}/4$
τ_C	thrust load coefficient
τ_{ch}	trim angle of mean chord of wetted arc of curved surface and the water's surface
τ_e	effective trim angle of planing surface
τ_1	angle of the tangent to the point of intersection of convex buttock, $B_{PX}/4$ and stagnation line at the bow as extended to the water's surface
φ	angle of water flow into a propeller
Ω	wave encounter number
ω	wave circular frequency
ω_e	frequency of wave encounter

Subscripts

e	refers to the engine
f	refers to the full-scale vessel
h	refers to water depth
m	refers to model scale (example: $R_{n,m}$ is Reynolds No. of model)

Symbols

$=$	equal
$\equiv$	identically equal
$\approx$	approximately equal
$\geq$	greater than or equal
$\leq$	less than or equal
∞	infinite water depth (deep water)
$\propto$	proportional
δ	numerical difference between two values
$\int$	integral

Conversion Factors

1 atmosphere = 14.7 lb/in^2
1 foot = 0.3048 m
1 horsepower (DIN) = 0.7355 kw
1 horsepower (SAE) = 0.7457 kw
1 g (gravitational acceleration) = 32.17 ft/sec^2
1 g (gravitational acceleration) = 9.806 m/sec^2
1 kilogram, kg (weight) = 0.4536 lb
1 kilogram, kg (force) - 9.806N
1 kilonewton, kN(force) = 224.8 lb_F
1 knot = 1.688 ft/sec
1 knot = 1.151 mi/hr
1 knot = 0.5144 m/sec
1 long ton = 2,240 lb = 1,016 kg
1 long ton = 35 ft^3 of salt water
1 long ton = 36 ft^3 of fresh water
1 metric ton (tonne) = 2,205 lb
1 metric ton (tonne) = 1,000 lit. of fresh water
1 metric ton (tonne) = 34.4 ft^3 of salt water
1 metric ton (tonne) = 35.4 ft^3 of fresh water
1 pound (force) = 0.004482 kN
1 pound (weight) = 2.205 kg

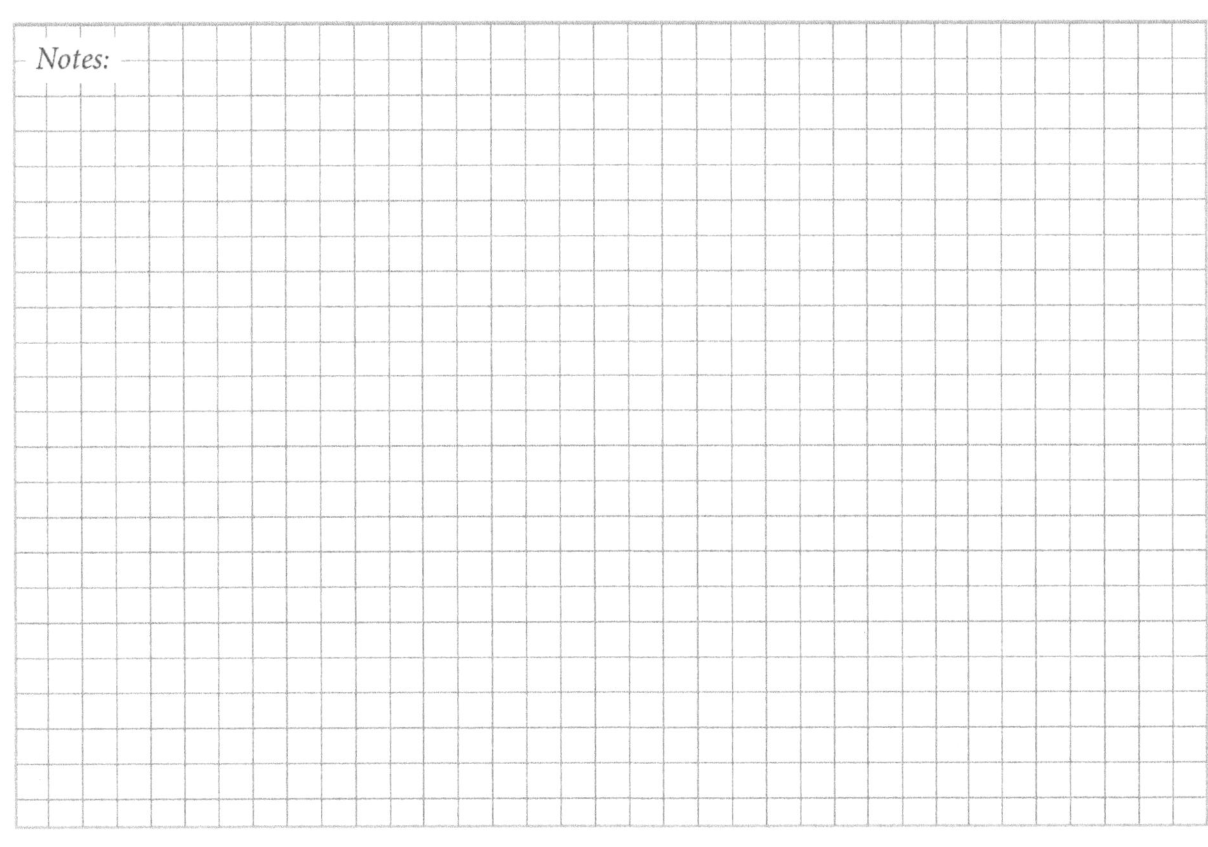

Abbreviations

AIAA	American Institute of Aeronautics and Astronautics
ASNE	American Society of Naval Engineers
ATTC	The American Towing Tank Conference
DTMB and TMB	David Taylor Model Basin
DTNSRDC/SPD	David Taylor Naval Ship Research and Development Center, Ship Performance Department
FAST	International Conference on Fast Sea Transportation
FPS	Foot-Pound-Second System of Units
HPMV	High Performance Marine Vehicles
HSMV	High Speed Marine Vessels
INA	Institution of Naval Architects
ITTC	International Towing Tank Conference
MACC	Multi-Agency Craft Conference
NACA	National Advisory Committee for Aeronautics
NASA	National Aeronautics and Space Administration
NAVSEA	Naval Sea Systems Command
NAVSEACOMBATSYSENGSTA	Naval Sea Combat Systems Engineering Station
NPL	National Physical Laboratory
NSRDC	Naval Ship Research and Development Center
NSWC Carderock Division	Naval Surface Weapons Center Carderock Division
RINA	The Royal Institution of Naval Architects
SDF	Standard Deviation Factor
SI	International System of Units
SIT-DL	Stevens Institute of Technology Davidson Laboratory
SIT-ETT	Stevens Institute of Technology Experimental Towing Tank
SNAME	The Society of Naval Architects and Marine Engineers
SSPA	Swedish Maritime Research Centre
VSM	Very Simple Model

Definitions and General Notes

Resistance

Discussions and numeric examples herein represent calm, deep salt water at 59°F (15°C) and, unless otherwise stated, are for either displacements of 100,000 pounds (45.4 mt) for planing boats or 1,102,500 pounds (500 mt) for semi-displacement vessels.

Hydrodynamics

For this book the term "dynamic" represents analysis at speed while analysis at zero speed is indicated as "static." The study of hydrostatics by naval architects has evolved over a long period of time, mostly focused at displacement conditions. This mature technology is thoroughly documented in textbooks and in schools of higher learning and is not included.

Dimensionless Speed

Various dimensionless speed coefficients have been used to document hydrodynamic trends for craft. In addition to velocity in knots, a commonly used dimensional speed, speed-length ratio, is widely used for reporting technology trends for displacement speeds when wetted length of vessels remain essentially constant. No dimensionless-speed coefficient is universally suitable for all technical analysis; however, one of the following is used in this text when appropriate for the technology being studied. F_{nL} is used herein in preference to speed-length ratio.

Volume Froude Number $F_{nV} = v/(g\nabla^{1/3})^{1/2}$ Equation A7-1

Beam Froude Number $F_{nB} = v/(gB_{PX})^{1/2} = C_V$ Equation A7-2

Length Froude Number $F_{nL} = v/(gL)^{1/2}$ Equation A7-3

Speed-Length Ratio $V/(L)^{1/2}$ (kt/ft$^{1/2}$) Equation A7-4

Depth Froude Number $F_{nh} = v/(gh)^{1/2}$ Equation A7-5

Transom Draft Froude Number $F_T = v/(gT_T)^{1/2}$ Equation A7-6

F_{nV} relates speed to volume of displacement and F_{nB} and F_{nL} respectively relate speed to the hydrodynamic beam and length. Thus, technology comparisons presented as a function of F_{nV}, F_{nB}, F_{nL} or F_{nh} from different sources must account for hull loading, length-beam ratio, and/or length-water depth ratio. The relationships between these coefficients follow, and a graphical representation for F_{nL} and F_{nV} is seen in Figure A7-1.

$F_{nB} = F_{nV} (\nabla^{1/3}/B_{PX})^{1/2}$ Equation A7-7

$F_{nV} = F_{nL}(L/\nabla^{1/3})^{1/2}$ Equation A7-8

$F_{nL} = 0.298[V/(L)^{1/2}]$ Equation A7-9

$F_{nL} = F_{nh}/(L/h)^{1/2}$ Equation A7-10

F_{nB} seems to best represent hydrodynamic phenomena at high speeds where the beam of the planing surface is constant, and the forebody is not in contact with the water surface. F_{nV} is often used to represent hydrodynamic phenomena when wetted length varies with speed and hydrodynamic support of the vessel's weight is significant. F_{nL} is representative when wetted length remains not less than 90% of its static length as speed varies, such as for sailboats and powered vessels having substantial hydrostatic support.

The discussion regarding F_{nV}, F_{nB}, and F_{nL} relates to dimensionless speed for significant aspects of hydrodynamics in regards to hull form. For propulsors a cavitation number (σ) must also be considered as a significant dimensionless speed coefficient.

Cavitation number (σ_H) based on vessel speed and depth of the propulsor below the water surface is most commonly used in reporting characteristics of propellers. $\sigma_{0.7R}$, based on the resultant axial and rotational velocity at the 0.7 radius of propeller blades, is generally used as being representative of condition for thrust loading, τ_C. An alternative cavitation number (σ_R) based on the resultant velocity of both rotational and axial flow at the tip of a propeller has use whenever heavily thrust-loaded applications must be considered.

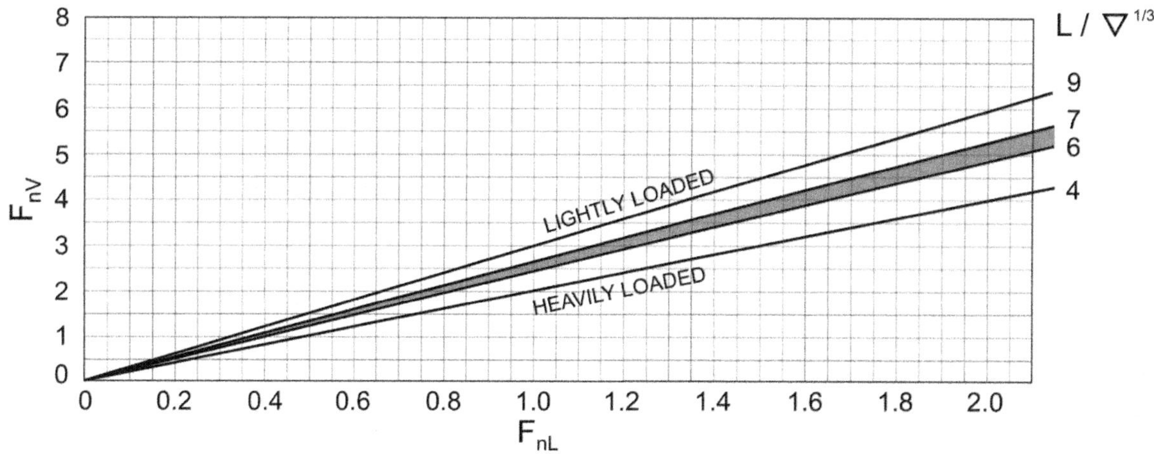

Figure A7-1: Relationship between F_{nL} and F_{nV} as a function of $L/\nabla^{1/3}$

$\sigma_H = (P_A + \rho g H - P_v)/1/2\ \rho v^2$ Equation A7-11

$\sigma_R = (P_A + \rho g H - P_v)/1/2\rho\ v_R^2$ Equation A7-12

The relationship between these two cavitation numbers is:

$\sigma_R = (\sigma_H J_T^2)/(1-W_T)^2(J_T^2 + \pi^2)$ Equation A7-13

Because no analytical relationship exists between Froude and cavitation numbers, Figure A7-2 provides a convenient reference for F_{nV} and σ_H for a range of vessel displacements and speeds.

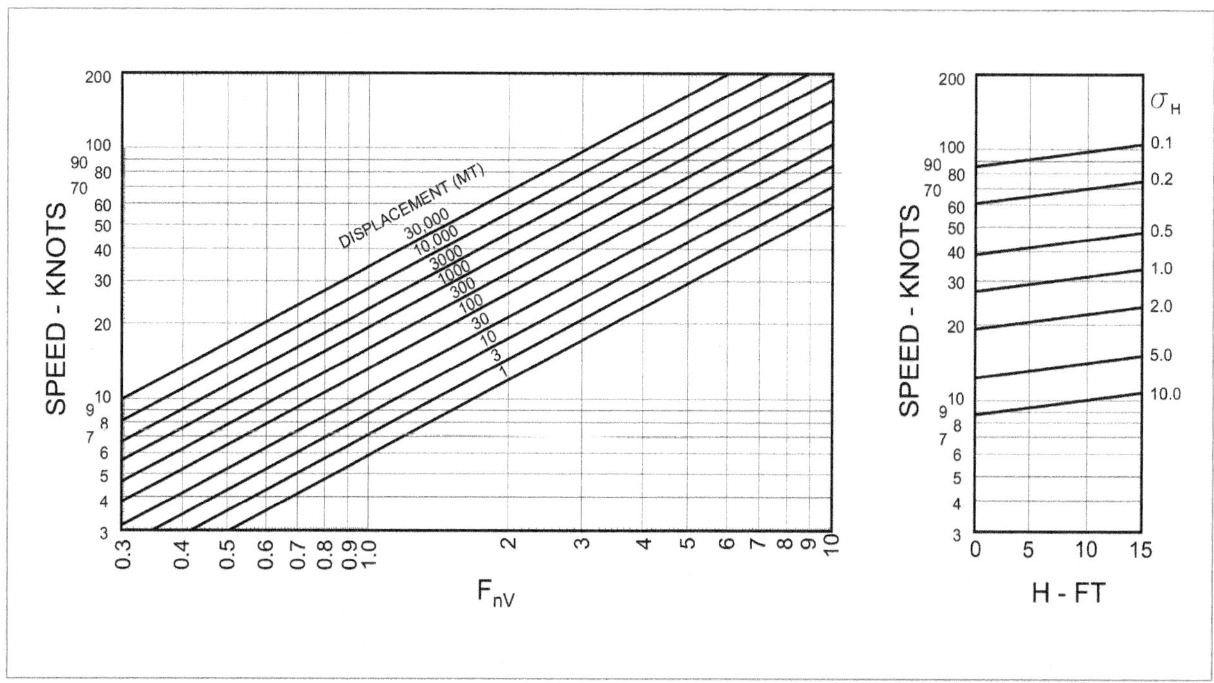

Figure A7-2: Relationships of F_{nV} and σ with speed, respectively, with displacement and depth of submergence

Relative Hull Loading

R/W, seakeeping, stability, and other important performance characteristics of craft are affected by hull loading. The following dimensionless and dimensional definitions are used in various references as quantitative expressions of loading for monohulls, demi-hulls of catamarans and amas (outriggers) of multi-hulls:

Slenderness ratio $L/\nabla^{1/3}$ Equation A7-14

Beam load coefficient $C_\Delta = \nabla/B_{PX}{}^3$ Equation A7-15

Area coefficient $A_P/\nabla^{2/3}$ Equation A7-16

Volume displacement coefficient $C_{DL} = \nabla/(0.1L)^3$ Equation A7-17

Displacement length ratio $\Delta/(0.01L)^3$ where Δ is in long tons/ft^3 Equation A7-18

$L/\nabla^{1/3}$ and C_{DL} relates hull loading (displaced volume) with respect to a function of length. C_Δ relates hull loading to a function of chine beam while $A_P/\nabla^{2/3}$ relates hull loading to a function of the projected bottom area of hard chine hulls.

The relationships between these coefficients are:

$L/\nabla^{1/3} = (L_P/B_{PX})C_\Delta{}^{1/3}$ Equation A7-19

$L/\nabla^{1/3} = 10(C_{DL})^{-1/3}$ Equation A7-20

$L/\nabla^{1/3} = 30.57/[\Delta/(0.01L)^3]^{1/3}$ (for salt water) Equation A7-21

$L_P/\nabla^{1/3} = 1.1[(A_P/\nabla^{2/3})(L_P/B_{PX})]^{1/2}$ (assuming $A_P = L_P[B_{PX}]0.83$) Equation A7-22

$\Delta/(0.01L)^3 = (1/35)[100/(L/\nabla^{1/3})]^3$ (for salt water) Equation A7-23

The displacement or weight of advanced craft is commonly reported either in the dimensions of metric tons (tonnes) or pounds (lb). Long tons (2,240 lb) and short tons (2,000 lb) are seldom used to dimensionally describe advanced craft weight.

Note: $L/L_P \approx 0.98$ is a reasonable approximation for a wide range of length-to-beam ratio planing hulls. For comparative purposes of performance vessels L and L_P may be assumed to be essentially equal. And with regard to calculating F_{nL}, assuming $L \approx L_P$ amounts to a difference of approximately one percent between using L_P vice L as the square root of length is involved.

Definition of Trim Angle

Many hours have been spent on technical discussions relative to the merits of geometric references of surfaces of planing hulls and the significant hydrodynamic trim relationship with respect to the water surface. Speed dependent trim and rise of a planing surface can be experimentally measured relative to arbitrary references for a specific hull shape. However, the full value of these data cannot be compared to data from other planing hull forms until arbitrary references are replaced with values for equivalent hydrodynamic significance. Table A7-1 shows examples of arbitrary references for the measure of hydrodynamically significant trim angles used in various documents.

There is no widely-accepted definition for the zero reference line from which dynamic trim is measured for planing craft. I prefer trim reference 3 in Table A7-1, that is a straight line representing the afterbody of the buttock at $B_{PX}/4$ for dynamic trim analysis and/or comparisons between hulls with varying shapes. It is, however, no longer hydrodynamically significant when any part of $B_{PX}/4$ having bow and/or stern curvature becomes wetted.

Viscous drag dominates at low angles and pressure drag dominates at high angles. For calm water operation, planing craft have an optimum dynamic trim for minimum drag, and this optimum angle has some dependence on aspect ratio and deadrise angle. For rough water, an important measure of quality of overall operations is vertical acceleration of the craft in response to waves. In general, vertical acceleration of planing craft is related to dynamic trim. The dynamic stability of a planing craft also has a strong relationship to trim angle.

Planing craft designers are continually involved in making trade-off decisions between operational objectives with dynamic trim being an important variable. The prediction of dynamic trim is the weakest link in technologies necessary for planing craft design. To make informed design decisions analytical methods need to be developed with trim prediction accuracy of at least +/- 0.2 degrees.

TABLE A7-1: EXAMPLES OF REFERENCES FOR THE MEASURE AND COMPARISON OF HYDRODYNAMICALLY SIGNIFICANT TRIM

TRIM REFERENCE	COMMENT
1. Baseline of lines drawing	Has no hydrodynamic significance. It is a geometric reference for developing a drawing of a three-dimensional shape.
2. Afterbody keel (Note 1)	It is part of the planing hull surface and influences hydrodynamic performance. The keel is only significant for constant deadrise hulls.
3. Afterbody buttock at $B_{PX}/4$ (Note 1)	It is part of the planing hull surface and influences hydrodynamic performance. The buttock at $B_{PX}/4$ is significant for varying deadrise hulls.
4. A straight line connecting the chine-bow profile intersection with the transom-keel intersection (Note 1)	It is analogous to the nose-tail line of an airfoil. Buttock curvature can be related to thickness to chord ratio of airfoil data. It is not easily related to analytical planing methodology.

Note 1: The geometry on the hull bottom, just forward of the transom, defining hook or rocker is considered to be appendages and is not to be used to represent the straight line of the afterbody of buttock $B_{PX}/4$. The geometry of hook, wedges, rocker, interceptors, etc., is addressed in this book on the topic of trim control.

Coordinate System

The coordinate system for angles and moments used herein is depicted in Figure A7-3.

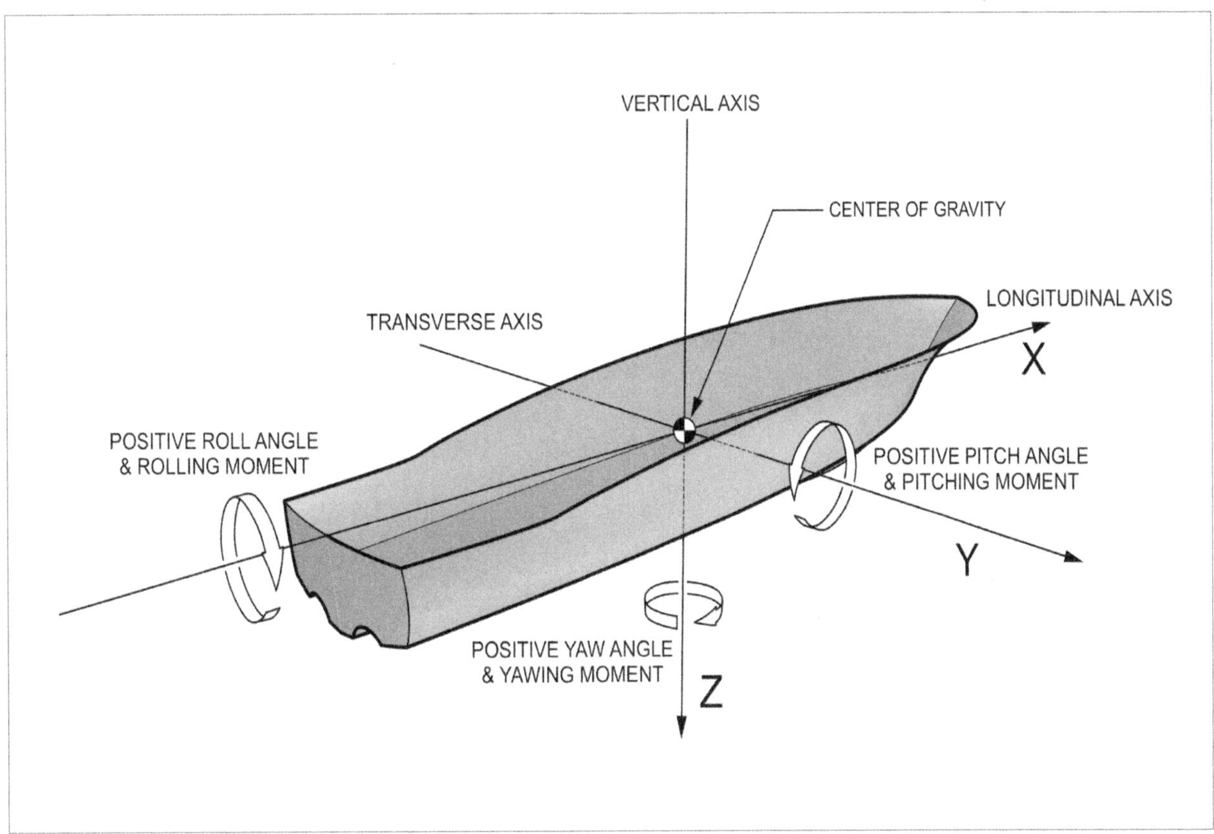

Figure A7-3: Coordinate system for angles and moments

Orientation of Vessels Relative to Seas

The relationship of vessel heading relative to sea direction is shown graphically in Figure A7-4.

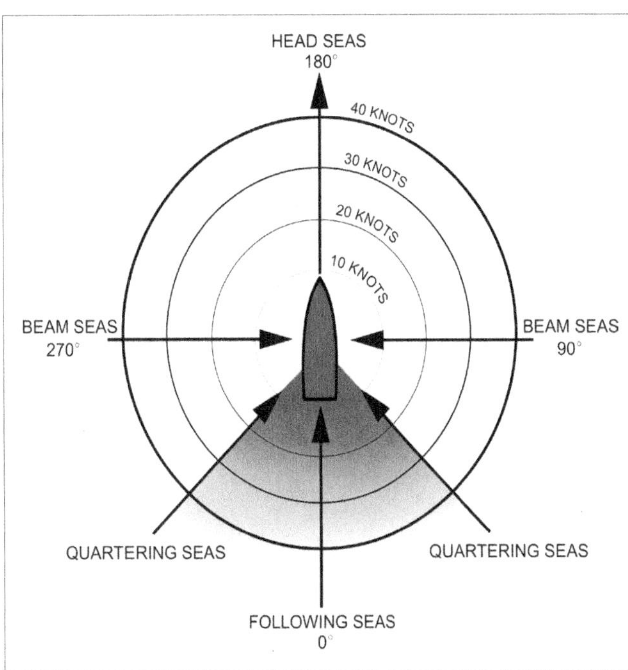

Figure A7-4: Orientation of vessel relative to sea direction

REFERENCES

Aage, C. (1971). "Wind Coefficients for Nine Ship Models," Hydro-Og Aerodynamisk Laboratorium (Hydro- and Aerodynamics Laboratory, Denmark), HyA Report No. A-3.

Abbott, I.H. and von Doenhoff, A.E. (1959). *Theory of Wing Sections*, Dover Publications, New York.

Abrahamsson, S. (1992). "35/38M Yacht – Study of Lift-Drag-Ventilating Properties of a Course-Stabilizing Fin," SSPA Report 6663-1 (Proprietary).

Akers, R.H., Hoeckley, S.A., Peterson, R.S. and Troesch, A.W. (1999). "Predicted vs. Measured Vertical-Plane Dynamics of a Planing Boat," FAST Conference.

Akers, R.H. (1999). "Dynamic Analysis of Planing Hulls in the Vertical Plane," SNAME, New England Section.

Allen, R.G. and Jones, R.R. (1978). "A Simplified Method for Determining Structural Design-Limit Presssures on High Performance Marine Vehicles," AIAA/SNAME Advanced Marine Vehicles Conference.

Allen, R.G. and Jones, R.R. (1977). "Considerations on the Structural Design of High Performance Marine Vehicles," SNAME, New York Metropolitan Section.

Allison, J. (1993). "Marine Waterjet Propulsion," *Transactions*, SNAME.

Anonymous, *High Speed Craft Human Factors Engineering Design Guide*, American Australian British Canadian Dutch-Working Group, Document ABCD-TR-08-01 V1.1, Free Download at: http://www.str.eu.com/human-factors-design-guide.php.

Baader, J. (1970). "Motorkreuzer und Schnelle Sportboote," Verlag Delius, Klasing & Co. (German Language).

Baader, J. (1951). "Cruceros Y Lanchas Veloces," Nautica Baader (Spanish Language).

Bailey, D. (1982). "A Statistical Analysis of Propulsion Data Obtained from Models of High Speed Round Bilge Hulls," RINA, Symposium on Small Fast Warships and Security Vessels.

Bailey, D. (1976). "The NPL High Speed Round Bilge Displacement Hull Series," *Maritime Technology Monograph*, No. 4.

Bailey, D. (1974). "Performance Prediction–Fast Craft," RINA, Occasional Publication No. 1.

Barlow, J.B., Rae, W.H. Jr., and Pope, A. (1999). *Low-Speed Wind Tunnel Testing*, John W. Ley & Sons, Inc., 3rd Edition.

Barnaby, K.C. (1960 & 1969). *Basic Naval Architecture*, Hutchinson Scientific and Technical, Third Edition and Sixth Edition.

Barnaby, S.W. (1891, 1900, 1908). *Marine Propellers*, London, E. & F.N. Spon, Third Edition, Fourth Edition, Fifth Edition, Revised.

Barry, C.D., Ghosh, D., Akers, R.H., and Ulak, A. (2002). "Implementation, Application and Validation of the Zarnick Strip Theory Analysis Technique for Planing Boats," RINA, High Performance Yacht Design Conference.

Beaufoy, M. (1834). *Nautical and Hydraulic Experiments, with Numerous Scientific Miscellanies*, Private Press of Henry Beaufoy.

Becker, C. (2007). "Database Investigation and Analysis," Donald L. Blount and Associates, Inc. Unpublished Document.

Benford, H. (2006). "Naval Architecture for Non-Naval Architects," SNAME.

Beveridge, J.L. (1971). "Design and Performance of Bow Thrusters," NSRDC Report 3611.

Beys, P.M. (1993). "Series 63 – Round Bottom Boats," SIT-DL Report 949.

Blanchard, U.J. (1952). "The Planing Characteristics of a Surface Having a Basic Angle of Dead Rise of 40° and Horizontal Chine Flare," NACA TN 2842.

Blount, D.L. and McGrath, J.A. (2009). "Resistance Characteristics of Semi-Displacement Mega-Yacht Hull Forms," International Conference on Design, Construction & Operation of Super and Mega Yachts, *International Journal of Small Craft Technology*, RINA.

Blount, D.L. and Funkhouser, J. (2009). "Planing in Extreme Conditions," HPMV, ASNE.

Blount, D.L., Schleicher, D.M. and Buescher, P.D. (2006). "Seakeeping: An Experimental Study of Accelerations and Motions of a High-Speed, Double-Chine Craft," International Symposium on Marine Design, RINA.

Blount, D.L. (2003). "Response of Monohulls Operating in Random Seas as Predicted from Analysis of a Collection of Model Test Data," Unpublished.

Blount, D.L., and Dawson, D. (2002). "Rudder Design for High-Performance Boats," *Professional BoatBuilder*, No. 78, August/September.

Blount, D.L. and Blount, Douglas (2002). "Powerboat Performance Tests," *Professional BoatBuilder*, October/November.

Blount, D.L. (1997). "Design of Propeller Tunnels for High-Speed Craft," FAST Conference.

Blount, D.L. (1997). "Tuning a Twin-Screw Rudder Installation," *Professional BoatBuilder*, No. 45, February/March.

Blount, D.L. and Bartee, R.J. (1997). "Design of Propulsion Systems for High-Speed Craft," *Marine Technology*, October.

Blount, D.L. (1994). "Achievements with Advanced Craft," *Naval Engineers Journal*, ASNE.

Blount, D.L. (1993). "Reflections on Planing Hull Technology," SNAME Southeast Section 5th Power Boat Symposium.

Blount, D.L. (1993). "Prospects for Hard Chine, Monohull Vessels," FAST Conference.

Blount, D., Grossi, L., and Lauro, G., (1992). "Sea Trials and Model-Ship Correlations Analysis of the High Speed Gas Turbine Vessel "DESTRIERO," HPMV Conference.

Blount, D.L. and Codega, L.T. (1992). "Dynamic Stability of Planing Boats," *Marine Technology*, January.

Blount, D.L., and Codega, L.T. (1991). "Designing for the Atlantic Challenge," Presentation to the Hampton Roads Section SNAME.

Blount, D.L., and Bjarne, E. (1989). "Design and Selection of Propulsors for High-Speed Craft," 7th Lips Propeller Symposium.

Blount, D.L. and Hubble, E.N. (1981). "Sizing Segmental Section Commercially Available Propellers for Small Craft," SNAME Propeller Symposium.

Blount, D.L. and Fox, D.L. (1978). "Design Considerations for Propellers in a Cavitating Environment," *Marine Technology*, April.

Blount, D.L. and Fox, D.L. (1976). "Small-Craft Power Prediction," *Marine Technology*, January.

Blount, D.L. and Hankley, D.W. (1976). "Full-Scale Trials and Analysis of High-Performance Planing Craft Data," *Transactions*, SNAME.

Blount, D.L., Stuntz Jr., G.. Gregory, D.L., and Frome, M.J. (1968). "Correlation of Full-Scale Trials and Model Tests for a Small Planing Boat," *Transactions*, RINA.

Blount, D.L. (1965). "Resistance and Propulsion Characteristics of a Round-Bottom Boat (Parent Form of TMB Series 63)," DTMB Report 2000.

Brix, J. (1993). "Manoeuvring Technical Manual," Seehafen-Verlag.

Burgh, N.P. (1869). "Modern Screw-Propellers Practically Considered," The Society of Arts, London, E & F.N. Spon.

Chambliss, D.B., and Boyd, G.M., Jr. (1953). "The Planing Characteristics of Two V-Shaped Prismatic Surfaces Having Angles of Dead Rise of 20° and 40°," NACA TN 2876.

Cimino, D., Tellet, D. (editors) (2007). "Marine Vehicle Weight Engineering," Society of Allied Weight Engineers, Inc.

Clark, D.J., Ellsworth, W.M., and Meyer, J.R. (2004). "The Quest for Speed at Sea," NSWC Carderock Division Technical Digest.

Clement, E.P. (1956). "Analyzing the Stepless Planing Boat, DTMB Report 1093.

Clement, E.P. and Blount, D.L. (1963). "Resistance Tests of a Systematic Series of Planing Hull Forms," *Transactions*, SNAME.

Clement, E.P., and Kimon, P.M. (1957). "Comparative Resistance Data for Four Planing Boat Designs," DTMB Report 1113.

Clement, E.P. (1957). "Scale Effect on the Drag of a Typical Set of Planing Boat Appendages," DTMB Report 1165.

Codega, L., and Lewis, J. (1987). "A Case Study of Dynamic Instability in a Planing Hull," *Marine Technology*, Vol. 24, No. 2.

Compton, R.H. (1986). "Resistance of a Systematic Series of Semi-Planing Transom-Stern Hulls," *Marine Technology*, October.

Crane, C.W. (1952). *Clinton Crane's Yachting Memories*, D. Van Nostrand Company, Inc.

Dawson, D., "Once Around the Design Spiral (1997), *Professional BoatBuilder*, October/November.

Dawson, D. and Blount, D.L. (2002). "Trim Control," *Professional BoatBuilder*, February/March.

DeGroot, D. (1956). "Resistance and Propulsion of Motorboats," (Translated by W.B. Hinterthan), DTMB Translation 244.

Denny, S.B. (1988). "Prediction of Thrust Limits for Partially Submerged Propellers," NAVSEA COMBATSY-SENGSTA Report No. 60-212.

Denny, S.B. and Hankley, D.W. (1985). "Prediction of Inclined Shaft Induced Propeller Face Cavitation," NAV-SEACOMBATSYSENGSTA Report No. 60-150.

Denny, S.B. and Feller, A.R. (1979). "Waterjet Propulsor Performance Prediction in Planing Craft Applications," DTRC Report SPD/0905/01.

Diehl, W.S. (1939). "The Application of Basic Data on Planing Surfaces to the Design of Flying-Boat Hulls," NACA TR 694.

Doctors, L.J. and MacFarlane, G.J. (2007). "A Study of Transom-Stern Ventilation," *International Shipbuilding Progress.*

Doctors, L.J. (2006). "A Numerical Study of the Resistance of Transom-Stern Monohulls," HPMV.

DuCane, P. (1950 & 1964). *High-Speed Small Craft*, Cornell Maritime Press, First Edition and Third Edition.

Durand, W.F. (1898). "Description of Steam Yacht Ellide and Her Speed Trials," *Marine Engineering*, December.

Egorov, I.T., Bunkov, M.M., and Sadovnikov, Y.M. (1978). "Propulsive Performance and Seaworthiness of Planing Vessels," (Russian language - limited availability of English as NAVSEA Translation No. 1965).

Faltinsen, O.M. (2005). *Hydrodynamics of High-Speed Marine Vehicles*, Cambridge University Press.

Fridsma, G. (1971). "A Systematic Study of the Rough-Water Performance of Planing Boats (Irregular Wave - Part II)," SIT-DL-71-1495.

Gabrielli, G. and von Karman, T. (1950). "What Price Speed?" *Mechanical Engineering*, October.

Gawn, R.W.L. and Burrill, L.C. (1957). "Effect of Cavitation on the Performance of a Series 16-Inch Model Propellers," *Transactions*, INA, Vol. 99.

Gertler, M. (1954). "A Reanalysis of the Original Test Data for the Taylor Standard Series," DTMB Report 806.

Gertler, M. (1947). "The Prediction of the Effective Horsepower of Ships by Methods in Use at the David Taylor Model Basin," DTMB Report 576.

Graff, W., Kracht, A. and Weinblum, G. (1964). "Some Extensions of D.W. Taylor's Standard Series," *Transactions*, SNAME.

Gray, H.P., and Allen, R.G., and Jones, R.R. (1972). "Prediction of Three-Dimensional Pressure Distributions on V-Shaped Prismatic Wedges During Impact or Planing," NSRDC Report 3795.

Gregory, D.L. and Dobay, G.F. (1973). "The Performance of High-Speed Rudders in a Cavitating Environment," SNAME Spring Meeting.

Grigoropoulos, G.J. and Loukakis, T.A. (1999). "Resistance of Double-Chine, Large, High-Speed Craft," *Bulletin de L'Association Technique Maritime et Aeronautique*, (ATMA) Vol. 99.

Grimsley, J.S. (2010). "Methodology to Quantify Vertical Accelerations of Planing Craft in Irregular Waves," PhD Thesis, Old Dominion University.

Hadler, J.B., Cain, K.M. and Singleton, E.M. (2009). "On the Effect of Transom Area on the Resistance of High-Speed Catamaran Hulls," FAST Conference.

Hadler, J.B., Kleist, J.L. and Unger, M.L. (2007). "On the Effects of Transom Area on the Resistance of Hi-Speed Mono-Hulls," FAST Conference.

Hadler, J.B., Gallagher, N.J., and VanHooff, R.W. (2003). "Model Resistance Testing in the Robinson Model Basin at Webb Institute," SNAME, New York Metropolitan Section.

Hadler, J.B., Hubble, E. Nadine, Allen, R.G., and Blount, D.L. (1978). "Planing Hull Feasibility Model – Its Role in Improving Patrol Craft Design," Symposium on Small Fast Warships and Security Vessels, RINA.

Hadler, J.B., Hubble, E.N. and Holling, H.D. (1974). "Resistance Characteristics of a Systematic Series of Planing Hull Forms Series 65," SNAME Chesapeake Section.

Hadler, J.B. and Hubble, E.N. (1971). "Prediction of the Power Performance of the Series 62 Planing Hull Forms," *Transactions*, SNAME, Volume 79.

Hadler, J.B. (1966). "The Predictions of Power Performance on Planing Craft," *Transactions*, SNAME, Volume 74.

Hadler, J.B., Wilson, C.J. and Beal, A.L. (1961). "Ship Standardization Trial Performance and Correlation with Model Predictions," *Transactions*, SNAME.

Hadler, J.B. (1958). "Coefficients for International Towing Tank Conference Correlation Line," DTMB Report 1185.

Hama, F.R., Long, J.D., and Hegarty, J.C. (1957). "On Transition from Laminar to Turbulent Flow," *Journal of Applied Physics*, Volume 28, No. 4.

Hamlin, C. (1989). *Preliminary Design of Boats and Ships*, Cornell Maritime Press.

Hammitt, A.G. (1975). *Technical Yacht Design*, Van Nostrand Reinhold Company.

Harbaugh, K.H. and Blount, D.L. (1973). "An Experimental Study of a High Performance Tunnel Hull Craft," *Transactions*, SNAME Spring Meeting.

Hatchell, E.G. and Wilson, R.C. (1988). "Margin Management and Procedures for U.S. Navy Small Craft," NAV-SEACOMBATSYSENGSTA Report No. 60-196.

Hoerner, S.H. (1965). *Fluid-Dynamic Drag*, Published by the Author.

Hofman, M., and Radojcic, D. (1997). "Resistance and Propulsion of Fast Ships in Shallow Water," Faculty of Mechanical Engineering, University of Belgrade (in Serbian).

Hoggard, M.M. and Jones, M.P. (1980). "Examining Pitch, Heave and Accelerations of Planing Craft Operating in a Seaway," High Speed Surface Craft Symposium.

Hoggard, M. (1979). "Examining Added Drag of Planing Craft Operating in a Seaway," Presentation to Hampton Roads Section/SNAME.

Hough, G.R. and Ordway, D.E. (1964). "The Generalized Actuator Disk," Second Southwestern Conference on Theoretical and Applied Technology.

Hoyle, J.W., Cheng, B.H., Hays, B., Johnson, B. and Nehrling, B. (1986). "A Bulbous Bow Design Methodology for High-Speed Ships," *Transactions*, SNAME.

Hubble, E.N. (1982). "Program PHFMOPT – Planing Hull Feasibility Model User's Manual," DTNSRDC/SPD-0840-01, Revised.

Hubble, E.N. (1980). "Performance Predictions for Planing Craft in a Seaway," DTNSRDC/SPD-0840-02.

Hubble, E.N. (1974). "Resistance of Hard-Chine, Stepless Planing Craft with Systematic Variation of Hull Form, Longitudinal Center of Gravity and Loading," NSRDC Report 4307.

Hubble, E.N. (1972). "Correlation of Resistance Test Results from Fixed- and Free-to-Trim Methods for a Dynamic-Lift Craft (Model 4667)," DTRC Report 3544.

Isherwood, R.M. (1973). "Wind Resistance of Merchant Ships," *Transactions*, RINA, Vol. 115.

ITTC (2006). "Testing and Extrapolation Methods, General Density and Viscosity of Water," ITTC 7.5-02-01-03 Rev. 01.

ITTC (2002) "Testing and Extrapolation Methods High Speed Marine Vehicles Resistance Test," ITTC 7.5-02-05-01, Rev. 1.

Judge, C.Q. (2013). "Comparisons Between Prediction and Experiment for Lift Force and Heel Moment for a Planing Hull," *Journal of Ship Production and Design*, February.

Kapryan, W. and Weinstein, I. (1952). "The Planing Characteristics of a Surface Having a Basic Angle of Dead Rise of 20° and Horizontal Chine Flare," NACA TN 2804.

Kapryan, W.J. and Boyd, G.J., Jr. (1955). "Hydrodynamic Pressure Distributions Obtained During a Planing Investigation of Five Related Prismatic Surfaces," NACA TN 3477.

Keuning, J.A., Gerritsma, J., and van Terwisga, P.F. (1993). "Resistance Tests of a Series of Planing Hull Forms with 30° Deadrise Angle and a Calculation Model Based on This and Similar Systematic Series," *International Shipbuilding Progress*, Vol. 40, No. 424.

Keuning, J.A. and Gerritsma, J. (1982). "Resistance Tests of a Series of Planing Hull Forms with 25 Degrees Dead-Rise Angle," *International Shipbuilding Progress*, Vol. 29, No. 337.

Kimon, P.M. (1957). "The Planing Characteristics of an Inverted V Prismatic Surface with Minus 10 Degrees Dead Rise," DTMB Report 1076.

Kinney, F.S. (1981 & 1962). *Skene's Elements of Yacht Design*, Dodd, Mead & Company, Seventh Edition, and Eighth Edition.

Koelbel, J.G. Jr. (1995). "Comments on the Structural Design of High Speed Craft," (With Discussions), *Marine Technology*, Volume 32, No. 2.

Korvin-Kroukovksy, B.V., Savitsky, D. and Lehman, W.F. (1949) "Wetted Area and Center of Pressure of Planing Surfaces," SIT-ETT Report No. 360.

Korvin-Kroukovsky, B.V. and Chabrow, R.R. (1948). "The Discontinuous Fluid Flow Past an Immersed Wedge," Davidson Laboratory Report SIT-DL-48-9-334.

Kowalyshyn, D.H. and Metcalf, B. (2006). "A USCG Systematic Series of High Speed Planing Hulls," *Transactions,* SNAME.

Kracht, A.M. (1978). "Design of Bulbous Bows," *Transactions,* SNAME.

Lasky, M.P. (1970). "Performance of High-Speed Naval Ships, Part II: Results of Resistance Tests in Smooth Water on Nine Hull Forms (LCB/LCG Effect)," NSRDC Report C-3311 (Approved for Public Release).

Leshnover, S. (1953). "An Experimental Study of the Effects of Buttock Curvature and Deadrise Distribution on the Diving Tendencies of Six Flying-Boat Forebodies during Landing," SIT-ETT Report No. 518.

Levi, R. (1971). *Dhows to Deltas*, Nautical Publishing Company.

Liljenberg, H. (1992). "35m Yacht Wind Load Measurements," SSPA Report 6663-8 (Proprietary).

Liljenberg, H. (1990). "Yacht Destriero Wind Load Measurement," SSPA Report 5904-8 (Proprietary).

Locke, F.W. (1948). "Tests of a Flat Bottom Planing Surface to Determine the Inception of Planing," Bureau of Aeronautics, Research Division DR Report No. 1096.

Lord, L. (1963). *Naval Architecture of Planing Hulls*, Cornell Maritime Press, Inc., Third Edition.

Maki, K.J., Doctors, L.J., Beck, R.F. and Troesch, A.W. (2005). "Transom-Stern Flow for High-Speed Craft," FAST Conference.

Mansoori, M. (2012). Private Correspondence.

Mapryan, W.J. and Boyd, G.M., Jr. (1955). "Hydrodynamic Pressure Distributions Obtained During a Planing Investigation of Five Related Prismatic Surfaces," NACA TN 3477.

Martin, M. (1978). "Theoretical Determination of Porpoising of High-Speed Planing Boats," *Journal of Ship Research*, Volume 22, No. 1.

Marwood, W.J. and Bailey, D. (1969). "Design Data for High-Speed Displacement Hulls of Round-Bilge Form," National Physical Laboratory, Report 99.

Marwood, W.J. and Bailey, D. (1968). "Transverse Stability of Round-Bottomed High Speed Craft Underway," National Physical Laboratory, Report 98.

Mathias, P.B. and Gregory, D.L. (1974). "Propeller Slipstream Performance of Four High-Speed Rudders Under Cavitating Conditions," NSRDC Report 4361.

Millward, A. (1979). "Preliminary Measurements of Pressure Distribution to Determine the Transverse Stability of a Fast Round Builge Hull," *International Shipbuilding Progress*, Volume 26, No. 297.

Misra, S.C., Gokarn, R.P., Sha, O.P., Suryanarayana, Ch., and Surech, R.V. (2012). "Development of a Four-Bladed Surface Piercing Propeller Series," *Naval Engineers Journal*, December.

Molland, A.F., Turnock, S.R. and Hudson, D.A. (2011). *Ship Resistance and Propulsion*, Cambridge University Press.

Molland, A.F., and Turnock, S.R. (2007). *Marine Rudders and Control Surfaces*, Elsevier Ltd.

Moore, W.L. and Hawkins, F. (1969). "Planing Boat Scale Effects on Trim and Drag (TMB Series 56)," NSRDC Technical Note No. 128.

Morabito, M.G. (2013). "Planing in Shallow Water at Critical Speed," *Journal of Ship Research*, June.

Motora, S. (1960). "On the Measurement of Added Mass and Added Moment of Inertia of Ships in Steering Motion," 1st Symposium on Ship Maneuverability, DTMB Report 1461.

Mottard, E.J. (1960). "Hydrodynamic Characteristics of a Planing Surface with Convex Longitudinal Curvature and an Angle of Dead Rise of 20°," NASA Technical Note D-180.

Mottard, E.J. (1959). "Effect of Convex Longitudinal Curvature on the Planing Characteristics of a Surface Without Dead Rise," NASA Memo 1-25-59L.

Mullar-Graf, B. (1997). "Dynamic Stability of High-Speed Small Craft," WEGEMT Association.

Murray, A.B. (1950). "The Hydrodynamics of Planing Hulls," *Transactions*, SNAME, Volume 58.

NATO Standard Agreement (STANAG) (2000). "Common Procedures for Seakeeping in the Ship Design Process," 4154, Edition 3.

Nicolai, L.M. (1984). "Fundamentals of Aircraft Design," METS, Inc.

Nordstrom, H.G. (1951). "Some Tests with Models of Small Vessels," SSPA Report No. 19.

O'Brien, T.P. (1969). "The Design of Marine Screw Propellers," Hutchinson Scientific and Technical, Third Impression.

Olofsson, N. (1996). "Force and Flow Characteristics of a Partially Submerged Propeller," PhD. Thesis, Chalmers University of Technology.

Oving, A.J. (1985). "Resistance Prediction Method for Semi-Planing Catamarans with Symmetrical Demihulls," Maritime Research Institute Netherlands (MARIN).

Parkinson, J.B. (1956). "Hydrodynamics of High-Speed Water-Based Aircraft," Symposium on Naval Hydrodynamics.

Payne, P.R. (1988). *Design of High-Speed Boats – Planing*, Fishergate, Inc.

Peach, R.W. (1963). "A Method for Determining Acceleration of a Ship," *International Shipbuilding Progress*, Vol. 10, No. 106.

Phillips-Birt, D. (1966). *Motor Yacht and Boat Design*, Adlard Coles Limited, Second Edition.

Pope, J.D. (1958). "The Planing Characteristics of a V-Shaped Prismatic Surface with 70 Degrees Dead Rise," DTMB Report 1285.

Radojcic, D. and Bowles, J. (2010). "On High Speed Monohulls in Shallow Water," 2nd Chesapeake Power Boat Symposium.

Radojcic, D., Rodic, T., and Kostic, N., (1997). "The Resistance and Trim Predictions for the NPL High speed Round Bilge Displacement Hull Series," Conference on Power, Performance and Operability of Small Craft, RINA.

Radojcic, D., (1988). "Mathematical Model of Segmental Section Propeller Series for Open-Water and Cavitating Conditions Applicable in CAD," SNAME Propellers '87 Symposium.

Raymer, D.P. (2006). "Aircraft Design: A Conceptual Approach," American Institute of Aeronautics and Astronautics, Inc., Fourth Edition.

Ridgely-Nevitt, (1967). "The Resistance of a High Displacement-Length Ratio Trawler Series," *Transactions,* SNAME.

Rijkens, A.A.K., Keuning, J.A. and Huijsman, S. (2011). "A Computational Tool for the Design of Ride Control Systems for Fast Planing Vessel," *International Shipbuilding Progress,* Vol. 58.

Riley, M.R., Coats, T. (2012). "Development of a Method for Computing Wave-Impact Equivalent Static Accelerations for Use in Planing Craft Hull Design," Chesapeake Power Boat Symposium.

Riley, M.R., Coats, T.W. (2012). "A Simplified Approach for Analyzing Accelerations Induced by Wave-Impacts in High-Speed Planing Craft," Chesapeake Power Boat Symposium.

Riley, M., Coats, T., Haupt, K. and Jacobson, D. (2011). "Ride Severity Index – A New Approach to Quantifying the Comparison of Acceleration Responses of High-Speed Craft," FAST Conference.

Riley, M.R., Coats, T., Haupt, K., and Jacobson, D. (2010). "The Characterization of Individual Wave Slam Acceleration Responses for High Speed Craft," 29th ATTC.

Riley, M. R., Haupt, K.D., and Jacobson, D.R. (2010). "A Generalized Approach and Interim Criteria for Computing $A_{1/n}$ Accelerations using Full-Scale High-Speed Craft Trials Data," Naval Surface Warfare Center.

Riley, M.R., Haupt, K.D., and Murphy, H.P. (2014). "An Investigation of Wave Impact Duration in High-Speed Planing Craft in Rough Water, Naval Surface Warfare Center.

Rose, J.C., Kruppa, C.F.L., and Koushan, K. (1993). "Surface Piercing Propellers – Propeller/Hull Interaction," FAST Conference.

Rose, J.C. and Kruppa, C.F.L. (1991). "Methodical Series Model Test Results," FAST Conference.

Rossell, H.E. and Chapman, L.B. (1949). *Principles of Naval Architecture,* Volume II, SNAME, Eighth Printing.

Savitsky, D. (2012). "The Effect of Bottom Warp on the Performance of Planing Hulls," Third SNAME Chesapeake Power Boat Symposium.

Savitsky, D. and Morabito, M. (2011). "Origin and Characteristics of the Spray Patterns Generated by Planing Hulls," Davidson Laboratory Technical Report SIT-DL-10-1-2882, January 2010 and *Journal of Ship Production and Design,* SNAME, May 2011.

Savitsky, D., DeLorme, M.F., and Datla, R. (2007). "Inclusion of Whisker Spray Drag in Performance Prediction Method for High-Speed Planing Hulls," *Marine Technology,* SNAME, Vol. 44, No. 1.

Savitsky, D. and Ward-Brown, P. (1976). "Procedures for Hydrodynamic Evaluation of Planing Hulls in Smooth and Rough Water," *Marine Technology,* Volume 13, No. 4.

Savitsky, D., Roper, J., and Benen, L. (1972). "Hydrodynamic Development of a High-Speed Planing Hull for Rough Water," 9th Symposium of Naval Hydrodynamics.

Savitsky, D. (1964). "Hydrodynamic Design of Planing Hulls," *Marine Technology,* Volume 1, No. 1.

Savitsky, D. and Neidinger, J. (1954). "Wetted Area and Center of Pressure of Planing Surfaces at Very Low Speed Coefficients," Davidson Laboratory Report SIT-DL-54-9-493.

Savitsky, D., and Ross, E.W. (1952). "Turbulence Stimulation in the Boundary Layer of Planing Surfaces – Part II Preliminary Experimental Investigation," SIT-ETT, Report 444.

Savitsky, D. (1951). "Wetted Length and Center of Pressure of Vee-Step Planing Surfaces," Stevens Institute of Technology, SIT-ETT Report 378.

Schleicher, D.M. (2008). "Regarding Small Craft Seakeeping," First SNAME Chesapeake Power Boat Symposium.

Schleicher, D.M. (2004). "Seakeeping – The Forgotten Requirement," MACC.

Schleicher, D. and Bowles, J. (2004). "Seakeeping Characteristics of a 100 Knot Yacht," RINA High Speed Craft Conference.

Schleicher, D.M. (2002). "Effective Use of the Transport Efficiency Concept," MACC.

Schmidt, R. (1936). "The Scale Effects in Towing Tests with Airplane-Float Systems," NACA TM 826.

Scholars, R.E. (1968). "An Investigation of the Performance Characteristics of a Long, Slender Hull," Trident Scholar Report, U.S. Naval Academy.

Sharqawy, M.H., Lienhard, J.H., and Zubair, S.M. (2010). "Therophysical Properties of Seawater: A Review of Existing Correlations and Data, Desalination and Water Treatment."

Shoemaker, J.M. (1934). "Tank Tests of Flat and V-Bottom Planing Surfaces, NACA-TN-509.

Shuford, C.L. Jr. (1958). "A Theoretical and Experimental Study of Planing Surfaces Including Effects of Cross Section and Plan Form," NACA Report 1355.

Sireli, E.M., and Goren, O. (2000). "The Effect of the Transom Stern on the Resistance of High-Speed Craft," IV Congress International Maritime Association of Mediterranean, Proceedings Vol. I.

Skene, N.L. (1925). *Elements of Yacht Design*, Yachting, Inc., Third Edition.

Skolnick, A. (1969). "A Structure and Scoring Method for Judging Alternatives," IEEE Transactions on Engineering Management.

Smith, R.M. (1924). *The Design and Construction of Small Craft*, Vizetelly & Co., Ltd.

Soletic, L. (2010). "Seakeeping of a Systematic Series of Planing Hulls," SNAME, Second Chesapeake Power Boat Symposium.

Sottorf, W. (1944). "Analysis of Experimental Investigations of the Planing Process on the Surface of Water," NACA TM 1061.

Springston, G.B., Jr. and Sayre, C.L. Jr. (1955). "The Planing Characteristics of a V-Shaped Prismatic Surface with 50 Degrees Dead Rise," DTMB Report 920.

Stoltz, J. and Koelbel, J.G., Jr. (1963). "How to Design Planing Hulls," *Motor Boating*.

Stromgren, C. (1995). "A Comparison of Alternative Bow Configurations," *Marine Technology*, July.

Sturtzel, W., Graff, W. (1963). "Investigation of Optimal Form Design for Round-Bottom Boats," Forschungsbericht des Landes Nordrhein-Westfalen, Nr. 1137 (German Language).

Sugai, K. (1964). "On the Maneuverability of the High Speed Boat," Bureau of Ships Translation No. 868, Translated from Transportation Technical Research Institute, Tokyo.

Tanaka, H., Nakato, M., Nakatake, K., Ueda, T. and Araki, S. (1991). "Cooperative Resistance Tests with Geosim Models of a High-Speed Semi-Displacement Craft," *Journal of the Society of Naval Architects of Japan*, Vol. 169.

Taravella, B., McKesson, C., Vorus, W. (2012). "A Very Simple Model for the Wake Hollow Behind a High-Speed Displacement Hull," *Naval Engineers Journal*, September.

Taylor, D.W. (1943). *The Speed and Power of Ships*, U.S. Govt. Printing Office (3rd Edition).

Taylor, D.W. (1910). *The Speed and Power of Ships*, John Wiley & Sons, Inc. (1st Edition).

Teale, J. (2003). *How to Design a Boat,* Sheridan House, Third Edition.

Toro, A.I. (1969). "Shallow-Water Performance of a Planing Boat," SNAME Southeast Section.

Tuthill, J. T. (1960). *An Advanced Hull and Propeller Design*, Fishing Boats of the World: 2, Food and Agriculture Organization of the United Nations (FAO).

van Berlekom, W.B. (1981). "Wind Forces on Modern Ship Forms—Effects on Performance," *Transactions*, NECI, Vol. 97, No. 4.

van Berlekom, W.B., Tragardh, P., and Dellhag, A. (1974). "Large Tankers—Wind Coefficients and Speed Loss Due to Wind and Sea," RINA.

Van Lammeren, W.P.A., Troost, L., Koning, J.G. (1948). *Resistance, Propulsion and Steering of Ships*, H. Stam-Haarlem-Holland Publishing.

Van Oossanen, P., Heimann, J., Henrichs, J., Hochkirch, K. (2009). "Motor Yacht Hull Form Design for the Displacement to Semi-Displacement Speed Range," FAST Conference.

Vorus, W.S. (1996). "A Flat Cylinder Theory for Vessel Impact and Steady Planing Resistance," *Journal of Ship Research*, June.

Vorus, W.S. (1991). "Forces on Surface-Piercing Propellers with Inclination," *Journal of Ship Research*, Vol. 35, No. 3.

Wagner, H. (1933). "Planing of Watercraft," NACA Technical Memo No. 1139, April 1948, (Translation of "Jahrbuch der Schiffbautechnik," Vol. 34).

Ward-Brown, P. (1980). "An Experimental Study of Aeration on Planing Surfaces," Davidson Laboratory, Technical Report 2153.

Ward-Brown, P. (1971). "An Experimental and Theoretical Study of Planing Surfaces with Trim Flaps," Davidson Laboratory Report SIT-DL-71-1463.

Walshe, D.E. (1983). "A Method for the Calculation of the Wind Resistance on High-Speed Marine Craft," NMI, Ltd. Report 168.

Weinstein, I. and Kapryan, W.J. (1953). "The High Speed Planing Characteristics of a Rectangular Flat Plate Over a Wide Range of Trim and Wetted Length," NACA TN 2981.

White, W. (1899 & 1900). "Review of Past Progress in Steam Navigation and Forecast of Future Development," *Marine Engineering*, November 1899 and January 1900.

Wood, K.D. (1949). *Airplane Design*, Ninth Edition.

Xu, L., Troesch, A.W., Vorus, W.S. (1998). "Asymmetric Vessel Impact and Planing Hydrodynamics," *Journal of Ship Research*, September.

Yeh, H.Y.H. (1965). "Series 64 Resistance Experiments on High-Speed Displacement Forms," *Marine Technology*, July.

Young, Y.L. and Kinnas, S.A. (2004). "Performance Prediction of Surface-Piercing Propellers," *Journal of Ship Research*, Vol. 28, No. 4.

Zarnick, E.E. (1979). "A Nonlinear Mathematical Model of Motions of a Planing Boat in Irregular Waves," DTNSRDC/SPD-0867-1.

Zarnick, E.E. (1978). "A Nonlinear Mathematical Model of a Planing Boat in Regular Waves," DTNSRDC Report 78/032.

Zseleczky, J. and McKee, G. (1989). "Analysis Methods for Evaluating Motions and Accelerations of Planing Boats in Waves," ATTC.

Credits for Illustrations

Graphic figures herein are formatted by the author and print quality digital files prepared by the DLBA staff other than as indicated in the following list.

Item	Figure or Appendix	Source
Photograph	Cover	Author
Photograph	Frontispiece	Will Cafnuk, Bravo Romeo Limited
Photograph	1-5	*Marine Engineering*, December 1898
Drawing	1-6	*Marine Engineering*, November 1900
Photographs	Chapter 2, Page 33	Author
Underwater photographs	2-8	BRIM
Model test results	2-9	BRIM
Resistance regression analysis	2-10 & A4-4	Hubble 1978
Drawing	2-30	Muller-Graf 1997
Graphs	3-7 & 3-8	Radojcic & Bowles 2010
Photograph	4-11 A&B	SSPA 1992
Graphs	4-12 & 4-13	SSPA 1992
Drawing	4-17	Kracht 1978 & Hoyle 1986
Drawing	7-14	Wagner 1948
Drawing	7-15	Allen & Jones 1978
Graph	7-16	Riley, Haupt & Jacobson 2010
Graph	7-17	Korvin-Koukovsky & Chabrow 1948
Photograph	9-1	Jim Smith Boats, Inc.
Etching	10-1	Beaufoy 1834
Drawings	Appendix 3	Blount & McGrath 2009
Drawings	A4-1 to A4-3	Hubble 1974
Graph	A4-5	Hubble 1982
Graphs	Appendix 5	Blount & Fox 1976

Author Biography

Donald L. Blount, P.E., founded Donald L. Blount and Associates, Inc., a Naval Architecture and Marine Engineering design office in 1988. The company provides marine design, engineering, and consulting services for high-speed special-purpose commercial, military, and recreational craft for domestic and international clients. Noteworthy vessels from this design office include the 67.7 m (222 ft) *Destriero*, which holds the non-refueled Atlantic crossing record in 1992 with an average speed of 53.1 knots (61 mph) and the 41m (135 ft) GT/MY *Fortuna*, the Royal Yacht of Spain, having a speed of 68-plus knots (78-plus mph).

Donald previously served as Head of the Department of the U.S. Navy's Combatant Craft Engineering Department (CCED) and had been employed at the David Taylor Model Basin prior to working at CCED. During his 15 years of employment at David Taylor, Donald conducted individual research and directed engineering programs relating to emerging technology in the field of hydrodynamics.

He has co-authored more than 50 papers and articles for technical societies in various countries. He is a registered professional engineer in two states and is a Fellow of both SNAME and RINA. He graduated from George Washington University in 1963 with a bachelor's degree in mechanical engineering.

Donald L. Blount

PO Box 55171

Virginia Beach, VA 23471

dblount2014@gmail.com

Index

A

B

C

D

E

F

G

H

I

J

K

L

M

N

O

P

Q

R

S

T

U

V

W

X

Y

Z

CPSIA information can be obtained at www.ICGtesting.com
Printed in the USA
BVOW07*0418141015

422404BV00012B/39/P

9 780989 083713